International Conference on Biomass (ICB 2016)

AA006082

Technology, Application, and Sustainable Development

IOP Conference Series: Earth and Environmental Science
Volume 65

Bogor, Indonesia
10 – 11 October 2016

ISBN: 978-1-5108-4172-7
ISSN: 1755-1307

Printed from e-media with permission by:

Curran Associates, Inc.
57 Morehouse Lane
Red Hook, NY 12571

Some format issues inherent in the e-media version may also appear in this print version.

Copyright© (2016) by the Institute of Physics
All rights reserved. The material featured in this book is subject to
IOP copyright protection, unless otherwise indicated.

Printed by Curran Associates, Inc. (2017)

For permission requests, please contact the Institute of Physics
at the address below.

Institute of Physics
Dirac House, Temple Back
Bristol BS1 6BE UK

Phone: 44 1 17 929 7481
Fax: 44 1 17 920 0979

techtracking@iop.org

Additional copies of this publication are available from:

Curran Associates, Inc.
57 Morehouse Lane
Red Hook, NY 12571 USA
Phone: 845-758-0400
Fax: 845-758-2633
Email: curran@proceedings.com
Web: www.proceedings.com

TABLE OF CONTENTS

IOP Conf. Series: Earth and Environmental Science **65** (2017) 011001 doi:10.1088/1755-1315/65/1/011001

Preface

The International Conference on Biomass (ICB) 2016 could be conducted by Surfactant and Bioenergy Research Center (SBRC) IPB in cooperation with Indonesian Biofuel Expert Association (IKABI), Lampung University, Villanova University, and Indonesian Biogas Association (AbgI) on 10-11 October 2016 in Bogor, Indonesia. This conference with the theme of **Sharing Knowledge and Experience on Biomass: Technology, Application and Sustainable Development** was the first scientific conference on biomass Indonesia. It was **expected** that biomass-themed conference can be conducted annually by raising current issues in biomass.

The committee accepted 55 of 86 papers which were presented in the conference and published in this ICB proceeding. These papers were divided into the following groups:

1. Technology,
2. Application, and
3. Environment.

Appreciations were given to Rector and Vice Rectors of IPB for their support to the conference, reviewers, supporters, and committee members who had worked dedicatedly to make the conference and the publication of this proceeding successful. Supports from Bogor Agricultural University, Indonesia Estate Crop Fund-Palm Oil (BPDPKS), PT Indocement Heidelberg Cement Group and Ministry of Research, Technology and Higher Education of the Republic of Indonesia were acknowledged. We tried our best to do paper editing by following the writing format used in ICB 2016. However, there might be some shortcoming editing is found in this proceeding. Therefore, inputs from readers and writers are expected so that editing can be done better in the future.

We hope this ICB 2016 proceeding brings benefits to all of us. Thank you for your attention and cooperation.

Bogor, April 2017
Organizing Team

Reviewer of ICB 2016

1. Prof. Basuki Wirjosentono
2. Dr. Dwi Setyaningsih
3. Prof. Dr. Erliza Hambali
4. Dr. *Farah Fahma*
5. Dr. Ika Amalia Kartika
6. Dr. Jupikely James Silip
7. Prof. Dr. Misri Gozan
8. Dr. Obie Farobie
9. Prof. Pudji Permadi
10. Dr. Taufik Djatna
11. Dr. Titi Candra Sunarti
12. Prof. Dr. Udin Hasanudin
13. Dr. Uju
14. Dr. Yohanes Aris Purwanto
15. Dr. Yandra Arkeman
16. Zuchra Helwani, PhD

Content from this work may be used under the terms of the Creative Commons Attribution 3.0 licence. Any further distribution of this work must maintain attribution to the author(s) and the title of the work, journal citation and DOI.
Published under licence by IOP Publishing Ltd

IOP Conf. Series: Earth and Environmental Science **65** (2017) 011002 doi:10.1088/1755-1315/65/1/011002

Peer review statement

All papers published in this volume of *IOP Conference Series: Earth and Environmental Science* have been peer reviewed through processes administered by the proceedings Editors. Reviews were conducted by expert referees to the professional and scientific standards expected of a proceedings journal published by IOP Publishing.

Content from this work may be used under the terms of the Creative Commons Attribution 3.0 licence. Any further distribution of this work must maintain attribution to the author(s) and the title of the work, journal citation and DOI.
Published under licence by IOP Publishing Ltd

A comparative study on pyrolysis characteristic Indonesia biomassa and low grade coal

G I Adhityatama[1], F Hanif[1], R B Cahyono[1], M Hidayat[1], T Akiyama[2]

[1]Department of Chemical Engineering, Gadjah Mada University, Jl. Grafika 2, Bulaksumur, Yogyakarta 55281, Indonesia
[2]Center for Advanced Research of Energy Conversion Materials, Hokkaido University, North 13 West 8, Kita-ku, Sapporo, 060-8628, Japan.

Email : rochimbakti@ugm.ac.id

Abstract. A comparative study on pyrolysis of biomass and low grade coal was conducted using a thermogravimetric analyzer. Each kind of biomass and coal has a characteristic pyrolysis behavior which is explained based on its individual component characteristics. All fuels experienced a small weight loss as temperatures approached 450K because of moisture evaporation. The coal had smallest total weight loss compared to biomass due to its high content of fixed carbon, suggesting that coal would produce high amounts of char and small amounts of volatile matter (*e.g.*, tar and gas). The biomass exhibits similar tendency regarding the decomposition process which is the hemicelluloses break down first at temperatures of 470 to 530K, cellulose follows in the temperature range 510 to 620K, and lignin is the last component to pyrolyzer at temperatures of 550 to 770K. The thermal decomposition of biomass consisted of two predominant peaks corresponding first to the decomposition of cellulose and, second, to the decomposition of lignin. Meanwhile, the coal exhibited only single peak because these fuels were predominantly composed of carbon. Based on the kinetic analysis, coal have the smaller activation energy (55.32kJ/mol) compared to biomass (range from 89.80-172.86 kJ/mol). Pyrolysis process also created more pore material in the solid product. These results were important for the optimization of energy conversion from those solid fuels. Biomass resulted lower solid product and higher tar product, thus would be suitable for liquid and gas energy production.

1. Introduction

Indonesia produce 147.7 million tons of biomass per year, equivalent to about 470 GJ/year. Biomass is a renewable energy source[1]. Indonesian biomass spread all over the country such as rice residue, wood, and oil palm residue. Biomass is a solution to substitute the excessive demand of non-renewable energy like fossil fuel. Burning fossil fuels release large amount of carbon dioxide into atmosphere causing environmental problem. Therefore, if biomass is used as an alternative fuel it will reduce the consumption of fossil fuel and also reduce the environmental pollution of greenhouse[2,3].

The energy source within the biomass can be obtained either through biological or thermochemical conversion. Biological conversion involves the usage of microorganisms such as anaerobic digestion and fermentation. Thermochemical conversion involves the application of heat and chemicals in the production of energy. Thermochemical conversion can be divided into combustion, gasification and

Content from this work may be used under the terms of the Creative Commons Attribution 3.0 licence. Any further distribution of this work must maintain attribution to the author(s) and the title of the work, journal citation and DOI.
Published under licence by IOP Publishing Ltd

pyrolysis. Among the thermochemical conversion, pyrolysis is considered the most efficient process due to its high feed to fuel ratio in comparison to combustion and gasification process[4].

Biomass pyrolysis can be converted into bio-fuels, adsorbent bio-char, and other useful chemicals. Kinetic is very important to understand reaction mechanism, predict the conversion and design reactor TGA measure the amount and rate of change in the weight of tested material as a function of temperature or time in atmospheric pressure. Various weight loss processes determined during the TGA reflect the physical and chemical structure changes during the conversion. Differential thermogravimetry (DTG) shows thermal behavior of material and also the optimum operating condition of the material pyrolysis. The kinetics of these thermal events has been determined by the application of the Arrhenius equation corresponding to the separate slopes of constant mass degradation[5,6].

Nomenclature

A: pre-exponential factor (min^{-1})
B : heating rate ($K\ min^{-1}$)
E_a : activation energy ($J\ mol^{-1}$)
K : rate constant (min^{-1})
n : order of reaction (-)
R : universal gas constant (=8.314 $J\ mol^{-1}\ K^{-1}$)
T : temperature (K)
t : heating time (minute)
X : conversion of sampe (-)
W : weight of sample (mg)

Subscript
i : initial state (at 105°C)
f : final state (at 800°C)

2. Material and experiment method

2.1. Material
The biomass residues of rice huck and ulin wood were obtained from agricultural residue in Bantul Yogyakarta while low grade coal was obtained from Kalimantan. The proximate analysis of the samples was shown in table 1. To avoid the effect of particle size, the sample were crushed and sieved to a granular size.

Table 1: Proximate analysis of solid fuels

Sample	Proximate analysis (%mass)		
	Fixed C	Volatile	Ash
Low grade coal	47.77	31.05	1.47
Rice husk	21.14	55.23	15.75
Palm kernel shell	30.78	57.52	2.21
Ulin wood	22.83	74.32	2.85

2.2. Thermogravimetric analysis (TGA)
The pyrolysis characteristics of the samples were examined using a thermogravimetric analyzer (TG\DTA). Sample was placed inside the TG where the weight was constantly measured. The functions of the TG were to measure and record the dynamics of sample weight loss with increasing temperature or time. Sample use for TG analysis is 11.788 mg for palm kernel shell, 8.541 mg rice husk, and 10.40 mg ulin wood. Heating rate was controlled at 10 °C/min, temperature range used in this experiment was 25 – 800 °C. Once the heating temperature reached 105 °C, it was held for 30 min

to completely remove moisture and provide a basis for analysis. Nitrogen was used as the carrier gas in the TG, so that the sample was paralyzed in an inert environment without oxygen.

2.3. Pyrolysis Kinetics
The non-isothermal kinetics for solid decomposition is usually written as follows:

$$\frac{dX}{dt} = kf(X) \tag{1}$$

where X is fuel conversion, and is given

$$X = \frac{Wi-W}{Wi-Wf} \tag{2}$$

In the above equation, W, W_i , and W_f represent the instantaneous, initial (at 105°C), and final (at 800°C) weights of the sample. The reaction rate constant k is expressed in terms of the Arrhenius equation as

$$K = A.exp(\frac{-Ea}{RT}) \tag{3}$$

and the function f(X) can be written as

$$f(X) = (1 - X)^n \tag{4}$$

Substituting Eqs. (3) and (4) into Eq. (1) gives

$$\frac{dX}{dt} = A.exp\left(\frac{-Ea}{RT}\right)(1 - X)^n \tag{5}$$

For a constant heating rate $\beta - \frac{dT}{dt}$, Eq.(5) can be rearranged to the following equation

$$\frac{dX}{dT} = \left(\frac{1}{\beta}\right)A.exp\left(\frac{-Ea}{RT}\right)(1 - X)^n \tag{6}$$

The integral method based on Coats and Redfern (CR) equation[7,8,9] is used in this work, and the approximate integration of Eq. (6) gives

$$ln\left[\frac{-ln(1-X)}{T^2}\right] = ln\frac{AR}{\beta Ea}\left[1 - \frac{2RT}{E}\right] - \frac{Ea}{RT}, if\ n = 1 \tag{7}$$

or

$$ln\left[\frac{-ln(1-X)^{1-n}}{(1-n)T^2}\right] = ln\frac{AR}{\beta Ea}\left[1 - \frac{2RT}{E}\right] - \frac{Ea}{RT}, if\ n \neq 1 \tag{8}$$

For most reactions, the value of $\frac{2RT}{E}$ is very small (i.e. $\frac{2RT}{E}$<<1), and thus the preceding two equations can be approximated by

$$ln\left[\frac{-ln(1-X)}{T^2}\right] = ln\frac{AR}{\beta Ea} - \frac{Ea}{RT}, if\ n = 1 \tag{9}$$

or

$$ln\left[\frac{-ln(1-X)^{1-n}}{(1-n)T^2}\right] = ln\frac{AR}{\beta Ea} - \frac{Ea}{RT}, if\ n \neq 1 \tag{10}$$

The plot of $ln\left[\frac{-ln(1-X)}{T^2}\right]$ versus $\frac{1}{T}$ becomes a linear line for $n = 1$; the plot of $ln\left[\frac{-ln(1-X)^{1-n}}{(1-n)T^2}\right]$ versus $\frac{1}{T}$ is also a linear line for $n \neq 1$. Accordingly, the apparent activation energy (Ea) and the apparent frequency factor (A) can be determined from the slope and intercept of the regression line, respectively

3. Result and Discussion

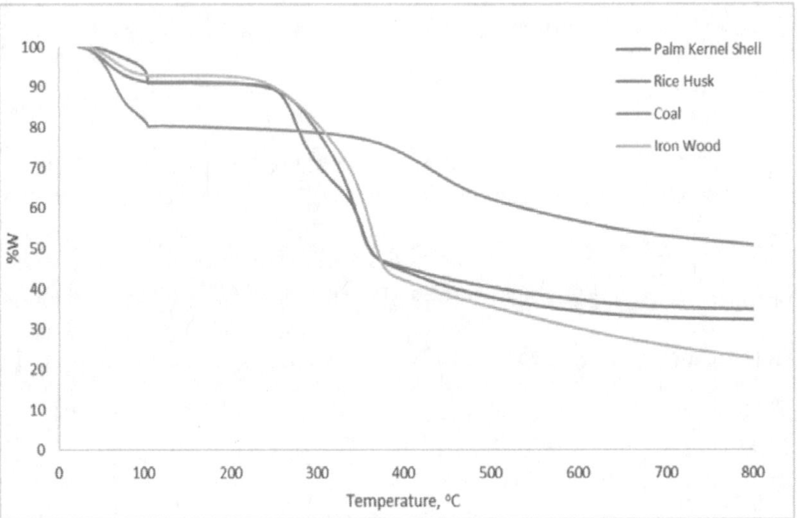

Figure 1. TGA Curves for Palm Kernel Shell, Rice Husk, Coal, and Ulin wood

For pyrolysis of palm kernel shell, rice husk, ulin wood, and coal, the percent weight change curves are shown in Fig. 1. It can be seen that the percent weight change of all samples are decreased with the increasing temperature. Ulin wood have the most percent weight loss from all the biomass sample and coal was the least. The remaining ulin wood after pyrolysis is 16.88% from the initial mass and coal is 35.87%. It indicating all sample are undergo thermal degradation and decomposition. Derivative of TG curve are shown in figure 2. Biomass shown slightly same curve pattern. It has 2 peak curve while coal only have one. The study of lignocellulose compounds is relevant to pyrolysis because decomposition these components occurs at different temperatures. The pyrolysis process divided into three main region: Moister content removal (<105 °C); degradation of hemicellulose (200 – 320 °C); lignin and cellulose decomposition (320 – 400 °C). Since coal doesn't have cellulose, hemicellulose and lignin the DTG curve only have one peak and have different pattern with biomass. Pyrolysis reaction end when the percent weight loss have horizontal pattern. Rice husk and palm kernel shell end at 600 °C, coal at 700 °C and Ulin wood at 750 °C.

Maximum decomposition hemicellulose temperature at 10 °C/min for rice husk is 295 °C, palm kernel shell 278.38 °C, and ulin wood 276.65 °C. Maximum decomposition cellulose and lignin temperature at 10 °C/min 349.47 °C for rice husk, 353.61 °C palm kernel shell, Ulin wood 362.82 °C.

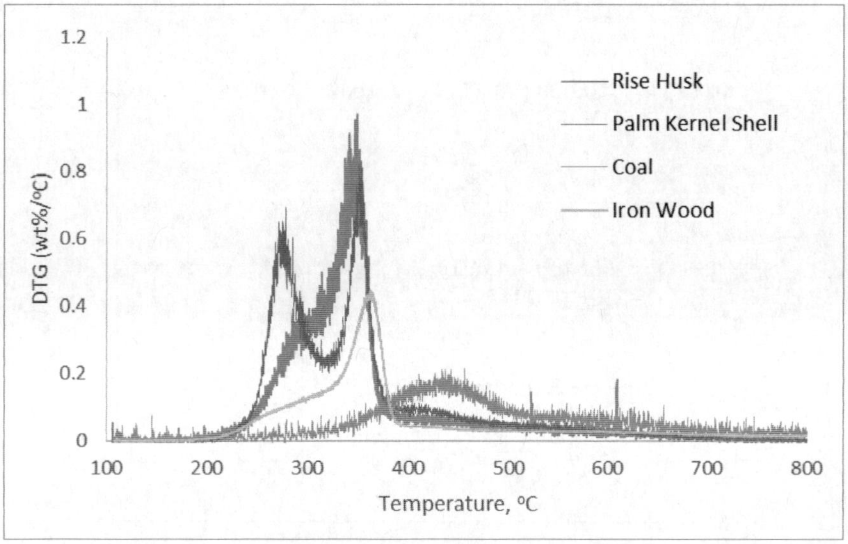

Figure 2. DTG Curves for Palm Kernel Shell, Rice Husk, Coal, and Ulin wood

Table 2 show the result of kinetic parameter obtained from the Arrhenius method for rice husk, palm kernel shell, ulin wood, and coal. All the data obtained at different value of n are fitted and the best-fit regression line that has highest value of correlation R^2 was shown in each figure. Coal has pyrolysis range from 594 – 779 °C activation energy 55.32 kJ/mol A = 610 R and the best-fitted regression n = 2.6 with correlation coefficient R= 0.99. Palm kernel shell the decomposition divided into two stage with activation energy value for the first region is 172.86 kJ/mol second stage 164.23 A for first stage and second stage is 5275810276075090.00 and 141694773043700.00 best-fitted regression n = 7 for first stage and n = 4 for the second stage which give coefficient correlation (R) 0.99 and 0.97. Rice husk activation energy for first and second stage is 89.80 and 108.5226 A for first and second stage 11963597.96 and 688935605.9 and the best-fitted regression n = 2.2 and 2.5 which give coefficient correlation 0.99 and 0.97. Ulin wood activation energy for first and second stage is 85.42635 and 135.0692 A for first and second stage 3745831.323 and 88315997175 and the best-fitted regression n = 3 and 4 which give coefficient correlation 0.98 and 0.97. Coal give least energy activation and ulin wood give most value of activation energy. It mean that coal is still a material which easiest to burn. Calculation using Arrhenius method shown that value of activation energy for ulin wood in first stage is the least among the other biomass.

Table 2: Kinetic parameters and characteristic temperatures for pyrolysis of biomass and coal

Sample	Pyrolysis range (°C)		Order	Ea (kj/mol)	R^2
	T_i	T_f			
Low grade coal	594	778	2.6	55.32	0.99
Rice husk	200	302	2.2	89.90	0.99
	302	385	2.5	108.52	0.97
Palm kernel shell	200	322	7	172.86	0.99
	322	384	4	164.23	0.97
Ulin wood	180	328	3	85.42	0.98
	328	400	4	135.06	0.97

As examples, the linear regresion for low grade coal and rice husk can be shown in the below

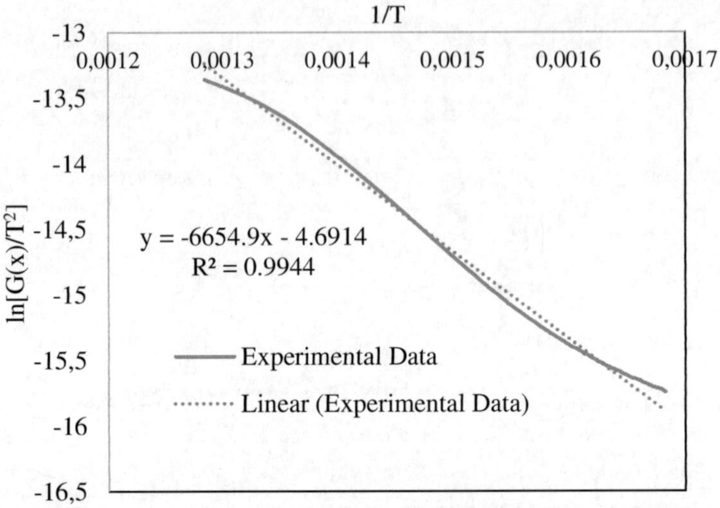

Figure 3. The linear regression of coal for calculation of activation energy

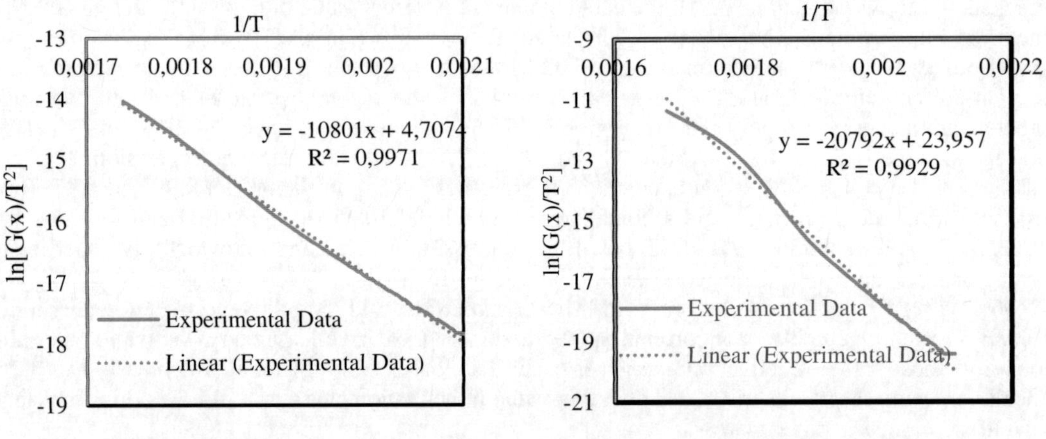

Figure 4. The linear regression of rice husk for calculation of activation energy

Figure 5 shows the comparison of SEM (Scanning Electron Microscopic) image between raw ulin wood residues before pyrolysis process and char products after pyrolysis process. It is seen that raw ulin wood there are no visible pores, while char products has visible pores.

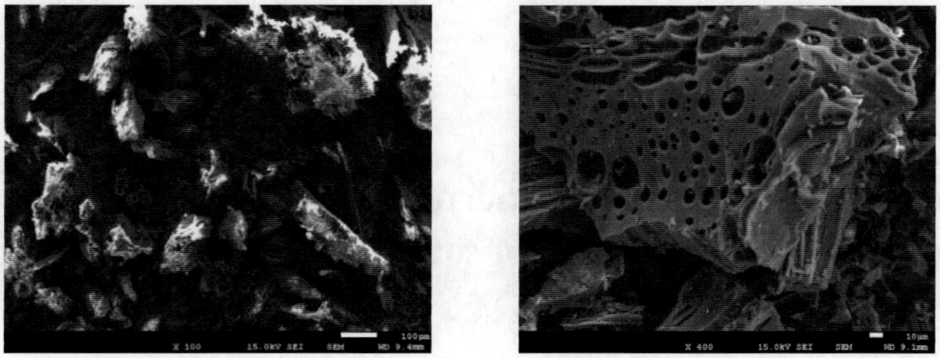

(a) (b)

Figure 5. The effect of the pyrolysis process on the SEM image of the ulin wood.

4. Conclusion

Based on TG-DTG analyses found that the biomass pyrolysis process occurs in two stages, the decomposition of hemicellulose and lignin decomposition while coal pyrolysis occurs only in one stage. This happens due to differences between the composition of biomass and coal. As a whole, coal have the smaller activation energy (55.32kJ/mol) compared to biomass (range from 89.80-172.86 kJ/mol).

References

[1] Zentrum fur rationell Energieanwendung und Umwelt GmbH (ZERU) 2000 Biomass in Indonesia-Business Guide

[2] Sami M, Annamalai K and Wooldridge M 2001Co-firing of coal and biomass fuel blends Prog Energy Combust Sci **27** 171–214

[3] Hamelinck C N, Faaij A P C, Den Uil H and Boerrigter H 2004 Production of FT transportation fuels from biomass; technical options, process analysis and optimisation, and development potential Energy **29** 1743–71

[4] Shafie S M, Masjuki H H and Mahlia T M I 2014 Rice straw supply chain for electricity generation in Malaysia: economical and environmental assessment Appl. Energy **135** 299-308

[5] Liu Q, Hu H, Zhou Q, Zhu S and Chen G 2004 Effect of inorganic matter on reactivity and kinetics of coal pyrolysis Fuel **83** 713–718

[6] Jakab E, Varhegyi G and Faix O 2000 Pyrolysis J. Anal. Appl. **56** 273–285.

[7] Guo J and Lua A C 2001 Kinetic study on pyrolytic process of oil palm solid wasteusing two step consecutive reaction model Biomass Bioenergy **20** 223–33.

[8] Sait H H, Hussain A, Salema A A and Ani F N 2012 Pyrolysis and combustion kinetics of date palm biomass using thermogravimetric analysis Bioresour Technol **118** 382–9

[9] Coats A V and Redfern J P 1964 Kinetic parameters from thermogravimetric data Nature **201** 68–69

A feasibility study of hydrothermal treatment of rice straw for multi-production of solid fuel and liquid fertilizer

S Samnang[1], P Prawisudha [1], A D Pasek[1]

[1]Bandung Institute of Technology, Department of Mechanical Engineering, Faculty of Mechanical and Aerospace Engineering, Bandung, Indonesia

Email : sann.samnang.itc@gmail.com

Abstract. Energy use has increased steadily over the last century due to population and industry increase. With the growing of GHG, biomass becomes an essential contributor to the world energy need. Indonesia is the third rice producer in the world. Rice straw has been converted to solid fuel by Hydrothermal Treatment (HT) for electricity generation. HT is a boiling solid organic or inorganic substance in water at high pressure and temperature within a holding time. HT converts high moisture content biomass into dried, uniform, pulverized, and higher energy density solid fuels. HT can effectively transport nutrient components in biomass into a liquid product known as fertilizer. This paper deals with an evaluation of hydrothermal treatment of rice straw for solid fuel and liquid fertilizer. An investigation of rice straw characteristics were completed for Bandung rice straw with various condition of temperature, biomass-water ratio, and holding time in the purpose to find the changes of calorific value for solid product and (N, P, K, and pH) for liquid product. The results showed that solid product at 225 °C and 90 min consists in a heating value 13.8 MJ/kg equal to lignite B. Liquid product at 225 °C and 90 min had the NPK content similar to that of micronutrients compound liquid fertilizer. The dried solid product should be useful for Coal Fire Power Plant, and the liquid product is suitable for plants. This research proves that hydrothermal process can be applied to rice straw to produce solid fuel and liquid fertilizer with adequate quality.

1. Introduction

Energy consumption has augmented gradually over the last century as world population has grown and more countries have become industrialized since 1900. Fossil fuel, especially crude oil, is currently the more noticeable energy source around the world. However, the reserves of fossil fuel are limited and will be reduced in the near future at its recent use rate.

Moreover, burning fossil fuels causes environmental issues including GHG emission, the main cause of climate change[1]. Because of the insufficient oil availability in the future, there is a good interest in discovery alternative energy sources. Due to the bad environmental impact of burning fossil fuels, the

Content from this work may be used under the terms of the Creative Commons Attribution 3.0 licence. Any further distribution of this work must maintain attribution to the author(s) and the title of the work, journal citation and DOI.
Published under licence by IOP Publishing Ltd

alternative energy source should be sustainable and environmental-friendly. Utilization biomass for energy production can have a balanced CO_2 production and consumption compared to CO_2 from burning fossil fuels. Nowadays, energy production from biomass is only a small production of the total energy production. The resources of biomass materials such as crops, grasses, wood, agricultural residues, and organic wastes are abundant, so there is a great potential to augment energy production from biomass materials.

Renewable energy plays a crucial role in Indonesia's national energy policies, especially as maximizing renewable resources helps strengthen the security of energy supply. On the other hand, Indonesia is recently exploiting only around 5% of its renewable energy capacity. The government intends to speed up the exploitation of renewable energy and to increase its share of primary energy to 23% by 2025. Renewable energy accounted for 33.4% of total primary energy supply (TPES) in Indonesia in 2012[2]. This amounts to 71.4 million tons of oil-equivalent (Mtoe) (282.3 million barrels of oil-equivalent), made up primarily of biofuels and waste biomass (25.3% of TPES). The remainder is from geothermal (7.6%), hydro (0.5%) and negligible levels of wind and solar power. Current capacity of on-grid biomass power plants in 2013 was 90.5 MW, mainly located in Sumatra Island (Northern Sumatra and Riau), Bali, West Java and Kalimantan. Biomass power plant feedstock in Indonesia is usually comprised of palm oil residues, municipal solid waste, sugar residues and other solid biomass, including rice residues, rubber wood and wood chip. Smaller quantities are available from other agricultural waste, such as logging residues, sawn-timber residues and coconut residues.

The rice production in Indonesia were increasing from years to years[3]. Indonesia is the world's third-largest rice producer and also one of the world's biggest rice consumers[4]. The country's rice area expanded from 11.4 million ha in 1995 to 13.2 million ha in 2010, which represented 24% of the total agricultural area. Rice yield increased slightly from 4.3 t/ha in 1995 to 5 t/ha in 2010. Rice production is heavily concentrated on the islands of Java and Sumatra; nearly 60% of total production emanates from Java alone. Data from the International Fertilizer Industry Association (IFA) reveal that the country's fertilizer consumption increased enormously from 0.14 million t in 1961 to 4.47 million t in 2009. FAO recorded that about 52% of all fertilizer use in Indonesia is for rice.

In 2008, about 620 million tons of rice straw were produced in Asia, and this quantity is increasing every year[5]. Burning of rice straw causes the effects on soil and atmosphere[6]. In Cambodia, sustainable rice straw management technologies are becoming more important as rice farming produces million tons of the material every year. Mechanization for rice production has increased resulting more rice straw being left in the field[7].

According to the above description, one of the appropriate solution is to invent rice straw to become a useful solid fuel which can be mixed with coal for Coal Fire Power Plant by hydrothermal process. It will benefit not only in solving GHGs emission problem but also decreasing the coal consumption in power plant.

The liquid organic fertilizer potentially holding solubilized nutrient is being rarely used. Now, the liquid residue of hydrothermal treatment is considered as wastewater, and there is a limited study on this liquid product. During HT, some inorganic and organic contents in rice straw may dissolve into the liquid phase. Thus HT can effectively transport nutrient components in biomass into a liquid product. This nutrient source might be used for organic fertilizer [8].

The objective of the study is to investigate the possibilities of using HT in producing solid fuel and bio-organic fertilizer. In this study, the influences of some operating parameters such as temperature and holding time to the product will be also investigated.

The hydrothermal process involves the application of heat and pressure in aqueous medium, similar to the natural processing of organic remains from plants or animals to become coal or crude oil we consume today. Hydrothermal treatment breaks the physical and chemical structure in the materials such as cellulose, hemicellulose, and lignin, and these biomasses are broken down into smaller and simpler molecules, see Figure 1 and Figure 2.

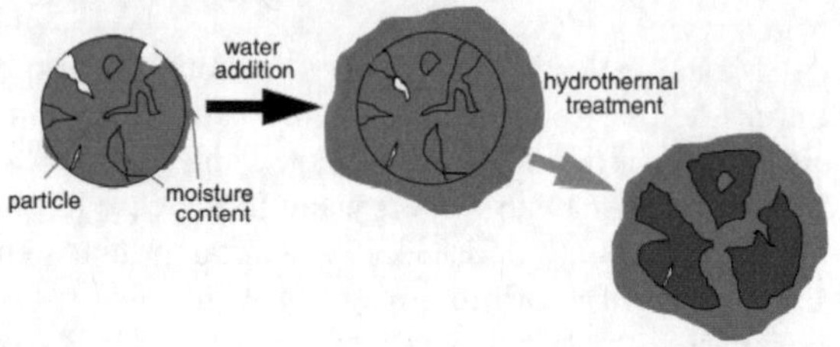

Figure 1. Visualization of particle breakage due to HT [9]

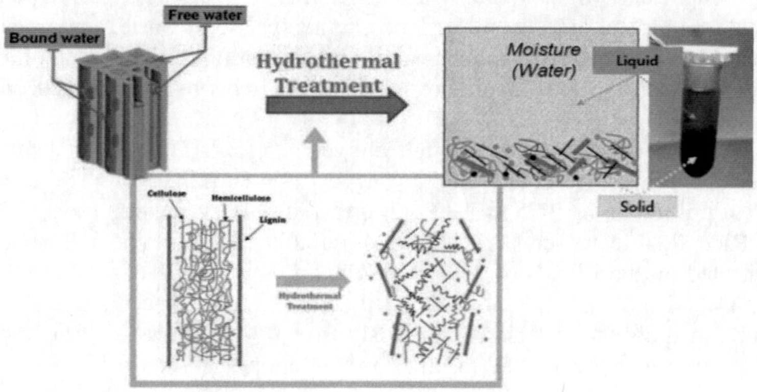

Figure 2. Physical structure change of biomass by HT [9]

2. Experimental Procedure

The rice straw samples used were obtained from the rice field around Bandung. The experimental apparatus setup consists of a 2-liter-capacity autoclave in Thermodynamic Laboratory, see Figure 3. In this experiment, 150 g of chopped rice straw and distilled water in 1:5 biomass-water ratio was put into the reactor. The reactor temperature was varied at 150, 175, 200 and 225 °C with the varied holding time of 30 min and 90 min. After having finished, the product was sent out from the reactor. The solid part was separated from the liquid by using a filtration.

Figure 3. Autoclave in Thermodynamic Laboratory

Hydrothermal process produces two kinds of product, see Figure 4 and Figure 5. Solid product and raw rice straw characteristics were analyzed in term of calorific value. The tests were done at chemical department. The total nitrogen in the liquid residue and the concentrations of P and K in the liquid residue are analyzed[10]. The micronutrient (N, P, K and pH) was defined for liquid products. The tests were done in Thermodynamic Laboratory. Two replications of N, P and K test were carried out.

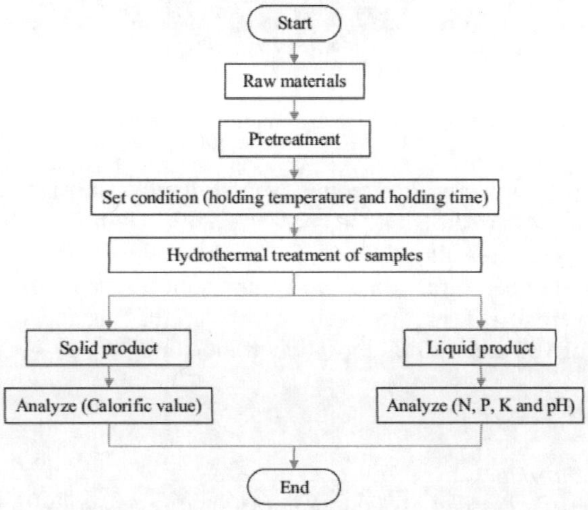

Figure 4. Hydrothermal treatment experimental flow chart

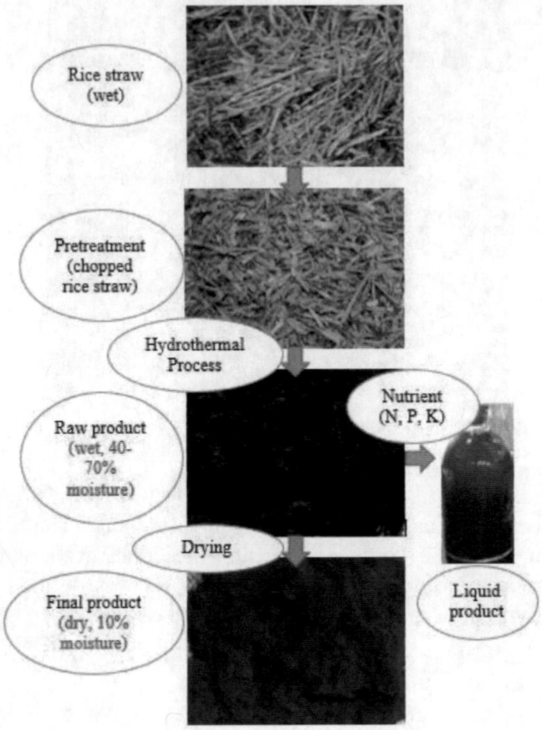

Figure 5. Raw rice straw conversion by employing the hydrothermal treatment

Solid product and raw rice straw were dried in the oven at 40 °C. The sample has been pulverized to pass 250-μm (No.60) sieve as prepared in accordance with Method D 346 or Method D 2013. Moisture was determined by establishing the loss in weight of the sample when heated under rigidly controlled conditions of temperature, time and atmosphere, sample weight, and equipment specifications. Samples were dried in the oven at 40 °C for surface moisture and temperature regulation between limits of 104 and 110 °C for inherent moisture[11].

3. Result and Discussion

3.1. Fuel properties of solid product

Higher heating value (HHV) is defined as the amount of heat released by the unit mass or volume of fuel (initially at 25 °C) once it is combusted and the products have returned to a temperature of 25 °C. It includes the latent heat of vaporization of water. HHV can be measured in a bomb calorimeter [12]. The calorific values of rice straw after hydrothermal treatment are presented in Table 1 based on ASTM D5865 standard in Chemical Department.

IOP Conf. Series: Earth and Environmental Science **65** (2017) 012002 doi:10.1088/1755-1315/65/1/012002

Table 1. Calorific value of rice straw after HT

Holding Temperature	Holding Time	Calorific Value (MJ/kg) (adb)
150 °C	30 min	10.55
	90 min	11.28
175 °C	30 min	10.96
	90 min	11.67
200 °C	30 min	12.92
	90 min	12.95
225 °C	30 min	11.88
	90 min	13.80
Raw rice straw		10.44

At 225 °C and 90 min, solid product contains a heating value 13.8 MJ/kg equal to lignite B which is less than 14.65 MJ/kg [13]. It reveals the possibility of using the product of HT as co-firing fuel with coal. According to Figure 6, the calorific value of solid products have increased in function of temperature and holding time after HT. If the holding time is increased from 30 min to 90 min, the average increment of the calorific value is about 0.84 MJ/kg. So, the holding time has a little effects on the caloric value. Moreover, if the temperature is increased from 150 °C to 225 °C, the calorific value increases from 10. 55 to 13.8 MJ/kg. Thus, the effect of temperature on the carlorific value is higher than that of holding time.

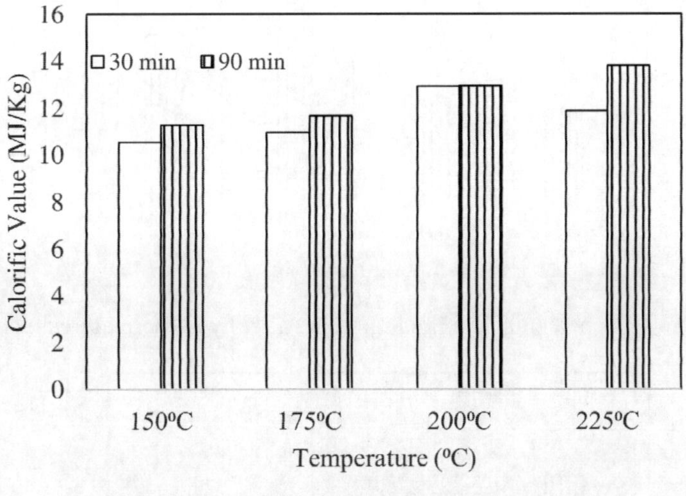

Figure 6. Variation of calorific value after hydrothermal treatment

3.2. Main nutrient elements in the liquid residue

The primary macronutrients utilized by plants in large amounts for crops growth are nitrogen, phosphorous and potassium. Table 2 reveals the main nutrient elements of liquid fertilizer after hydrothermal treatment.

Table 2. Main nutrient elements of liquid fertilizer after HT

Holding Temperature	Holding Time	Main nutrient elements			
		N(%)	P(%)	K(%)	pH
150 °C	30 min	0.02±0.00	0.09±0.02	1.35±0.31	5.6
	90 min	0.20±0.16	0.16±0.02	1.28±0.24	5.2
175 °C	30 min	0.29±0.20	0.14±0.06	1.45±0.31	5.5
	90 min	0.22±0.17	0.14±0.07	4.66±0.07	4.8
200 °C	30 min	0.30±0.26	0.12±0.03	5.28±0.82	4.4
	90 min	0.44±0.08	0.12±0.06	5.03±0.63	4.9
225 °C	30 min	0.28±0.24	0.12±0.04	6.87±0.03	4.8
	90 min	0.35±0.21	0.15±0.06	9.48±1.28	4.6

Data show mean±standard error

The percentage of N fluctuates when the holding temperature and holding time increase, see Figure 7. The percentage of P has increased when the reaction temperature and holding time have increased, see Figure 8. The percentage of K increases when the treatment temperature and holding time increase, see Figure 9.

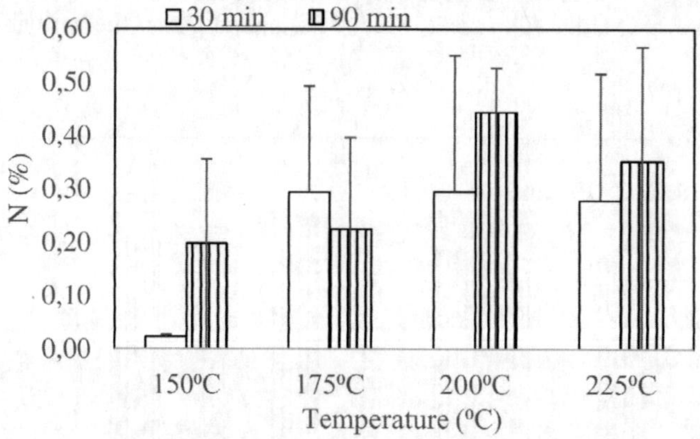

Figure 7. Variation of nitrogen percentage in liquid product after hydrothermal treatment

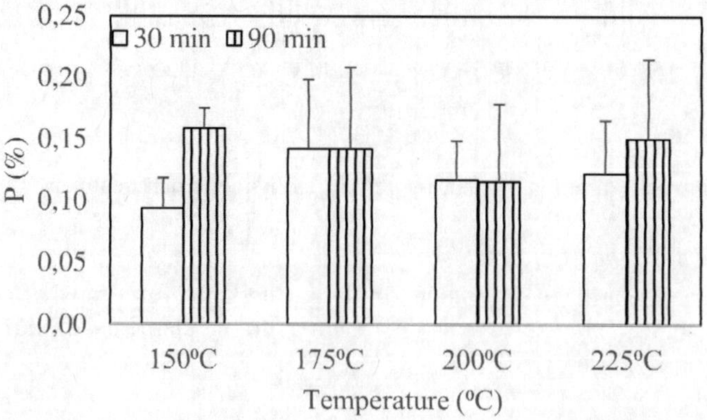

Figure 8. Variation of phosphoric percentage in liquid product after hydrothermal treatment

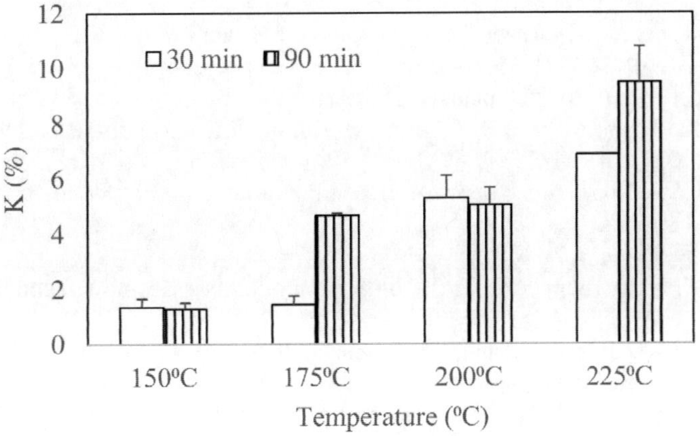

Figure 9. Variation of potassium percentage in liquid product after hydrothermal treatment

As shown in Figure 7, Figure 8 and Figure 9, the holding time had little effects on the percentage of N, P and K. However, variation of reaction temperatures provides the significant differences of the percentage of N, P and K.

Table 3. Representative commercial liquid fertilizer [10]

Types	N(%)	P(%)	K(%)
Urea compound liquid fertilizer	2-16	1-14	1-10
Organic liquid fertilizer	1.5-12	1-12	2-11.5
Mircronutreints compound liquid fertilizer	1.5-12	0.8-4	2-12

Percentage of N, P and K in Table 2 are considerably low compared with that of commercial liquid fertilizer in Table 3. However, this liquid product can be used as the liquid fertilizer because the value of N, P and K are almost equal to the micronutrients compound liquid fertilizer in the condition of 225 °C temperature and 90 min holding time.
The optimum value of pH at which plants can take up nutrients to the root zone and the foliage best is ranging from 5.5 to 8.5 [8]. The pH of sewage sludge after HT is equal 5.17 so it can applied directly to plants and soil because pH is not much low [10]. At (150 °C and 30 min), (150 °C and 90 min), and (200 °C and 90 min), pH of liquid products are good for crops, see Table 2.

4. Conclusion

From the carried out experiment in this study, it was found that hydrothermal treatment can produce solid fuel and liquid fertilizer from rice straw at the same time. The best operating parameter is at 225 °C temperature and 90 minutes holding time, producing solid fuel with calorific value 13.8 MJ/kg and liquid fertilizer containing the compositions of N, P and K with value 0.57%, 0.22% and 8.2%, respectively. The temperature has significant effect while the holding time has the lower effect for the calorific value, N, P, and K. Calorific value and soluble nutrients are linearly increased as the reaction temperature and holding time are increased.

Acknowledgements

The research is supported by the research fund from Decentralization Project, Ministry of Higher Education of Indonesia.

References

[1] Cheng J 2010 *Biomass to Renewable Energy Processes* (Boca Raton: CRC Press)
[2] Anonim 2015 *Indonesia 2015* (Paris: International Energy Agency)
[3] Anonim 2012 *Ricepedia* (http://ricepedia.org/indonesia)
[4] Idrees K 2015 *World Knowing* (http://worldknowing.com/top-10-largest-rice-producing-country-in-the-world/)
[5] Anonim 2008 *IRRI* (http://irri.org/our-work/research/value-added-rice/rice-straw-and-husks)
[6] Anonim 2009 *Recycling of Rice Straw*
[7] Anonim 2016 *IRRI* (http://news.irri.org/2016/05/cambodia-assesses-new-technologies-for.html)
[8] Nurdiawati A *et al.* 2015 Evaluation of hydrothermal treatment of empty fruit bunch for solid fuel and liquid organic co-production. *Energy Procedia* **79** 226-232.
[9] Yoshikawa K and Prawisudha P 2014. Hydrothermal treatment of municipal solid waste for producing solid fuel *Application of Hydrothermal Reactions to Biomass Conversion, Green Chemistry and Sustainable Technology* pp. 355-383.
[10] Jambaldorj G, Takahashi M, and Yoshikawa K 2007 Liquid fertilizer production from sludge by hydrothermal treatment *Proceedings of International Symposium on EcoTopia Science* **07** 605-608
[11] ASTM Interntional. 1981 *Annual Book of ASTM Standards Gaseous Fuels, Coal and Coke; Atmospheric A.* (USA: ASTM International)
[12] Basu P 2010 *Biomass Gasification and Pyrolysis, Practical Design and Theory* (USA: Elsevier Inc.)
[13] Matuszewski M 2012 *Quality Guidelines for Energy System Studies, Detailed Coal Specifications* (USA: National Energy Technology Laboratory)

A pilot-scale study of wet torrefaction treatment for upgrading palm oil empty fruit bunches as clean solid fuel

M H Gusman[1], P N E Sastroredjo [1], P Prawisudha[1], T Hardianto[1] and A D Pasek[1]

[1] Faculty of Mechanical and Aerospace Engineering, Bandung Institute of Technology, Bandung, Indonesia

Email : muhammadhanifgusman@gmail.com

Abstract. Less utilized empty fruit bunch (EFB) is seldom used as solid biofuel due to its high alkali content that potentially cause ash deposit called slagging and fouling. This phenomenon could harm biomass-fired power plant equipment. Some pre-treatment of EFB is needed to reduce EFB ash deposit potential. The effect of wet torrefaction pre-treatment in laboratory scale was successfully proven in decreasing slagging and fouling potential while increasing EFB calorific value that could fulfill clean solid fuel criteria. This research focuses on wet torrefaction process that conducted on a pilot scale with the capacity of 250 liters. It was found that wet torrefaction process can improve the product's calorific value up to 9.41% while reduce its ash content down to 1.01% comparing to the raw EFB. The reduction of ash content also leads to the reduction of slagging and fouling tendency that presents in terms of alkali index. Alkali index is a quantitative method that can be calculated after obtaining metal oxides fraction on solid fuel. Metal oxides could be obtained by using energy dispersive x-ray spectroscopy.

1. Introduction

As one of the largest renewable energy resources, it is expected that biomass could substitute the utilization of fossil fuel. Biomass stockpile, obtained from agricultural plant by-product process, is abundant and produced continually, less utilized, and nearly free [1]. In the other hand, biomass as an energy resources cannot directly utilized due to its high moisture content, bulk volume, low heating value, and also high alkali content [2]. High moisture content and bulk volume will decrease the combustion performance inside the furnace. Subsequently, low heating value makes biomass no suitable enough to substitute fossil fuel. Whereas high alkali content can bring harm risks of ash deposit called slagging and fouling [3]. Therefore, biomass needs a pre-treatment process to fulfill clean solid fuel condition.

Indonesia plays important role in palm oil industry. In period 2015/2016, Indonesia alone contributes over 32 million tonnes of palm oil and accounting for over than 54 percent of world's total palm oil production [4]. Referring to the process flow diagram, it was found that only 21-23 wt. % oil palm produced from process plant and the rest is left as biomass residue [5]. The largest biomass solid residue produced from palm oil mill process is palm oil empty fruit bunch with the amount of 23-25 wt. % from each unit of fresh fruit bunch. Palm oil empty fruit bunch (EFB) is termed as palm oil bunches waste residue generated that has loosen its fruit after sterilization and threshing process [2].

Content from this work may be used under the terms of the Creative Commons Attribution 3.0 licence. Any further distribution of this work must maintain attribution to the author(s) and the title of the work, journal citation and DOI.
Published under licence by IOP Publishing Ltd

One of the most promising pre-treatment process to obtain clean solid fuel criteria is wet torrefaction. Wet torrefaction or so-called hydrothermal carbonization is one of thermal degradation conversion process in a subcritical water as its medium [6]. Wet torrefaction aims to decompose lignocellulosic component in biomass such as hemicellulose, cellulose and lignin through hydrolysis between 180 to 280 °C [7].

Similar to other biomass, empty fruit bunch (EFB) has high alkali content that potentially cause problem such as ash deposition on biomass-fired plant equipment. Potassium element in alkali content that present in EFB act as principal bonding agents in most deposits [3]. Previous research has proved that wet torrefaction process could effectively remove chlorine content [8] and potassium content [9]. By date, wet torrefaction experiments to produce clean solid fuel on a pilot scale reactor by using quantitative criteria has not been conducted.

In this research, the authors employed an experimental investigation on wet torrefied solid products that has been proceed by using pilot scale reactor (250 liters). Indonesian EFB was chosen as the raw material. This research focuses on finding process parameters that could increase high heating value while decrease ash deposits potential. Calorific value were obtained by using adiabatic oxygen bomb calorimeter while major ash metal oxides composition were obtained by using energy dispersive x-ray spectroscopy.

2. Materials and methods

2.1. *Materials*
Sterilized and pressed empty fruit bunches (EFB) were used in this research. The EFB were taken from PT Perkebunan Nusantara VIII, Kertajaya, Banten, Indonesia. It was found that the EFB had 39.08% of moisture content (as received) and no further treatment is employed to the EFB.

2.2. *Wet torrefaction experimental process*
Wet torrefaction experiments were conducted by using 250 L pilot-scale reactor, consist of a body covered by Rockwool insulation, an oil heater, steam input from boiler, and direct contact steam condenser. During pilot-scale experiments, 15 kg of feedstock were supplied to the reactor with biomass-to-water ratio (BWR) of 1 : 3 and 1 : 5. 150 and 175 °C were chosen as the holding temperature, with variation of holding period of 30 minutes and 90 minutes.
EFB feedstock was loaded and mixed with certain amount of water, resulting water-to-biomass ratio parameter. Then, reactor heated until reaching the holding temperature. Holding temperature will be held for certain holding period. After completion, treated products were taken out from the reactor and dried at 40 °C for subsequent experiments and analysis.

2.3. *Analytical procedures*

2.3.1. *Drying process*
The drying process was tested referring to ASTM D 3173. The samples were dried naturally until the mass difference reached 0.01%. The samples crushed until it could pass through a 60-mesh sieve.

2.3.2. *High heating value analysis*
The calorific value of products was tested referring to ASTM D 5865. The preparation process of the samples referring to section 2.3.1. Parr adiabatic oxygen bomb calorimeter apparatus is used to determine the HHV of untreated EFB and all solid products derived from wet torrefaction experiment.

2.3.3. Physical composition analysis
The physical composition analysis of product was tested referring to ASTM D 3172. The preparation process of the samples referring to section 2.3.1. Leco TGA - 601 apparatus is used to determine the ash content of untreated EFB and all solid products derived from wet torrefaction experiment.

2.3.4. Ash composition analysis
The untreated EFB and all products derived from wet torrefaction experiment were heated inside 815 °C furnace. Then, ash composition of all samples were evaluated. The ash composition analysis was determined by using Energy Dispersive X-Ray Spectrometry (EDX) – 720 Shimadzu referring to IK-paskal-05-01-5B/3 standard.

2.3.5. Alkali index calculation
The empirical calculation were applied by using percentage of potassium oxide and the amount of ash present in the untreated EFB and wet torrefied solid products. The alkali index is showing the alkali oxides contribution to the potential level of slagging and fouling. Equation 1 describe the amount of alkali resulted the energy produce by solid fuel utilization [10].

$$Ai\ \left(\frac{lb\ alkali}{MM\ BTU}\right) = \frac{1\times10^{6}}{BTU/lb} \times \%ash \times \%alkali\ on\ ash \qquad (1)$$

Alkali index below 0.4 lb/MM Btu is considered a fairly low slagging risk. Values between 0.4 and 0.8 lb/MM Btu will probably slag with increasing certainty of slagging as 0.8 lb/MM Btu value is approached. Above 0.8 lb/MM Btu, the fuel is virtually certain to slag and foul [10].

3. Results and Discussion

3.1. Calorific Value Analysis
Table 1 shows that all wet torrefied solid products had an increasing of HHV compared to raw material. According to Figure 1, the increased process temperature and holding period affected to HHV. This phenomenon can most likely happened due to lignocellulosic decomposition during the process.

Table 1. HHV of raw material and wet torrefied solid products on pilot-scale reactor

Process Parameters	HHV (J/g)
Raw material	17,252
150 °C – 30 minutes – BWR 1 : 5	17,311
150 °C – 90 minutes – BWR 1 : 5	17,531
175 °C – 30 minutes – BWR 1 : 5	17,853
175 °C – 30 minutes – BWR 1 : 3	18,560
175 °C – 90 minutes – BWR 1 : 5	18,874

Lignocellulosic substances decomposition process plays significant roles in the increased of HHV. Lignocellulosic is mainly composed of three polymers: (1) cellulose, (2) hemicellulose, and (3) lignin. Hydrothermal experiment has been conducted to observe hemicellulose decomposition temperature by using wheat bran, spent grain, corn cobs, and eucalyptus [11]. It was found that at 150 °C – 160 °C, the solid products shown similar structure to xylose. Xylose is a substance derived from xylan through hydrolysis process. It can be concluded that hemicellulose is poorly resistant to hydrolysis and starts to decompose in the temperature of 150 °C into its monomers.

The increased of HHV has a significant correlation with lignin fraction. Lignin has the largest HHV among the other lignocellulosic substituent and lignin is not ease to decompose due to its higher degradation temperature. At 175 °C followed by maximum holding period, the increased of HHV was found at its largest. This effect can be occurred because at higher operating temperature, the time required to hydrolyze hemicellulose into its monomers is shorter, which mean at the same holding period, solid products yield from 175 °C shown plenteous decomposed substances compared to solid products yield from 150 °C.

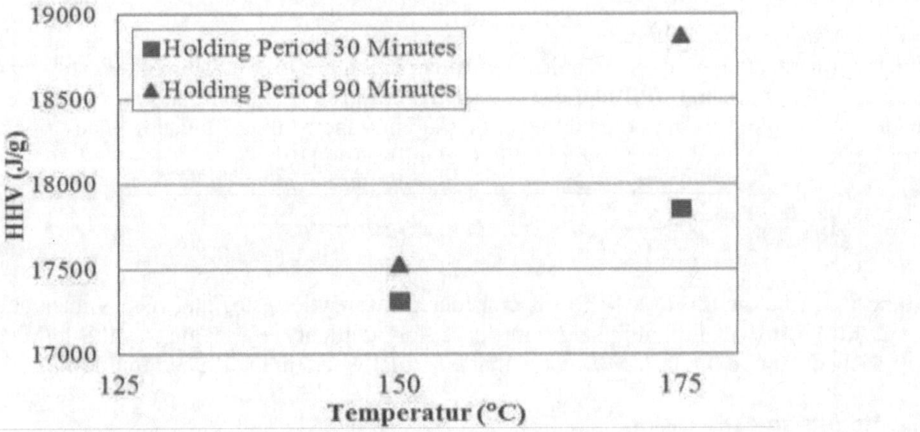

Figure 1. Temperature and holding period effect on HHV (BWR 1 : 5)

The increased of HHV, not only can be reached with the increased of holding period and temperature. Figure 2 showing that lower water to biomass ratio could lead to higher HHV. Feedstock with less amount of water also has less EFB fraction that submerged. Less water used during the process may lead to partial dry torrefaction condition.

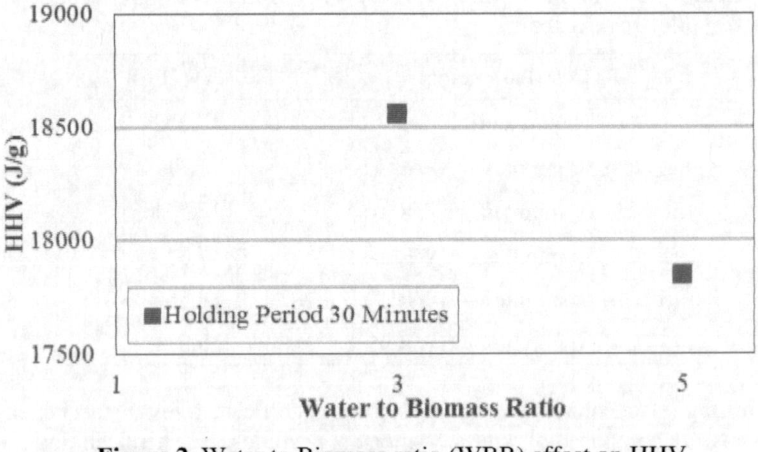

Figure 2. Water to Biomass ratio (WBR) effect on HHV

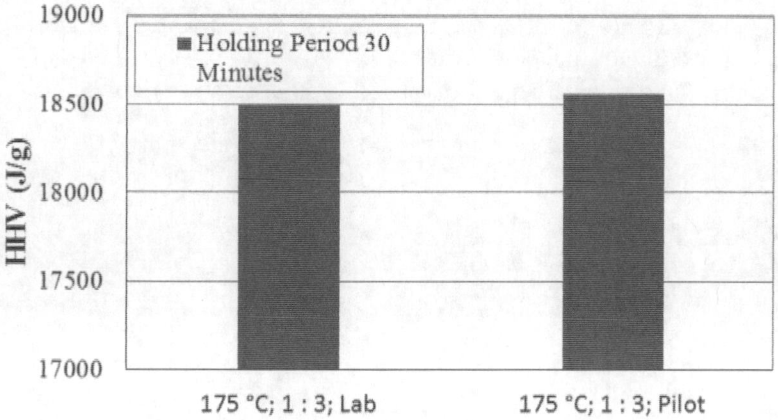

Figure 3. HHV comparison between laboratory scale and pilot scale at the same process parameter

The feasibility process of wet torrefaction in larger scale should be investigated by comparing several criteria in both lab-scale and pilot-scale. This step of study allows investigation of a product and process on an intermediate scale before commercially applied in the industrial-scale production. Characteristics of treated products in pilot-scale reactor can be used for further apparatus design development as well as predicting wet torrefied solid products behavior.

Upscaling study were done by comparing several criteria such as HHV and ash composition. From the previous experiment that has been conducted by using 2.5 L reactor, the HHV of solid product obtained from laboratory scale experiment in the temperature of 175 °C, BWR 1 : 3, and holding period 30 minutes reached 18,502 J/g, thus had slightly lower value than the pilot scale. This difference happened due to the time required to reach the desired holding temperature.
On laboratory scale experiment, heat energy source provided by electric heater with rated power of 1.5 kW. The time needed to reach 175 °C from 35 °C is approximately 25 – 27 minutes. Meanwhile on pilot-scale experiment, the time needed to reach 175 °C from 164.1 °C is 130 minutes, which means pilot-scale reactor has lower heat energy input. This phenomenon leads to hemicellulose degradation through hydrolysis reaction that considerably occurred before holding temperature is reached.

3.2. Physical composition analysis
According to Table 2, the ash fraction in the solid products are showing fluctuated result. This could be happened due to two reasons. The first one is the unknown rate of ash dissolution into the water comparing to other substances. Water act as a solvent for biomass substances, organic and inorganic substances that dissolved in a high rate and high amount must be evaluated. The second one is the presence of impurities in the solvent in form of metal oxides and/or metal hydroxides.

Table 2. Ash Fraction of raw material and wet torrefied solid products

Process Parameters	Ash (%db)
Raw Material	5.62
150 °C – 30 minutes – BWR 1 : 5	6.62
150 °C – 90 minutes – BWR 1 : 5	4.61
175 °C – 30 minutes – BWR 1 : 5	10.36
175 °C – 30 minutes – BWR 1 : 3	12.24
175 °C – 90 minutes – BWR 1 : 5	14.32

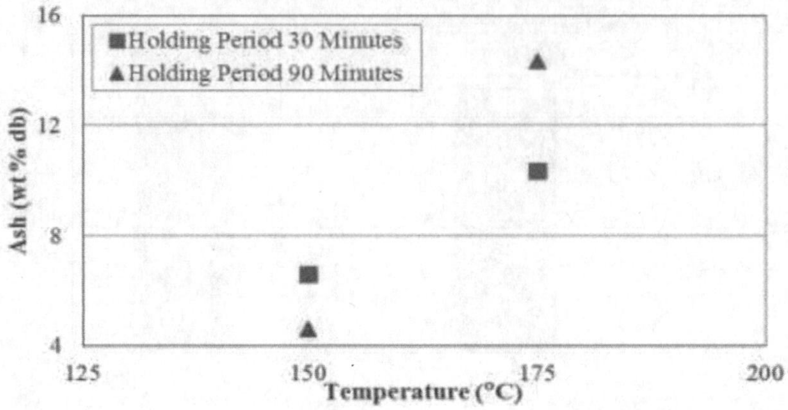

Figure 4. Temperature and holding period effect on ash fraction (BWR 1 : 5)

According to Figure 4, the increased of operating temperature leads to the increase of ash fraction in the solid product. Meanwhile, correlation between holding period and ash fraction is yet to be concluded. According to Figure 5, the ash fraction in the solid product reduced along with the addition of water. The addition of water will soak most fraction of the EFB and enlarge the amount of impurities that could dissolved into the water.

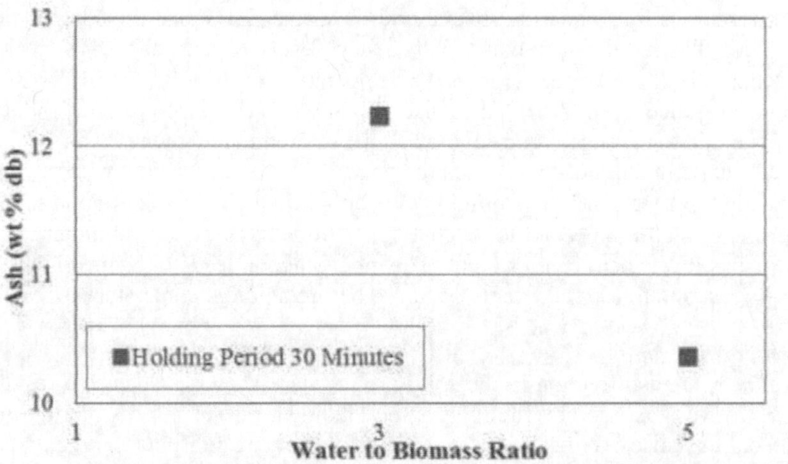

Figure 5. Water to biomass ratio (WBR) effect on ash fraction

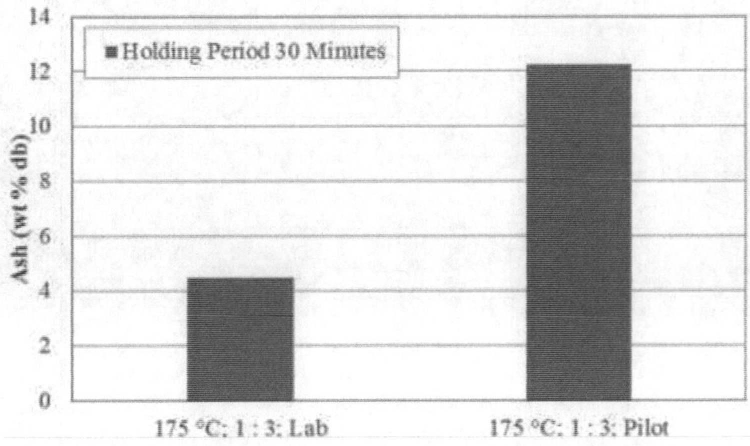

Figure 6. Ash fraction comparison between laboratory scale and pilot scale at the same process parameter

According to Figure 6, pilot scale at the temperature of 175 °C, holding period of 30 minutes, and BWR of 1: 3 showing higher ash fraction comparing to the solid product derived from the laboratory scale. The ash fraction derived from laboratory scale solid product was found at 4.47% (dry basis). This happened due to the amount of water that soaked the biomass. At pilot scale experiment, BWR of 1 : 3 did not soaked the biomass at all. Meanwhile, at the laboratory scale, BWR of 1 : 3 has been soaked the biomass. On pilot scale, there are a lot of solid biomass fraction that did not exposed to water directly and continuously. So, the only way to eliminate ash from the biomass is through metal evaporation mechanism, which the mechanism needs higher operating temperature. So that the process will be similar to dry torrefaction process, which is not using water for its thermal decomposition and inorganic matter solvent.

3.3. Empirical calculation of slagging and fouling tendency

Table 3. Metal oxides composition of raw material and wet torrefied solid products

Compound	Raw Material	150 °C - BWR 1 : 5		175 °C - BWR 1 : 5		150 °C - BWR 1 : 3
		30 Minutes	90 Minutes	30 Minutes	90 Minutes	30 Minutes
SiO_2	39.842	31.463	27.919	29.122	33.39	25.268
K_2O	18.204	7.852	6.813	10.803	6.810	4.703
CaO	21.665	16.437	15.294	15.579	11.914	10.075
Fe_2O_3	2.695	34.635	41.303	35.326	39.102	52.396
SO_3	5.353	3.662	3.0246	3.373	3.378	2.918
P_2O_5	10.719	4.321	4.275	4.527	4.064	3.330
MnO	0.073	0.778	0.775	0.966	0.719	0.524
TiO_2	0.103	0.852	0.596	0.304	0.623	0.785

The ash fraction and metal oxides composition in Table 3 were used to evaluate the alkali index of raw material and wet torrefied solid products based on semi-quantitative analysis. Based on the empirical formula calculation, it was found that raw material have a very high potential of slagging and fouling.

IOP Conf. Series: Earth and Environmental Science **65** (2017) 012003 doi:10.1088/1755-1315/65/1/012003

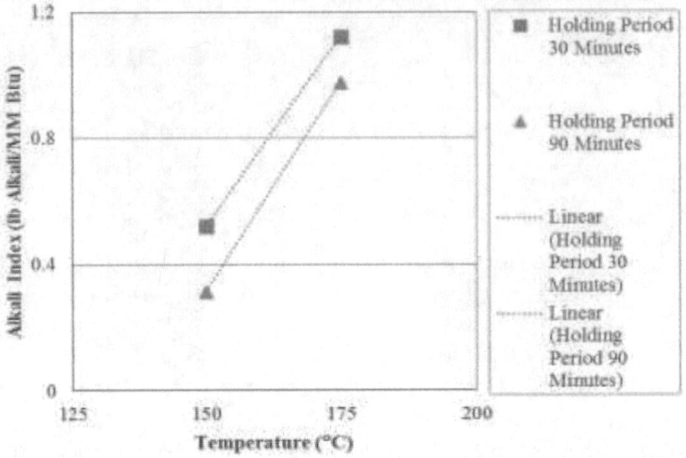

Figure 7. Operating temperature and holding period effect on alkali index of wet torrefied solid products

Based on Table 4, raw material showing a very high slagging and fouling tendencies. Meanwhile, there are three parameters that showing reduction level of slagging and fouling potential when compared with the raw material. The first one is at the operating temperature of 150 °C, holding period of 30 minutes, and biomass to water ratio of 1 : 5 that showing high level of slagging and fouling tendency. The second one is at the operating temperature of 150 °C, holding period of 90 minutes, and biomass to water ratio of 1 : 5 that showing low level of slagging and fouling tendency. The third one is at the operating temperature of 175 °C, holding period of 30 minutes, and biomass to water ratio of 1 : 3 that showing high level of slagging and fouling tendency. According to Figure 7, the alkali index is increasing along with the increasing of operating temperature. In every temperature, longer holding period resulting in lower alkali index. This probably happened because longer holding period leads to longer dissolution reaction.

From the previous research, wet torrefaction experiment on laboratory scale reactor has been conducted to obtain the alkali index of wet torrefied solid products. The comparison of the alkali index is shown throughout Table 5. Based on Table 5, it can be concluded that wet torrefied solid products derived from pilot scale reactor is showing smaller value of alkali index, which means that the products will have smaller tendencies of slagging and fouling potentials.

Table 4. Alkali index of raw material and wet torrefied solid products

Process Parameters	Alkali Index (lb Alkali/MM Btu)	Level of Severity	Annotation
Raw Material	1.022	Very High	Not Selected
150 °C – 30 Minutes – 1 : 5	0.520	High	Selected
175 °C – 30 Minutes – 1 : 5	1.120	Very High	Not Selected
175 °C – 30 Minutes – 1 : 3	0.511	High	Selected
150 °C – 90 Minutes – 1 : 5	0.314	Low	Selected
175 °C – 90 Minutes – 1 : 5	0.975	High	Not Selected

Table 5. Alkali index of raw material and wet torrefied solid products from pilot scale reactor and lab scale reactor

Process Parameters	Alkali Index (lb Alkali/MM Btu) from Pilot-Scale Reactor	Alkali Index (lb Alkali/MM Btu) from Lab-Scale Reactor
Raw Material	1.022	1.022
175 °C – 30 Minutes – 1 : 3	0.511	0.642

The HHV, ash fraction, and alkali index of pilot scale solid product are compared in Table 6. The solid product yield from temperature 150 °C, holding period of 30 minutes, and BWR of 1 : 5 and also temperature 175 °C, holding period of 30 minutes, and BWR of 1 : 3 are showing significant result because it shown an increased HHV and reduced level of slagging and fouling potential compared to the raw material. Solid product yield from temperature 150 °C, holding period of 90 minutes, and BWR of 1 : 5 is much more preferred because it shown an increased HHV, reduced of ash fraction, and decreased slagging and fouling potential down to the lowest level.

Table 6. Raw material and wet torrefied solid products overall comparison

Process Parameters	HHV (J/g)	Ash (%db)	Alkali Index (lb Alkali/MM Btu)	Level of Severity	Annotation
Raw Material	17,252	5.62	1.022	Very High	Not Selected
150 °C – 30 Minutes – 1 : 5	17,311	6.62	0.520	High	Selected
175 °C – 30 Minutes – 1 : 5	17,853	10.36	1.120	Very High	Not Selected
175 °C – 30 Minutes – 1 : 3	18,560	12.24	0.511	High	Selected
150 °C – 90 Minutes – 1 : 5	17,531	4.61	0.314	Low	Selected
175 °C – 90 Minutes – 1 : 5	18,874	14.32	0.975	High	Not Selected

4. Conclusion

The comparison of the trend and result for laboratory scale and pilot scale on wet torrefaction using empty fruit bunch (EFB) have been evaluated. This study has shown the relation of operating parameters (temperature, holding time, and biomass-to-water ratio) to the calorific value, also the slagging and fouling tendency. The relation of operating parameter that has been obtained will be useful to get preferable operating conditions choice on the industrial level. Also with the comparison, we can assure the result similarity between laboratory scale and pilot scale through diagrams and trend lines.

The result shown that the increasing of temperature and holding time with less used of water, affected on the increasing of calorific value. On the same operating parameter, pilot-scale reactor result a slight higher calorific value due to its reactor heat rate. For overall characteristics that has been identified, holding temperature of 150˚C, holding period of 90 minutes and BWR of 1 : 5 is chosen as the best operating parameter to produce solid product with low level of ash severity and an increasing value of HHV comparing to the raw material. The ash severity level successfully drop from very high down to low level while the HHV increase from 17,252 J/g up to 17,531 J/g.

Acknowledgement
This research is supported by Insentif SINAS 2016 (Insinas 2016) Indonesian government research fund.

References

[1] Aghamohammadi N *et al.* 2016 An investigation of sustainable power generation from oil palm biomass: a case study in sarawak *J. Sustainability* **8** 416

[2] Kerdsuwan S and Laohalidanond K 2011 *Renewable energy from palm oil empty fruit bunch* In: *Renewable Energy - Trends and Application* ed M Nayeripour and M Kheshti (Croatia: InTech) chapter 6 pp 123-50

[3] Miles T R *et al.* 1995 *Alkali deposits found in biomass power plants* (Golden: National Renewable Energy Laboratory) p 8

[4] Abdullah N and Sulaiman F 2013 *The oil palm wastes in Malaysia* In *Biomass now - sustainable growth and use* ed Miodrag Darko Matovic (Croatia: InTech) chapter 3 pp 75-99 (http://wwwintechopencom/books/biomass-now-sustainable-growth-and-use/the-oil-palm-wastes-in-malaysia)

[5] United States Department of Agriculture 2017 *World Agricultural Production* (appsfasusdagov/psdonline/circulars/productionpdf)

[6] Jain A, Balasubramanian R and Srinivasan M 2016 Hydrothermal conversion of biomass waste to activated carbon with high porosity: A review *Chem. Eng. J.* **283** 789-805

[7] Reza M T *et al.* 2014 Hydrothermal carbonization of biomass for energy and crop production *J. Applied Bioenergy* **1** 11-29

[8] Yoshikawa K and Prawisudha P 2014 *Hydrothermal treatment of municipal solid waste for producing solid fuel* In: *Application of hydrothermal reaction to biomass conversion* ed F Jin (Berlin: Springer) chapter 13 pp 355-83

[9] Novianti S *et al.* 2015 Low-potassium fuel production from empty fruit bunches by hydrothermal treatment processing and water leaching *J. Energy Procedia* **75** 584-589

[10] Spellman F R and Bieber R M 2012 *Environmental Health and Science Desk Reference* (Plymouth: The Scarecrow Press) p 111

[11] Kabel M *et al.* 2002 Hydrothermally treated xylan rich by-products yield different classes of xylo-oligosaccharides *Carbohydrate Polymers* **50** 47-56

An identification and characterization of biodiesel fatty acid based by using dielectric sensor

Rahmawati[1], T Djatna[2], E Noor[2] and Irzaman[3]

[1]Post Graduate Program of Agroindustrial Technology, Bogor Agricultural University, Bogor ; Department of Electrical Engineering, Lhokseumawe State Polytechnic, Aceh, Indonesia.
[2]Department of Agroindustrial Technology, Bogor Agricultural University, Bogor, Indonesia.
[3]Department of Physics, Bogor Agricultural University, Bogor, Indonesia.

E-mail: rahmawati.gunawan@gmail.com

Abstract. The fatty acids composition is identified by a gas chromatography mass spectrometer (GC-MS), then it is characterized in saturated and unsaturated components. This paper investigates biodiesel fatty acid by using dielectric constant measurements and focuses on dielectric sensor based on biodiesel chemical properties characterization. The objectives of this paper are identification fatty acids and determination of correlation dielectric properties and biodiesel fatty acid characterization. The proposed method is dielectric constant by using capacitance sensor are applied to determine the response dielectric sensor from the fatty acid composition. Sixteen fatty acid methyl esters were identified and two characterizations the amount of saturated and unsaturated fatty ester fractions. The model parameter was determined by regression analysis for estimating the relationships among fatty acid content and dielectric properties. The results show that measurements of electrical properties, successfully used for the characterization of fatty acids. The dielectric constant of biodiesel was found increasing as the saturated decreases. This relationship becomes calibration for the assessment of the quality of biodiesel based on the dielectric sensor. The model reveals that the fatty acid composition affects the value of the biodiesel dielectric and show that dielectric sensor potentially to handle for characterization of biodiesel fatty acid content.

1. Introduction

Biodiesel is a fuel based ester (Fatty Acid Methyl Ester: FAME) must meet the requirements of the specification to be used in diesel engines. A mixture of different fatty acid methyl esters is commonly referred to as biodiesel, which is a renewable alternative fuel. Triacylglycerol (oil or fat) transformed to form esters and glycerol esters are produced are called biodiesel [1,2,3]. ASTM International (American Society for Testing and Materials) defines biodiesel as a mixture of long-chain monoalkylic esters from fatty acids. The fatty acids identification was needed for production to determine the biodiesel quality. The composition of fatty acid content, structure feature and the length of carbon atoms determine the properties of physics and chemistry biodiesel. However, the physicochemical analysis requires high costs and time. This study presents the use of a dielectric

Content from this work may be used under the terms of the Creative Commons Attribution 3.0 licence. Any further distribution of this work must maintain attribution to the author(s) and the title of the work, journal citation and DOI.
Published under licence by IOP Publishing Ltd

sensor and a quick and inexpensive method to correlate the physicochemical variables to the dielectric constant of the fatty acids biodiesel characterization. The change in dielectric constant of FAME from different biodiesel production will be presented. This research wants to apply the application of dielectric measurements to identified saturated degree of biodiesel fatty acid content, thus, the relation between the electrical properties and the degree of saturated were analyzed. The proposed method allows reducing costs in the characterization of fatty acids and the reduction in analysis time. In addition, the method allows an assessment of the quality of biodiesel based on fatty acids composition. As a commercial fuel, biodiesel quality is analyzed using sophisticated equipment to make sure it meets the requirements of international standards and to be stored in the long term. The quality of biodiesel based on physicochemical properties depending on the distribution of fatty acid [4], [5]. Generally, gas chromatography (GC) has been the most widely used method for the analysis of biodiesel owing to its generally higher accuracy in quantifying minor components but not typically used for online measurement [6]. However, the accuracy of GC analyses can be influenced by factors such as baseline drift, overlapping signals, derivative samples and aging of standards and samples. Furthermore, gas chromatography is a measurement that is not real-time and requires expensive equipment costs that are not easily applied to process monitoring and control in small and medium scale industries [7].

Most of the important parameters of the end product biodiesel is a fatty characterization of mono-alkyl esters, fatty acids, glycerol, and derivatives. Feature structure, the length of the carbon atom, and the degree of saturation of fatty acids determines the physical properties of biodiesel such as viscosity, flash point (FP), cetane number, density, high heating values [5], [8]–[12]. It takes the alternative technologies that are a simple, non-invasive and low cost to evaluate and monitor the production process and the quality of biodiesel. Romano and Sorichetti 2011, the parameters that determine the quality of biodiesel can be classified into two major groups, the first includes public properties (density, viscosity, flash point, cloud point, pour point, cetane number, numbers neutralization), the second describes the chemical composition and purity of a mixture of fatty acid esters (alcohol content, ester content, the content of mono, di and triglycerides, total quantity of glycerol, iodine value).

Some measurement techniques such as the implementation of electrochemical impedance spectroscopy [14], dielectric spectroscopy [15], have managed to characterize the physical properties of biodiesel based on the dielectric constant, resistivity and relaxation time or frequency. Dielectric spectroscopy has been successfully used for monitoring the production and characterization of biodiesel, these measurements give a quantitative indication of the advance of the purification process and the conversion to biodiesel [13]. Permittivity and Conductivity are used to characterize feedstocks from different origins and biodiesel content estimation of biodiesel blend [7], [16], to characterize the different stages of the production process during the purification process and in the final product [17], [18], and also for the characterization of FAME [19]. Although in the literature there are references on the dielectric of biodiesel properties there is comparatively less information on correlation dielectric sensor with fatty acid in the final product. The physical and chemical fuel properties of biodiesel fundamentally depend on the fatty acids distribution of the triglyceride obtained from the raw material used for biodiesel production [5], [8]. Fatty acid content such as higher degree of unsaturation of biodiesel fuels led to a longer ignition delay and, consequently, a more retarded start of combustion [20]. Dielectric measurements are relevant for the production and characterization of biodiesel since the refractive index of FAME depends essentially on the electronic polarization of the molecules of fatty acid esters. The reduction of the dielectric losses of treated oil is affected due to the elimination presence of contaminants [16], [17]. The differences in dielectric properties among tested oils generally correspond to the degree of unsaturation as evidenced by iodine values (IV), increase with the degree of unsaturation or the number increasing of double bonds of unsaturated fatty acids in oils [21]–[23]. Since dielectric properties of materials are closely corelated with saturated/unsaturated fatty acids in oils and therefore, researchers properly designed and calibrated electrical instruments, which can be directly used to determine the saturated/unsaturated fatty acids in oils accurately. The technique is reliable enough for predicting fatty acid in biodiesel.

In this paper, we report results from the investigation of fatty acid content in biodiesel production used the dielectric sensor. Quantitative characterization and quality biodiesel based on fatty acid composition were analyzed using Gas Chromatography, which the amount and type of fatty acids determine the physical properties of biodiesel [24]. Dielectric sensors use capacitance to measure the dielectric constant of a surrounding medium. The sensitivity of the capacitive sensor is affected by the dielectric constant of the target material being detected. The permittivity of materials is most commonly expressed in relation to the permittivity of free space and therefore named relative permittivity (dielectric constant). The same dielectric models applied to biodiesel with the content and composition of fatty acids that are different. The dielectric sensor is currently in widespread use in biodiesel production since it makes possible the characterization of the sample and the detection of moisture and other contaminants that impact on the quality of the feedstocks and the final product. In this paper, biodiesel classification is determined by dielectric constant measurements in the range from 100 Hz to 100K Hz, for samples of biodiesel at room temperatures. As an alternative biodiesel quality testing, dielectric properties require little independent evaluation.

The electrical properties are capable of measurements in both laboratory and industrial environments, so very potential applied to the automated process control from online measurement systems [19]. The objectives of this paper are identification fatty acids by Gas Chromatography and determination of correlation dielectric properties and characterization of biodiesel fatty acid. This relationship becomes calibration for the assessment of the quality of biodiesel based on the dielectric sensor. The results presented in this work show that dielectric sensor is potentially useful for these purposes, and this technique is fast, non-destructive, real time and reasonable cost.

2. Materials and methods

2.1. Materials

Samples of biodiesel used are biodiesel feedstock RBDPO and used cooking oil. Testing was performed using three samples in different production systems. The percentage composition of fatty acids in the biodiesel was determined by the gas chromatography-mass spectrometer (GCMS), and the dielectric properties were determined by LCR meter.

2.2. Biodiesel Characterization

Biodiesel Characterization by using GC-MS analysis showed that major fatty acid components in all esters were saturated and unsaturated fatty acid. The retention time and peak height/area were used in qual*itative and quantitative deter*mination of the biodiesel components. The presence of saturated fatty acid in the obtained biodiesel leads to high viscosity, high cetane number and better biodiesel stability [25]. The dielectric constant values of biodiesels were mainly affected by unsaturated fatty acids. Further- more, the dielectric constant of biodiesels increased with increasing the degree of unsaturation or the number increasing of double bonds of unsaturated fatty acids in biodiesel [23]. The higher level of unsaturated fatty acid reduces fuel quality, because of its easy oxidation [26] and saturated fatty acids such as 16:0 or 18:0 are more stable than unsaturated ones like 18:1, 18:2 and 18:3, respectively, which decreases the fuel quality [27].

Generally, the main components of each biodiesel were five FAMEs methyl palmitate (hexadecanoic acid, methyl ester; CAS No. 112-39-0), methyl stearate (octadecanoic acid, methyl ester; CAS No. 112-61-8), methyl oleate (octadecenoic acid, methyl ester; CAS No. 112-62-9), methyl linoleate (octadecadienoic acid, methyl ester; CAS No. 112-63-0), and methyl linolenate (octadecatrienoic acid, methyl ester; CAS No. 301-00-8). Three other FAMEs were present in minor amounts methyl

myristate (tetradecanoic acid, methyl ester; CAS No. 124-10-7), methyl arachidate (eicosanoic acid, methyl ester; CAS No. 1120-28-1), and methyl behenate (docosanoic acid, methyl ester; CAS No. 929-77-1) [25], [28]. The composition of biodiesel fatty acids is varying from source to source and usually contains methyl esters of C-12 to C-24 fatty acids having quite appreciable quantities of unsaturated (C18:1 and C18:2) fatty acids [29].
Quantitative analysis is used to identify the amount of fatty acid in biodiesel because biodiesel quality is reflected in the composition of fatty acids [30]. Identification of FAME components of biodiesel was analyzed by using GC-MS (ASTM E202). Gas chromatography to analyze the content of fatty acid using Agilent 6890 GC Method. The oven program was given as follows initial temperature 60 ^{o}C post temperature 80 ^{o}C maximum temperature 325 ^{o}C. The previous studies have done calculations fatty acid composition of biodiesel based on measurements GCMS [29], [31].

2.3. Dielectric Properties Measurement
The testing dielectric of biodiesel using a sensor with two parallel electrodes is done by fixing in PCB plate. Dielectric measurements on a sample of biodiesel using LCR meter Lutron 8014. The tests were carried out and verified with the various reference fuel in 3 sample reference biodiesels from RBDPO and cooking oil. Some previous researchers showed that measurements of electrical properties, successfully used for the characterization of fatty acid methyl ester (FAME) based on mixtures of fatty acid methyl esters from different vegetable oils [19], [32], [33].

A capacitor is a device that stores electric charge. The simplest example of a capacitor consists of two conducting plates of area A (A = 4 cm x 4 cm), which are parallel to each other, and separated by a distance d (d = 2.5 mm) as shown in Figure 1. Electrical properties of the samples were determined from dielectric constant in the frequency 100 Hz and at room temperature.

The capacitance C (farads) of a parallel plate capacitor of area A (m^2), the separation between the plates separated by empty space (or air) is:

$$C = \varepsilon_0 \frac{A}{d} \qquad (1)$$

where ε_0 is the *permittivity of free* space (8.85 x 10^{-12} *F/m*, farads/meters), d is the separation between the plates (meter). If the empty space is replaced by a dielectric, capacitance increases [34], [35]. The dielectric constant of material is complex and proportional to the material electrical permittivity and free space permittivity.

$$K = \frac{\varepsilon}{\varepsilon_0} \qquad (2)$$

$$K = \varepsilon_r + i\varepsilon_i \qquad (3)$$

where K is the complex dielectric constant, ε is the electrical permittivity, ε_0 is the *permittivity of free* space, ε_r is the real part of the complex dielectric constant will be referred to as the dielectric constant of material and ε_i is the imaginary part of the complex dielectric constant can be called loss factor, dissipation factor or dielectric loss [36]–[40]. The permittivity of materials is most commonly expressed in relation to the permittivity of free space and therefore named relative permittivity (dielectric constant).

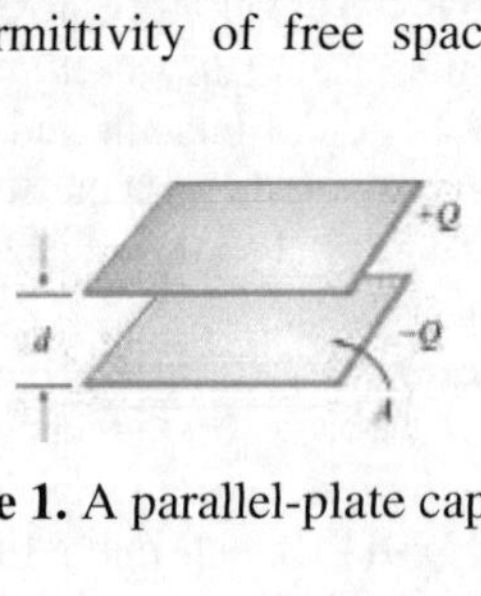

Figure 1. A parallel-plate capacitor

3. Results and Discussion

In order to understand the fatty acid content and dielectric constant of biodiesel. The contents of fatty acids were analyzed using GCMS, identify the fatty acid content consisting of saturated and unsaturated. Furthermore, the correlation analysis of dielectric properties and chemical composition based on the amount of saturated and unsaturated.

3.1. Chemical Composition of Biodiesel Using GC-MS

GC-MS analysis was used the chemical composition of the three biodiesel fuel samples for profiling of various fatty acid methyl esters. Twelve peaks were observed in the area picture; each peak corresponds to a fatty acid methyl ester and was identified from the library match software (Willey09TH). The fatty acid chromatogram showing the different components present in each biodiesel sample.

Each FAME was present in different proportions in the three samples. The physicochemical properties of biodiesel are influenced by the structural features of fatty acid, chain length, the degree of unsaturation, and branching of the chain [8]. Table 1 gives the comparative analysis of saturated and unsaturated properties of fatty acids. The result shows that the biodiesel contains saturated fatty acid 63.97%, 61.93%, 55.23%, and unsaturated fatty acid 36.03%, 38.075%, 45.062% respectively for sample 1 (S1), sample 2 (S2) and sample 3 (S3). The characteristics of these biodiesel samples fall within a narrow band. Methyl ester contains a higher percentage of saturated fatty acid, with a lower percentage of unsaturated from GC analysis showed which clearly proved that the prepared esters have a very good fuel property [25]. The influence of these proportions on the dielectric properties will be discussed in the sections that follow.

3.2. Dielectric constant of Biodiesel

Dielectric properties of biodiesel samples were measured at three different total amount of saturated and unsaturated fatty acid. In the dielectric characterization of materials, simple measurement is capacitance and then converted to the dielectric constant. Capacitance measurement of materials and air using LCR meter, in which the tube is filled to biodiesels and measurement of air in the empty tube. Calibration of measuring instruments in the short and open circuit. Based on the results of measurements of air capacitance, constant dielectric obtained on the empty cell is 1,005. Constant dielectric was calculated by the formula 1 and 2. The dependence of the dielectric constant on the frequency and saturated fatty acid content in this work are illustrated in Figure 2. The dielectric constant of all samples decreased with increasing frequency at individual saturated fatty acid. Generally, the dependence of the dielectric constant of biodiesels on frequency as well as overall values, are similar to those obtained for other agricultural commodities including those used in other biofuel studies was also reported [39], [41]–[44].

Prediction content of saturated FAME is possible using the measurement of dielectric because it has different properties in each polar fatty acid component. This observation is in close agreement with earlier results, i.e., [25] observed comparative analysis of saturated properties with unsaturated properties of fatty acids and [23] reported the effect of oil composition on dielectric properties. Saturated FAME of biodiesel FAME composition for predicting physical properties of biodiesel from its FAME composition using regression equation. Dielectric measurements can be used to predict the FAME content of saturated fatty acids in biodiesel because have different polar properties. Using the regression equation by previous researchers that the physical properties of biodiesel and biodiesel fatty acid composition have an excellent correlation [45], [46].

Figure 3 shows the dependence of dielectric constant on saturated fatty acid content at 100 Hz, 120 Hz, 1K Hz, 10 K Hz and 100 K Hz. This observation is appropriate with earlier results [23,37] reported the differences in dielectric properties among tested oils generally correspond to the degree of unsaturation. Fatty acid composition of biodiesels is another determining factor of dielectric constant. Dielectric constant increases with the decreases in the degree of saturation for all frequencies, which increases in the degree of saturation as the electron rich double bond has a negative charge and tends to oscillates to a certain extent, with the changing electric field.

The saturated fatty acids contain single bonds between carbon atoms of parent chain of molecule. Each carbon atom connected with four other atoms in molecule. Since all are single covalent bonds which are sigma bonds, C-C bonds are very strong and difficult to cleave under normal conditions. The unsaturated fatty acids have multiple covalent bonds like double or triple covalent bonds between carbon atoms of parent chain which have pi-bonds. Pi-bonds are weaker than sigma bonds so pi-bonds can easily cleaved. The ability for the π electron to move freely between carbon atoms gives rise to a high electrical conductivity [33], [47]. The permittivity depends on dielectric constant ε′, which is related to capacitance of a substance and its ability to store electrical energy.

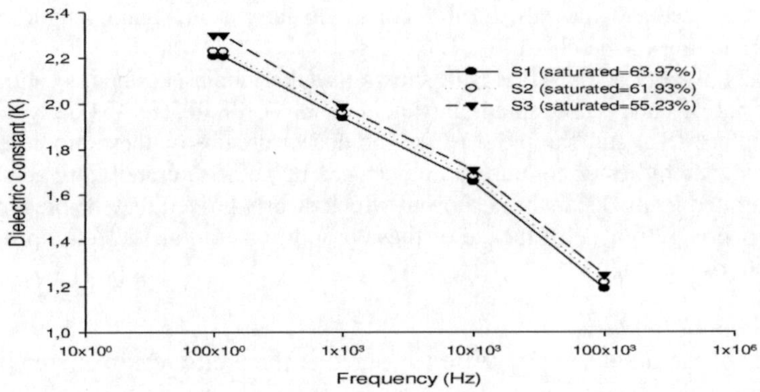

Figure 2. The dielectric constant as function of frequency for different saturated fatty acid contents

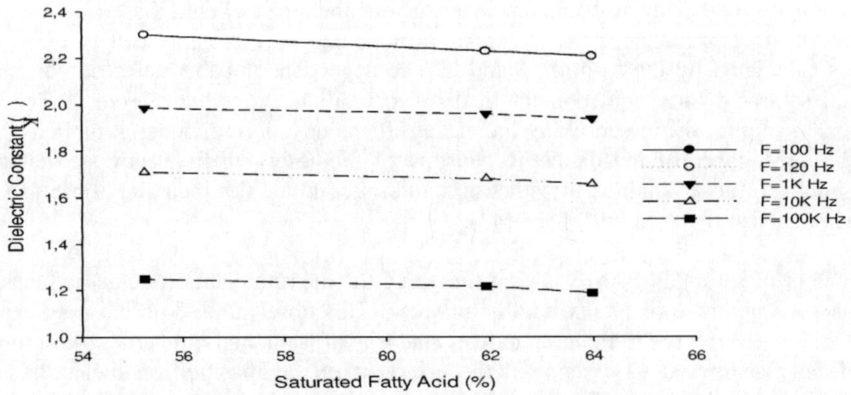

Figure 3. The dielectric constant as a function of saturated fatty acid

Table 1. Measured values of fatty acid in biodiesel

Type	Fatty acid	Peak area (%)		
		S1	S2	S3
Unsaturated	1. Oleic	35.466	34.698	41.701
	2. Palmitoleic		3.021	
	3. Eicosenoic	0.564	0.356	1.854
	4. Linoleic			0.719
	5. Octadecadienal			0.467
	6. Eicosapentaenoic			0.321
Total unsaturated		36.030	38.075	45.062
Saturated	1. Caprylic		0.215	0.383
	2. Lauric	0.567	1.247	1.041
	3. Palmitic	45.84	45.521	40.210
	4. Stearic	10.037	8.806	6.756
	5. Arachidic	1.674	0.427	1.860
	6. Myristic	4.734	4.553	4.332
	7. Margaric		1.156	
	8. Hexadecanoic	0.489		
	9. Docosanoic	0.328		0.358
	10. Lignoceric	0.299		0.290
Total Saturated		63.970	61.925	55.230

The dielectric properties of biodiesels were correlated with their different compositions using a statistical model. The best subset regression was used to determine which parameters affect the dielectric properties of biodiesels such as dielectric constant. The parameters include percent of fatty acids which are the main components in biodiesel. For saturated FAME, there is a nonlinear decrease in dielectric properties with increasing saturated fatty acid content. The plot of the predicted saturated fatty acid versus the actual output saturated fatty acid are depicted in Figure 4. The saturated fatty acid of FAME was given as the dependent variable Y, while the dielectric properties (capacitance and dielectric loss) were given as variable x. Regression diagnostic performs a multilinear regression of the response in Y on the predictors in x described as the following multiple regression equations:

$$Y=77.86+0.05\ x_1 - 90.38\ x_2 \qquad (4)$$

where Y is the predicted saturated fatty acid (%) of biodiesel, x_1 is capacitance and x_2 is a dielectric loss of dielectric properties of biodiesel. A high correlation coefficient (R^2:0.981) indicates the excellent correlation between the saturated fatty acid and the dielectric properties of biodiesel.

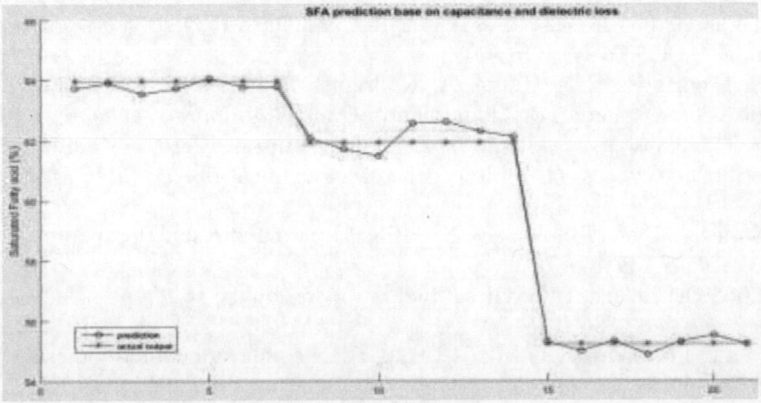

Figure 4. The saturated fatty acid prediction base on dielectric properties

Several kinds of biodiesel derived from RBDPO and cooking oil were selected to verify the practicability of the dielectric measurement. Their typical FAME compositions are shown in Table 1 base on GC-MS. The predicted saturated fatty acid of these biodiesels is also plotted in Fig. 4. The desired result is observed namely, that all of the predicted values are in the range of the experimental values. The results are consistent with earlier results, it is reported that dielectric constant for vegetable oil, soybean oil is higher than palm olein suggesting that dielectric increase with degree of unsaturation [23], [48]. It seems that the technique is reliable not only in preliminary tests for specific biodiesel origin. This indicates that the present correlative model is a simple and practical method to give a satisfactory predicted saturated fatty acid of FAME composition of a biodiesel from its dielectric properties without arduous and expensive experimental determination.

The measurements of dielectric properties, successfully used for the characterization of fatty acids and the dielectric sensor potentially to handle for characterization of biodiesel fatty acid content. In practice, in the case of an important number of potential effects, this could be achieved from the computational viewpoint by training. Need more observation by using another parameter such as temperature to develop a handy measurement for fatty acid identification.

4. Conclusion

Fatty acid characterization using dielectric sensor showed that the increasing dielectric constant of biodiesel was found as the saturated decreases. The results show that measurements of electrical properties, successfully used for the characterization of fatty acids. For future works, it is required to embed this approach to a handy sensor that integrates dielectric probe with identification tool.

Acknowledgements

This research has been supported by the BPPDN Grant, the Ministry of Research Technology and the Higher Education Republic of Indonesia in Domestic Doctorate Scholarship Program.

References

[1] Ghanei A E, Moradi R, Taherpour G R and Kalantari R 2011 Variation of physical properties during transesterification of sunflower oil to biodiesel as an approach to predict reaction progress *Fuel Processing Technology* **92** 1593–8

[2] Leduc S, Natarajan K, Dotzauer E, McCallum I and Obersteiner M 2009 Optimizing biodiesel production in India *Appl. Energy* **86** S125–S131

[3] Randazzo M L and Sodré J R 2011 Exhaust emissions from a diesel powered vehicle fuelled by soybean biodiesel blends (B3-B20) with ethanol as an additive (B20E2-B20E5) *Fuel* **90** 98–103

[4] Gerhard K, Van Gerpen J, Haas M J, Foglia T A and Dunn R O 2005 *The Biodiesel Handbook* (Champaign : AOCS Press)

[5] Ramos M J, Fernández C M, Casas A, Rodríguez L and Pérez A Influence of fatty acid composition of raw materials on biodiesel properties *Bioresource Technology* **100** 261–8

[6] Rachmanto T, Allanson D, Matthews D and Jenkinson I 2014 Monitoring of Biodiesel Transesterification Process Using Impedance Measurement *Int. J. Mater. Mech. Manuf.* **2** 265-71

[7] Corach J, Sorichetti P and Romanoa S 2015 Electrical and ultrasonic properties of vegetable oils and biodiesel *Fuel* **139** 466–71

[8] Knothe G 2005 Dependence of biodiesel fuel properties on the structure of fatty acid alkyl esters *Fuel Process. Technol.* **86** 1059–70

[9] Fakhry E M and El Maghraby D M 2013 Fatty acids composition and biodiesel characterization of *Dunaliella salina Journal of Water Resource and Protection* **2013** 894–9

[10] Mariod A, Klupsch S, Hussein I H and Ondruschka B 2006 Synthesis of alkyl esters from three

unconventional sudanese oils for their use as biodiesel *Energy and Fuel* **2** 2249–52

[11] Giwa S O, Adekomaya S O, Adama K O and Mukaila M O 2015 Prediction of selected biodiesel fuel properties using artificial neural network *Fronties in Energy* **9** 433–45

[12] Cheenkachorn K 2006 Predicting properties of biodiesels using statistical models and artificial neural networks *As. J. Energy Env*, **7** 299–306

[13] Romano S D and Sorichetti P A 2011 Standards for fuel characterization in *Dielectric Spectroscopy in Biodiesel Production and Characterization* (New York : Springer) pp 29–41.

[14] Pereira T C, Conceição C A F, Khan A, Fernandes M S, Raquel M.T. Ferreira, E. P. Marques, and A. L. B. Marques, 2016 Application of electrochemical impedance spectroscopy: A phase behavior study of babassu biodiesel-based microemulsions *Spectrochimica Acta Part A: Molecular and Biomolecular Spectroscopy* **168** 60–4

[15] Schönhals A and Kremer F 2003 Analysis of dielectric spectra in *Broadband Dielectric Spectroscopy* (Springer : Heidelberg)

[16] Corach J, Sorichetti P A, and Romanoa S D 2016 Permittivity of biodiesel-rich blends with fossil diesel fuel : Application to biodiesel content estimation *Fuel* **177** 268–73

[17] Sorichetti P A and Romanoa S D 2005 Physics and Chemistry of Liquids : An Physico-chemical and electrical properties for the production and characterization of Biodiesel *Phys. Chem. Liq. An Int. J.*, 43 37–41

[18] Gonzalez Prieto L E, Sorichetti P A and Romanoa S D Electric properties of biodiesel in the range from 20 Hz to 20 MHz . Comparison with diesel fossil fuel *Int. J. Hydrogen Energy* **33** 3531–7

[19] Corach J, Sorichetti P A and Romanoa S D 2011 Electrical properties of mixtures of fatty acid methyl esters from different vegetable oils *Int. J. Hydrogen Energy* **37** 14735–9

[20] Benjumea P, Agudelo J R and Agudelo A F 2011 Effect of the degree of unsaturation of biodiesel fuels on engine performance, combustion characteristics, and emissions *Energy and Fuels* **25** 77–85

[21] Rudan-Tasič D and Klofutar C 1999 Characteristics of vegetable oils of some slovene manufacturers *Acta Chim. Slov.* **46** 511–21

[22] Pace W E, Westphal W B and Goldblith S A 1968 Dielectric properties of commercial cooking oils *J. Food Sci.* **33** 30–36

[23] Hu L, Toyoda K and Ihara I 2008 Dielectric properties of edible oils and fatty acids as a function of frequency, temperature, moisture and composition *J. Food Eng.* **88** 151–8

[24] Gangadhara R and Prasad N B L 2016 Studies on optimization of transesterification of certain oils to produce biodiesel *Chem. Int.* **2** 59–69

[25] Murugesan A, Subramaniam D, Avinash A, and Nedunchezhian N 2013 Quantitative and qualitative analysis of biodiesel – an in-depth study A *Int. J. Ambient Energy*, 36 37–41

[26] Monyem A and Van Gerpen J H 2001 The effect of biodiesel oxidation on engine performance and emissions *Biomass and Bioenergy* **20** 317–25

[27] Tyson K S 2001 *Biodiesel Handling and Use Guidelines* U.S. Department of Energy

[28] Ott L S and Bruno T J 2008 Variability of Biodiesel Fuel and Comparison to Petroleum-Derived Diesel Fuel : Application of a Composition and Enthalpy Explicit Distillation Curve Method *Energy & Fuels* **22** 2861–8

[29] Chopra A, Tewari A K, Vatsala S, Kumar R, Sarpul A S and Basu B 2011 Determination of Polyunsaturated Fatty Esters (PUFA) in Biodiesel by GC / GC – MS and 1 H-NMR Techniques *J Am Oil Chem Soc* **88** 1285–96

[30] Koria L and Nithya G 2012 Analysis of Datura stramonium Linn . biodiesel by gas chromatography - mass spectrometry (gc-ms) and influence of fatty acid composition on the fuel related characteristics *Journal of Phytology* **4** 6–9

[31] Basumatary S, Barua P and Deka D C 2013 Research Article Identification of chemical composition of biodiesel from Tabernaemontana divaricata seed oil *Journal of Chemical and Pharmaceutical Research* **5** 172–9

[32] M'Peko J C, Reis D L S and De Souza J E 2013 Evaluation of the dielectric properties of biodiesel fuels produced from different vegetable oil feedstocks through electrochemical impedance spectroscopy *Int. J. Hydrogen Energy* **38** 9355-9

[33] Spohner M 2012 A Study of the Properties of Electrical Insulation Oils and of the Components of Natural Oils *Acta Polytech.* **52** 100–5

[34] Onimisi M Y and Ikyumbur J T 2015 Comparative Analysis of Dielectric Constant and Loss Factor of Pure Butan-1-ol and Ethanol *Am. J. Condens. Matter Phys.* **5** 69–75

[35] Eskiner M, Bär F, Rossner M, Munack A and Krahl J 2015 Determining the aging degree of domestic heating oil blended with biodiesel by means of dielectric spectroscopy *Fuel* **143** 327–33

[36] Seyfried M S and Murdock D 2004 Measurement of Soil Water Content with a 50-MHz Soil Dielectric Sensor *Soil Sci. Soc. Am. J.*, **68** 394-403

[37] Muley P D and Boldoe D 2013 Investigation of microwave dielectric properties of biodiesel components *Bioresour. Technol.* **127** 165–74

[38] Korostynska O, Blakey R, Mason A and Al Shamma'a A 2013 Physical Novel method for vegetable oil type verification based on real-time microwave sensing *Sensors Actuators A. Phys.* **202** 211–6

[39] Nelson S O 2010 Fundamentals of Dielectric Properties Measurements and Agricultural Applications', *J. Microw. Power Electromagn. Energy* **44** 98–113

[40] Blakey R T and Morales-Partera A M 2016 Microwave dielectric spectroscopy – A versatile methodology for online, non-destructive food analysis, monitoring and process control *Eng. Agric. Environ. Food* **9** 264–73

[41] Nelson S O 2008 Dielectric properties of agricultural products and some applications *Res. Agric. Eng.* **54** 104–112

[42] Guo W, Shaojin W, Tiwari G, Johnson J A and Tang J 2010 Temperature and moisture dependent dielectric properties of legume flour associated with dielectric heating *LWT - Food Science and Technology* **43** 193–201

[43] Skierucha W, Wilczek A and Szyplowska A 2012 Dielectric spectroscopy in agrophysics *Int. Agrophys* **26** 187–197

[44] Fennell L P and Boldor D 2013 Dielectric Characterization of the Seeds of Invasive Chinese Tallow Tree *J. Microw Power Electromagn Energy* **47** 237–50

[45] Tong D, Hu C, Jiang K and Li Y 2011 Cetane Number Prediction of Biodiesel from the Composition of the Fatty Acid Methyl Esters *Journal of the American Oil Chemists' Society* **88** 415–23

[46] Sivaramakrishnan K and Ravikumar P 2012 Determination of cetane number of biodiesel and it's influence on physical properties *ARPN Journal of Engineering and Applied Sciences* **7** 205–11

[47] Littlejohn S D 2014 *Electrical Properties of Graphite Nanoparticles in Silicone* (New York : Springer)

[48] Sairin M A, Nizar N N A, Aziz S A, and Rokhani F Z 2016 Study of dielectric permittivity and fatty acid composition for fats and oil in wide frequency spectroscopy measurement at 0.5-50 GHz *Proc. Int. Conf. Sens. Technol. ICST* November 2016

IOP Conf. Series: Earth and Environmental Science **65** (2017) 012005 doi:10.1088/1755-1315/65/1/012005

Application of diethanolamide surfactant derived from palm oil to improve the performance of biopesticide from neem oil

F N Nisya[1], **D Prijono**[1,2] **and A Nurkania**[1]

[1]Surfactant and Bioenergy Research Center (SBRC), Bogor Agricultural University, Jl. Pajajaran No.1. Bogor, Indonesia
[2]Department of Plant Protection, Bogor Agricultural University, Bogor ,Indonesia

Email : fifin.nisya@gmail.com

Abstract. The purpose of this research was to improve the performance of organic pesticide derived from neem plant using diethanolamide surfactant (DEA) derived from palm oil in controlling armyworms. The pesticide was made of neem oil. Neem oil is a neem plant product containing several active components, i.e. azadirachtin, salanin, nimbin, and meliantriol which act as a pesticide. DEA surfactant acts as a wetting, dispersing and spreading agent in neem oil pesticide. The neem oil was obtained by pressing neem seeds using a screw press machine and a hydraulic press machine. DEA surfactant was synthesized from methyl esters of palm oil olein. Pesticide formulation was conducted by stirring the ingredients by using a homogenizer at 5,000 rpm for 30 minutes. Surfactant was added to the formulation by up to 5%. Glycerol, as an emulsifier, was added in to pesticide formulations of neem oil. The efficacy of the pesticides in controlling armyworms fed soybean leaves in laboratory was measured at six concentrations, i.e. 10, 13, 16, 19, 22, and 25 ml/L. Results showed that the neem oil used in this study had a density of 0.91 g/cm^3, viscosity of 58.94 cPoise, refractive index of 1.4695, surface tension of 40.69 dyne/cm, azadirachtin content of 343.82–1.604 ppm. Meanwhile, the azadirachtin content of neem seed cake was 242.20 ppm. It was also found that palmitic (31.4%) and oleic (22.5%) acids were the main fatty acids contained in neem oil. As the additive material used in neem oil in this study, diethanolamide surfactant had a pH of 10.6, density of 0.9930 g/cm^3, viscosity of 708.20 cP, and surface tension of 25.37 dyne/cm. Results of CMC, contact angle, and droplet size analyzes showed that diethanolamide surfactant could be added into insecticide formulation by 5%. Results of LC tests showed that on *Spodoptera litura* the LC_{50} and LC95 values were 13 and 22 ml/L, respectively. Neem oil was found to inhibit the development of *Spodoptera litura* and its larval molting process.

1. Introduction

Surfactant is a compound which has an ability to reduce surface tension of a medium. It can be produced biochemically or in a synthetically chemical way. It is able to reduce surface tension because it has both hydrophilic (like water) and hydrophobic (like fat/oil) groups. These polar and non-polar parts of surfactant make it able to make water and oil mix well in a mixture. According to [1], in general, the functions of a surfactant are to reduce surface and interfacial tension, to increase the stability of dispersed globules, and to control the formation type of emulsion, such as oil in water (O/W) or water in oil (W/O). Today, surfactant is used as a component of adhesive material, coagulating agent, wetting agent, foaming agent, emulsifier, penetrating agent, and dispersing agent [2]. Diethanolamide surfactant is a surfactant whose molecules have no charges and its hydrophilic and hydrophobic prop-

Content from this work may be used under the terms of the Creative Commons Attribution 3.0 licence. Any further distribution of this work must maintain attribution to the author(s) and the title of the work, journal citation and DOI.
Published under licence by IOP Publishing Ltd

erties are resulted from the existence of ester oxygen and hydrocarbon ester groups. Hydrocarbon groups consist of carbon-carbon and carbon-hydrogen bonds. In organic molecules, these bonds are non-polar bonds and hydrophobic. As these parts get longer, the solubility of the molecules in water is low. According to [3], alcohol and ester groups consist of polar molecules. This property makes ester oxygen groups hydrophilic. Diethanolamide can be produced in two ways, namely through (1) the reaction between methyl ester and diethanolamide and (2) reaction between fatty acid and diethanolamide.

Insecticide from neem oil has been long known and used by farmers to organically control pests in plants. Among other plant insecticides available in the market today, neem oil is the one which has low toxicity for beneficial organisms. This has made it a promising plant insecticide [4]. The target pests of neem oil include *A. culicifacies* [5], *Ceraeochrysa claveri* [6,7] *Diaphorina ri* [8], *Mamestra brassicae* [9] *Nilaparvata lugens Stal* [10], *Pieris brassicae* [11[, *Limnoperna fortunei* (Dunker 1857) [13], and *Spodoptera frugiperda* [13].

Neem oil is extracted from seeds of neem plants. Neem plants belong to the family of *Meliaceae* and are found to be limited in Indonesia and the Philippines-Azadirachta excelsa Kack [14]. Today, neem oil is well known all over the world as a source of phytochemicals for human health and pest control. It contains at least 100 active components but triterpenoid, better known as limonoids, is the main component and azadirachtin which has a pesticide effect is the most important component. Azadirachtin has a molecular weight of 720 g/mol [4]. In Indonesia, neem plants are found alongside the north coast of Java, stretching from Indramayu to Banyuwangi. In addition, they are also found in Nganjuk, Jombang, Blitar, Ponorogo, Madiun, Bojonegoro, Bondowoso, Gianyar, Negara, and East Lombok [15].

One of the disadvantages of neem oil is that it is less stable under the field condition as a result of its high photodegradation level, short staying time, and slow killing power compared to non-plant pesticides [16, 17, 18]. Environmental factor and method of extraction gave significant effects on the composition of neem oil. As a result, there is no standard active material in the composition of this plant insecticide and this restricts the use of it in agricultural pest control [19,20,21]. Therefore, in its application, an additive material which can stick, wet, and disperse it to the all parts of leaves, is required to make pest control more effective. One of the additive materials that can be used for this purpose is diethanolamide (DEA) surfactant. This study was aimed at improving the performance of organic pesticide made of neem seed oil by using DEA surfactant from palm oil in controlling *Spodoptera litura* pests.

2. Materials and Methods

2.1 The process of neem oil production and the analysis of its physicochemical properties

Dried neem seeds were pressed by using a screw press. The cake produced from this pressing process was repressed to take the remaining oil in the cake out. The crude oil obtained from this pressing process was left alone for 24 hours to separate the sediment which was still found in the crude oil. The sediment was then pressed by using a hydraulic press to take the oil out from the sediment. This crude oil was then analyzed for its azadirachtin content and physicochemical properties. Measurements were taken on azadirachtin content, refractive index, optical rotation, density, viscosity, surface tension, solubility, and contact angle. The analysis of physicochemical properties of neem oil was done at the laboratories of Surfactant and Bioenergy research Center of Bogor Agricultural University and Center of Agro industries, Indonesian Ministry of Industry.

2.2 The process of DEA surfactant production and the analysis of its physicochemical properties
DEA surfactant was produced by using an amidation reactor of 25 L/batch scale. The surfactant was synthesized by reacting methyl ester of palm oil with diethanolamide at the mole ratio of 1:2 by using 1% NaOH 30% as a catalyst at 140°C and the speed of 100-200 rpm for 3-5 hours.

2.3 Determination of optimum concentration of DEA surfactant as an additive material in the formulation of plan insecticide from neem oil.
This trial was done to obtain the concentration of DEA surfactant that could dissolve neem oil to form a stable emulsion as indicated by its droplet size analyzed by using a ICC 50 HD microscope (1000 x magnification), had small contact angle as analyzed by using a Contact Angle Analyzer Phoenix 300, and had a small surface tension as analyzed by a Critical Micelle Concentration (CMC) method. The common way of CMC determination is done by measuring the surface tension of surfactant solution as the function of concentration. DEA surfactant was dissolved in neem oil in the concentration of 1, 2, 3, 4, 5, 6, 7, 8, 9, and 10%. The surface tension of these surfactant solutions was measured by using a Spinning Drop Tensiometer at Surfactant and Bioenergy Research Center, Bogor Agricultural University.

2.4 Formulation of neem insecticide by using DEA surfactant
Insecticide formulation process was done by using a homogenizer. The formulation was made by mixing 90-95% neem oil, 1-10% DEA surfactant (in accordance with DEA surfactant concentration test results), and 1-5% supporting material to improve the stability of pesticide emulsion. The formulation process was done for 30-60 minutes at the speed of 5,000-6,000 rpm. Glycerol as much as 3% was used as the supporting material in this study.

2.5 Assessment of the effects of lethal concentration of neem formula on S. litura larvae
The resulted insecticide formula was tested at six concentration rates (10, 13, 16, 19, 22, and 25 ml/l) expected to cause death of 15-95% tested insects. Each insecticide formula was diluted with distilled water containing surfactant at the concentration similar to that of each neem oil formula. As a control, distilled water containing surfactant at the concentration similar to that contained in each formula was diluted to reach the concentration of 25 ml/l. For each treatment, 60 instar III larvae of *Spodoptera litura* were put into 6 petri dishes (10 in each petri dish). The tissue paper put in the lower and upper parts of each dish was wetted every day in order to maintain the moist of the leaves in the dishes. On day-2 and 4, treatment or control leaves were added. Observation on the mortality of tested insects was done on day-9.

3. Results

3.1 Neem oil production and the analysis of its physicochemical properties
The extraction process of neem oil was done by using screw and hydraulic pressing methods resulted in an oil yield of 12.3%. This analysis was aimed at assessing the properties of neem oil produced. The physicochemical properties of neem oil are listed in Table 1. It was shown that neem oil produced in this study had 0.91 g/cm^3 density, 58,94 cPoise viscosity, and 1.4695 refractive index. As a comparison, at 25^{0}C water and olive oil have viscosity of 0.890 and 81 cPoise, respectively.

Neem oil produced in this study was found to have a surface tension of 40.69 dyne/cm which was considered to be higher than those (25-35 dyne/cm) of commercial pesticides available in the market. High level of surface tension of neem oil has made the direct application of this oil in the field was less effective as neem oil could not dissolve in water nor spread evenly to the surface of leaves. High surface tension also made neem oil applied to the plants easy to fall down or evaporate as it could not attach to the leaves. This made the spraying of this oil, particularly during rainy season, a waste as the oil was easily washed away by the rain.

Table 1. Physicochemical properties of neem oil

No.	Physicochemical properties	Value
1	Density	0.91 g/cm^3
2	Iodine number	74.5 g Iodine/100 g
3	Saponification number	189 mg KOH/g
2	Viscosity	58.94 cPoise
3	Surface tension	40.69 dyne/cm
4	Refractive index	1.4695
5	Optical rotation	Not measurable/dark
6	Azadirachtin content of neem oil	343.82—1,604 ppm
7	Azadirachtin content of neem seed cake	242.20 ppm
8	Fatty acid composition (%)	
	Butyric (C4)	0.98
	Lauric (C12)	7.29
	Palmitic (C16-0)	31.4
	Stearic (C18-0)	9.45
	Oleic (C18-1)	22.5
	Linoleic (C18-2)	10.2
	Linolenic (C18-3)	1.94

Results of the HPLC analysis showed that the azadirachtin content of neem oil was 343.82–1,604 ppm while that of neem seed cake resulted from the pressing process was 242.20 ppm. These results were lower than the azadirachtin content found by [22], namely 0.1-0.5% of dried neem seed weight. Meanwhile, palmitic (31.4%) and oleic (22.5%) acids were found as the main fatty acids of neem oil.

3.2 *Physicochemical properties of DEA surfactant*

DEA surfactant produced in this study was found to have pH of 10.6, density of 0.9930 g/cm^3, viscosity of 708.20 cP, and surface tension of 25.37 dyne/cm. These were in line with what was found by [23]. The pH level of diethanolamide used by Kirk-Othmer was 9-10. The value of surface tension of DEA surfactant produced in this study indicated that this surfactant could lower the surface tension of water by 61-65% from 72 dyne/cm.

3.3 Determination of optimum concentration of DEA surfactant as an additive material in the formulation of plant insecticide from neem oil

a. CMC Analysis on surface tension

Determination of optimum concentration of DEA surfactant by using a CMC analysis in order to assess the highest reduction in the surface tension of neem oil. It was shown that the addition of DEA surfactant at the concentration of 5 and 6% could lower the surface tension of neem oil from 40.69 to 23.82 dyne/cm. The addition of DEA surfactant higher than 6% was found to make the surface tension of neem oil increase back. According to [24], the addition of surfactant to a solution would lower the surface tension of the solution. After a certain concentration was reached, the surface tension would stay constant even though the concentration of surfactant was increased. The adding of surfactant to the amount that exceeded this concentration would make the surfactant aggregate to form micelles. The concentration when miselles start to form is known as the Critical Micelle Concentration (CMC). Surface tension will keep lowering down until CMC is reached. Once CMC is reached, surface ten-

IOP Conf. Series: Earth and Environmental Science **65** (2017) 012005 doi:10.1088/1755-1315/65/1/012005

sion will remain constant indicating that the interfacial is saturated and the formed micelles are in their dynamic equilibrium state with their monomers.

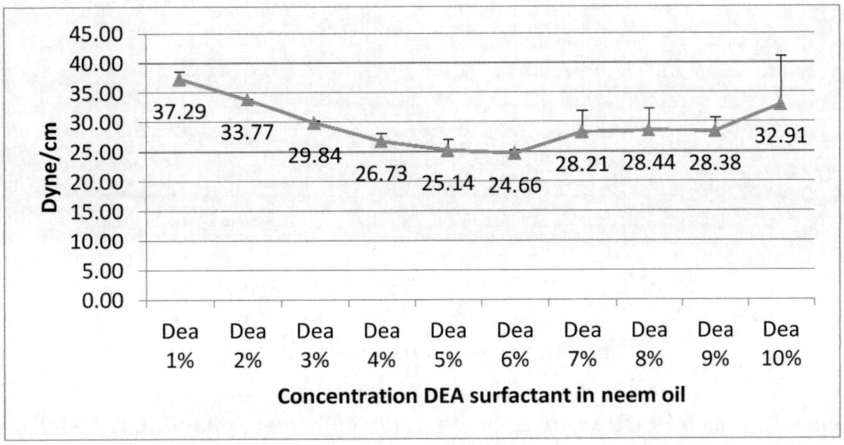

Figure 1. Surface tension values of neem oil at various concentrations of DEA surfactant

CMC is a standard parameter in the characterization of surfactant solution [25] as, in general, it indicates the minimum concentration needed to obtain surfactant association structure. A number of studies on the methods or techniques of CMC determination and factors affecting CMC of various types of surfactants have been done. The common methods or techniques used to determine CMC are based on surface tension [26], capillary electrophoresis [27] and optical density [28]. Theories have also been developed to explain the occurrence of CMC [29,30,31].

b. Analysis of Contact Angle

The analysis of contact angle was done to determine the contact angle of formulation which was diluted to 1%. This analysis was necessary to assess the ability of DEA surfactant in spreading neem oil on the surface of leaves. The graph of the decrease in contact angle of neem oil is depicted in Figure 2.

A contact angle is an angle formed between the surface of tested material (leaves) and neem oil dropped onto the surface of the tested material. The mean contact angles resulted from the addition of DEA surfactant to neem oil solution were about 17.82 to 52.52°. It is shown in the above graph that the addition of DEA surfactant at a concentration higher than 5% resulted in smaller contact angles. This indicated that the solution of neem oil was able to spread well on the surface of the leaves resulting in very small contact angles. Insecticide formulation solution was expected to have small contact angles as small as possible to make it able to attach to and spread over the surface of targeted objects. A contact angle close to 0° shows that neem oil droplets do not only attach to but also spread over well on the leaves. A contact angle close to 90° indicates that neem oil droplets are able to attach to the leaves but they do not spread well on the leaves. A contact angle close to 180° indicates that neem oil droplets do not attach to the leaves at all and even they directly slip down from the leaves.

The mixture of neem oil and surfactant with small contact angle indicated that the mixture spread over the leaves well when it is applied by farmers in the field. Therefore, DEA surfactant produced in this study could be utilized as a spreading and wetting agent in insecticide formula. Solution containing surfactant spreading all over the leaves could penetrate and spread over the fine hair on the surface through a capillarity force, while the droplets of solution containing solution with no surfactant did not spread but only attached to the surface of leaves [32].

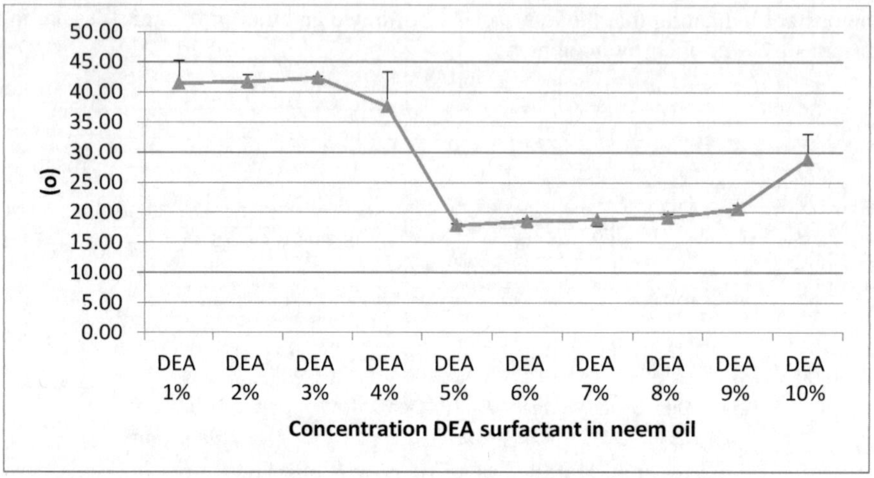

Figure 2. Effects of DEA surfactant concentration in neem oil on contact angles

c. Analysis of droplet size

This analysis was done to determine the droplet size of the emulsion formed by the mixture of neem oil and DEA surfactant diluted in water until the formula concentration was 1%. The analysis was done to ensure that the emulsion system of formulation product had a micro size. Small-sized droplets made the emulsion more kinetically stable so that sedimentation and creaming during storage could be avoided [33]. The graph showing the decrease in droplet size as a result of the addition of DEA surfactant produced in this study is depicted in Figure 3.

Emulsion is a preparation containing two phases, usually water and oil, which are not mixed. In an emulsion, a liquid is dispersed into droplets and the other liquid is stabilized with an emulsifier or a suitable surfactant [34]. An emulsion system is generally easy to get damaged by the addition of energy or by the passing by of time. This problem can be overcome by making the size of droplets smaller and by using a stabilizing agent such as surfactant. Small-sized droplets can be reach through the use of appropriate ratio of surfactant and oil and right temperature. In this study, it was found that the mixture of neem oil and DEA surfactant diluted in water to a concentration of 1% had mean droplet size of 1.61 to 8,17 µm. Smaller size droplets were obtained when higher concentration of DEA surfactant was added into neem oil.

As stated by [35], the length of hydrocarbon chains of a surfactant also affects the performance of surfactant in emulsion formation. A surfactant with long hydrocarbon chains increases the protective role of steric in minimizing the possibility of particle or droplet mergers. Olein, as a synthetic feedstock of DEA surfactant in this study, is composed dominantly of long chain fatty acids. Therefore, it was found in this study that the size of droplets formed was smaller as more DEA surfactant was added.

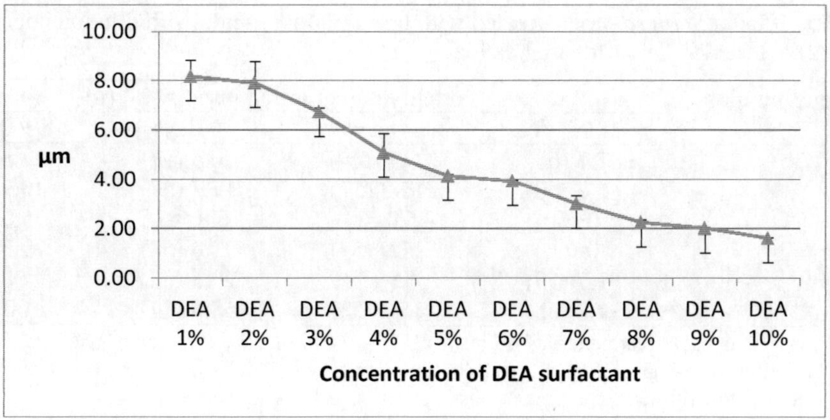

Figure 3. Effects of DEA surfactant concentration in neem oil on droplet size

3.3 Lethal Concentration Test of Selected Insecticide Formula

Results of determination of optimum concentration of DEA surfactant as an additive material in the formulation of plant insecticide from neem oil showed that DEA surfactant concentration of 5% was the most effective in lowering the surface tension and contact angle of neem oil. This determination was made as no statistical differences were found in the effectiveness of DEA surfactant concentration of 5 and 6% in lowering the surface tension and contact angle of neem oil.

Results of the application test of insecticide formula of neem oil by using 5% DEA surfactant showed that the mortality rates of tested insects at 2, 4, 6, and 9 days after treatment (DAT) were about 0 - 1.8%, 5 - 15.2%, 6.8 - 55.8%, and 22.8 - 100%, respectively. The effect of neem formula treatment on mortality rates significantly increased at 9 DAT. This was caused by the existence of active substances, particularly azadirachtin, which inhibited insect development and this effect was observable long after the treatment was given. Larvae which fail to molt can live for significantly long time before they finally die [36].

Based on the values of LC_{50} and LC_{95} at 9 DAT and results of probit analysis of mortality rates of tested larvae, the neem formula by using DEA surfactant was $y = -3.13 + 7.28\ x$, with LC_{50} and LC_{95} values of 13 and 22 ml/L, respectively. This result was lower than [13] who found that neem oil of *Azadirachta indica juss* could kill *Spodoptera frugiperda* Smith at mortality rates of 83.33 +/- 0.83% and 89.58 +/- 0.90% at 4 and 6 DAT, respectively. Differences in efficacy might be caused by differences in types of pests and methods and techniques of insecticide production. Most neem product used in this study was neem oil which was obtained by pressing neem seeds. A solvent was then added to extract the active substances of the pesticide. Different methods and techniques applied in neem oil production resulted in different concentration of active substances and biological efficacy [37].

Table 2 Mortality of larva *S. litura* larvae treated with neem oil insecticide with dilution concentration of 10-25 ml/l at 2, 4, 6, and 9 days after application

Formulation concentration	Mortality rate of tested insects (%) (n)[a]			
(ml/l)	2 days	4 days	6 days	9 days
10	1.7 (60)	5.1 (59)	10.5 (57)	22.8 (57)
13	1.7 (60)	12.3 (57)	17.5 (57)	43.9 (57)
16	0 (60)	5.0 (60)	6.8 (59)	55.9 (59)
19	0 (60)	11.9 (59)	22.4 (58)	79.3 (58)
22	1.7 (59)	6.9 (58)	29.3 (58)	94.8 (58)
25	1.8 (55)	15.2 (53)	55.8 (52)	100 (52)

[a]Number of tested insects in each treatment was 60 instar III larvae of *S. litura*. Figures in bracket (n) show the number of larvae which were found back at a certain observation time. Figures less than 60 indicate that there are larvae which were not found back at the time of observation for unknown reasons.

Insecticide of neem oil is a systemic insecticide [38]. The work mechanisms of neem oil include inhibiting feeding, affecting hormone function in juvenile stages, reducing ecdysone, deregulating growth, altering development and reproduction, suppressing fertility, sterilizing, repelling oviposition, and disrupting molting processes [39]. Based on the complex active substances they contain, derivative products of neem oil can be used as antifeedants, growth regulators, sterilants, anti-oviposition agents, and repellents [40].

4. Conclusions

Neem oil used in this study had a density of 0.91 g/cm^3, viscosity of 58.94 cPoise, refractory index of 1.4695, surface tension of 40.69 dyne/cm, azadirachtin content of neem seed oil of 343.82 – 1,604 ppm, and azadirachtin content of neem seed cake of 242,20 ppm. The main fatty acid composition consisted of palmitic (31.4%) and oleic (22.5%) acids. As an additive substance in neem oil insecticide, the ethanolamide surfactant used in this study had a pH of 10.6, density of 0.9930 g/cm^3, viscosity of 708.20 cP, and surface tension of 25.37 dyne/cm. based on the results of CMC analysis and the values of contact angle and droplet size, the most effective concentration of DEA surfactant used as an additive in neem oil insecticide was 5%. The LC_{50} and LC_{95} values on *Spodoptera litura* were 13 and 22 ml/L, respectively. Neem oil worked by inhibiting the development of *Spodoptera litura* and its larval molting process.

Acknowledgement
Financial support from Ministry of Research, Technology, and Higher Education through the National Strategic Excellence Research Scheme was acknowledged.

References

[1] Rieger M M 1985 *Surfactant in Cosmetic. Surfactant Science Series* (Marcel Dekker Inc., New York)

[2] Jungermann E 1979 *Bailey's Industrial Oil and Fat Products* **Vol 1**, 4th ed (John Willey and Son, New York)

[3] Fessenden R J and Fessenden J S 1986 *Kimia Organik*. Edisi Ketiga Jilid 2 (Erlangga)

[4] Campos E V R, de Oliveira J L, Pascoli M, de Lima R, Fraceto L F 2016 Neem oil and crop protection: From now to the future *Frontier in Plant Scince* **7** 1-8

[5] Chandramohan B, Murugan K, Madhiyazhagan P, Kovendan K, Kumar P M, Panneerselvam C 2016 Neem by-products in the fight against mosquito-borne diseases: biotoxicity of neem cakefractions towards the rural malaria vector *Anopheles culicifacies* (Diptera: Culicidae) *Asian Pac.J.Trop. Biomed* **6** 472–76

[6] Scudeler E L, Garcia A S G, Padovani C R and Santos D C 2013 Action of neem oil (*Azadirachta indica* A. Juss) on cocoon spinning in *Ceraeochrysa claveri* (Neuroptera: Chrysopidae) *Ecotoxicol. Environ. Saf.* **97** 176–82

[7] Scudeler E L and dos Santos D C 2013 Effects of neem oil (*Azadirachta indica* A. Juss) on midgut cells of predatory larvae *Ceraeo chrysaclaveri* (Navás,1911) (Neuroptera:Chrysopidae) *Micron* **44** 125–32

[8] Weathersbee A A and Mc Kenzie C L 2005 Effect of a neem biopesticide on repellency, mortality, oviposition, and development of diaphorina citri (homoptera: psyllidae) *Fla. Entomol* **88** 401–7

[9] Seljåsen R and Meadow R 2006 Effects of neem on oviposition and egg and larval development of *Mamestra brassicae* L: dose response, residual activity, repellent effect and systemic activity in cabbage plants *Crop Prot.* **25** 338–45

[10] Senthil-Nathan S, Choi M Y, Seo H Y, Paik C H and Kalaivani,K. 2009 Toxicity and behavioral effect of 3b,24,25-trihydroxycycloartane and beddomeilactoneonthericeleaffolder *Cnaphalocrocis medinalis* (Guenée) (Lepidoptera:Pyralidae) *Ecotoxicol. Environ.Saf.* **72** 1156–62

[11] Hasan F and Shafiq Ansari M 2011 Toxic effects of neem-based insecticides on *Pierisbrassicae* (Linn.). *Crop Prot.* **30** 502–7

[12] Pereyra P J, Rossini G, Darrigran G 2012 Toxicity of neem's oil, a potential biocide against the invasive mussel Limnoperna fortunei (Dunker 1857). *Annals of the Brazilian Academy of Sciences* **84** 4 1065-1071

[13] Tavares W S, Costa M A, Cruz I, Silveira R D, Serrã J E and Zanuncio J C 2010 Selective effects of natural and synthetic insecticides on mortality of *Spodoptera frugiperda* (Lepidoptera: Noctuidae) and its predator *Eriopis connexa* (Coleoptera: Coccinellidae). *J. Environ.Sci.Health B* **45** 557–561

[14] Pankaj S, Lokeshwar T, Mukesh B, Vishnu B 2011 Review of neem (*Azadirachta indica*): thousand problems one solution. IRJP **212** 97-102

[15] Subiyakto 2002 Pemanfaatan serbuk biji nimba (*Azadirachta indica* A. Juss) untuk pengendalian serangga hama kapas. Perspektif **11** 9-17

[16] Isman M B 2006 Botanical insecticides, deterrents, and repellents in modern agriculture and an increasingly regulated world. *Annu. Rev.Entomol.* **51** 45–66

[17] de Oliveira J L, Campos EVR, Bakshi M, Abhilash P C and Fraceto L F 2014 Application of nanotechnology for the encapsulation of botanical insecticides for sustainable agriculture: prospects and promises. *Biotechnol. Adv.* **32** 1550–1561

[18] Miresmailli S and Isman M B 2014 Botanical insecticides inspired by plant–herbivore chemical interactions. *TrendsPlant Sci.* **19** 29–35

[19] Ghosh A, Chowdhury N and Chandra G 2012 Plant extracts as potential mosquito larvicides. *Indian J.Med.Res.* **135** 581–98

[20] Tangtrakulwanich K and Reddy G V P 2014 Development of insect resistance to plant biopesticides: anoverview, in *Advancesin Plant Biopesticides*, ed.D.Singh (New Delhi: Springer) 47–62

[21] Siegwart M, Graillot B, Blachere Lopez C, Besse S, Bardin M, Nicot P C 2015 Resistance to bio-insecticides or how to enhance their sustainability: a review. *Front. PlantSci.* **6** 381

[22] Sukrasno 2003 *Nimba Tanaman Obat Multifungsi*. (Jakarta: Agromedia Pustaka) 81 p

[23] Kirk-Othmer 1968 *Encyclopedia of Chemical Tehnology* **Vol 3** (New york : John Wiley and sons)

[24] Genaro R A 1990 *Remington's Pharmaceutical Science* 18th ed (Macle Printing Company, Easton-Pennsilva, USA)

[25] Wang Hai-Bo and Liu De-Shan 2003 CMC of Nonyphenol Polyoxyehylene Ethers in Oil Phases and Problems Concerned, Chemical Journal of Microemulsion-Gel Glass Containing Copper Nitrate *J. Dispersion Science and* Technology **15** 4 513-42

[26] Zulauf M, Furstenberger U, Grabo M, Jaggi P, Regenass, M and Rosenbusch 1989 Methods in Enzymology vol 172ed S Fleischer and B Fleischer B (Academic Press, San Diego) p 528

[27] Nakamura H, Sano A and Matsuura K1998 Determination of Critical Micelle Concentration of Anionic Surfactants by Capillary Electrophoresis Using 2-Naphthalenemethanol as a Marker for Micelle Formation. *Analytical Sciences* **14** 379-82

[28] Chung J J, Lee S W and Chol J H 1991 Salts Effects on Critical Micelle Concentration and Counterion Binding of Cetylpyridinium Bromide Micelles *Bull Korean Chem. Soc.* **12** 4

[29] Tanford C 1973 *The hydrophobic Effect, Formation of Micelles and Biological Membranes* (John Wiley, New York) p. 84

[30] Mukherjeee P and Mysels K J 1971 *Critical Micelle Concentrations of Aqueous Surfactant Systems* (Nath. Stand. 36, Washington) p. 51

[31] Mukherjeee P and Mysels K J 1979 *Reporting Experimental Data Dealing With Critical Micellization Concentrations (c.m. c's) of Aqueous Surfactant Systems* (Pergamon Press, Oxford) p 1085-1089

[32] Yu Y, Zhu H, Frantz J M, Reding M E, Chan K C, Ozkan H E 2009 Evaporation and coverage area of pesticide droplets on hairy and waxy leaves *J Biosyst Eng*. **104** 324 – 334

[33] Solans C, Izquierdo P, Nolla J, Azemar N and Garcia-Celma M J 2005 Nano-emulsions *Current Opinion in Colloid and Interface Science* **10** 102-10.

[34] Anief M 2000 *Ilmu meracik obat. Yogyakarta* (Gajah Mada University Press)

[35] Voigt R 1995 Buku Pelajaran Teknologi Farmasi Edisi kelima (Soewandhi, Dr. Rer. Nat. S. N. Apt., penerjemah) Yogyakarta: Gadjah Mada University Press 165-225, 568-571, 577-578.

[36] Prijono D and Hassan E 1993 Laboratory and field efficacy of neem (*Azadirachta indica* A. Juss) extracts against two broccoli pests / Wirksamkeit von Niemextrakten (*Azadirachta indica* A. Juss) gegen zwei Schädlinge an Brokkoli im Laboratorium und Freiland *Journal of Plant Diseases and Protection* **100** 4 354-370

[37] Roychoudhury R 2016 Neem products *Ecofriendly Pest Management for Food Security*, ed.Omkar (Amsterdam:Elsevier) 545–562

[38] Cox C 2002 Pyrethrins/pyrethrum insecticide fact sheet. *J. Pestic.Reform* **22** 14–20

[39] Brahmachari G 2004 Neem–anomnipotent plant: aretrospection. *Chembiochem* **5** 408–421

[40] Gonzalez-Coloma A, Reina M, Diaz C E, Fraga B M and Santana-Meridas O 2013 Natural product-based biopesticides for insect control R*eference Modulein Chemistry, Molecular Sciences and Chemical Engineering*, ed. J.Reedijk (Amsterdam: Elsevier)

IOP Conf. Series: Earth and Environmental Science **65** (2017) 012006 doi:10.1088/1755-1315/65/1/012006

Assessment of agricultural biomass potential to electricity generation in Riau Province

P Papilo[1], I Kusumanto[1], K Kunaifi[2]

[1]Departement of Industrial Engineering, UIN Suska Riau University, JL HR Subrantas No. 155 , Pekanbaru, Indonesia
[2]Departement of Electrical Engineering, UIN Suska Riau University, JL HR Subrantas No. 155 , Pekanbaru, Indonesia

Email : pilo1244@gmail.com

Abstract : Utilization of biomass as a source of electrical power is one potential solution that can be developed in order to increase of the electrification ratio and to Achieve the national energy security. However, now it is still difficult, to Determine the amount of potential energy that can be used as an alternative power generation. Therefore, as a preliminary step to assess the feasibility of biomass development as a power generation source, an analysis of potential resources are required, especially from some of the main commodities, both of residues of agriculture and plantation. This study aims to assessing the potential of biomass-based supply from unutilized resources that can be Obtained from the residues of agricultural and plantations sectors, such as rice straw and rice husk; Dry straw and chaff of rice; corn stalks and cobs; stalks of cassava; and fiber, shell, empty fruit Bunches, kernels and liquid wastes in the palm oil factories. More research is focused on the theoretical energy potential measurements using a statistical approach which has been developed by Biomass Energy Europe (BEE). Results of the assessment has been done and showed that the total theoretical biomass energy that can be produced is equal to 77,466,754.8 Gj year -1. Theoretically, this potential is equivalent to generate electricityof year 21,518,542.8 MWh -1.

1. Introduction

National electricity consumption in the period of 2000-2012 continued to increase with an average growth rate of 6.2% per year. It was lower than coal (9.9%) and LPG (13.5%). It also led to the national electrification ratio still at the value of 75.8% in 2012. Higher value of other ASEAN countries such as Singapore (100%), Malaysia (99.4%), Philippines (89.7%), and Vietnam (97.6%) showing that electrification rate of Indonesia is still low [1].

Furthermore, in electricity supply of year 2012, the power plants owned by the Indonesian State Electricity Company still dominated with a share of over 73% (32.9 GW), power plants owned by Independent Power Producer (IPP) in the range of 17% (7.4 GW), and the rest is met by power plants owned by Private Power Utility (PPU), power plant of non-fossil Operating Permit, and rental power plants with a share of the range of 10% (4.5 GW) [1].

Meanwhile, especially in the province of Riau, the electrification ratio has reached 77.56%. However, in some districts, seven regencies / cities in Riau province have lower electrification ratio-than the national average [2]. Certainly, it should be considered, especially by local government, in order to overcome problems of the growing demand for electricity in the future.

Content from this work may be used under the terms of the Creative Commons Attribution 3.0 licence. Any further distribution of this work must maintain attribution to the author(s) and the title of the work, journal citation and DOI.
Published under licence by IOP Publishing Ltd

To address the problems of the demand for electricity which continues to increase, it is necessary to utilize various alternative sources. An alternative source is through the utilization of the remaining agricultural products, plantation or residual forest products such as biomass.

Currently, biomass has been the most important source of energy in every region of the world [3]. Biomass has the potential to be one of the main energy source in the future, and the modernization of bioenergy systems suggested as an important contributor in the development of sustainable energy in the future, especially for sustainable development in industrialized countries and in developing countries [4]. As a result, there will be mobilization of the provision of biomass on a large scale as an effort to fulfill the energy needs in each region [5].

Biomass is the term for any organic material derived from plants (including algae, trees, and plants). Biomass is produced by green plants that converts sunlight into plant material through photosynthesis. Biomass resource can be considered as an organic material, in which the solar energy stored in chemical bonds. When the bonds between adjacent carbon, hydrogen and oxygen molecules are broken by digestion, combustion, or decomposition, then it releases their stored chemical energy [6]. Exploiting the potential of biomass as a source of electrical energy has started to be developed in some countries in the world. Such as in China, with the potential of available biomass, allowing them to generate electrical energy with a capacity of 30 GW [7]. Similarly, in the EU, even the demand of biomass raw material exceeds the supply capability can be provided for the needs of power plants [8]. Biomass basically can be divided into three major groups, wood biomass, non-wood biomass and secondary fuels [9]. On the other hand, Biomass Energy Europe divides biomass into four categories: (1) forest biomass and waste, (2) energy crops, (3) agricultural waste, and (4) organic waste [10].

Efforts to address the problems of the community energy requirements continue to be implemented by the provincial government. Start with drafting the rules and policy of regional energy management, consolidation of cross-sectors, the budget provision that is sufficient for the construction of various center of electrical energy generation, and foster cooperation with various industry parties in and outside the country to jointly develop technologies supplying alternative energy in accordance with the potential which exists.

Basically, the potential of new and renewable energy, especially biomass energy in the province of Riau are available in large enough quantities. Sources of potential energy are from agricultural products waste, forest products waste, industrial waste, and urban sewage coming from households. One of agricultural potentials to be utilized as an alternative energy source in the province of Riau is palm oil.

However, the underlying issue is the level of the overall potential of the various sources of biomass energy in Riau Province that cannot be known certainly. Lack of availability of data, undeveloped knowledge and technology utilization of biomass energy, cause portions of the utilization of biomass energy for the effort to fulfill the energy needs of society and increase the portion of renewable energy in the national mixed energy is still quite low.

The lack of data on potential energy source causes doubts to investors in investing their capitals for alternative energy development efforts in the province of Riau. The complexity of the existing problems, could impede utilization development efforts of biomass energy in the province of Riau. Therefore, it is necessary to carry out assessment of the potential of biomass energy in the province of Riau as a first step. This study will be the baseline to determine the direction of policy on the use and development of biomass energy in the province of Riau.

The purpose and scope of this study include: 1) Obtain the value of biomass energy potentials from two main sources, from the rest of the agricultural products which are distinguished into two groups: *First*, the rest of the agricultural product of horticulture crops such as straw and rice husks, leftover corn cobs, and cassava stem. *Second*, biomass potential can also be measured from the waste products of oil palm plantations that are divided into empty fruit bunches of palm oil, palm shells, kernel fibers, and liquid waste of POME (Palm Oil Mill Effluent). In this study, the biomass potential obtained is still a theoretical potential. 2) Obtain the value of the electricity that is a conversion of whole agricultural biomass energy. 3) Obtain the classification of agricultural biomass energy potential of any region in the province of Riau. Classification is divided into three levels, high, medium, and low potential.

2. Literature Reviewe

According to the Biomass Energy Europe, there are five types of potential sources of biomass energy: theoretical, technical, economical, implementation, and sustainable implementation. Because this study is a preliminary study, the research focuses on the theoretical potential. Figure 1 shows an illustration of the first four potentials and are described in the following sections [10].

Theoretical potential is the maximum number of total land biomass that is theoretically available for bioenergy production with basic biophysical limits. The theoretical potential is usually expressed in Joules primary energy, which is the energy contained in the biomass raw material, and has not been processed. Primary energy is converted into secondary energy, such as electricity, liquid fuels, and gaseous fuels. In the case of crops and forest biomass, theoretical potential describes the maximum productivity under theoritical optimal management by considering constraints such as soil conditions, temperature, solar radiation, and precipitation.

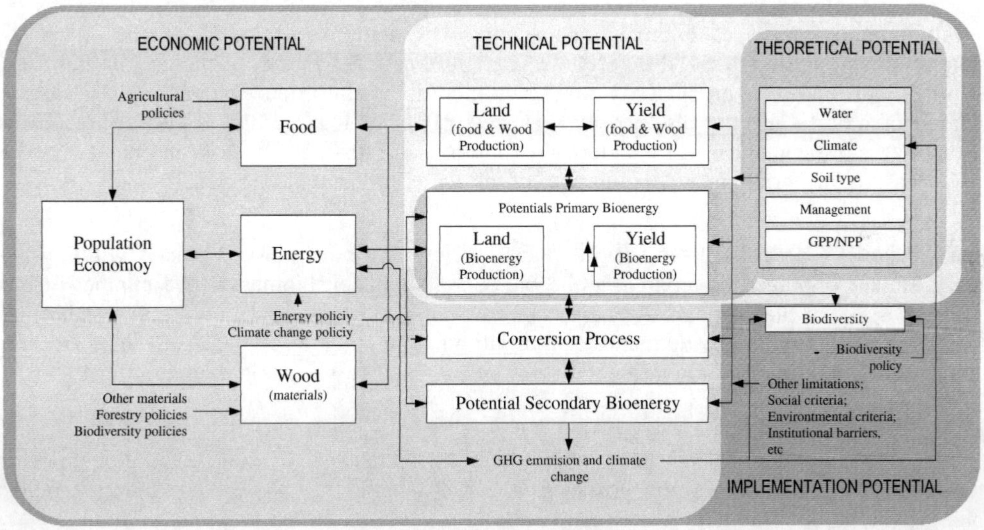

Figure 1. Biomassa potential classifications [10].

Technical potential is part of the theoretical potential that is available under the techno-structural condition with current available technology (eg harvesting techniques, infrastructure and accessibility, and processing techniques). The technical potential also considers the spatial conditions related to land use (eg for the production of food, feed, and fiber) including ecological aspects (ie natural reserves) and constraints due to the possibility of non-technical use. The technical potential is usually expressed in Joule primary energy, but sometimes also expressed in a secondary unit for energy.

Economic potential is part of the technical potential that meets the criteria of economical profits under certain conditions. The economic potential generally refers to the secondary bio energy, although sometimes the primary bio energy is also considered. The final results of the assessment of the economic potential is in the form of Supply Curve (IDR /tons).

Potential implementation is part of the economic potential that can be applied to both certain time period and the real socio-political, includes obstacles (and incentives) of economic, institutional and social. Potential implementation is focused on the feasibility or impact of economic, environmental, and social from bioenergy policies.

Theoretically, there is the fifth type of potential biomass sources that is the sustainable implementation potential. Actually, that is not a new type of potential, but it is a combination of the criterias of environmental sustainability, economic, and social assessment of biomass sources. It describes that the sustainability criteria acts as a filter on the theoretical, technical, economical and implementation potential aimed to the sustainable implementation potential. Depending on the type of potential, the sustainability criteria can be applied into different levels. For example, to get the technical potential, criteria and environmental constraints are integrated to limit the available area and / or results that can be achieved.

2.1 Biomass Potential Energy Measurement

Information on the production and use of biomass energy is usually difficult to obtain because of the lack of long-term data. Even if available, the data and information is usually inaccurate and too focused on a specific location. This is usually caused by traditional biomass that is often considered as part of the non-formal economy so it is not included into the activities of statistical agencies. In turn, although biomass plays an important role in developing countries, the planning, management, production, distribution, and use of biomass rarely get the attention of policy makers and energy planners [9].

Determining the potential of biomass energy involves many factors such as the complexity of production and consumption, difficulty in determining the energy sources of biomass, the sustainability of long-term productivity, and economic aspects in production and use. In addition, the utilization of biomass energy also involves aspects of technology, ecology, and social. Three key issues in determining analysis of the potential of biomass are distinguish between potential and actual supply, measure the variation, and the number of units used in the measurement [9].

Biomass is usually collected from various plant species. The potential of biomass is the number of biomass sources produced by an area, while the actual supply of biomass is the number of biomass sources accessed for the purposes of energy conversion. The limitation factors to access biomass are topography, law or local regulation, and local traditions. The potential of biomass can be measured by conducted three main methods of geospatial technology, field surveys, and modeling [12].

a. Geospatial Technology

Including remote sensing (RS), geographic information systems (GIS), and global positioning systems (GPS). Remote sensing images such as aerial and satellite photos, provide an efficient and reliable way to monitor biomass resources over time.

This technique is used to estimate the growth in the stock of biomass and productivity estimates. RS provides an effective-cost way to gather the necessary information in outlying and inaccessible area. Remote sensing images are also the only practical approach to analyze land use and land cover changes in large economic scale, regional, and global. The resulting patterns can be studied by comparing images obtained at different times. Data derived from remote sensing imagery, such as land use / cover, altitude, and temperature of the surface, are further used in geographic information systems (GIS).

b. Field Survey

The field survey is used to collect data as part of an evaluation on a specific location. Usually, a field plot (size may vary) is selected as representative of the type of vegetation in an area of study, and parameters such as stem diameter, tree height, or the dimensions of the crown for resources of the forest and density, altitude, and the phenology development for easily measured plants are measured. Further extrapolation results in a larger area are used to develop equations that predict the availability of biomass. This sampling technique provides the most accurate estimation of biomass resources at specific locations, but not practical for large scale. The survey takes a long time, a lot of employees, therefore it is expensive, even with today's satellite communications technology (GPS). Currently, the field survey in the assessment of biomass resources is used when other methods are not effective or when the ability to use other methods does not exist.

c. Modelling

Model is a framework that is simplified and designed to visualize a system or process that often uses mathematical techniques to facilitate the calculation or prediction. The complexity of the model and modeling techniques depends on the needs of assessment and data availability. Models can be as simple as an extrapolation of data that are measured using statistical methods, or as complex to balance process (set up as a separate module) to get the resource characteristics.

Both static (analytical) and dynamic (simulation) models are used in the assessment of biomass resources. Static model describes a mathematical system, in an equation, and can be built in a spreadsheet. For example, estimating the number and cost of crop residues (straw rice for example) with set a value (usually average) for several variables such as crop production, the resulting residues, labor costs, and the price (chemical, fertilizer, fuel, and planting).

3. Methodology

The study was conducted within the scope of Riau Province which consists of 12 districts and cities. In general, the data used was secondary data obtained from annual reports to the relevant agencies, including the Department of Agricultural, Department of Agriculture and Horticulture, Department of Mines and Energy and the Central Bureau of Statistics in the province of Riau.

This research included activities that consist of the studying the literature, surveying through secondary data collection on several relevant agencies, analyzing the data, and mapping classification. The data collected included all aspects within the scope of the purposes of the study. In general, the biophysical data can be collected from some types of exist agricultural commodities. Biophysical data included land resources and biological resources of food crops, including plantation crops.

Methods and approaches used in analyzing biomass potential referred to the guidelines developed by the Biomass Energy Europe [11]. The assessment of the level of potential energy biomass was based on the availability of such resources, and conducted using statistical methods. This method used statistical data based on land use, crop yields, crop production, and from the literature. Statistical data were then combined with the conversion factor, such as the yield per hectare, residues to crop factor, etc. These factors was based on an assessment / opinions of expert, field study, or review of the literature. In addition, the assumptions was carried out to determine the portion of biomass that can be used for energy production, by considering the needs for the land use for other purposes.

Furthermore, the biomass potential had been obtained in previous measurements would be converted into electrical energy units (MWh). Hierarchical Cluster Analysis approach was carried out to determine the classification of the energy potential of each region. In this method, the region cluster is formed due to the similarity (proximity) between clusters. In this study, cluster region was formed based on the potential of each commodity. Each region was divided into three levels of potential clusters: high, medium, and low potential. The range of potential biomass energy value of each type of commodity was different, depend on the range of the potential value of each commodity.

The use of this method is limited to the knowledge of the potential energy in theory only. To find a more realistic potenial of energy, assessment processes needs to be done technically, economically and implementation. For that, it will require other supporting data as well as the number of population, the number of target customers will get a power supply, the data shrinkage during distribution, the conversion process or other data related to financing as the cost of production, transportation and electricity tariffs.

The following section outlined the measurement method of the biomass potential, and the classification method of biomass potential of each region.

a. Statistics Method –Primary Agricultural Residue (PAR)

Most important primary agricultural residues of available agricultural biomass for bioenergy was straw and rice straw. Parameters that influenced the potential of straw / rice straw was the area covered by these plants and the amount of straw / rice straw produced per hectare or per ton of plant. Other types of residues that should be included in the residual category was the primary product of the cultivation process (eg, pruning fruit trees).

The theoretical method of annual crop residues, such as rice, was estimated based on the cultivated area and agricultural production (AP) in ton per hectare, for each species of plant and the average ratio between the product and the residue (PtR). Primary agricultural residues (*THP_PAR*) was calculated by:

$$THP\_PAR = \sum(CA_i x AP_i x PtR_i x Av_i) \quad (1)$$

where: ***THP_PAR*** is an agricultural residues primary (eg, straw, rice straw), in tons, ***CAi*** is the cultivated area for *i*-th crop, in hectares (ha), ***APi*** is the agricultural production for the *i*-th crop, in tons per hectare (tons / ha), ***PtRi*** is product-residue ratio for *i*-th crop, while ***Avi*** is the availability of residues for *i*-th crop according current harvest system.

Tons unit of the theoretical potential of agricultural residues was converted to energy units by multiplying the potential (in tons / year) with a lower calorific value of a particular residue (GJ / ton_{dry}). Energy value was calculated using a low heating value of 14.7 GJ / tonnes with a water content of about 20% [9]. The ratio of product-residue by plant species (*PtRi*) was amounted to 1.757 [13]. The availability of residues for *i*-th crop according to current harvesting system (*Avi*) was assumed equal to 1.

In oil palm plantation commodities, palm oil biomass portion was determined based on the net energy balance proposed by Hambali [14]. The processing of fresh fruit bunches (FFB) produced from a plantation, will produce palm oil, kernel, empty fruit bunches, fiber, shell, and liquid waste or POME. Processing the oil at a refinery industry will produce Refined Bleached Deodorized Palm Oil (RBDPL) and Palm Fatty Acid Destillated (PFAD). Furthermore, RBDPL through fractionation processes will produce Refined Bleached Deodorized Palm Olein (RBDPO) and Refined Bleached Deodorized Palm Stearin (RBDPS).

However, the energy potential was only measured in the palm oil biomass which consisted of empty fruit bunches, shells, fibers, kernel, and POME with a portion of each energy of 21%, 6.4%, 14.4%, 5%, and 58, 3%, respectively [14]. In general, the net energy balance of palm oil can be described in Figure 2.

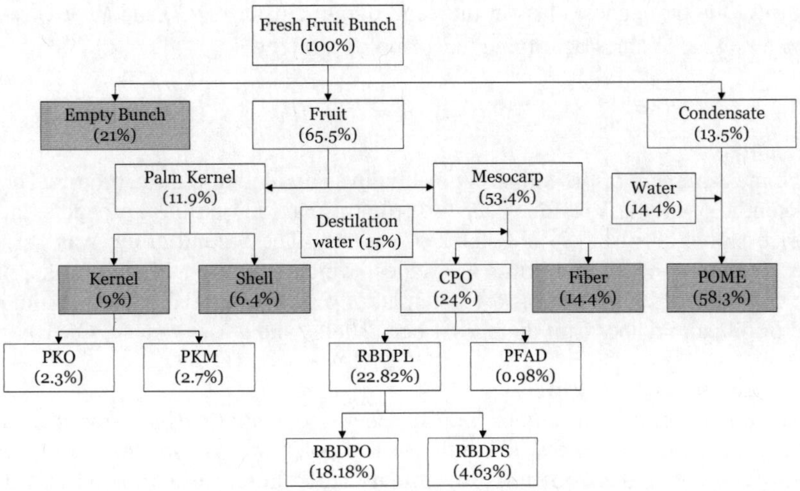

Figure 2. Palm oil mass balance [14].

b. Statistics Method– Secondary Agricultural Residue (SAR)

Secondary agricultural residues (SAR) were generated and collected from companies that process agricultural plant parts harvested to produce the food / feed. In some European countries, the company must report the volume and utilization of residues they produce to the local statistics agency. However, if the direct statistics are not available, the methodology for assessment using the following equation :

$$P_{ti} = Cr_i \; x \; PtSR_i \tag{2}$$

where: $\boldsymbol{P_{ti}}$ is the theoretical potential of secondary agricultural residues of *i*-th crop (tons / year), $\boldsymbol{Cr_i}$ is the production of *i*-th crop (tons / year), and $\boldsymbol{PtSR_i}$ is the ratio between products and secondary residues of *i*-th crop.

c. Hierarchical Cluster Method

Cluster is a method of distribution data into meaningful subgroups, when the number of subgroups and other information about their composition may not be known [15]. Classification can also be described as an activity to divide a set of objects into multiple objects that resemble each other in the same class while the different objects will form another class [17]. Hierarchical Cluster Analysis is a cluster method that produces a unique set of categories or groups by variables of sequentially couples, groups, or variables [18]. Hierarchical clustering methods is processed to the stage of generating sequence of partitions, each corresponds to a different number of clusters [19].

In multivariate data analysis, sampling units is generally represented as a multidimensional space in which the distance between pairs of points is defined as some function of observed sample value [20]. According to [16], determining the distance between clusters can be performed with agglomerative method. In this method, determining the distance between two possible clusters is built with a single n clusters, one for each point, and successively joining the pair nearest cluster to form a new cluster.

The general equation of Aglomerative Algorithm was developed by Lance and William [21] in which the difference between the new class is the union of Ci u Cj and some other class Ck that is defined as:

$$\begin{aligned} d(C_i \cup C_j, C_k) = {} & \alpha_i d(C_i, C_k) + \alpha_j d(C_i, C_k) + \beta d(C_i, C_k) + \gamma |d(C_i, C_i) - \\ & d(C_i, C_k)| + \delta_i h(C_i) + \delta_j h(C_j) + \varepsilon h(C_k) \end{aligned} \tag{3}$$

In this equation, *h(Ci)* is the highest class in the dendrogram cluster of *Ci*, and $\theta \equiv(\alpha_i, \alpha_j, \gamma, \delta_i, \delta_j, \varepsilon)$ is a set of parameters which values determine the grouping strategy.

4. Results

4.1 Biomass Potential

The results of the measurement of the biomass potential was divided into three groups. The first group described the potential of biomass residues derived from primary agricultural products, including rice straw, dryland rice straw, corn stalks, and stems of cassava. The second group was the potential of biomass sourced from secondary agricultural residues, which consisted of rice husks, dryland rice husks, and corn cobs. The third group was the potential of biomass derived from oil palm plantations, which consisted of oil palm empty fruit bunches, fibers, shells, and waste water (POME).

a. Agricultural Residue Biomass Potential.

Table 1, describes the assessment results of biomass energy potential derived from primary residues on four agricultural commodities. Based on the theoretical potential assessment, it is known that there were three regions with the highest potential for primary agricultural residue biomass, which were to Rokan Hilir by 3,666,063 Gj (21.48%), Indragiri Hilir by 3,116,287 Gj (18.26%) and Kampar by 2,082,552 (12.20%).

Table 1. Theoritical potential of primary agricultural residue biomass

Districts	Paddy Rice Straw		Dry Rice Straw		Corn Stem		Cassava Stem		Energy Total/ Districts (Gj/yr)
	Production (Ton/yr)	Energy (Gj/yr)	Production (Ton/yr)	Energy (Gj/yr)	Production (Ton/yr)	Energy (Gj/yr)	Production (Ton/yr)	Energy (Gj/yr)	
Bengkalis	24,626	632,429	2,624	43,681	469	16,214	4,837	170,649	862,974
Indragiri Hilir	121,681	2,829,124	1,952	22,620	5,844	160,711	3,167	103,832	3,116,287
Indragiri Hulu	13,650	505,463	4,155	87,659	841	31,706	7,285	219,071	843,900
Kampar	29,346	910,835	9,656	431,177	3,146	142.991	19,552	597,549	2,082,552
Kuansing	44,275	1,255,925	13	224	389	22,123	4,677	153,155	1,431,427
Pelalawan	37,475	997,481	309	10,621	18,361	566,949	5,388	161,629	1,736,681
Rokan Hilir	157,959	3,512,679	385	12,140	774	34,005	3,383	107,239	3,666,063
Rokan Hulu	19,812	488,308	30,743	852,281	1,267	51,244	5,597	153,281	1,545,114
Siak	27,032	720,660	414	9,737	491	17,025	4,676	159,014	906,436
Dumai	694	112,428	3,620	85,837	225	9,987	6,505	160,310	368,563
Pekanbaru	32	744	6	207	1,312	44,523	12,348	217,247	262,720
Meranti	5,419	177,809	0	0	78	6,689	2,065	57,773	242,270
Total	482,001	12,143,883	53,877	1,556,184	33,197	1,104,166	79,480	2,260,750	17,064,985

Where: Primary Residues – Product Ratio (*PtRi*) for straw= 1.757; corn dan cassava = 2. Residues available (Avi) =1,

While Table 2, shows the potential of secondary biomass from agricultural residues in each district in the province of Riau. Based on the assessment results of potential had been conducted, it is known that there were two most potential regions, Rokan Hilir with theoretical potential of agricultural residues for biomass energy of 1,299,584 Gj (28.84%) and Indragiri Hilir with a potential of 1,033,295 Gj (22.93%).

Based on the four types of existing agricultural potential, it can also be known that the rice paddy yield had the largest biomass potential that was 75% of total primary and secondary agricultural biomass potential. Whereas dryland rice, corn, and cassava were respectively only 9%, 6%, and 10% of the overall potential of the existing agricultural biomass. (Figure - 3). Overall total theoretical potential agricultural biomass, which can be produced from residues of paddy rice, dry rice, corn and cassava, was 12,571,642 Gj.\

Table 2. Theoritical potential of secondary agricultural residue biomass

Districts	Hull of paddy rice		Hull of dry rice		Corn hump		Energy Total/ District (Gj/yr)
	Production (Ton/yr)	Energy (Gj/Th)	Production (Ton/yr)	Energy (Gj/yr)	Production (Ton/yr)	Energy (Gj/yr)	
Bengkalis	24,626	201,687	2624	21,491	469	1664	224,842
Indragiri Hilir	121,681	996,567	1952	15,987	5844	20,740	1,033,295
Indragiri Hulu	13,650	111,794	4155	34,029	841	2985	148,808
Kampar	29,346	240,344	9656	79,083	3146	11,165	330,592
Kuansing	44,275	362,612	13	106	389	1381	364,099
Pelalawan	37,475	306,920	309	2531	18,361	65,163	374,614
Rokan Hilir	157,959	1.293,684	385	3153	774	2747	1,299,584
Rokan Hulu	19,812	162,260	30,743	251,785	1267	4497	418,542
Siak	27,032	221,392	414	3391	491	1743	226,525
Dumai	694	5684	3620	29,648	225	799	36,130
Pekanbaru	32	262	6	49	1312	4656	4968
Meranti	5419	44,382	0	0	78	277	44,658
Total	482,001	3,947,588	53,877	441,253	33,197	117,816	4,506,657

Where : Secondary Residues – Product Ratio (*PtRsi*) for hull of paddy rice = 0.7; corn hump (*PtRsi*) = 0.3

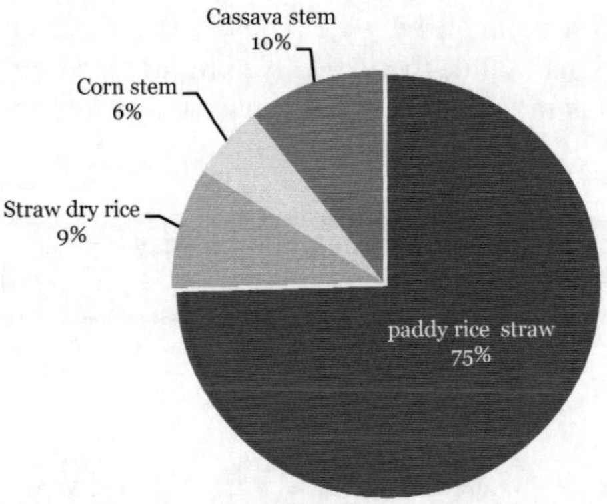

Figure 3. Pie diagram of agricultural biomass potential

b. Palm Oil Biomass Potential.

Riau Province is one of areas that has the largest area of palm oil plantations in Indonesia. According to statistics of the Agricultural Department of Riau Province (2013), the area under palm oil plantations in Riau Province is 2,401,460 million hectares and spread over 12 districts. 461,007 hectares of oil palm plantations from the existing total area, has produced a total production of 4,477,081 tonnes.

The largest community-owned palm plantation area is in Indragiri Hilir, with a plantation area reaches 580,627 Ha (24.1%). Indragiri Hilir regency could produce as many as 644,098 tonnes of palm oil during 2012. Other area in Riau Province that has the quite large resource potential of oil palm is in Rokan Hulu and Kampar regency with palm oil plantations respectively reach 256,696 ha (10.7%) and 289,156 ha (12.1%). During 2012, production of those regencies had reached respectively 579,011 tonnes (12.9%) and 573,338 tonnes (12.8%) of fresh fruit bunches (FFB) [22].

Based on Table 3, it can be seen that in total, biomass energy can be generated from the potential of palm oil plantations in Riau Province was 30,786,475.29 Gj / year. The largest potential could be produced from the utilization of fiber (coir) of palm oil, which was 8,800,474.52 GJ (27.9%), and the utilization of oil palm bunches with theoretical energy potential value of 8,402,587.5 Gj / year (27.2%).

IOP Conf. Series: Earth and Environmental Science **65** (2017) 012006 doi:10.1088/1755-1315/65/1/012006

Table 3. Theoretical potential of palm oil biomass (community – owned)

Districts	Fiber (Gj/yr)	Shell (Gj/yr)	Bunch (Gj/yr)	Kernel (Gj/yr)	POME (Gj/yr)	Total Potential (GJ/yr)
Bengkalis	696,520.02	443,782.24	680,493.90	394,489.22	277,995.25	2,493,280.62
Indragiri Hilir	1,237,312.26	88,343.75	1,208,843.13	700,778.62	493,836.38	4,429,114.13
Indragiri Hulu	468,047.81	298,212.97	457,278.57	265,089.02	186,807.36	1,675,435.73
Kampar	1,101,382.30	701,737.05	1,076,040.76	623,791.74	439,583.98	3,942,535.82
Kuansing	356,608.68	227,210.41	348,403.52	201,973.06	142,329.74	1,276,525.41
Pelalawan	937,326.98	597,210.49	915,760.16	530,875.46	374,106.18	3,355,279.26
Rokan Hilir	1,006,951.70	641,571.33	983,782.90	570,308.93	401,894.81	3,604,509.68
Rokan Hulu	1,112,280.13	708,680.51	1,086,687.84	629,963.97	443,933.52	3,981,545.98
Siak	1,076,645.58	685,976.24	1,051,873.20	609,781.57	429,711.05	3,853,987.65
Dumai	150,740.87	96,043.36	147,272.50	85,375.36	60,163.73	539,595.82
Pekanbaru	6437.27	4101.46	6289.16	3645.89	2569.25	23,043.02
Meranti	450,220.93	286,854.71	439,861.86	254,992.38	179,692.29	1,611,622.18
Energy Total/Biomass	8,600,474.52	5,479,724.51	8,402,587.50	4,871,065.22	3,432,623.54	30,786,475.29

Where: Primary Residues – Product Ratio(*PtRi*) of fiber = 0.170; shell= 0.065; bunch= 0.230; kernel = 0.064 and POME = 1.217)

Table 4 shows that theoretically, total energy potential of palm oil from private plantations could be produced in the province of Riau was 25,108,837.3 Gj / year. The largest energy potential could be produced from the utilization of oil palm fiber, which was 7,014,320.2 Gj / year (27.9%). Palm oil bunches waste of private plantations could, theoretically, produce energy by 6,852,928.8 Gj / year (27.2%). The biomass potential was in the district of Kampar with a total potential biomass energy of 5,162,942.44 Gj / year (20.6%).

Table 4. Theoretical potential of palm oil biomass (private-owned)

Region	Fiber (Gj/yr)	Shell (Gj/yr)	Bunch (Gj/yr)	Kernel (Gj/yr)	POME (Gj/yr)	Energy Total / Region(Gj/yr)
Bengkalis	322,192.04	205,282.12	314,778.77	182,480.45	128,593.37	1,153,326.75
Indragiri Hilir	911,274.38	580,611.28	890,307.00	516,120.00	363,708.06	3,262,020.71
Indragiri Hulu	364,440.59	232,200.45	356,055.24	206,408.83	145,455.62	1,304,560.73
Kampar	1,442,313.69	918,958.80	1,409,127.72	816,885.63	575,656.60	5,162,942.44
Kuantan Singingi	392,245.15	249,915.90	383.220.04	222,156.54	156,552.98	1,404,090.61
Pelalawan	1,378,480.79	878,288.16	1,346,763.53	780,732.48	550,179.60	4,934,444.55
Rokan Hilir	670,186.95	427,004.33	654,766.72	379,574.91	267,485.18	2,399,018.11
Rokan Hulu	786,799.34	501,302.99	768,695.99	445,620.86	314,027.55	2,816,446.73
Siak	592,029.15	377,206.70	578,407.24	335,308.54	236,290.82	2,119,242.45
Pekanbaru	58,751.86	37,433.29	57,400.05	33,275.39	23,449.06	210,309.65
Meranti	95,606.25	60,914.77	93,406.46	54,148.67	38,158.39	342,234.54
Energy Total/Biomass	7,014,320.20	4,469,118.80	6,852,928.80	3,972,712.30	2,799,557.20	25,108,637.30

Where: Primary Residues – Product Ratio(*PtRi*) of fiber = 0.170; shell= 0.065; Bunch= 0.230; kernel = 0.064 and POME = 1.217)

Based on the overall analysis, every potential area of biomass and the potential of each type of biomass can be summarized in Table 5.

Table 5. Recapitulation of biomass energy potential

Agricultural Commodity	Type of Biomass	Energy (Gj/Yr)	%	Higest Potential District
Wet Rice	straw, hull	16,091,472.1	20.7	Rokan Hilir
Dry Rice	straw, hull	1,997,436.9	2.5	Rokan Hulu
Cassava	Stem	2,260,750.2	2.9	Kampar
Corn	stem, knob	1,221,983.1	1.6	Pelalawan
Palm oil (community owned)	fiber, shell bunch, kernel, POME	30,786,475.3	39.7	Indragiri Hilir
Palm oil (private owned)	fiber, shell, bunch, kernel, POME	25,108,637.3	32.4	Kampar
Energy Total of Biomass		77,466,754.8		Kampar

4.2 Electricity Potential

As shown in Table 5, it can be seen that the overall theoretical potential of biomass could be produced from two groups of agricultural commodities and plantations in Riau Province was 77,466,754.8 Gj. This is equal to 21,518,542.8 MWh of electrical energy (where 1 Gj = 0.28 MWh).

4.3 Region Claster Biomass Potential

Based on cluster analysis has been conducted, six types of clusters were formed which were cluster of areas based on the potential of biomass palm oil, clusters of rice biomass potential, cluster of corn biomass, cluster of cassava biomass, cluster of crops biomass (which is a combination of rice, corn, and cassava biomass), and clusters of the total biomass of agricultural commodities. Each type of clusters was classified into three levels of potential: high, medium, and low potential. The potential level value of each cluster type was different, depending on the range of values in a type of cluster.

Using Hierarchical Cluster Method and the calculation of proximity analysis, level value of each cluster type had been obtained. The values range of each potential level of each cluster type, are summarized in Table 6.

Table 6. Interval of potential level each claster

Level	Palm oil biomass (Gj/yr)	Rice Biomass (Gj/yr)	Corn Biomass (Gj/yr)	Cassava Biomass (Gj/yr)
High	9,105,478	4,821,656	632,112	597,549
Middle	3,646,607	1,754,634	154,156	161,629
Low	365,278	1262	17,878	57,773

The agglomeration process of each parameter formed dendogram of each cluster type as presented in Figure 4.

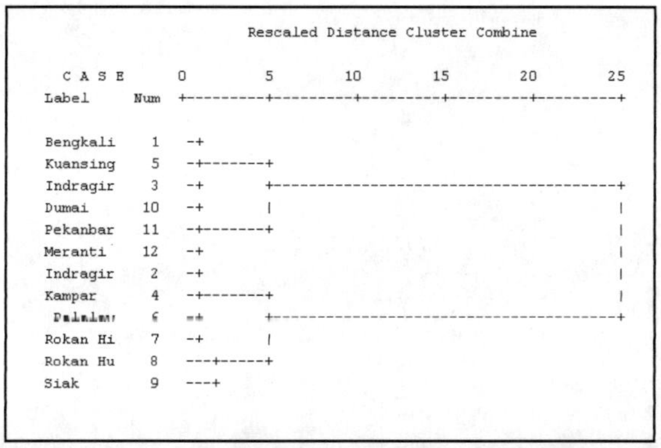

Figure 4a. Cluster Dendrogram of Palm Oil Biomass

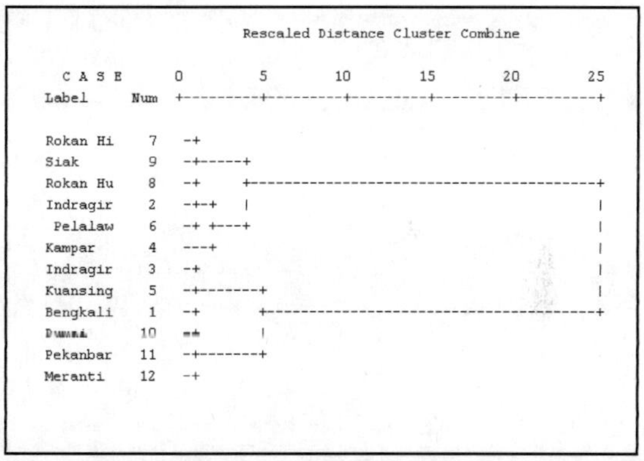

Figure 4b. Cluster Dendrogram of Rice Biomass

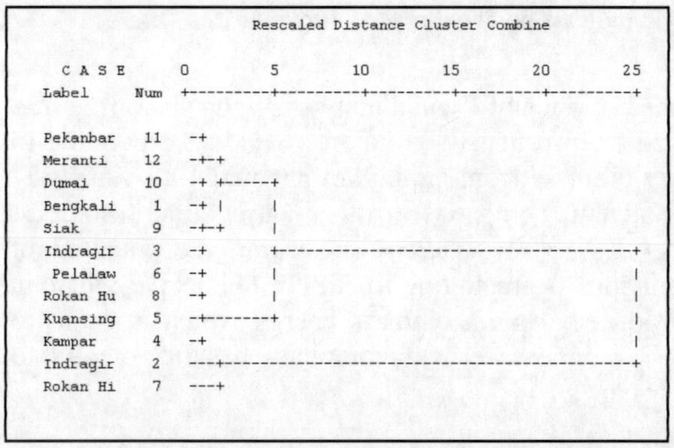

Figure 4c. Cluster Dendrogram of Corn Biomass

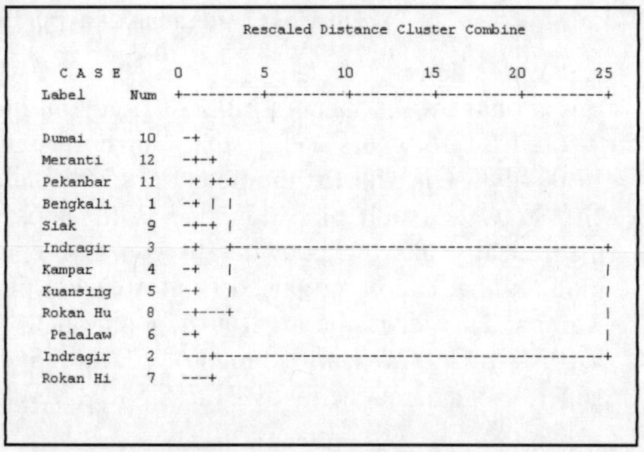

Figure 4d. Cluster Dendrogram of Cassava Biomass

Figure 4. Cluster Dendrogram of Biomass Potential

In the following section (see Figure 5), refers to the data of biomass potential of each region and the results of the cluster analysis, map of the biomass cluster in the province of Riau could be illustrated. Four maps were formed, consisted of cluster map of the palm oil biomass potential, cluster map of the potential of the rice biomass, cluster map of the potential of corn biomass, and potential map of cassava biomass.

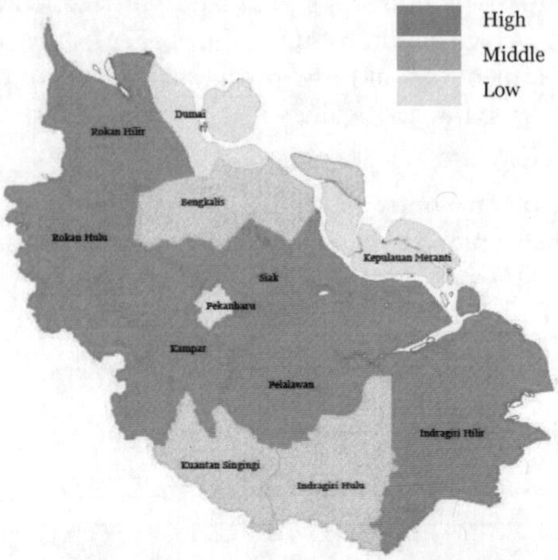

Figure 5a. Cluster map of palm oil biomass

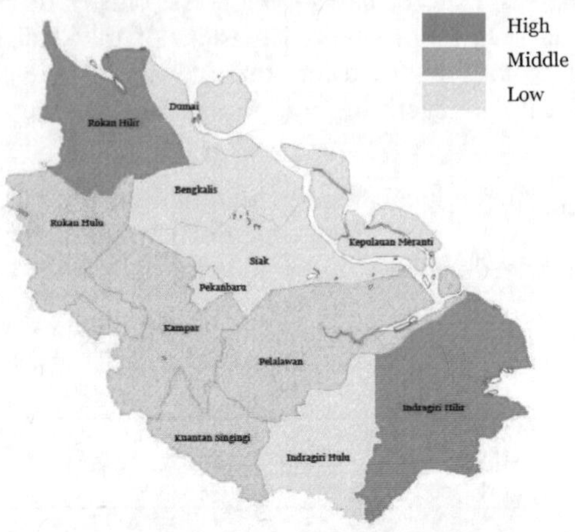

Figure 5b. Cluster map of rice biomass

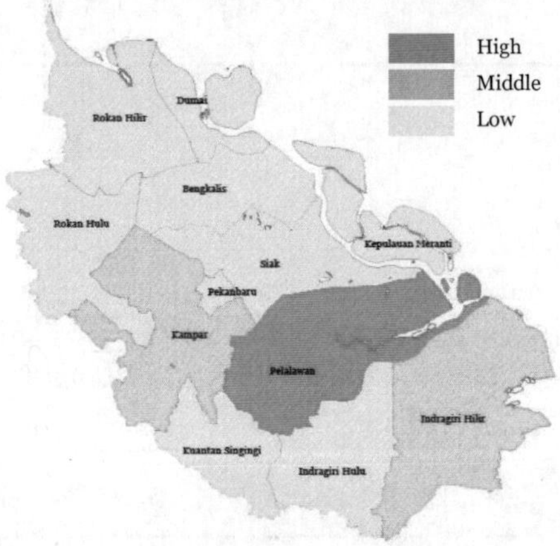

Figure 5c. Cluster map of corn biomass

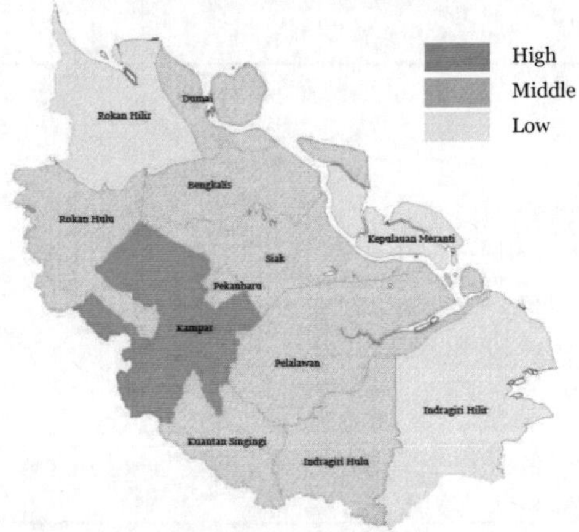

Figure 5d. Cluster map of cassava biomass

Figure 5. Cluster map of agricultural biomass potential in Riau Province

5. Conclusions

Based on the assessment of biomass potential in the province of Riau, a number of conclusions: ***First***, theoretically biomass energy that can be produced from primary agricultural residues amounted to 17,064,985 Gj, whereas the potential of secondary biomass from agriculture amounted to 4,506,657 Gj. Straw and hull of paddy rice is the most potential compared to other agricultural resources, theoretically has the potential to generate energy by 16,091,472.33 Gj. ***Second***, the potential of biomass that can be produced from oil palm plantations is amounting to 55,895,112.56 Gj, wherein, Kampar District is the area most potential to produce oil palm biomass energy which is equal to 9,105,478 Gj. ***Third***, with a total of 77,466,754.8 Gj biomass energy, theoretically has the potential to generate electrical energy by 21,518,542.8 MWh / year.

References

[1] [AAAT] Agency for Assessment and Application of Technology 2014 Outlook Energy Indonesia 2014
[2] General Directorate of Electricity of Energy 2013 Bioenergi Bulletin Edision 2014
[3] Thran D T Seidenberger J Zeddies R. Offermann 2010 Global biomass potentials — Resources, drivers and scenario results *J. Ener. For. Sust. Dev* **14** 200-2005

[4] Berndes G Hoogwijk M & Broek RVD 2003 The contribution of biomass in the future global energy supply: a review of 17 studies *J. Biom and Bioen* **25** 1–28

[5] Welfe A Gilbert P and Thornley P 2014 Increasing biomass resource availability through supply chain analysis *J. Biom and Bioen* **70** 249-266

[6] McKendry P. 2002. Energy production from biomass (part 1): overview of biomass *J. Biores Tech* **83** 37–46

[7] Xingang Z Zhongfu T & Pingkuo L 2013 Development goal of 30 GW for China's biomass power generation: Will it be achieved? *J. Ren and Sust. En. Rev* **25** 10-317

[8] Bertrand V Dequiedt B & Cadre E L 2014 Biomass for electricity in the EU-27 : Potential demand,CO2 abatements and break even prices for co-firing *J. En. Pol.* **73** 631–644.

[9] Calle F, Rosillo P, Groot S L, Hemstock and J Woods 2007 The Biomass Assessment Handbook: Bioenergy for a Sustainable Environment (London: Earthscan)

[10] Biomass Energy Europe 2010a *Harmonization of biomass resource assessments, Vol I: Best Practices and Methods Handbook* (Freiburg-Germany : BEE)

[11] Biomass Energy Europe 2010b. *Methods & Data Sources for Biomass Resource Assessments for Energy* (Freiburg-Germany : BEE)

[12] APEC 2008. Survey of Biomass Resource Assessments and Assessment Capabilities in APEC Economies. www.nrel.gov/docs/fy09osti/43710.pdf [acceded July 10, 2016].

[13] Koopmans A and J Koppejan 1997 Agricultural and Forest Residues - Generation, Utilization and Availability. Wood Energy Conservation Specialists - Regional Wood Energy Development Programme in Asia http://wgbis.ces.iisc.ernet.in/energy/HC270799/RWEDP/acrobat/presidues.pdf [acceded July 15, 2016]

[14] Hambali E A Thahar and A Komarudin 2010 The Potential of Oil Palm and Rice Biomass as Bioenergy Feedstock *7th Biomass Asia Workshop 2010* Jakarta Indonesia

[15] Hartigan J A 1975 *Clustering Algorithms* (New York: Wiley)

[16] Hartigan J A 1985 Statistical Theory in Clustering *J. Class* 2 pp 63-76

[17] A.D.Gordon 1987 A Review of hierarchical classificatio *J. Roy.Stat. Soc*. Series A (General) **150** 119-137

[18] Bridges 1966 Hierarchical cluster analysis *Psychological Reporis*(Southern Universities Press) **18** 851-854

[19] Fraley and Raftery 1998 How many clusters? Which clustering method? Answers via model-based cluster analysis *The. Comput. J* **41** 578-588

[20] Gower dan Ross 1969 Minimum spanning trees and single linkage cluster analysis *J. Roy. Stat. Soc*. Series C (Applied Statistics) **18** 54-64

[21] Plantation Office of Riau Province 2013 Statistik Perkebunan Provinsi Riau Tahun 2012

[22] Lance G N and William W T 1967 A general theory of classificatory sorting strategies *Comput. J.* **9** 373–380

IOP Conf. Series: Earth and Environmental Science **65** (2017) 012007 doi:10.1088/1755-1315/65/1/012007

Benefits of using biogas technology in rural area: karo district on supporting local action plan for greenhouse gas emission reduction of north sumatera province 2010-2020

N Ginting

Faculty of Agriculture, University of Sumatera Utara, North Sumatera, Indonesia

Email : nurzainahginting@gmail.com

Abstract. Indonesia committed to reduce its greenhouse gas (GHG) by 26% in 2020. At the UNFCCC (Conference of the United Nation Framework Convention on Climate Change) held in Paris in December 2015 Indonesia committed to reduce GHG; one way by promoting clean energy use for example biogas. Agricultural industry produces organic waste which contributes to global warming and climate change. In Karo District, mostly the people were farmers, either horticulture or fruit and produces massive organic waste. Biogas research was conducted in Karo District in May until July 2016 used 5 biodigesters. The purpose was to determine benefits of using biogas technology in order to reduct GHG emissions. The used design was Completely Randomized Design (CRD) with treatments: T1 (100% cow feces), T2 (75% cow feces + 25% horticultural waste), T3 (50% cow feces + 50% horticultural waste), T4 (25% cow feces + 75% horticultural waste) and T5 (100% horticultural waste). Parameter research were gas production, pH and temperature. The research result showed that T1 produced the highest methane ($P<0.05$) compared to other treatments while T2 produced methane higher ($P<0.05$) compared to T4 or T5. There was no difference on methane production between T4 and T5. As conclusion application of biogas on agricultural waste supported local action plan for greenhouse gas emission reduction of North Sumatera Province 2010-2020. From horticultural waste, there were 2.1 x 10^6 ton CO_2 eq in 2014 which were not calculated in RAD GRK (Regional Action Plan for Greenhouse Gas Emissions Reduction).

1. Introduction

Associated with Indonesia was rated as the biggest four GHG emiter together with India, China, USA, Indonesia National Development Planning Agency formulated such policy in order to reduce greenhouse gases in Indonesia. In addition, President of Republic Indonesia committed to reduce greenhouse gases 26% by 2020 by its own effort and another 15% by international support[1].

In bottom level, such policy were implemented by Article 6 paragraph (1) of Presidential Regulation of the Republic of Indonesia Number 61 Year 2011 on the National Action Plan for Greenhouse Gas Emission Reduction. It was stated that to reduce the Greenhouse Gas emissions in each province, the Governor of each province should compose the Regional Action Plan for Greenhouse Gas Emissions Reduction (RAD GRK). In defenition, RAD GRK was a document with the objective to reduce GHG

Content from this work may be used under the terms of the Creative Commons Attribution 3.0 licence. Any further distribution of this work must maintain attribution to the author(s) and the title of the work, journal citation and DOI.
Published under licence by IOP Publishing Ltd

emissions pursuant to national commitments until 2020 in the form of policy directions, strategies and programs include activities[2].

To achieved the target, RAD GRK need to supported by the local government, the private sector and communities. In North Sumatera Province, associates with composing on the Local Action Plan for Greenhouse Gas Emissions Reduction on North Sumatera Province of 2010-2020, Governor of North Sumatera province stipulated several regulations, for examples scope of activities of RAD GRK which comprises agriculture, forestry and peat land, energy and transportation, industry and waste management. Moreover, international support such as JICA and USAID supported in the form of guiding target achievement. The Regional Action Plan for reducing emissions of GRK (RAD-GRK) in North Sumatra Province concentrated in (1) Identification of areas and activities that have the potential as the source/ absorption of GRK emissions (2) Identification of emissions and projection of GRK period (years 2010-2020) (3) Formulation of the mitigation actions that have the potential to reduce GHG emissions from agriculture; industry; forestry and peatland; energy, transportation, and waste[3].

There are 6 types of gases which are consisted in GHG, namely Carbon Dioxide (CO_2), Dinitro Oxide (N_2O), Methane (CH_4), Sulfurheksaflorida (SF_6), Perflorokarbon (PFCs) and Hidroflorokarbon (HFCs). Any activities which produces any kind of these gases is classified as GHG emissions. activities that contribute to GHG emissions could be energy-based emissions, emissions from the production process and product use, land-based emissions, and emissions from waste[4].

A wide range of industrial and economic activity contributes to GHG emissions for example agriculture. In Karo District, mostly the people are farmers, either horticulture or fruit and produces tremendous organic waste. Typically in Indonesia only small percentage of waste are transported to landfill and mostly occurs in major cities[5].

As mentioned above, RAD GRK concentrates for example in formulation of mitigation actions that have the potential to reduce GHG from agriculture waste. However mitigation in RAD GRK on waste is focused on how the waste to be transported to landfill[3]. As waste from agriculture especially horticulture have a potential to be converted into energy through biogas technology, mitigation policy in RAD GRK should be improved. Biogas reduce CH4 emissions from organic waste while biogas also produced renewable energy [6]. The objective of this study was to analyze energy potential from agricultural waste in Karo District associated with improving RAD GRK policy in North Sumatera Province.

2. Materials and Methods

Organic waste which were cow waste and horticultural waste were used as an input into 5 biogas units. In Karo District, the highest percentage of horticultural vegetation were kinds of cabbage and oranges. In this research, horticultural waste which were used for examples cabbage, china cabbage, broccoli combined with orange waste.

The research method was used Completely Randomized Design (RAL). HRT (Hydraulic Retention Time) 45 days. Treatments were as follows : T1 (100% cow feces), T2 (75% cow feces + 25% horticultural waste), T3 (50% cow feces + 50% horticultural waste), T4 (25% cow feces + 75% horticultural waste) and T5 (100% horticultural waste). Parameter research were gas production, tempperature and pH. There were five replications and 5 laboratories scale biogas digesters were used (capacity 500 liters). Gas production was measured by a simple gas meter with 1 atm pressure. Figure 1 shows prototype of gas production measured.

3. Results and Discussion

3.1. Gas Productions

Gas production from this study was presented in table 1. There were differences in gas production between T1 and other treatments. Data analysis was then followed by Duncan Test where T1 and T2 showed gas production were significantly different (5%) from other treatments. Higher gas production should occured also on other treatments [6], any agricultural waste has a biogas potential. However, accumulation of rotten oranges in T3, T4 and T5 reduced pH of substrates. A pH of 6 to 7 facilitated good environment for microorganisms in growing. Gas production were measured by gas meter with 1 atm pressured. Gas meter prototype was presented in figure 1.

Table 1. Gas Production (CH_4) : input cow and horticultural waste

Treatments	Replication					TOTAL	Mean
	1	2	3	4	5		
T1	128,14	126,87	152,58	153,2	160,86	721,65	144.33a
T2	96,01	98,22	107,24	112,07	107,24	520,78	104,16b
T3	80,57	88,97	76,6	76,6	68,94	391,68	78, 33c
T4	60,43	61,28	52,77	37,28	38,3	250,06	50,01d
T5	62,35	22,98	61,28	45,45	45,96	238,02	47,6d
			Total			2122,19	

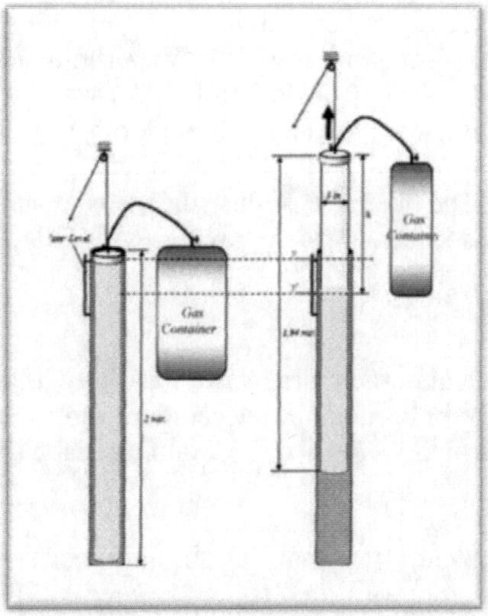

Figure 1. Gas Meter Prototype

3.2. Temperature

In this research, data of temperature were taken as temperature as part of climate effected gas production[6]. Temperature in this research was about 19°C which was same as temperature of environment. Location of the research was in highland area where ambient temperature was 22 - 23°C in July and August thus temperature of underground where biodigester was set up was below 20°C and automatically effect temperature inside biodigester. Three temperature ranges in the anaerobic digestion process which were 1) < 20°C; 2) 20 - 45°C; 3) 45 - 60°C[7] and optimal temperature was 35°C [8]. High gas production were occured in July, August and September where temperature were 19 - 21°C [6]. Extreme temperature affected methane bacterium thus cause gas production stopped.

3.3. pH

pH in this research was took from samples in the outlet of biodigesters/slurry. pH were range from 4 to 7. pH is influenced by combination of substrates materials. Lower pH affect microorganisms in growing and pH of 6 to 7 are preferable. Lower pH such 4 may affect microorganisms metabolism thus affect gas production. It was important to measure pH in order to ensure that there was no disturbance on gas production thus gas production could ceased due to acidification or alkaline condition [9]. pH of biogas slurry is a critical point whether slurry could be used as fertilizer. In this research slurry from T 1 and T2 only were suggested be used as liquid fertilizer as pH are range from 6 to 7. Standard National Of Indonesia on pH of liquid fertilizer were ranged from 4 to 9 however in order to facilitate habitate for a wider range of good microorganisms, better to apply pH as 6 to 7 which were pH for normal soil. A wider range of good microorganisms in soil will create a good competition with pathogen microorganims thus create a healthy soil.

3.4. Benefits of Biogas for Local Farmers

There were three major benefits by using bigas technology, e.i. CH_4 as source of renewable energy, slurry as liquid fertilizer and clean environment [10]. Slurry as fertilizer had been proven by research which was conducted by Ginting and Mustamu in 2012. Application of biogas slurry as 250 ml could be an alternative of 2.5 g NPK as early fertilizer on the growth of Spinach Plant (*Amaranthus tricolor*) [11]. Either livestock manure or agro by-product which are utilized for example through biogas technology is ecologically promoting as cleanliness of the environment is one of biogas technology production [12]. In Malaysia, Biogas technology give benefits such as tremendous renewable energy, fertlizer and could enhancing public health whereas anaerobic digestion reduces the impacts of waste on the environment [13].

This research is followed by dissemination of research result on local farmers in Karo District. 12 biodigesters from PE plastic bag with 0.02 thick were used as biodigester. Kinds of cow waste combined with agricultural waste except orange were used by local farmers. Orange waste which is rotten orange is not used as it is decrease pH thus decrease microorganism metabolism thus decrease gas production. Rotten orange should processed in composting with addition of $CaCO_3$ to increase pH.

3.5. Benefits of using biogas technology in Karo District on supporting of local action plan for greenhouse gas emission reduction of North Sumatera Province 2010-2020

It was mentioned in the beginning of RAD GRK concentrates on (1) Identification of areas and activities that have the potential as the source/ absorption of GRK emissions. (2) Identification of emissions and projection of GRK period (years 2010-2020) (3) Formulation of the mitigation actions that have the potential to reduce GHG emissions from agriculture; industry; forestry and peatland; energy, transportation, and waste.

In RAD GRK document of North Sumatera Province, from agriculture, it was mentioned that GHG from horticulture was not calculated as the horticulture area was not significant thus GHG emited also was not significant. There were 6 major emiter activity in agriculture, namely palm oil mill, irrigated rice field, rubber plants, food crops, urea in oil palm plants and livestock. While based on share of agriculture on GRDP (table 2), it was known that horticulture together with crops influenced GRDP from agricultural sector. It concluded from this fact that activities in horticulture were tremendous moreover the highlands in North Sumatera Province were potential for the development of horticulture crops. Associated with RAD GRK where mitigation for waste was to be dumped onto landfill, in case of horticultural waste it was better to be processed by using biogas technology. Meanwhile in order for biogas technology to be used, it need to be supported by the local government and communities, i.e. farmers.

Table 2. Share of Agriculture, Processing Industries and Others to GRDP at Current Price, 2015 in North Sumatera Province

Sector/Subsector	2015
Agriculture	23.03
• Horticulture	8.05
• Crops	9.87
• Livestock	2.54
• Forestry	1.23
• Fishery	2.34
Processing Industries	23.02
Others Sectors	52.12

Table 3. GHG emission from Agricultural Sector (Local Action Plan for Greenhouse Gas Emissions Reduction of North Sumatera Province 2010-2014)

Emission from Agricultural Sector (Year)	**Emission from Agriculture Sector (tCO_2eq)**
2010	9.322.588,52
2011	9.412.212,01
2012	9.551.101,32
2013	9.641.232,02
2014	10.521.102,21

From table 3, it could be concluded that the emissions increase significantly from 2006 to 2014. So mitigation should be taken place and based on this research mitigation could be by using biogas technology on waste from horticulture which has not been calculated. By capturing the gas emitted with biogas technology, CH_4 emissions were considerably reduced. Combustion of CH_4 will produce CO_2. As methane was a greenhouse gases with a global warming potential 28 times greater than CO_2 [14], combustion of CH_4 would create benefits such energy and gases with lower global warming potential [15].

Statistic of North Sumatera Provinces 2014, showed that in Karo District there were 256,735 ton horticulture production which cabbage was significantly produced, i.e. 64,305 ton. In general, waste from horticultural product could be as much as 1.5 times than the product itself. So it could be concluded that there were 96,457 ton cabbage waste in Karo District in 2014.

From the research, the best result on biogas production were on T1 and T2 while T1 only contained with cow feces and in T2 contained 75% cow feces and 25% horticultural waste. As it is horticultural waste which would be calculated, so T2 would be used as based of calculation. From 96,457 ton horticultural waste mostly cabbage, it would be produced about 1,415 x 10^9 CH_4. Capturing methane was equal to cutting emission of methane thus minimizing global warming [6]. Methane had a global warming potential 28 times greater than CO_2 [14]. From this research, it concluded that by capturing methane from horticultural waste mostly cabbage, there were global warming potential as much as 39.62 x $10^9 CO_2$ eq.

4. Conclusions

RAD GRK North Sumatera Province should be improved since it was not included waste from agricultural sector, i.e. horticulture. Eventhough land size of horticulture was less significantly than other sector but it produced tremendous product followed by tremendous waste which produced CH_4. From horticultural waste mostly cabbage waste alone, it could be assumtion that there were global warming potential as much as 1,415 x 10^9 CH_4 or 39.62 x 10^9CO_2 eq in 2014 which could be cut. By capturing CH_4 through biogas technology, CH_4 emissions are considerably reduced.

Acknowledgements

The author gratefully acknowledges that financial support from Republic of Indonesia: Ministry of Research, Technology and Higher Education

References

[1] Bappenas 2011 *Pedoman Umum Penyusunan Rencana Aksi Daerah Penurunan Emisi Gas Rumah Kaca Draft 2 – 21 Mei 2011* (Jakarta: Ministry of National Development Planning/National Development Planning Agency).

[2] Endah M 2013 *Update on Indonesia's Climate Change Mitigation and Adaptation Plan* (Jakarta: Ministry of National Development Planning/National Development Planning Agency)

[3] The Government of North Sumatra Province 2012 *Local Action Plan for Greenhouse Gas Emissions Reduction of North Sumatra Province 2010-2010*

[4] IPCC 2008 2006 *IPCC Guidelines for National Greenhouse Gas Inventories*, Prepared by the National Greenhouse Gas Inventories Programme, Ed. H S Eggleston, K Miwa, N Srivastava and K Tanabe (Japan: IGES)

[5] Department of Spatial and Redential of North Sumatera 2015 *Annual Report*

[6] Balde H, Andrew C V, Stephen D B, Claudia W R, Anna C, Raymond L D and Douglas J M 2016 Methane emissions from digistate at an agricultural biogas plant. *Bioresource Technology* **216** 914 – 922.

[7] Cantrell, Keri B, Thomas D, Kyoung S R and Patrick G H 2008 Livestock waste-to-bioenergy generation opportunities *Bioresource Technology* **99** 7941-7953

[8] Usama Z, Craig F, Paramod P and Shulin C 2008 Evaluation of a new fixed-bed digester utilizing large media for flush dairy manure treatment *Bioresource Rechnology* **99** 8619-8625.

[9] Ware A and Niamh P 2016 Biogas from cattle slaughterhouse waste: Energy recovery towards energy self-sufficient industry in Ireland *Renewable Energy* **97** 541-549.

[10] Ginting N 2012 Supporting policy to succeed the regional action plan (RAD) greenhouse gas (GHG) emissions in North Sumatra *Innovation: J of Politics and Policy* **9**

[11] Ginting N and Mustamu N 2012 The application of biogas sludge as organic fertilizer on the growth of spinach plant (*Amaranthus tricolor*). *Procs. Annual International Conference Syiah Kuala University* in 2012 (Banda Aceh: The 8th IMT-GT *Uninet Biosciences Conference*) pp 154-156

[12] Francesco C, Cesare C, Giulio d N d V and Massimo D d 2015 Technical and financial analysis of a cogeneration plant fueled by biogas produced from livestock biomass *Energy Procedia* **82** 666-673

[13] Abdeshahian P, Jeng S L, Wai S H, Haslenda H, and Chew T L 2016 Potential of biogas production from farm animal waste in Malaysia *Renewable and Sustainable Energy Reviews* **60** 714-723

[14] Myhre G *et al.* 2013 *Climate Change 2013: The Physical Science Basis*. Contribution of Working Group I to the Fifth Assessment Report of the Intergovernmental Panel on Climate Change *Anthopogenic and natural radiative forcing*. Ed. Stocker T F *et al.* (London: Cambridge University Press).

[15] Van S P 2007 *Biomass Energy* (Belanda : Centre Technology for Sustainable Development TDO, Technische Universiteit Eindhoven)

Bioindustry development based on citronella essential oil to meet the needs for renewable energy: A review

M Rizal[1], S Wiharna[1] and A Wahyudi[1]

[1]Indonesian Spices and Medicinal Crops Research Institute, Jl. Tentara Pelajar No. 3, Bogor, Indonesia

Email: molide2012@gmail.com

Abstract. In 2015, national domestic fuel consumption is already above 1.5 million barrels per day, while production is under 800,000 barrels per day. The gap between production and consumption will be considerably widened, serious efforts is needed to save the use of petroleum and also look for alternative replacement with renewable natural energy. Two approaches that can be taken: First, save the use of fossil fuel directly by using citronella essential oil-based bioadditive by 20 percents, equal to at least Rp. 55.2 trillions of national petroleum subsidies. Second, encourage increased utilization of biofuel mixed with that bioaditives that gradually reduce dependence on fossil fuels while developing machines which will fully operated with biofuels. Development of Sustainable Agricultural Bioindustry (SAB) system by integrating crops (candle nut, citronella) with livestock (dairy cattle) in a specific region. could contribute to : a) production of biodiesel and bioadditive feedstocks, b) production foodstuffs like beef and/or milk, c) utilitization of non-productive land, d) employment, by absorbing large number of farmer, e) increase the farmers income, f), biogas that can be used to meet daily household energy needs, and g) environmental conservation and sequestration of carbon emissions.

1. Introduction

Increasing need in energy demand which parallel with the development of the human population with all their needs are faced with the facts for depletion of the world's energy reserves, especially fossil energy. This conditions globally, sooner or later, have sparked increasingly fierce competition for dominance and the search for foddil energy resources. For Indonesia, which is located in the equator with a tropical climate, the problem can not be ignored, because it has been stated that Indonesia's oil proven reserve is 3.7 milyard barrels that will be exhausted in 10 years [1].

In 2015, domestic fuel consumption is already above 1.5 million barrels per day, while production is under 800,000 barrels per day. Indonesia ranks as 24th producer of petroleum in the world. To meet these needs, the Government of Indonesia has to import 350,000 - 500,000 barrels per day [2]. The gap between production and consumption of the national fuel are considerably widened so that serious efforts is needed to fulfill most of the fuel demand through utilization of renewable energy. This effort is hoped could help ease the burden of fuel subsidies in the state budget by reducing the consumption of domestic and imported fuel.

Realizing this, before emerging energy crisis which will have a wide impact politically, economically and socially, the Government through the Directorate General of New, Renewable Energy and Energy

Content from this work may be used under the terms of the Creative Commons Attribution 3.0 licence. Any further distribution of this work must maintain attribution to the author(s) and the title of the work, journal citation and DOI.
Published under licence by IOP Publishing Ltd

Conservation (EBTKE), Ministry of Energy and Mineral Resources, has established policies and facilitated various parties to carry out research and development efforts for utilization energy of biomass (bioenergy), whose raw materials are available in abundance in the sheer number continuously in the archipelago, especially biofuels. Biofuel can be determined as a fuel from organic resources including plants and animals [3]. Biofuel has a specific characteristic that is renewable, whereas it can be produced by using raw material which can be grown or developed. There are a variety of biofuels potentially available, but the main biofuels being considered globally are biodiesel and bioethanol. Bioethanol could be use as a substitute for gasoline and biodiesel as a substitute for diesel oil.

On the other side, applications of biofuel nowadays still facing some problems because it is not fully compatible with recently available machinery technologies as they were designed to work with petroleum fuels. Nevertheless, by adding a small number of bioadditive which was derived from citronella essential oils into the fuel tank then the effect of clogging, heating and knocking in the engine can be minimized.

This paper examines the ideas how to couple the upstream-downstream production and utilization of biofuels in an integrated manner in order to fulfill part of the national needs for renewable energy.

2. Citronella Based Additive: Prospect and Contribution in the Utilization of Renewable Energy

Citronella (*Cymbopogon nardus* L, family Gramineae) is one kind of essential oil producing plants that can grow in Indonesia (Figure 1). Citronella oil was obtained from distillation of citronella grass leaves, and Indonesian citronella oil in world trade has known as "Java Citronella Oil". Citronella crop acreage in 2014 was 18,660 ha [4]. Citronella production centers area in Indonesia were located in Aceh, West Sumatra, West Java, Central Java, East Java and East Kalimantan [5].

Figure 1. Citronella plantation at Manoko Experimental Garden, Lembang, West Java [6]

Citronella oil containing about 35-45% citronella compounds, 85-90% geraniol, 11-15% citronellal, 3-8% geraniol acetate, 2-4% citronellal acetate and slightly contains sesquiterpenes as well as other compounds. Citronella oil is widely used as an industrial raw material of soap, toothpaste and medicines. Citronellal and geraniol are generally used for the manufacture of esters, such as hydroxy citronellal, geraniol acetate, and synthetic menthol which have the nature of a more stable and widely used in the perfume and other fragrance industries. Hydroxy citronellal is important for high-quality soaps and perfumes, while menthol flavors to the basic ingredients of cough medicine, toothpaste and as dessert. Citronella oil is also used as a plant-based pesticides such as fungicides, bactericides, insecticides and nematicides [7]. At lately, citronella oil also developed as bioadditive for petroleum based fuel oil [8, 6].

Citronella oil-based industry is very potential to be developed as Indonesia has a comparative advantage in the supply of raw material. The development of this essential oils industry will create a multiplier effect by increasing the income of essential oil crop farmers, and creating employment in the field of agroindustry [6]. If a few years ago, Indonesia ranks as the second largest citronella oil producing countries after China, then Indonesia is currently the largest producer in the world because of declining acreage and production in China.

One of the benefits of citronella oil which is highly prospective for development in the future is utilization of this oil as a main component for fuel additive (bioadditive). Bioadditive is made of a mixture of several essential oils, main component is citronella oil, which are native to Indonesia, easy to cultivate in the country and does not interfere with human food needs.

Bioadditive can be used easily by added to the fuel tank, it can be mixed with any composition of gasoline and diesel oil, at 1 ppm concentration, have the physical and chemical properties similar to gasoline or diesel fuel, it can be degraded easily (biodegradable). Bioaditive serves as a catalyst, enhance the combustion process in the engine, so that the resulting combustion emissions environmentally friendly and does not add to the accumulation of carbon dioxide in the atmosphere that can further reduce the effects of global warming or known with zero CO2 emission [8].

Two approaches that can be taken in implementing citronella essential oil-based bioadditive for lowering national fossil fuel consumption: ***First,*** save the use of fossil fuel directly by using bioadditive on commercial machines. If the use of citronella-based bioadditive simply able to economize on the use of fuel by 20%, then if the quota of national fuel subsidies in 2015 is estimated to reach 46-47 million KL [9], equal to Rp. 276 Trillion, there will be a saving of fuel subsidy by Rp. 55.2 trillion. Laboratory test results on the Lemigas Laboratory and the vehicles that use bio fuel additive products showed a decrease in exhaust emissions/opacity of the vehicle are quite significant, this means that they would help in the global warming reduction program. ***Second,*** encourage increased use of biofuel mixed with bioadditive that can reduce fossil fuels usage while developing machines which will fully operated with biofuels. It was considered that biofuel produced today were not optimum working on gasoline and diesel machines used by today's society and industry. Some of the problems faced is the clots of biofuel in the engine combustion chamber, the crust of dirt that interfere with engine performance and increased engine temperature. Utilization of citronella bioadditive will improve biofuel engine performance whiles maintaining machines and extend engine life.

There are some advantages in using citronella based bioadditive, among other things: 1) It is a renewable energy, 2) cultivation of the raw material is not difficult because the plants easily adapt to the climate and local soil, 3) Use of raw materials for the additive does not conflict with the interests of provision of food or products vital in community life, 4) reduce exhaust emissions (especially reducing black smoke) in comparing with fossil fuels without the addition of these additives, and 5) environmentally friendly [8].

A series of tests in the field using a variety of gasoline and diesel fuelled engines show that citronella-based bioadditive is able to save petroleum consumption: Gasoline 10-30%, Solar 10-40%. These figures show that bioadditive utilization will obtain a significant decrease in petroleum based fuel consumption. Sucipto stated that in addition to the more complete combustion in the engine it will obtain better power and exhaust emissions are more environmentally friendly [8]. Further assessment is technically and economically necessary to provide opportunities of additive use as a blending component of gasoline and diesel oil for supporting fuel consumption efficiency in the transportation sector, power plant generators, military equipments and other industries.

The better prospects for the future is the use of this citronella bioadditive in supporting the above mentioned machines to be operated with biofuels, i.e bioethanol as a substitute for gasoline and biodiesel as a substitute for diesel oil. There is also an important need to develop and mass production of machines and vehicle that could be fully operate with biofuel. Collaboration among academicians and industrialists were needed to bring the dream into reality. Indonesia then could play an important role by producing these machines as well as their renewable biofuel for energy independence and energy security.

3. Sustainable Agricultural Bioindustry: Components and Products for Production of Bioenergy

Bioindustry term is used for integrated system where from single bio raw material through multiple process to produce multiple valuable products [10]. The bioindustry term used in this paper is intended to applied on Sustainable Agricultural Bioindustry (SAB), the new paradigm to acommodate transformation of the development orientation from fossil-based raw materials into renewable resources (biological resources). Thus, the role of agriculture is not only a major producer of materials food, but a producer of biomass feedstock biorefinery to produce food, feed, fertilizer, fiber, energy, pharmaceutical products, chemical and other bioproduct. SAB systems in principle how to manage and/or make an optimum use of all biological resources including biomass and/or waste organic farming, for the welfare of the community in an ecosystem in harmony. In this scheme, agricultural waste and biomass utilized by applying bio-technology and genetic engineering process to produce tangible output of agricultural products for final consumption as stock feed, processed food, feed, energy, bioproduct. It also includes environmental services to public and private and also has economic value added. It was expected that from agricultural bioindustry will occur economic participation and distribution of added value, which will be beneficial socially and politically [11].

SAB can be developed in a farmer level, farmer's group, or by large agricultural companies in a specific region. The development area can be tangible ecological region with a particular landscape, or in a village-level administrative region, district or a wider area. The SAB systems designed so that a sustainable way of bioprocess which took place in the region are able to produce range of products such as biomass, nutrients, water, gas and energy independently with minimum external input. The type of raw material used is the entire biomass, agricultural waste and the results derived from the use of crop, animal, other organism. Besides aiming to reduce yield loss, agriculture bioindustry characterized by reprocessing and recycling of waste and the waste processed, biorefinery where an integration processing with multi input multi output as well as the joint input-joint output, the agricultural bioindustry can also be utilized for the expansion of the product and industry (Product / industry widening). From this SAB system, we could have produced the main products or by-products of high economic value or strategic value, including citronella oil based bioadditive which became the topic of discussion in this paper.

SAB components developed for the production of biofuels and bioadditive consists of: candle nut pant *(Reutealis trisperma,* family Euphorbiaceae*)* cultivated in strip farming with citronella plant, cattles, biogas fermentor, candlenut oil processor, and citronella oil distillator. Candlenut tree is able to produce 5 ton of oil / ha and its use as a biofuel is not a conflict as foodstuffs such as palmoil (Figure 2). Pranowo *et al.,* that productivity level of candle nut can reach 8-9 tons of crude oil, equivalent to 6-8 tons of biodiesel/ha/year, make as a strategic commodity associated with government programs to find alternative sources of renewable energy [12]. Besides as a biofuel feedstock, candlenut also useful as conservation plants because it has deep and broad roots [13, 14], even in degraded land [12]. This plant has a good adaptability to grow in various regions in Indonesia. Meanwhile citronella oil production obtained is 480 kg /ha/yr [6].

Figure 2. Candle nut plantations for conservation [13]

Each of these plants are cultivated following their Good Agricultural Practices (GAP) procedures and processed into candlenut oil and citronella oil in accordance with their Good Manufacturing Practices (GMP), the dregs material from citronella distillation waste is used as feed for cattle (beef or dairy cattle) (Figure 3), while livestock manures are used for the biogas production process and further processed into compost which is returned to the plantation to fertilize candlenut and citronella plants. Biogas produced in this bioindustry system can be used for domestic use (cooking and lighting) or used as a fuel for the distillation of citronella. Other benefit of this SAB system are production of foodstuffs such as meat or dairy milk. Analysis of the citronella grass residual distillates indicate that the nutritional content are balanced with grass and a little much of better than rice straw (Sukamto and Syukur, 2015) [6]. Meanwhile, fresh milk produced 12 litres/cow/day, the quality still meet the required standards SNI-3141.1.:2011 about fresh milk [15] and are safe for public consumption as complied with Indonesian National Standard on Testing for Microbial Contamination on dairy Products [16, 17].

Figure 3. Utilization of Citronela distillation waste as cattle feed [18]

In terms of the environment, citronella-based SAB have positive impacts as one way to use marginal lands, also for post-mining and post-disaster land reclamation activity. Combination of annual with perennial plants like citronella and candlenut stands in which the harvesting activities does not damage the land must also contribute as a carbon sink to the surrounding environment.

Indonesia as the third largest forest area in the world, can play an important role to reduce world carbon emissions through carbon sinks [19]. Development of biofuel and essential oil crops that are integrated with cattle in SAB system is expected to be one of the efforts to preserve the forest, through the planting (afforestation) in the non-forest area (degraded land). As well as doing repairs the damaged forest areas (degraded forest) by way of reforestation with both types of plants.

4. Implementation and Benefits for the National Energy Requirements

The implementation of technological innovations and products that have been reinvented by the best sons and daughters of this nation have a great potential to combine in an inclusive SAB system in a region expected to fulfill the need for a national energy by producing and utilizing renewable energy. Development of citronella-based SAB system has the following benefits [8]:

a. Saving national fuel subsidies through utilization of citronella bioadditive

b. Increased use of Biofuel (BBN) combined with citronella bioadditive on commercial machines

c. Utilization of non-productive land
Use of biofuel additives widely and continuous direct need input source sustainable raw materials as feed stock, for it takes the availability of land for cultivation. The utilization of marginal land (non-productive) become a great potential for growing essential so that it can be a source of raw materials to meet the needs of this bio-fuel additive. Assuming national fuel consumption in 2015 reached 46 million kiloliters, it need 46,000 litres bioadditive (applied at 1 ppm concentration), then it can be use non-productive land as much as 95,833 ha for citronella plantation to cover this need.

d. Employment
The need of land used to meet the needs of raw materials related to cause a major impact on employment. The positive impacts that occur after the non-productive land is developed for planting effort citronella is absorbed a large number of farmers. As an illustration is in 1 ha of citronella plantation takes at least 4 farmers who can work from planting to harvesting. With as many as 95,883 ha of land use is needed additional workforce of 383,532 farmers.

e. Increasing farmers' income
The positive impact that followed as a result of land use is increasing farmer incomes. For comparison farmer per day minimum wage of Rp 50,000, - then in one month (25 days) pay farmers to Rp 1.500.000, -. By cultivating citronella in 1 ha, farmers get as much as 150 kg lemongrass oil, with lemongrass oil price per kg Rp 150.000, then the farmers' income to Rp 22.500.000, - / month. After deducting the production cost by 30% the number of farmers' income to Rp 15.750.000, - / month. If in 1 ha of citronella takes at least four farmers to make the cultivation of citronella then every farmer would get additional income of Rp 3.937.500, - / person, meaning that there is an additional income of farmers amounted to 262.5%, which means there is an increased level of life of farmers [8]. This scheme were calculated based on 10 hectares citronella plantation, in order to get adequate number of citronella grass to be distillated every day. Considering that citronella plants were best distilled after 6 months, and then should be cutted for ditillation at every 3 consecutive months, so we propose for each one hectare of citronella farm there shoul be installed 1 distillation unit with capacity of 300 kgs citronella grass.

f. Food source
In integration with cattle, SAB acting as a source of food, such as milk and/or beef in a program of self-sufficiency and food security. The results of the research in Lembang, West Java show that the integration of planting citronella minimum area of 9 hectares is able to provide supplies forage for 30 dairy cows continuously [6]. Rizal *et al.* reported that dairy milk produced on that system were proven have suitable quality and safe [17].

g. Biogas
Livestock manures generated in SAB area could be processed further to into biogas that can be used to meet daily household energy needs.

h. Environmental conservation and carbon sink
In addition to functioning as a plant conservation, candlenut also has a very big role in environmental conservation. Since candlenut produce a large number of leaves, it also play a role in supplying oxygen to the environment in the sheer number is quite large, as well as absorption of CO_2 from the environment and water uptake by the rootssupply oxygen to the environment was certainly quite large, as well as absorption of CO2 from the environment and water uptake by the roots (Jurnalasia, 2015) [20].

5. Conclusion
Sustainable Agricultural Bioindustry approaches can be used to produce bioadditive and biofuel, as well as foodstuffs, in a specific region to meet a portion of national renewable energy needs.

References
[1] Ditjen Migas 2016 RI punya cadangan minyak 21,5 miliar barel yang belum bisa diambil (http://finance.detik.com/energi/d-3217723/ri-punya-cadangan-minyak-215-miliar-barel-yang-belum-bisa-diambil) October 28th 2016
[2] Indonesia Investments 2016 Minyak bumi (http://www.indonesia-investments.com/id/bisnis/komoditas/minyak-bumi/item267) October 6th 2016
[3] Putrasari Y, Praptijanto A, Santoso W B, and Lim O 2016 Resources policy and research activities of biofuel in Indonesia: A review *Energy Reports* **2** 237–245
[4] BPS 2016 Luas areal tanaman perkebunan rakyat menurut jenis tanaman 2000-2014 Biro Pusat Statistik Jakarta (https://www.bps.go.id/linkTabelStatis/view/id/1669) October 8th 2016
[5] Ditjenbun 2012 Statistik perkebunan indonesia: citronella (Jakarta: Direktorat Jendral Perkebunan) p 131
[6] Sukamto, C Syukur, Syafaruddin, and I M Trisawa 2015 Citronella penghasil minyak atsiri dan sumber pakan ternak *Booklet* (Bogor: Pusat Penelitian dan Pengembangan Perkebunan) p 21
[7] Oroojalin F, Kasra-Kermen R, Azizi M, and Bassimi M R 2010 Phytochemical composition of the esential oils froms three apiaceae spesies and their antibacterial effect on food-borne pathogen *Food Chemistry* **120** 765-770.
[8] Sucipto J 2015 Atsiri sebagai bahan bakar: sumber energi baru dan terbarukan, potensi dan pemberdayaan ekonomi rakyat p 10
[9] BPH Migas 2016 Kuota BBM subsidi 2015 ditetapkan sebesar 46-47 juta KL (http://www.bphmigas.go.id/bbm/kuota-bbm-subsidi-2015-ditetapkan-sebesar-46-47-juta-kl) October 6th 2016
[10] Lakitan B 2014 Sistem pertanian bioindustri berkelanjutan *Workshop Pengembangan Sistem Bioindustri* Jakarta 26-27 September 2014
[11] Kementan 2014 Pertanian-bioindustri berkelanjutan: Solusi pembangunan indonesia masa depan *Strategi Induk Pembangunan Pertanian 2015-2045* (Jakarta Kementerian Pertanian) p 210
[12] Pranowo D, Herman M, and Syafaruddin 2015 Potensi pengembangan kemiri sunan *(Reutealis trisperma* (Blanco) Airy Shaw) di lahan terdegradasi *Perspektif* **14** 87-101
[13] Herman M, Syakir M, Pranowo D, Saefudin and Sumanto 2013 Kemiri sunan (*reutealis trissperma* (blanko) airy shaw): tanaman penghasil minyak nabati dan konservasi alam (Bogor: IAARD Pres) p 91
[14] Wahyudi A, A Aunillah, dan M Syakir 2010 Potensi dan prospek bahan bakar nabati sebagai pendukung desa mandiri energi. *Warta Penelitian dan Pengembangan Tanaman Industri* **17** (1)
[15] BSN 2016 SNI-3141.1.:2011 Susu Segar Bagian -1: Sapi (http://sisni.bsn.go.id/index.php/sni_main/sni/detail_sni/11914) October 31th, 2016

[16] BSN 2008 SNI 2897 Metode pengujian cemaran mikroba dalam daging, telur dan susu serta hasil olahannya (http://sisni.bsn.go.id/index.php/sni_main/sni/detail_sni/7779) October 31th, 2016

[17] Rizal M, Sukamto and Sujianto 2016. Mutu susu sapi segar dengan pakan limbah seraiwangi di kebun percobaan Manoko, Lembang, Bandung *Konferensi Nasional Minyak Atsiri* Aceh 19 Oktober 2016

[18] Sukamto and Djazuli M 2011 Limbah serai wangi potensial sebagai pakan ternak *Warta Penelitian dan Pengembangan Pertanian* **33** pp 10-12

[19] Mulyani S 2014 Hutan sebagai penyedia jasa lingkungan (http://srimuliyani.blogspot.co.id/2014/01/hutan-sebagai-penyedia-jasa-lingkungan.html) October 31th 2016

[20] Jurnalasia 2015 Kebutuhan produksi biodiesel serap pasar kemiri sunan (https://www.jurnalasia.com/bisnis/agribisnis/kebutuhan-produksi-biodiesel-serap-pasar-kemiri-sunan/)

IOP Conf. Series: Earth and Environmental Science **65** (2017) 012009 doi:10.1088/1755-1315/65/1/012009

Biomass of cocoa and sugarcane

Siswanto[1], Sumanto[1], R S Hartati[1] and B Prastowo[1]

[1]Indonesia Center for Estate Crops Research and Development, Bogor, Indonesia

E-mail: siswantos2002@yahoo.com

Abstract. The role of the agricultural sector is very important as the upstream addressing downstream sectors and national energy needs. The agricultural sector itself is also highly dependent on the availability of energy. Evolving from it then it must be policies and strategies for agricultural development Indonesia to forward particularly agriculture as producers as well as users of biomass energy or bioenergy for national development including agriculture balance with agriculture and food production. Exports of biomass unbridled currently include preceded by ignorance, indifference and the lack of scientific data and potential tree industry in the country. This requires adequate scientific supporting data. This study is necessary because currently there are insufficient data on the potential of biomass, including tree biomasanya detailing the benefits of bioenergy, feed and food is very necessary as a basis for future policy. Measurement of the main estate plants biomass such as cocoa and sugarcane be done in 2015. Measurements were also conducted on its lignocellulose content. Tree biomass sugarcane potential measured consist of leaves, stems and roots, with the weight mostly located on the stem. Nevertheless, not all the potential of the stem is a good raw material for bioethanol. For cocoa turned out leaves more prospective because of its adequate hemicellulose content. For sugarcane, leaf buds contain a good indicator of digestion of feed making it more suitable for feed.

1. Introduction

Biomass is organic material produced by photosynthetic processes, either in the form of products or effluent. Examples of biomass include plants, trees, grass, potatoes, agricultural waste, forest waste, urban sewage, feces and animal manure. The use of biomass as a renewable energy source is a way out of human dependence on fossil fuels. National energy policy through the issuance of Presidential Decree No. 5 of 2006 on the National The energy policy set targets for biofuels use to more than 5% of national energy consumption by 2025. The policy followed by Presidential Instruction No. 1 in 2006, which among other things instructed the Minister of Agriculture to encourage the provision of the plant including the facilitation of the provision of seeds and seedlings, counseling, and integrating the activities of development and post-harvest of plant material to support the provision of biofuels. Modernization of agriculture ahead, including agro-industry and agricultural mechanization will increasingly play a role in agriculture development in the future and also adds to the strategic environment of the agricultural sector. Agricultural mechanization that most of its operations rely on the availability of fuel-based oil and fossil energy (fuel) is increasingly scarce will face a big problem. Bioenergy, including biofuels which its raw materials from agricultural biomass are abundant in agricultural and forestry sectors, be an alternative that should be considered to address the problem.

Content from this work may be used under the terms of the Creative Commons Attribution 3.0 licence. Any further distribution of this work must maintain attribution to the author(s) and the title of the work, journal citation and DOI.
Published under licence by IOP Publishing Ltd

IOP Conf. Series: Earth and Environmental Science **65** (2017) 012009 doi:10.1088/1755-1315/65/1/012009

In Indonesia, hundreds of million tonnes of agricultural waste such as straw, bran, litter, sugarcane, oil palm empty fruit bunches and others generated annually. Most of the waste is untapped or more that just burned in the land. As the solid fuel, agricultural waste and other biomass can be effectively optimized in a way to convert them into fuel gas through a reactor. A study that the appointment of gaseous fuels from biomass materials containing a very high calorific value which is 3.5 to 5.5 MJ Nm3, which consist of carbon monoxide (25% v / v) and hydrogen (15-20% v / v) , This gas can be burned by a burner and generate temperatures reached more than 1000°C which can be used for heating in industrial applications[1].

Waste utilization is expected to reduce the problem and generates revenue by making the high added value of the waste. The problem recently, the selection of waste biomass which has lignocellulosic as raw material for bioethanol still has obstacles to the process of hydrolysis to produce glucose. However, the hydrolysis of lignocellulose to glucose and xylose can be done using the enzyme β-glucosidase[2]. Sunarti and Richana conducted a study for lignocellulosic ethanol from corn cobs with a yield of 14.22%[3]. Technically the technology is here, so what is needed is kind and how many raw materials are available. Therefore the main plant biomass plantation trees need to know including the potential for bioenergy as well as in support of agricultural development bioindustry forward.

Search alternative materials that do not compete with food and feed is necessary and urgent to think about. This occurs because the technology development of first generation biofuels require or utilize the results of the primary agriculture or plantations, such as palm oil, coconut oil, castor oil, molasses, cassava and others. On the other hand, the available agricultural and forestry waste as a source of lignocellulose to be a promising source of energy. Waste utilization is expected to reduce environmental problems and generates revenue by making the high added value of the waste. There are currently more potential calculated or known biomass is the only potential macro, therefore to the next, the tree biomass and its potential for bioenergy needs to be known, so it will known quantitatively the usage of biomass as organic material, feed or for bioenergi that will not disrupt the environmental sustainability of agriculture.

Based of that explanation the objective of this research are to know the biomass trees and the potential biomass of estate crops (cocoa and sugarcane) as a source of bioenergy feedstock. The output will be gained are the biomass trees and potential biomass of cocoa and sugarcane plants for bioenergy supportting agriculture bioindustry, and the outcome are the ease and provided a quantitative basis which is a clear potential for the development and utilization of biomass from estate crops and its economic viability as a source of bioenergy raw materials in support of agricultural development bioindustry.

2. Methodology

The research will be conducted through a methodology:

- A literature study on the activities (including research) that has been done by others so that can be arranged as a biomass tree of plantation.
- Field studies and measurements.

Estimation of biomass can be using Allometri with mathematical models[4,5]. The activities are not using Allometri, but with a destructive method, by direct measurement as much as possible components of plant biomass through field studies. Measurement of lignocellulose or lignin and cellulose, which are useful for the determination of the potential of biomass for the manufacture of bioethanol and other bioindustry products[6,7]. Lignin and cellulose content analysis performed in the laboratory of Agricultural Research Agency in Cimanggu, Bogor.

Field studies conducted to determine and obtain the data included for reconfirmation literature, as well as perform the necessary measurements if for no libraries or secondary data that support the tree industry. As an example of the biomass weight data of stems, roots, branches and other. Measurement for the field is required to obtain those data, by pulling, cutting and weighing individual components of the biomass.

- Interview

Interviews were conducted only if necessary, to confirm and to obtain information regarding the allotment of biomass, based on the experience of farmers in the field. An example is the use of cocoa biomass that during this outcome of events pruning of old leaves, and how far the benefits of the biomass. As well as the methodology to a literature study on the activities), this activity depends on the activities of field studies and measurements) on this methodology.

Field data collection include:

- The biomass of plant parts, including: leaf, rod stands, branch / Twig Trimming, stem from demolition, skin and flesh, crude oil (yield juice), roots and others. Derivative type of tree biomass can also be developed in accordance with field observations.
- Criteria for potential : There is a chance utilized as fertilizer, feed, energy (burnt or conversion into oil, etc.), but the real number relative precedence which utilized for bioenergy (eg flowers fall was not recorded because it was too little). It will be confirmed by desk-study and field confirmation (statement of farmers etc)

To know the characteristics and potential of biomass, analyze were carried out to several parameters which include moisture content, ash content, fiber content including lignin, hemicellulose and cellulose.

2.1. Water Content

Aluminum cup dried in an oven for 15 minutes and cooled in a desiccator for 10 minutes and weighed (A). The sample is weighed as much as ± 2 g in the cup (B). Cup and its contents were dried in an oven of 100 ° C for 6 hours. Grail was transferred to a desiccator and then cooled and weighed. Cup and its contents dried again to obtain a constant weight (C).

$$\text{Calculation: Water content } (\%) = \left[\frac{B-(C-A)}{B}\right] x100\% \quad (1)$$

2.2. Ash Content

Prepared dish to ashing, and then dried in an oven for 15 minutes and then cooled in a desiccator and weighed (A). The sample is weighed as much as ± 3 g in the cup (B), and then burned in the smoke chamber until no smoke anymore. Furthermore, the ashing in an electric furnace at a temperature of 400-600 ° C for 4-6 hours until a white ash or has a fixed weight. Ash along with the cup cooled in a desiccator and weighed (C).

$$\text{Calculation: Ash content } (\%) = \frac{C-A}{B} x100\% \quad (2)$$

2.3. Crude fiber content

Five (5) g sample put in a 500 ml Erlenmeyer flask was then added 100 ml of H2SO4 0 325 N and boiled for approximately 30 minutes. Added another 50 ml 1:25 N NaOH and boiled for 30 menit. In hot conditions is filtered with Whatman No. 40 after a known weight of the dry. Filter paper used consecutively washed with hot water, 25 ml of H2SO4 and 95% ethanol. Then dried in an oven at 100-110°C until constant weight. The filter paper is cooled in a desiccator and weighed.

$$\text{Crude fiber content } (\%) = \frac{\text{weight of dry sludge}}{\text{Weight of sample}} \text{ x } 100\% \quad (3)$$

2.4. Lignin content

A sample of 1 g was weighed in a 250 ml Erlenmeyer flask then added 20 ml H2SO4. Subsequently allowed to stand for 2 hours and shuffled slowly. Samples were then added 250 ml of distilled water, heated in a water bath at a temperature of 100 0 C for 3 hours. Further filtering using filter paper of known weight (A). Erlenmeyer and funnel rinsed with distilled water 3 times. Filter paper along diovenkan residue at 105 0C for 12 hours or at a temperature of 500 ° C for 24 hours. Cooled and weighed filter paper weight (B). Filter paper with residue diabukan with muffle furnace at a temperature of 600 ° C for 34 hours. Then cooled and weighed (C).

$$\text{Lignin content } (\%) = \frac{B-A-C}{Weight\ of\ sample} x100\% \quad (4)$$

Note:
A = weight of filter paper (g)
B = weight of filter paper and the residue after the oven (g)
C = weight of ash (g)

2.5. Contents of hemicellulose and cellulose

2.5.1. Hemicellulose content

Content of NDF (Natural Detergent Fiber)

Sample of A g, put in a 500 ml beaker and added with a solution of NDS. NDS solution consists of the following chemicals: 1 l of distilled water; Sodium sulfate 30 g; EDTA 18.81 g; Sodium Borate 10 H2O 6.81 g; NaHPO4 2etoksi anhydrous 4.5 g and 10 ml of pure ethanol. Then weighing G3 glass filter (B). Samples were added to NDS solution is filtered with the aid of a vacuum pump, rinsed with hot water and acetone. The screening results were dried in an oven of 105 ° C, after which it entered in esikator for an hour, then do last weighing (C).

$$\%\ \text{NDF} = \frac{C-B}{A} x100\% \quad (5)$$

Note :
A = weight of sample (g)
B = weight glass filter (g)
C = weight of glass filters and samples after oven (g)

2.5.2. ADF Content (Acid Detergent Fiber)

Sample of A g, was put into a beaker and added with 50 ml solution of ADS. ADS solution consisting of: H2SO4; CTAB (cethyle trimethyl ammonium bromide). Samples were added to the solution is heated for an hour on the electric bath. Filtering is done with the aid of a vacuum pump with a glass filter also uses weighted (B) .Washing was done with acetone and hot water. Drying and screening results enter into the oven. After it was put back into the desiccator to cool and weighed (C).

$$\%\ \text{ADF} = \frac{C-B}{A} x100\% \quad (6)$$

Note :
A = weight of sample (g)
B = weight glass filter (g)
C = weight of glass filters and samples after oven (g)

Hemicellulose content of =% NDF -% ADF (7)

2.5.3. Cellulose content

The residue ADF (C) inside the glass filter is placed on the tray containing water high about 1 cm. Then added H2SO4 as high as ¾ of the glass filter and left for 3 hours while stirring .Filtering carried out with the aid of the vacuum pump also uses glass filter. Washing is done with acetone and hot water. Drying and screening results enter into the oven. After it is put back into desiccator to cool and weighed (D).

$$\% \text{ Cellulose} = \frac{D-C}{A} x100\% \quad (8)$$

Note :
A = weight of sample (g)
D = weighting of filter glass and ADF residues after the oven (g)
C = weighting of filter glass and ADF residues early (g)

3. Results and Discusion

Collecting of data from the cocoa plant biomass and sugarcane have been carried out successively in the location: Jatitujuh, Majalengka and Sumedang for sugarcane and Sukabumi to cocoa. Samples of the plant are plants belonging to local farmers.

3.1.Weight of Biomass

3.1.1.Biomass of Sugarcane Plant

Weight measurements of sugarcane biomass was performed in sugarcane plantation belonging to farmer in Palasa village Kertajati District of Majalengka regency and subdistrict Cibuluh Ujung Jaya village Sumedang regency, West Java. Samples are taken from sugar cane plantation with PKP (center to center) 120 cm, is a PC plant (plan cane) planting in 2014, varieties grown is BL (Bululawang). Yield of sugar cane is sold in the form of fresh stem to PG Jati Tujuh Majalengka. Observations in sugarcane age of 12 months gained an average number of stems per clump 5.6, plant height 310 cm, stem diameter 1.6 cm, number of fresh leaves per stem 9. While for the sugar cane 6 months age was obtained the number of stems per clump 6, trunk diameter of 1.7 cm, high plant 172 cm with the number of leaves is 12 sheets.

The measurement results of biomass weight of sugarcane are mostly on the stem/rod, either of aged 6 months and 12 months (table 1). The weight of the stem is 84.76 to 92.0% of the weight of the sugar cane crop biomass as a whole. Although later processed into bagasse and its nira, bagasse weight is still the greatest weight in comparison with the leaves and roots.

Figure 1. Sugarcane age of 12 months and the process of observation cane biomass

Table 1. Weight of Sugarcane Biomass Plant

Plant part	Weight per plants		Hemicellulose	Cellulose	Utilization by Farmers			
	Green (kg)	Green %	(%)	(%)	Energy	Feed	Fertilizer	others
6 months Age								
leaves	0.82	13.5761	11,45-14,01	12,57-14,50	V	V	V	
Rod/stem	5.12	84.76821	10,93	10,51	V			
Root	0.1	1.655629	9,31	13,80	V			
Weight/clump	6.04	100						
Weight/ha	120.800							
12 months Age								
leaves	0.64	6.808511	12,00-12,63	11,67-14,38	V	V	V	
Rod/stem	8.65	92.02128	14,70	11,84	V			
Root	0.11	1.170213	11,54	15,67	V			
Weight/clump	9.4	100						
Weight/ha	188.000							

Note: Trunk have crushed and dried.

To be converted into bioethanol is actually the biomass which contain high hemicellulose and cellulose. The roots are high cellulose but low weight. So on sugarcane biomass is the stem which became the major known so far is utilizing its bagasse. Shoot of sugarcane apart from the small quantities, this biomass is suitable for feed, because the ADF (acid detergent fiber), which is an indicator of forage digestibility is lower than stems and roots. Fiber digestibility of feed ingredients will be strongly influenced by the content of the cell wall constituent materials such as fibers, lignin, cellulose and hemicellulose and the indicators include a low content of ADF[8]. In the future utilization of sugarcane shoots biomass preferably for livestock feed as long as it has been practiced by farmers in Indonesia.

3.1.2.Biomass of Cocoa Crop

First sampling of 5 cocoa plants aged of 30 years old was carried out in Ubrug village, Warung Kiara District, Sukabumi. Second sampling of 5 cocoa plants aged of 5 years old was carried out in Pakuhaji village, Parung kuda district, Sukabumi. Potency of cocoa plant biomass including roots, leaves, stem, twigs, and fruit waste of cocoa (pod husks). Assumptions planting space of 4 x 4 m, so the population per hectare = 625 trees. Until now the use of pod husks on cocoa plantations as fertilizer for the crops by stacked under cocoa plants. Pod husks of cacao weighing up to 75% of the entire weight of the fruit, so it can be said that the main waste processing cocoa fruit is the pod (fruit shell). Each of the cocoa plant produce average of 30 fruits/year with a weight of 300-500 g / fruit. Cocoa fruit produced about 7.5 tonnes/ha /year. Weight pod husks reached 75% whole weight of the fruit or the equivalent of 5.6 tonnes/ha/year[9].

In cacao plant, the biomass mostly of twigs, small branches, branches to the main stem, respectively 7.89%, 26.93% and 32.11% as well as a branch root, tap root consecutive weighs 6.87 % and 3.47% (table 3). When summed weight is approximately 77% of the weight of plant biomass aged 30 years and about 74% for plant biomass of 5 years old cocoa plant. This number can be obtained when the rejuvenation of plantations, while its leaves are usually for feed materials as well as organic fertilizer.

Table 2. The weight of the Biomass of Cocoa Crop Age 5 Years

Plant part	Weight per plant (kg)				Hemicellulose (%)	Cellulose (%)	Utilization by farmers			
	Green (kg)	Green %	Dry %	Dry (kg)			Energy	Feed	Fertilizer	Others
5 years Age										
Weight of dry leaves	0.85	1.56	3.10	0.85			v		v	
Weight of green leaves	8.45	15.46	21.89	6.01	9,80-14,97	9,04-12,85	v	v	v	v
Weight of small twigs	6.64	12.14	12.25	3.36	9,23	10,63	v			
Weight of branches (2)	16.11	29.48	26.44	7.26	10,19	11,19	v			
Weight of main stem	13.43	24.57	23.84	6.55	9,81	21,19	v			
Tap roots	5.72	10.47	8.40	2.31	12,79	5,51	v			
Secondary roots	1.89	3.45	3.46	0.95	7,66	24,63	v			
Fruits	1.58	2.89	0.61	0.17	7,92	11,51	v	v	v	v
Weight per plant	54.66	100.0	100.0	27.46						
Weight plant per ha (625 plt/ha)	34,16			17,16						

Table 3. The weight of the Biomass of Cocoa Crop Age 30 Years

Plant part	Weight per plant (kg)				Hemicellulose (%)	selulose (%)	Utilization by farmers			
	Green (kg)	Green %	Dry %	Dry (kg)			Energy	Feed	Fertilizer	Other
30 years Age										
Weight of dry leaves	3.94	5.23	10.92	3.94			v		v	
Weight of green leaves	7.74	10.27	11.20	4.04	7,31-8,71	23,13-25,29	v	v	v	v
Weight of small twigs	6.76	8.98	7.89	2.84	3,28	23,62	v			
Weight of branches	23.10	30.68	26.93	9.71	3,90	27,45	v			
Weight of main stem	23.28	30.92	32.11	11.58	3,24	30,77	v			
Tap roots	5.61	7.45	6.87	2.48	3,05	25,32	v			
Secondary roots	2.77	3.68	3.47	1.25			v			
Fruits	2.10	2.79	0.62	0.22			v	v	v	v
Weight per plant	75.29	100	100	36.05						
Weight plant per ha (625 plt/ha)	47,056			22,531						

3.2. The Content of the Lignocellulose Biomass Plant

3.2.1. The content of Lignocellulose Biomass of Sugarcane

Sugarcane waste made up of young leaves, old leaves, stems, roots. In this event be reviewed if any changes during planting (the effect of plant age). It has been observed when sewage plants aged 6 and 12 months. Observations fibers with properties derivatives are Neutral detergent fiber (NDF) and acid detergent fiber (ADF). Lignin, hemicellulose and cellulose.

Table 4. The content of the lignocellulosic biomass sugarcane age 6 months

Waste material	% Fiber	% ADF	% NDF	% Lignin	% Hemicellulose	% Cellulose
6 months age						
Old leaves	32.12	15.16	26.84	6.84	11.68	13.60
Old leaves no.2	33.34	14.15	28.94	4.64	14.79	13.91
Old leaves no.3	32.51	12.51	27.36	6.22	14.85	11.44
Old leaves no.4	30.33	13.73	28.47	4.62	14.74	10.97
Old leaves no.5	33.2	15.17	29.17	6.24	14.00	12.96
Average of old leaves	32.3	14.14	28.15	5.71	14.01	12.57
Young leaves	31.19	13.84	25.29	5.24	11.45	14.50
Stem	30.16	21.11	32.04	8.72	10.93	10.51
Root	34.12	24.42	33.73	11.01	9.31	13.80

Table 5. The content of the lignocellulosic biomass sugarcane age 12 months

Waste material	% fiber	% ADF	% NDF	% Lignin	% Hemicellulose	% Cellulose
12 months age						
Old leaves 1	30.82	15.32	28.11	7.12	12.79	10.91
Old leaves 2	30.71	17.16	29.63	5.8	12.47	12.44
Average of old leaves	30.76	16.24 20.24	28.87 28.87	6.46	6.46 12.63 8.63	11.67 15.75
Young leaves	32.49	15.32	27.32	6.11	12.00	14.38
Stem	38.06	19.75	34.45	11.52	14.70	11.84
Root	38.85	19.27	30.81	11.64	11.54	15.67

Van der Meer and Van Es, 2001 reported that the digestibility of feed ingredients will be greatly influenced by the fiber content of the cell wall constituent materials such as fibers, lignin, cellulose and hemicellulose.

Crude fiber is divided into two parts: neutral detergent fiber (NDF) and acid detergent fiber (ADF). neutral detergent fiber (NDF) is the fraction of a cell wall composed of cellulose, hemicellulose and lignin. The content of NDF linked to the consumption of feed, because all the components take up space and slow rumen digestibility, lower NDF content can be consumed more feed.

The content of ADF is an indicator of forage digestibility, because the content of lignin is part of a faction that can be digested. NDF is always greater than the ADF, for the ADF does not contain hemicellulose.

The results showed that the leaf has a sugar cane waste for NDF lower than stems and roots (table 4 and table 5). Likewise to ADF which is an indicator of forage digestibility lower than sugarcane leaf stems and roots. Meanwhile, the longer the time of planting, the NDF and ADF increased. It shows that sugarcane waste from the leaves, stems and roots contain lower NDF and ADF is less than 50 then all can be used for animal feed, and young and old cane leaves better than stems and roots.

Hemicellulose found in almost all plants, especially in food crop waste such as corn cobs, sugarcane bagasse, rice straw, wheat bran and cotton seed. These materials containing hemicellulose between 16-

IOP Conf. Series: Earth and Environmental Science **65** (2017) 012009 doi:10.1088/1755-1315/65/1/012009

40%, this observation is smaller that ranges from sugar cane waste hemicellulose 9.31-14,7% [10]. The older the plant hemicellulose tends to wane while cellulose increased.
Hemicellulose can be used for several products including:

- Xylitol sugar, through the process of hydrolysis of hemicellulose/xylan into xylose, then hydrogenated into xylitol. Xylitol has advantages over sugar (sucrose) is a low-calorie sweetener, is much lower glycemic index that does not raise blood sugar in the body, so it is good for diabetics. Xylitol can be used to make products such as chewing gum and others [11]. Besides, xylitol is currently widely used for toothpaste because it can strengthen the gums [12].
- hemicellulose in the form of xylan can produce furfural which can be used as an industrial solvent oil, solvent reactive phenol resins, disinfectants as well as starting materials for producing various chemicals and other polymers [13,14].
- As a substrate to produce xylanase enzyme, which is very useful in the food and feed industry. Xylanase for food can be used for animal feed mixes, syrup purification and manufacture of sugar xylose, while for non-food which is to process paper bleaching [15,16]. In the paper-bleaching process used thermostable xylanase and alkali resistant pH (6). In addition, to the hydrolysis of lignocellulose in the production of bioethanol.

The observation of sugarcane waste cellulose ranges from 10.51 to 15.67%. The young leaves and old leaves at the age of 6 and 12 months is not much different from hemicellulose and cellulose content. Based on its NDF and ADF, the young leaves and old leaves are excellent for feed and food. For raw materials of xylitol sugar and furfural is also a high potential since the requirement as those raw materials is a material with a higher hemicellulose content of 12%. According to the rules UNCTAD / Gatt, 1979 is recommended for raw materials are materials that contain a minimum of 12-20% hemicellulose / xylan. In addition to sugar as a food product can also be used as raw material of functional foods, as probiotics that are cello-oligosaccharides (SOS) and xilo-oligosaccharides (XOS).

With the content of cellulose and hemicellulose were higher then sugar cane waste is also potential for second-generation bioethanol. Utilization of lignocellulose to ethanol in addition to not disrupt the food supply can solve environmental problems and improve the efficiency and competitiveness of waste. Key of the production process of bioethanol from the lignocellulosic raw material, pretreatment is necessary to reduce. Hydrolysis of lignocellulose can also be done with acids and bases, combined with high temperature or high pressure. Cellulose is hydrolyzed by means of chemical and enzymatic cellulase is to produce glucose, while hemicellulose to produce xylanase xylose. Glucose and xylose are fermented into ethanol.

3.2.2. The Content of Lignocellulose of Cocoa Biomass

Table 6. The content of lignocellulose of cocoa biomass age 5 years

Cocoa 5 years age	% Fiber	% ADF	% NDF	% Lignin	% Hemicellulose	% Cellulose
Branches	32.64	14.23	24.42	11.26	10.19	11.19
Tap roots	30.45	15.46	28.25	12.15	12.79	5.51
Old leaves	28.63	12.17	27.14	4.62	14.97	9.04
Young leaves	25.41	12.48	22.28	2.76	9.80	12.85
Roots	41.72	28.25	35.91	9.43	7.66	24.63
Stem	44.41	35.14	44.95	13.41	9.81	21.19
Dry fruit	26.27	15.57	23.49	6.84	7.92	11.51
Twigs	29.13	20.29	29.52	9.27	9.23	10.63

Table 7. The content of lignocellulose of cocoa biomass age 30 years

Cocoa 30 Years age	% Fiber	% ADF	% NDF	% Lignin	% Hemicellulose	% Cellulose
Twigs	37.52	23.17	26.45	10.62	3.28	23.62
Leaves	33.11	20.41	27.72	5.67	7.31	23.13
Young leaves	31.45	18.23	26.94	3.45	8.71	25.29
Branches	45.07	25.27	29.17	13.72	3.90	27.45
Roots	39.48	30.18	33.23	11.11	3.05	25.32
Stem	48.73	30.23	33.47	14.72	3.24	30.77

Fiber of cocoa waste is higher than sugar cane waste but lower than palm oil waste. Fiber ranged between 25.41-48.73% (table 6 and table 7). Stems and roots had high levels of fiber and cellulose, whereas hemicellulose low. The stem and roots are not feasible for feed. Physically observations may also be difficult in enzymatic hydrolysis because it was too hard. Young and old leaves and tap roots still feasible for feed. As for bioethanol process leaves, dried fruit and twigs are eligible.

4. Conclusions and Recommendations

The biomass tree of sugar cane which measured potential detailed of leaves, stems and roots, with the weight mostly on the stem at around 84.76 to 92.0% of the weight of the sugar cane biomass as a whole. The leaves or shoots are few in number and are more suitable for feed because the content of ADF (acid detergent fiber) is quite low and is lower than the stems and roots. This is an indicator of forage digestibility of sugarcane leaves are suitable for feed.

The biomass tree of cocoa plant consists of leaves, fruits, small twigs, branches to the main trunk and roots. Biomass mostly of small branches, twigs branch to the main trunk and the total weight is around 77% of the biomass weight of plant age of about 30 years and about 74% for the plant of approximately 5 years age. Cocoa biomass contain fiber is higher than sugar cane biomass but lower than palm oil biomass. Fiber ranged between 25.41-48.73%. Stems and roots had high content of fiber and cellulose, whereas hemicellulose is low. The roots and stems were not eligible for the feed. Based on physical observations may also be difficult hydrolyzed enzymatically for being too hard. Young and old leaves still contain hemicellulose from 9.8 to 14.97% and cellulose from 9.04 to 12.85%, which is adequate for the manufacture of next bioethanol. Young and old leaves and tap roots still feasible for feed, while for bioethanol eligible process that leaves, dried fruit and twigs.

5. Suggestion

For other major estate plants that have the opportunity to bioenergy feedstock ie. coconut, rubber, coffee and jatropha should be immediately observable also to complement the data already gathered from the results of activity in 2015.

Acknowledgement

Thanks to researchers group of Indonesia Center for Estate Crops Research and Development as well as other friends who have helped the completion of the research and this report.

References

[1] Rajeev J and Rajvanshi A K 1997 Sugarcane leaf-bagasse gasifier for industrial heating application *Biomass and Bioenergy* **13** 141-146
[2] Koesnandar 2001 Biokonversi selobiosa langsung menjadi etanol menggunakan Ko-immobilisasi sel *Lipomyces starkeyi* dan *Saccharomyces cerevisiae* secara Fed-Batch *J Mikrob Ind* **6** 15-18

[3] Sunarti T C and N Richana 2006 Bioethanol production from corncob residue by two successive fermentations *Final report of Osaka Gas Foundation of International Cultural Exchange Research Grant* FY 2005/2006

[4] Sah J P, M S Ross, S Koptur and J R Snyder 2004 Estimating aboveground biomass of broadleaved woody plants in the understory of Florida Keys pine forests *Forest Ecology and Management* **203** 319–329

[5] Makungwa S D, Abbie C, David L S, George Y K-P and Iain H W 2013 Allometry for biomass estimation in Jatropha trees planted as boundary hedge in farmers' fields *Forests* **4** 218-233

[6] Richana N and Bambang P 2012 Teknologi biofuel generasi kedua: bioetanol dari lignoselulosa tandan kosong kelapa sawit *Warta Penelitian dan Pengembangan Pertanian* **34** 19-20

[7] Bambang P dan Richana N 2014 *Biofuel generasi 1, Generasi 2 dan Generasi 3* (Jakarta: IAARD Press)

[8] Van Der Meer J M and A J H Van Es 2001 Optimal degradation of lignocellulosic feeds by ruminants and *in vitro* digestibility tests *Proceedings of a Workshop, Degradation of Lignocellulosics in Ruminant and Industrial Processes* March 17-20 1986 Lelystad, Netherlands pp 21-34

[9] Wulan S N 2001 Kemungkinan pemanfaatan limbah kulit buah kakao (*Theobroma cacao*, L) sebagai sumber zat pewarna (B-Karoten) *Jurnal Teknologi Pertanian* **2** 22 – 29

[10] Jaeggle W 1975 Integrated production of furfural and acetic acid from fibrous residues in a continous process *Escher Wyss News* **2** 1-15

[11] Anonymous 2004 Alternative sweeteners: a balancing act *J Asia Pacific Food Industries* September pp 51-54

[12] Saha B C and Bothast R J 1998 Enzymology of xylan degradation [http:// www.usda.gov/ttich/tektran/data/000009/20/0000092095htmk-2.9kb 18 Dec 98]

[13] Sjostroom E 1995 *Wood Chemistry Jilid II* (Yogyakarta: UGM Pres)

[14] Mansilla HD, Baeza J, Urzua S, Maturana G, Villasenor J and Duran N 1998 Acid-catalysed hydrolysis of rice hull: Evaluation of furfural production *J Bioresource Technol* **66** 189-193

[15] Ruiz-Arribas A, Fernandez-Abalos J M, Sanches P, Gardu A L and Santamaria R I 1995 Over production, purification and biochemical characterization of xylanase I (xys 1) from *Streptomyces halstedii* JM8 *Appl And Environ Microbiol* **61** 2414 - 2419

[16] Stassen H E 1995 Small-scale biomass gasifier for heat and power; a global review *The World Bank* pp 49-50.

Biomass potential resources identification in Togean Islands, Central Sulawesi

A Bunyamin[1] and D Purnomo[1]

1 Agro-Industrial Technology Department, Padjadjaran University, Bandung, Indonesia

Email: anas.bunyamin@unpad.ac.id

Abstract. Togean Islands is one of remote area in Central Sulawesi Province, Indonesia. Togean has been already well known for its great underwater scenery which fascinating many foreign tourists stay there. The large number of visits to Togean doesn't mean at the same time it brings much improvement to local economy. People in Togean was used to live with limited utilities. Water and electricity are the two major problems that have been faced by the communities for many years. On the other hand, Togean has a very good potential for the development of biomass as a renewable energy source. This paper evaluated the potency of each resources using some parameters including availability, social support, technology feasibilities and sustainability aspect. Biomass potential resources that were investigated are hardwoods and forestry product, agro-industrial waste and by-products, and also household waste. Advanced analysis has concluded that the most feasible resources that eligible to be considered as future biomass resources is household waste followed by agro-industrial and agricultural waste then hardwood and forestry products.

1. Introduction

Togean Islands is an islands located in the middle of the Tomini bay and administratively belongs to the region of Central Sulawesi. This islands consist of 494 island and grouped into six districts which belonged Tojo Una-Una Regency. The districts are Batudaka, Una-Una, Togean, Talatako, Walea Besar and Walea Kepulauan.

Content from this work may be used under the terms of the Creative Commons Attribution 3.0 licence. Any further distribution of this work must maintain attribution to the author(s) and the title of the work, journal citation and DOI.
Published under licence by IOP Publishing Ltd

IOP Conf. Series: Earth and Environmental Science **65** (2017) 012010 doi:10.1088/1755-1315/65/1/012010

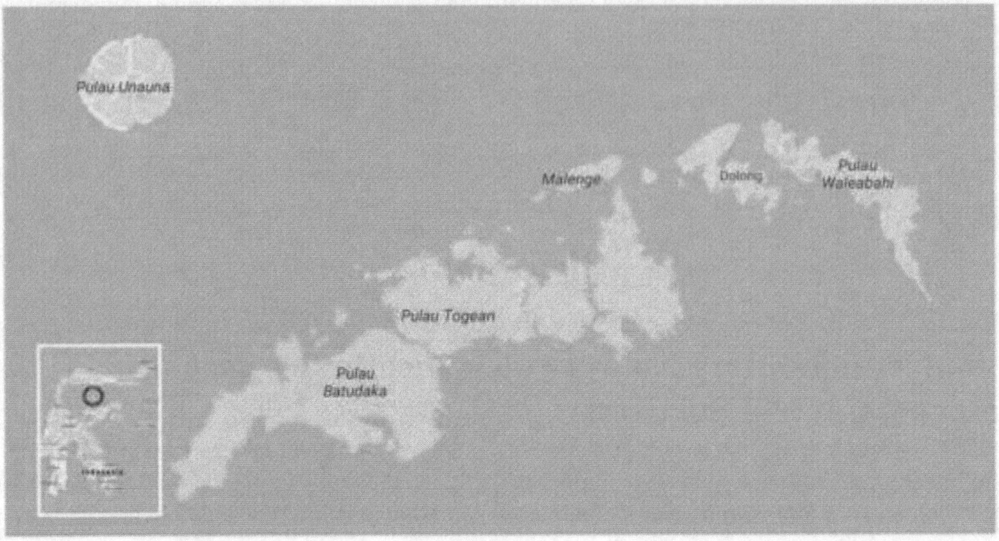

Figure 1. Map of Togean Islands (source: maps.google.com)

Comparing to other Indonesian area, Togean islands was classified as Frontier, Outermost and Disadvantaged islands in Indonesia. One of the main problems in such type areas is the lack of facilities and infrastructure so that economic development is lagging and left behind. Togean consisting of land and sea territory with a population of 30,000 peoples where 72.01% of these populations have an average spending 400-749 thousand per capita per month. This value when compared with the population on the island of Java is very much different. The value will also be smaller for a specific area, as they have more limited job and works. Most of them depend on fishing activity to survive.

One of the main factor that being the causes of the prosperity problems in Togean is the lack of infrastructure facilities which is needed as a part of production activity. Electricity and water as the main utility for production activity is a major obstacle that faced by the community.

In Togean islands, electricity come from diesel power generators, only operate for 18 hours a day from 6.00pm to 12.00am. This conditions affects the whole activities of the people. As one example, because the fishing activity is the main activity, then the fish is the main commodity that have a huge potency to be used as raw material for food products. However, the six-hour electricity power cut make the performance of related equipment such as freezers, refrigerator and also ice maker becomes unable to function properly. The quality of preserved fish finally could not match the standard of fish product.

Closely related to the lack of electricity supply is the availability of clean water. At some point near Wakai, the largest city in Togean, there is a water source that sufficient enough to supply Wakai City and some areas around. However, there are many people live in small islands around the main island that require the electricity as a power to distribute water to the rest of the region. The lack of electricity and water supply should be handled carefully in order to let the activities of community run well.

As a resource-rich islands, Togean should be able to develop and growth into an independent island of the electric needs. Some of the available resources are forest products including hard wood and other forest products (forestry residue, grass, etc.), Agroindustrial activity waste and household waste. The objectives of this study is to examine the possibility of development of biomass as an alternative energy feedstock for producing electricity in Togean. This study will show data results of an exploratory study and a review of some of the existing literature.

2. Methods

Some of the commodities in the islands Togean has been investigated as a potential source of biomass development. The simple method Multi Criteria Decision Making is used to defined most feasible commodities based on determined parameters. Several factors are the subject of assessment is the continuity and availability of raw materials, supporting facilities, the availability of human resources, regulatory support, the level of difficulty of application and alternative technology that already mastered. Each parameter weighted by the extent of its influence on the total score of each alternative source. Assessment done by the related stakeholders which have special interest with the development of alternative energy in Togean, including academics, businesses, governments and communities.

3. Results and Discussion

3.1. Forestry Products and Residue

Most of the islands of Togean is a region with a high proportion of forest where its products can be developed well. Unfortunately, the limitations of sustainability make the use of the forest must be maintained in order not to damage the existing forest ecosystems. Most of the forests in Togean located in mountainous terrain. Extensive mountainous terrain in six districts in Togean can be seen in Figure 1.

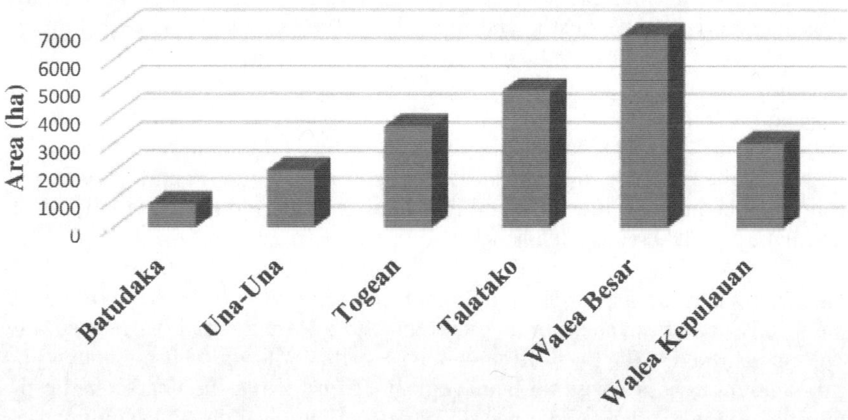

Figure 2. Mountainous Area in Togean Islands. source: BPS - Statistics of Tojo Una-Una Regency [1]

The forest area in Togean, is a conservation area which is named as Togean Islands National Park, so the utilization of forest timber and non timber product must consider the principles of environmental sustainability.

Forests in all districts in the region Togean are available in a very good condition. Although it was not designated as production forests, but forests also already become a part of the daily life of the community and the forest produce forestry residue that can be used as biomass feedstock. Nevertheless, the forestry residue which can be used as raw material for biomass production will be available in a number that is not equal with forestry residue produced from production forest, and that is why these kind of forestry residue often not to be considered [2] .

Technological difficulties may be encountered due to the material characteristics of the hardwood. Some alternative technologies that can be applied to obtain the energy content of hardwood, among others, based on thermochemical conversion such as direct combustion [3], charcoal production [4], pyrolisis [5,6] and gasification [7]; biological; fermentation [8,9] and anaerobic digestion [10,11]. Direct combustion is the application of technology with lowest level of difficulty, compared with other technologies. Nevertheless, most of the conversion rate only reached 20% of the total energy contained would cause the application of this technology to be very wasteful. It is feared, these technology will lead to a reduction in forest area significantly. Application that allows to obtain large amounts of energy recovery is the advanced technology which, undeniable, has much bigger cost consequences. This coupled with the difficulty of obtaining skilled personnel (HR) who mastering the technology.

Considering the regulation, the development of biomass energy using forestry products as raw material seems to be hampered by conservation forest regulatory itself.

3.2. Agro-industrial and Agricultural Waste

Agroindustry and agricultural waste is one of the high potential raw materials that can also be used as biomass feedstock to generate energy. Some of the activities that are carried out Togean people as Agroindustry activities include manufacture of palm sugar, processing of plantation products and also routine agricultural and farming activities.'

Availability of Agroindustry and agriculture waste in the region of Togean varies depending on the activity of the peoples. Agroindustry and agriculture is the second work activity after fishing. The number of Agroindustry and agriculture activities is largely determined by the availability of agricultural land and raw materials of Agro-industrial activity in each district.

Agro-based industrial activity is still relatively rare compared to agricultural activities. Thus, the quantity of agricultural waste will be greater than the waste of Agroindustry. Availability of agricultural waste in Togean relatively large because many agricultural activities associated with the presence of crop production such as cocoa, coconut, cloves. Each of these commodities either during the planting, maintenance and harvesting of agricultural waste such as stems and leaves that can be used as biomass feedstock.

Utilization of these organic materials as biomass feedstock will not reduce environment quality and people's daily activities. The more agricultural and agro-industrial waste used, the cleaner activity will be achieved. The part of tourism aspects that still not well maintained in Togean Islands is people awareness on waste management. Even though the characteristic of organic waste doesn't harm the sea, land and other environment, but when the huge amount of the waste disposed directly to the environment, it will become a serious threat. Amount of agricultural waste was calculated based on the type of each plant. Yields of each agricultural (ton per units area) were varied depend on the plants and activities [2] . Potency identification of biomass production from agricultural waste and by-products were conducted as already been perfectly reported by Wang *et al.*[12] . Coconut, Clove, Cocoa and Coffee was the most cultivated in Togean. Each productivity in 2015 displayed as follows.

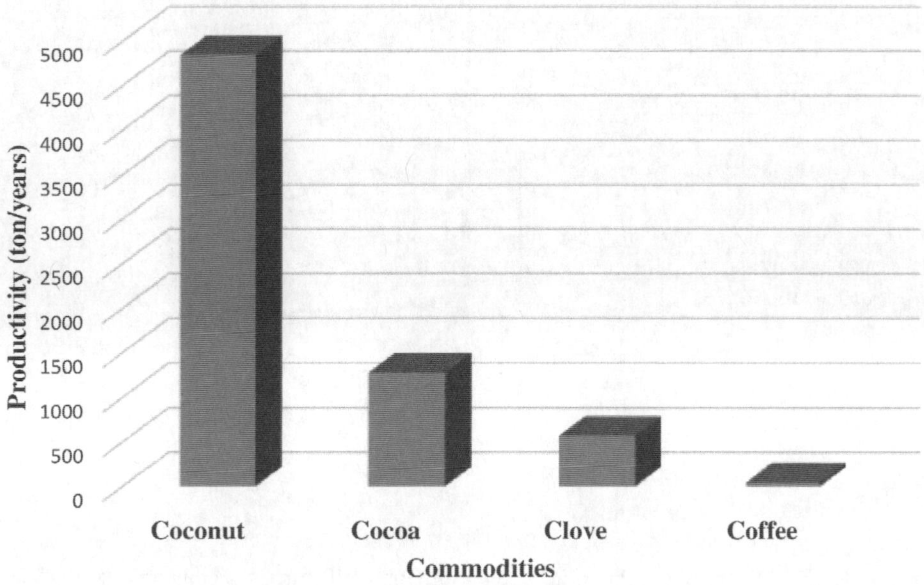

Figure 3. Annual productivity of Togean Crops [1]

Assuming the agricultural waste and residue was only 50% for coconut and coffee, at least we will get 402,286.6-ton biomass feedstock a year. These number equal with more than 1.1-ton production scale per day along the year continuously. We still couldn't find that scale for biomass production in Indonesia. However, the amount of resulted energy will be varied depend on the type of applied technology and the characteristic and composition of the feedstock.

As in forestry residue processing, biomass production using agricultural waste and residue also offered many modifications. In fact, some of agricultural residue will be easily to be converted than hardwood. Some initial decomposition is often to be founded than hardwood.

Current regulation fully supports the use of agro-industry and agricultural waste as the use of those both types of biomass sources will reduce environmental damage.

3.3. Household Waste

The opportunities in developing biomass feedstock that also reasonable to be considered came from the use of household waste. In addition to produce biomass, the use of household waste will reduce more pollution that has been happening in the area of Togean islands.

The availability of household waste which can be used as biomass feedstock is highly dependent on the number of population in an area. With a population of approximately 30,000 people in Togean, wherein assuming that each family produces an average of 0.33 kg of household waste per capita per day [13], then at least Togean will produce household waste by 3.3 tons per day. The next possible obstacles faced is the geographical situation of the islands. This islands consist of almost 500 island that complicate the collection of household waste to be processed into biomass.

3.4. Potency analysis

Each raw material then being analyzed to determine which material that can be considered as the most potential source for biomass development in Togean. Information gathered from stakeholders involved survey in the Togean islands and related experts resulted in the following results. It is seen that the most potential to be used as raw material for biomass development is Agroindustry and agricultural waste and household waste.

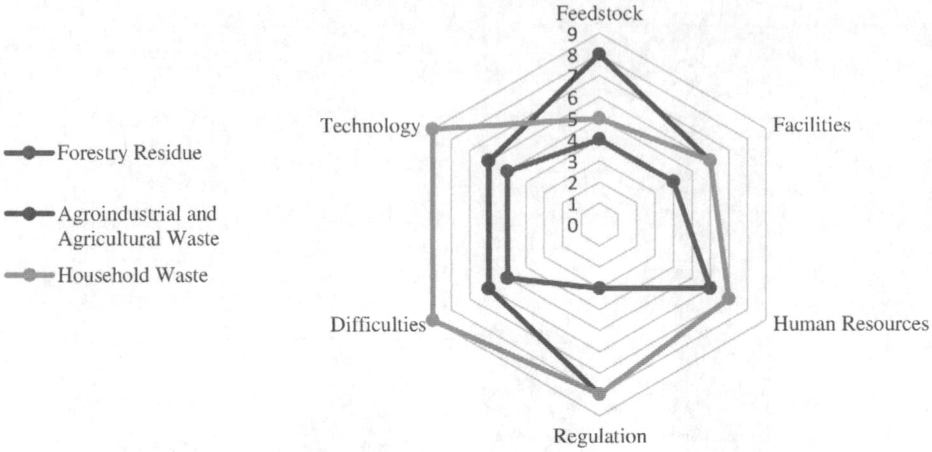

Figure 4. Radar Diagram of Potential Development of Biomass Feedstock from Various Materials

To refine the analysis, the analysis then continued by weighting each parameter according to the significance of its effect on the success of the development of biomass in Togean. The proportion are defined in accordance with the agreement of stakeholders is the availability of raw materials (25%), the availability of facilities (10%), capability of human resources (20%), supporting regulations (15%), difficulties level of application (10%) and technology support (20%). After further analysis conducted based on previous percentage of each parameter, the result shows that household waste ranks first as the most potential raw material for biomass development.

Figure 5. Assessment Result of Each Alternative Raw Materials

The development of biomass-based household wastes considered to be the most potential biomass feedstock comparing to the others as it only requires simple technology to convert into biomass and also available in largest quantity.

4. Conclusion

Selection of local raw materials as biomass feedstock development will differ in one place to another. Conditions in the local area (social, technology, sustainability, etc.) must be the primary consideration in the development of biomass as a renewable energy source. Biomass development in the islands area would not be similar to the development in the region of the main island.

Acknowledges
We would like express our grateful thanks to The Coordinating Ministry for Culture and Human Development who already fully support this program.

References

[1] BPS - Statistics of Tojo Una-Una Regency (2016) *Tojo Una-Una Regency in Figures*
[2] Koopmans A and Koppejan J (1998) The Resource Base September 28 2016 (http://wwwfaoorg/docrep/006/AD576E/ad576e00pdf)
[3] Wang B L *et al.* 2016 The contributions of biomass burning to primary and secondary organics: a case study in pearl river delta (PRD) China *Science of the Total Environment* **569–570** 548–556
[4] Feuerbacher A Siebold M Chhetri A Lippert C and Sander K 2016 Increasing forest utilization within Bhutan's forest conservation framework: The economic benefits of charcoal production *Forest Policy and Economics* **73** 99–111
[5] Rajić N *et al.* 2013 Hardwood lignin pyrolysis in the presence of nano-oxide particles embedded onto natural clinoptilolite *Microporous and Mesoporous Materials* **176** 162–167
[6] Stefanidis S D, Kalogiannis K G, Iliopoulou E F, Michailof C M, Pilavachi P A and Lappas A A 2014 A study of lignocellulosic biomass pyrolysis via the pyrolysis of cellulose hemicellulose and lignin *Journal of Analytical and Applied Pyrolysis* **105** 143–150
[7] Daystar J, Reeb C, Gonzalez R, Venditti R and Kelley S S 2015 Environmental life cycle impacts of cellulosic ethanol in the Southern US produced from loblolly pine eucalyptus unmanaged hardwoods forest residues and switchgrass using a thermochemical conversion pathway *Fuel Processing Technology* **138** 164–174
[8] Kotarska K, Świerczyńska A and Dziemianowicz W 2015 Study on the decomposition of lignocellulosic biomass and subjecting it to alcoholic fermentation: Study on the decomposition of lignocellulosic biomass *Renewable Energy* **75** 389–394
[9] Zabed H Sahu J N Boyce A N & Faruq G (2016) Fuel ethanol production from lignocellulosic biomass: An overview on feedstocks and technological approaches *Renewable and Sustainable Energy Reviews* **66** 751–774
[10] Ge X, Xu F & Li Y 2016 Solid-state anaerobic digestion of lignocellulosic biomass: Recent progress and perspectives *Bioresource Technology* **250** 239–249
[11] Zhong Y, Ruan Z, Zhong Y, Archer S, Liu Y and Liao W 2015 A self-sustaining advanced lignocellulosic biofuel production by integration of anaerobic digestion and aerobic fungal fermentation *Bioresource Technology* **179** 173–179
[12] Wang X, Yang L, Steinberger Y, Liu Z, Liao S and Xie G 2013 Field crop residue estimate and availability for biofuel production in China *Renewable and Sustainable Energy Reviews* **27** 864–875
[13] Dhokhikah Y, Trihadiningrum Y & Sunaryo S 2015 Community participation in household solid waste reduction in Surabaya Indonesia *Resources Conservation and Recycling* ***102*** 153–162

Bioplastic production from cellulose of oil palm empty fruit bunch

Isroi[1], A Cifriadi[2], T Panji[1], Nendyo A Wibowo[3] and K Syamsu[4]

[1] Indonesian Research Institute for Biotechnology and Bioindustri, No.1, Taman Kencana Street, Bogor – Indonesia
[2] Indonesian Rubber Research Institute, No.1, Salak Street, Bogor – Indonesia
[3] Indonesian Industrial and Beverage Crops Research Institute, Pakuwon Main Street Km 2 Parungkuda, Sukabumi – Indonesia
[4] Department of Agroindustrial Technology, Faculty of Agricultural Engineering and Technology, Bogor Agricultural University, Bogor –Indonesia

Email : isroi93@gmail.com

Abstract. Empty fruit bunch is available abundantly in Indonesia as side product of CPO production. EFB production in Indonesia reached 28.65 million tons in 2015. EFB consist of 36.67% cellulose, 13.50% hemicellulose and 31.16% lignin. By calculation, potential cellulose from EFB is 11.50 million tons. Cellulose could be utilized as source for bioplastic production. This research aims to develop bioplastic production based on cellulose from EFB and to increase added value of EFB. Cellulose fiber has no plastic properties. Molecular modification of cellulose, composite with plasticizer and compatibilizer is a key success for utilization of cellulose for bioplastic. Main steps of bioplastic production from EFB are: 1) isolation and purification of cellulose, 2) cellulose modification and 3) synthesis of bioplastic. Cellulose was isolated by sodium hydroxide methods and bleached using sodium hypochlorite. Purity of obtained cellulose was 97%. Cellulose yield could reach 30% depend on cellulose content of EFB. Cellulose side chain was oxidized to reduce hydroxyl group and increase the carboxyl group. Bioplastic synthesis used glycerol as plasticizer and cassava starch as matrix. This research was successfully producing bioplastic sheet by casting method. In future prospects, bioplastic from EFB cellulose can be developed as plastic bag and food packaging.

1. Introduction

Indonesia is the largest oil palm producers in the world, where its production predicted as about 31 million metric tons of oil palm in 2015 [1]. Crude oil palm (CPO) is extracted from the fruits and the lignocellulosic residual remains as oil palm empty fruit bunch (EFB). Accumulation of EFB in the mill is about 28.65 million metric tons per year. EFB has low commercial value and constitutes a disposal problem due to its large quantity. Conventionally, EFB is burned, disposed of in landfills, or composted to organic fertilizer. In order to prevent air pollution and other environmental problems, burning of EFB is forbidden. It is therefore of importance to optimally utilize EFB in order to solve these problems and at the same time utilize the resource for valuable products.

EFB is composed of 40.37% cellulose, 20.06% hemicellulose and 23.89% lignin [2]. Having high cellulose content, EFB has high potential as a source for cellulosic derived products, such as cellulose

Content from this work may be used under the terms of the Creative Commons Attribution 3.0 licence. Any further distribution of this work must maintain attribution to the author(s) and the title of the work, journal citation and DOI.
Published under licence by IOP Publishing Ltd

fiber [3], nano cellulose [4], glucose [5], xylose (Zhang *et al.*, 2012) and ethanol [7]. Cellulose also has others potential uses as a source for bioplastic production. Cellulose has no plasticity feature. Uses of cellulose in bioplastic production need some modification. Derivate of the cellulose that has been used in bioplastic synthesis was cellulose nano crystals (CNC) [8], nano fiber cellulose (NFC) [9], cellulose acetate butyrate [10], cellulose acetate [11] and bio-PE [12]. Cellulose derivate was mixed composed with other biopolymers matrix in order to increase physical properties or as a filler, such as starch and polylactic acid (PLA).

The demand for bioplastic increasing along with the rising concern towards environmental problems caused by pretroleum-based plastic. Global production capacity of bioplastic increased by 38% per year during 2003-2007 and predicted will be reach 3.45 million tonnes in 2020. Cellulose and cellulose derivate are about 11% from the global capacity [12]. Cellulose from EFB could be used as source for bioplastic production. Potential cellulose from EFB are 11.5 million tonnes in Indonesia making EFB as good source for bioplastic. This research aims to develop bioplastic production technology from EFB. Cellulose are isolated using sodium hydroxide and sodium hypochlorite. Cellulose obtained from EFB are modified in order to increase physical properties of the bioplastic.

2. Material and Methods

2.1. Oil Palm Empty Fruit Bunches

OPEFB was collected from an oil palm mill in Bangka Island, Indonesia. Fresh EFB were chopped and sun dried until the moisture content was less than 10%. Dried EFB chopped into small pieces about 5 cm long and stored in a container at room temperature prior to the experiment. It was analyzed for its lignin, cellulose, and hemicelluloses contents.

2.2. Cellulose isolation and purification

Dried EFB was delignified by sodium hydroxide in the dosage of 1 g NaOH/100 g EFB with consistency 10%. The digester heated until pressure inside the digester was 6 bar for 5 h. Pulp than washed with water to remove NaOH residue and black liquor. The pulp was beaten in laboratory beater to get pulp with freeness level of 300. Sodium hypochlorite (5.25% in water w/w) were used for cellulose purification. 100 gr of pulp (o.d. basis) was treated in a flask containing 3000ml of deionized water with 33.5 g sodium hypochlorite at 70-75°C. The addition of sodium hypochlorite was continued at 2 h intervals until the cellulose become white. The cellulose was left in acidified condition for 12 h before washing. Cellulose was washed three times using deionized water.

2.3. Oxidation of Cellulose

5 gr EFB cellulose (o.d. basis) was added into 1000ml Erlenmeyer glass. The cellulose than was impregnated by 500 ml of hydrogen peroxide with various concentrations (0%, 3%, 6% and 9%). The pH of the solution maintained at 11 using 0.1M NaOH or 0.1M HCl solution, as depending on the condition. The cellulose suspension was continuously shaken with electric shaker for 24 h. In the of reaction, the pulp filtered, washed for at least four time and then dried prior to subsequent treatment and analysis.

2.4. Cellulose-starch bioplastic preparation

Bioplastic composite of cellulose-starch composite was prepared by solution casting and evaporation process using cassava starch as the polymers matrix and glycerol as plasticizer. 30 gr of cassava starch was suspended in 1000 ml distilled water and heated at 60°C for 15 minutes for gelatinization. Cellulose solution was added slowly to the gelatinized starch and stir until all cellulose mixed well in the gelatinized starch. The amount of cellulose was 0% (C0), 12.5% (C1), 25% (C2), 37.5% (C3), 50% (C4) and 75% (C5). Glycerol addition was 0%, 12.5%, 25%, 37.5% and 50%. The mixture was cooled and cast on acrylic plates and air dried. The film produced was peeled of and kept in zipper bag and stored in desiccator.

2.5. Analytical methods

The cellulose, hemicellulose, and lignin of the OPEFB were determined according to the Chesson-Datta Method [13] and TAPPI Standard. A Fourier transform infrared (FTIR) spectrometer (Impact 410 iS10, Nicolet Instrument Corp.) was used for determining changes in the structure of the EFB, pulp, cellulose and bioplastic film according to the method described in reference [7]. Each spectrum was obtained with an average of 32 scans and resolution of 4 cm−1 from 600–4,000 cm−1. The spectrum data was controlled by Nicolet OMNIC 4.1 (Nicolet Instrument Corp.) software and analyzed by eFTIR® (EssentialFTIR, Operant LLC). Bioplastic film micromorphology was visualized using light microscope with 100x – 400x magnification.

The carboxylic acid content of all pulp and cellulose samples was determined by a titration technique described elsewhere [14]. An air-dry cellulose sample equivalent to 0.5-1.0 g was weighted into a 200ml glass- stopper flask. 100 ml of calcium acetate solution were added. The flasks were shaken overnight, and then the suspension of fibers was filtrated. The color indicator murexide was used as as a metalchromic indicator. The pH value of the filtrate was adjusted to 12 by the addition of 0.1 M sodium hydroxide solution. The decrease in concentration of calcium acetate solution after contact with the fibers was determined by titration technique. Solution of 0.1 M EDTA was used as titrant. Tensile properties which include the tensile strength and Modulus were determined using the ASTM D882-02 method. Films were cut manually into 1.5x10 cm and then stretched using at a crosshead speed of 12.5 mm/min. Testing conditions include relative humidity of 50±5 and temperature of 23±2°C. At least 7 samples per treatment were tested and values were averaged.

3. Result and Discussion

3.1. Characteristic of empty fruit bunch

The initial content of EFB, pulp and cellulose presented in Table 1. EFB used in this research has high lignin content and higher than reported in other references [15]. These differences could arise depending on the source of the EFB, the historical treatments of the material prior to the analysis and the analytical method. Cellulose is the highest constituent of the EFB, then followed by lignin and hemicellulose. EFB will ready to degrade by microbes when exposes to the air on disposal field and lignocellulosic component will be decreased. EFB samples that have already decayed has low hemicellulose and cellulose content.

EFB strand has unique structure when examined under the microscope as shown in Figure 1. EFB strand and smaller fiber covered by a ball-like structure. These round-shape spiky is called silica bodies [15]. The silica bodies attached to circular craters which are spread relatively uniform over the strand. Extensive treatment of the EFB; such as washing, hammering and crushing could dislodge the silica bodies.

Tabel 1. Lignocellulose content of the native EFB, pulp and cellulose

Parameter	EFB	Pulp	Cellulose
Ekstraktif/ How Water Soluble (HWS) (%)	17.79	-	-
Cellulose (%)	36.67	65.43	70.17
Hemicelulosa (%)	13.50	16.68	27.12
Lignin (%)	31.16	17.70	03.38
Ash (%)	0.89	0.18	0.16

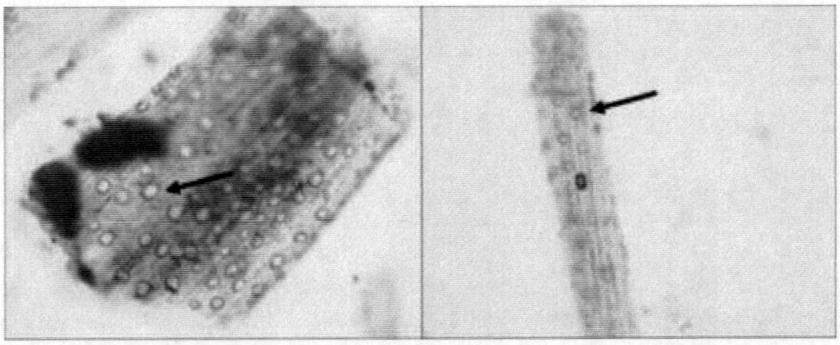

Figure 1. Surface of the EFB strand covered by silica bodies (arrow).

Lignin are more soluble in alkaline condition, but cellulose not soluble in alkaline. EFB treatment by NaOH in high pressure and temperature will decrease the lignin content. Lignin content of the EFB pulp was 56.8% lower than to the initial content. Consequently, cellulose content of the EFB increased almost two times than the initial content. Hemicellulose content relatively stable after NaOH digesting. Sodium hypochlorite common used for pulp bleaching. Sodium hypochlorite oxidized lignin in the pulp. Lignin content after bleaching the EFB pulp reduced significantly.

Structural changes of the EFB, pulp and EFB cellulose were analyzed using FTIR, which reflect the changes in the functional group (Figure 2). Two spectra band has been decreased after cellulose isolation form EFB. O-H valence vibration at wavenumber 3338 cm^{-1} and symmetric CH_2 vibration were decrease. Spectra band in the range of 1750 to 1200 cm-1 also decreased. Decreasing of the several FTIR band correlated with the decreasing of the lignocellulosic component. Assignment of FTIR band published in other reference [7].

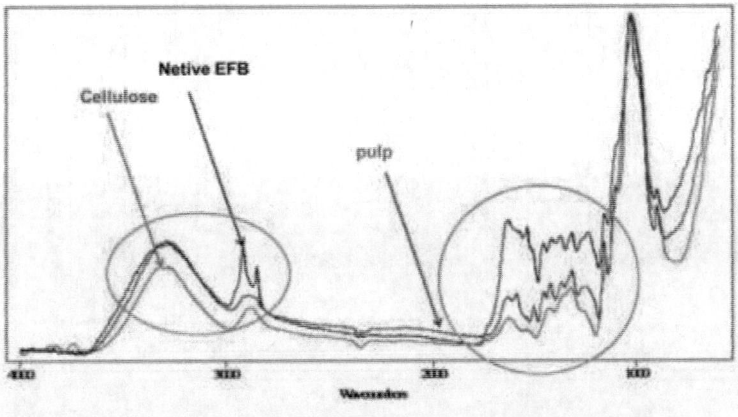

Figure 2. FTIR spectra of the native EFB, pulp and cellulose (bleached pulp).

Crystallinity of the lignocellulose could be predicted using a ratio from the specific FTIR band [16, 17]. The total crystallinity (TCI) and lateral order (LOI) indices, obtained from the 1420/893 and 1375/2900 cm^{-1} absorbance ratios respectively, were used to study the crystallinity changes. LOI and TCI values of the EFB, pulp and EFB cellulose shown in Figure 3. The LOI values was decrease after delignification and pulp bleaching, otherwise the TCI values was relatively constant. Delignification and bleaching of lignocellulose tend to decrease the crystallinity [18]. In this case, LOI was more sensitive to predict the crystallinity of the EFB and its derivate.

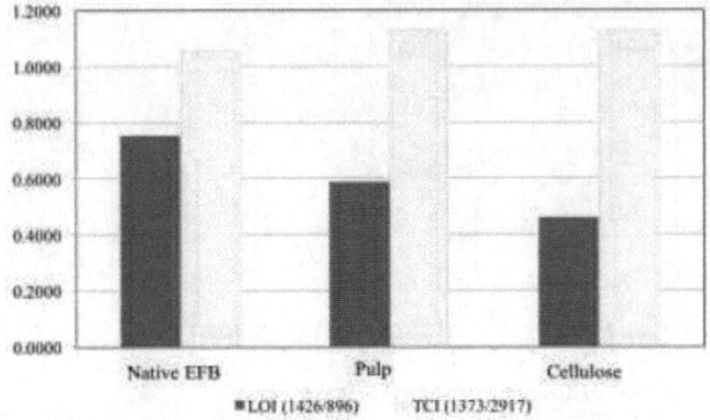

Figure 3. Determined lateral order (LOI) and total crystallinity (TCI) infrared crystallinity indices of native EFB, pulp and cellulose.

3.2 Oxidation of EFB cellulose

Hydrogen peroxide is a strong oxidant and common used in paper industry. Oxidation of the lignocellulose are applied in order to change the reactivity due to an alteration in the number and allocation of functional groups. Changes of the functional group of the cellulose evaluated using FTIR analysis as shown in Figure 4. Some peak increased and the other was constant. Peak at wavenumber 1722 cm^{-1} and 1633 cm^{-1} increased. These peaks assigned as CO stretch unconjugated from carboxyl groups and CO conjugated, respectively. Oxidation of the cellulose could increase the carboxyl group of the cellulose [14,17]. Oxidation may be break the covalent carbon to carbon bond and made new carboxyl group [19].

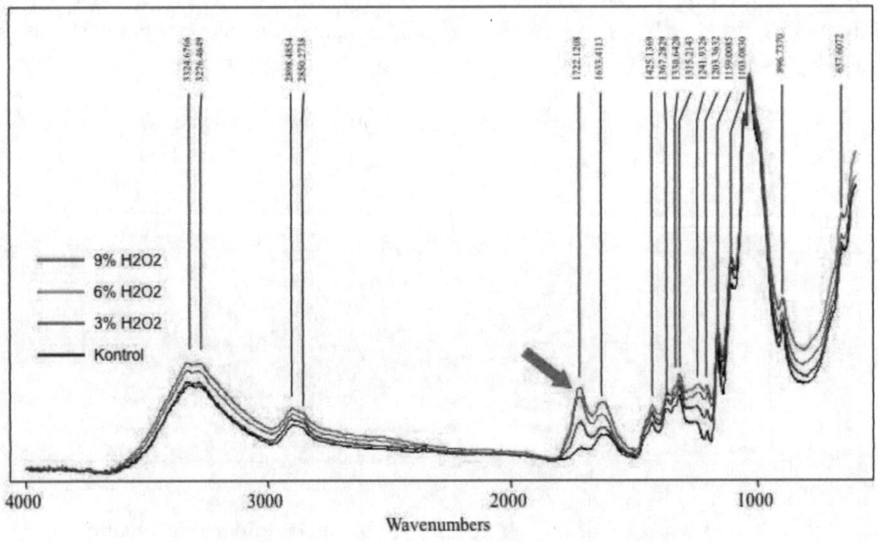

Figure 4. FTIR spectra of the oxidized EFB cellulose with different H_2O concentration.

Oxidation of the cellulose also will reduce the crystallinity as indicate in LOI value. LOI value decreased with the increasing of the percentage of hydrogen peroxide (Figure 5). Damage of the cellulose strand increased by increasing the percentage of the hydrogen peroxide and some strand could be broken. Amorphous region of the cellulose strand is more sensitive to attack. Breaking of the cellulose may be occur in the amorphous region. Damage in crystalline region suspected to be the cause of reducing the crystallinity of the cellulose.

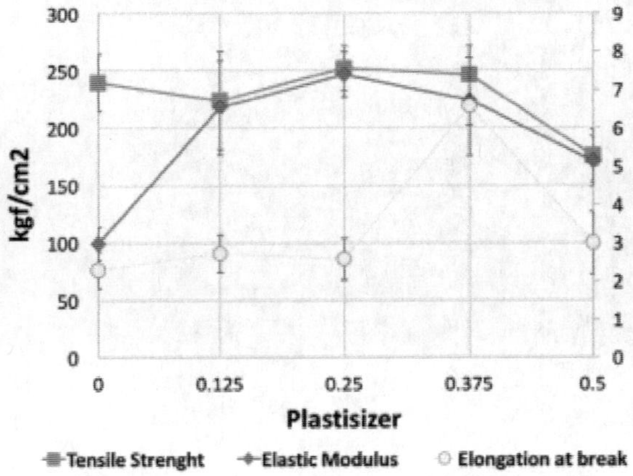

Figure 5. Lateral order index (LOI) of the EFB cellulose oxidized in different H_2O_2 concentration: O% (Control), 3%, 6% and 9%.

3.3 Physical characteristic of bioplastic

Tensile strength, elastic modulus and elongation at break of the cellulose bioplastic composite was shown in Figure 6. Tensile strength was little reduced by 12.5% plasticizer addition and then little bit increase until 37.5%. Elastic modulus increase significantly when plasticizer added and remain constant until 0.5% addition. Elongation at break almost constant until 25% plasticizer addition and then increase significantly at 37.5%. Addition glycerol more than 60% will reduce the elongation at break. Reducing of these characteristic because composite become more loose between matrix and fiber. These evidence also reported in reference that higher glycerol addition will reduce the physical properties [20].

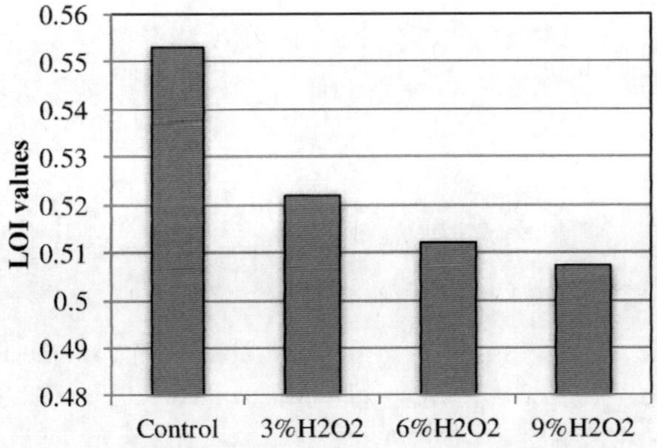

Figure 6. Tensile strength, elastic modulus and elongation at break of the bioplastic cellulose composite.

Cellulose arrangement in the bioplastic composite evaluated using light microscope and shown in Figure 7. Starch bioplastic is transparent and no fiber could be shown under the light microscope. Addition 12.5% cellulose in the bioplastic composite shown more gab between cellulose fiber. Addition more cellulose will increase the density of the fiber inside the bioplastic composite.

Figure 7. Cellulose arrangement in bioplastic with different percentage from: 0% until 75%.

Opacity of the bioplastic film shown in Figure 8 and Figure 9. Addition of the cellulose will reduce the opacity of the bioplastic film. Cellulose fiber is less transparent than starch. Higher density of the cellulose fiber as shown in Figure 7 will prevent the light and reduce the opacity of the bioplastic composite.

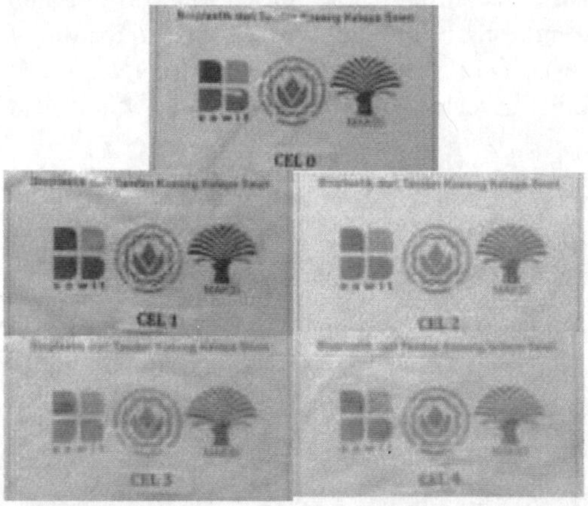

Figure 8. Transparency of the EFB bioplastic with different cellulose content: 0%, 12.5%, 25%, 37.5%, 50% and 75%.

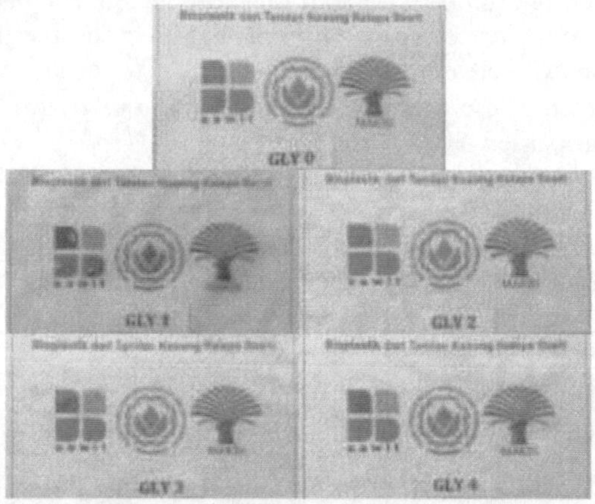

Figure 9. Transparency of the EFB bioplastic with different glycerol content: 0%, 12.5%, 25%, 37.5% and 50%.

4. Conclusion

Lignocellulosic content of EFB was 36.67% cellulose, 13.50% hemicellulose and 31.16% lignin. Delignification of EFB by sodium hydroxide and subsequence bleaching by sodium hypochlorite was remove lignin and increase percentage of holocellulose more than 97%. Oxidation of the EFB cellulose reduce the crystallinity and increase carboxyl content of the EFB cellulose. Bioplastic composite successfully made from EFB cellulose with cassava starch as a matric and glycerol as plastisizer.

Acknowledgements
The authors thank to Indonesia Estate Crop Fund for Palm Oil who funded this research with Contract Number PRJ – 32/DPKS/2015. The authors also thank Siti Ropikoh, Anggun Rahman, Dewi and Fikri Al Baiquni for their contribution in laboratory work.

References

[1] Dirjenbun 2015 *Tree Crop Estate Statistics of Indonesia 2014-2016* Available at: http://ditjenbunpertaniangoid/tinymcpuk/gambar/file/statistik/2016/SAWIT 2014-2016pdf
[2] Isroi 2015 Biological Pretreatment of Oil Palm Empty Fruit Bunches *2nd International Symposium on Integrated Biorefinary ISIBio* Bogor
[3] Khalid M, Ratnam CT, Chuah TG, Ali S and Choong TSY 2008 Comparative study of polypropylene composites reinforced with oil palm empty fruit bunch fiber and oil palm derived cellulose *Materials & Design* **291** 173–178
[4] Lani NS, Ngadi N, Johari A and Jusoh M 2014 Isolation characterization and application of nanocellulose from oil palm empty fruit bunch fiber as nanocomposites *Journal of Nanomaterials* **2014** 9
[5] Hamzah F, Idris A and Shuan TK 2011 Preliminary study on enzymatic hydrolysis of treated oil palm Elaeis empty fruit bunches fibre EFB by using combination of cellulase and $β$ 1-4 glucosidase *Biomass and Bioenergy* **353** 1055–1059
[6] Zhang D, Ong YL, Li Z and Wu J C 2012 Optimization of dilute acid-catalyzed hydrolysis of oil palm empty fruit bunch for high yield production of xylose *Chemical Engineering Journal* **181** 636–642
[7] Isroi Mofoluwake I and Taherzadeh M J 2014 Effect of fungal and phosphoric acid pretreatment on ethanol production from oil palm empty fruit bunches OPEFB *Bioresource Technology* **165**
[8] Arrieta MP, Fortunati E, Dominici F, López J and Kenny JM 2015 Bionanocomposite films based on plasticized PLA--PHB/cellulose nanocrystal blends *Carbohydrate Polymers* **121** 265–275
[9] Siró I and Plackett D 2010 Microfibrillated cellulose and new nanocomposite materials: A review *Cellulose* **173** 459–494
[10] Park H-M, Misra M, Drzal LT and Mohanty AK 2004 "Green" Nanocomposites from Cellulose Acetate Bioplastic and Clay: Effect of Eco-Friendly Triethyl Citrate Plasticizer *Biomacromolecules* **56** 2281–2288
[11] Grunert M and Winter WT 2002 Nanocomposites of cellulose acetate butyrate reinforced with cellulose nanocrystals *Journal of Polymers and the Environment* **101** 27–30
[12] Shen L, Worrell E and Patel M 2010 Present and future development in plastics from biomass *Biofuels Bioproducts and Biorefining* **41** 25–40
[13] Chesson A 1981 Effects of sodium hydroxide on cereal straws in relation to the enhanced degradation of structural polysaccharides by rumen microorganisms *Journal of the Science of Food and Agriculture* **328** 745–758
[14] Fras L and Stana-Kleinschek K 2002 Quantitative determination of carboxyl groups in cellulose by complexometric titration *Lenzinger* pp 80–88 (http://wwwlenzingcom/fileadmin/template/pdf/konzern/lenzinger_berichte/ausgabe_81_2002/LB_2002_Fras_16_evpdf)
[15] Law K-N, Daud WRW and Ghazali A 2007 Morphological and chemical nature of fiber strands of oil palm empty-fruit-bunch OPEFB *BioResources* **23** 351–362
[16] Hurtubise F G and Krässig H 1960 Classification of fine structural characteristics in cellulose by infared spectroscopy Use of potassium bromide pellet technique *Analytical Chemistry* ACS Publications **32** 177–181
[17] Nelson ML and O Connor RT 1964 Relation of certain infrared bands to cellulose crystallinity and crystal lattice type Part II A new infrared ratio for estimation of crystallinity in celluloses I

and II *Journal of Applied Polymer Science* **83** 1325–1341

[18] Åkerholm M, Hinterstoisser B and Salmén L 2004 Characterization of the crystalline structure of cellulose using static and dynamic FT-IR spectroscopy *Carbohydrate Research* **3393** 569–578

[19] Ma P U, Law K, Daneault C and Ma P 2009 *Influence of Oxidation on Intrinsic Fiber Strength* vol 43 pp 387–92

[20] Collinson SR and Thielemans W 2010 The catalytic oxidation of biomass to new materials focusing on starch cellulose and lignin *Coordination Chemistry Reviews* **25415** 1854–1870

[21] Carvalho A J F, Curvelo A A S and Agnelli J A M 2000 Wood pulp reinforced thermoplastic starch composites *International Journal of Polymeric Materials* **51** 1–16

IOP Conf. Series: Earth and Environmental Science **65** (2017) 012012 doi:10.1088/1755-1315/65/1/012012

Combustion quality analysis of briquettes from variety of agricultural waste as source of alternative fuels

S Suryaningsih[1], O Nurhilal[1], Y Yuliah[1], C Mulyana[1]

[1] Department of Physics, Mathematic and Natural Science Faculty, Padjadjaran University, Bandung, Indonesia

Email : sri_56@ymail.com,

Abstract. The increasing in world population and the industrial sector led to increased demand for energy sources. To do this by utilizing the agricultural waste as a fuel source of alternative energy in the form of bio briquette. The aim at this study was to obtain data onto the characteristics of a wide variety of biomass briquettes from waste agricultural industry. The basic ingredients used are biomass waste from coconut husks, sawdust, rice husks and coffee husks. Each of these biomass residues are dried, crushed, then mixed with starch adhesives. This mixture is molded and dried using sunlight. Each type of briquettes was characterized and analyzed the physical-chemical properties, including calorific value, water content, fixed carbon content and the results were compared with charcoal and coal that was used as fuel in public. The results showed that bio briquettes from coconut husks get the highest calorific value of 4,451 cal/g.

1. Introduction

Biomass is carbon based complex polymers and is composed of a mixture of organic molecules containing hydrogen, oxygen, and also small quantities of other atoms. Bio-briquettes are compacted combustible material created from biomass used as a form of fuel for heating or cooking in industry and household. Bio-briquettes could be a solution to both global and local efforts in the search for new, renewable energy that are clean, sustainable and environmentally friendly resources. By practicing this technology will bring not only environment and economic sustainability, but also social and political stability, signifying the true long-term sustainability [1].

Limited land for agriculture, such as food crops, herbaceous plants, industrial forest and the increasing land use of human settlement led to a specific land use for energy crops becomes difficult. Another alternative to be used as biomass briquettes is to utilize industrial waste based on existing agricultural industry.

Some of the major agricultural industry in Indonesia produces solid waste that has not been in exploitation. Coconut husks, sawdust, rice husks and coffee husks are the basic material that selected in this study. Each of the briquettes produced from these chosen raw materials is characterized and analyzed and present in this paper as a consideration stuff for the industry.

Content from this work may be used under the terms of the Creative Commons Attribution 3.0 licence. Any further distribution of this work must maintain attribution to the author(s) and the title of the work, journal citation and DOI.
Published under licence by IOP Publishing Ltd

IOP Conf. Series: Earth and Environmental Science **65** (2017) 012012 doi:10.1088/1755-1315/65/1/012012

2. Methods

2.1. Choosing of raw material
Selection of raw materials for biomass briquettes were considered in the availability of the waste in the major agricultural industry in Indonesia and has not been much in exploitation. These wastes include coconut husks, rice husks, coffee husks and sawdust. Especially for sawdust, waste used is a type of acacia tree *(acacia denticulosa)*, this species have been selected for fast-growing and widely sold in the market in the form of timber as a pre-construction material in the building construction.

2.2. Preparation of raw material
All selected raw materials are dried using the sun to reduce the moisture content, drying time between 2 to 8 hours depending on ambient temperature and humidity. Except sawdust, all the raw material are crushed into smaller parts to be easily inserted and arranged into the pyrolysis furnace.
Carbonization process is performed separately for each raw material. Turning on the stove over medium heat and insert raw material in the pan for about 30 to 180 minutes until it becomes charcoal. After becoming the charcoal, then cooled at room temperature and then crushed into fine particles. Fine charcoal, then filtered with a 120 mesh screen.

2.3. Fabrication of bio-briquettes
The additive substances are made from 20 g of starch (tapioka) in 80 g water solution to achieve a concentration of 4% additives. Mix the starch and water and heated to become an adhesive solution. Then the adhesive evenly mixed and stirred with 100 g of charcoal raw material.

Figure 1. A picture of cylindrical molding block

Figure 2. A picture of hydraulic compression unit

Figure 3. A picture of formed bio-briquettes

The dough is then printed on a cylindrical molding block with 2 cm in diameter (Figure 1), for 10 g per print and pressed with a pressure of 1000 kg/cm^2 (Figure 2), and keeping dwell time at optimum value of 40 seconds [2]. Figure 3 shows the bro-briquettes before drying process. Ideally bio-briquettes are then dried in the sun for 1 to 2 weeks, but for the needs of this experiment the bio-briquettes are dehydrated in an oven at a temperature of 55 °C for almost 24 hours.

2.4. Characterization of bio-briquettes

2.4.1 Bulk Density
Each type of bio-briquettes have 10 samples, the mass and dimension are taken individually. The bulk density was obtained by using this equation:

$$\rho = m/v \quad (1)$$

where: ρ = bulk density, m = mass of bio-briquettes, v = cylindrical volume of bio-briquettes.

2.4.2 *Proximate Test*

Proximate Test analysis is a procedure to determine the components inside the briquettes, including moisture content, volatile matter, ash content and fixed carbon.

The moisture content *MC (%)* inside the briquette can be evaporated by heating at temperature 100 °C -105 °C for 60 minutes. The equation used to calculate the lost of weight as the moisture content is:

$$MC(\%) = [m_2 - m_3/m_2 - m_1] \times 100 \quad (2)$$

where: m_1 = weight of empty cup, m_2 = weight of empty cup + sample, m_3 = weight of empty cup + sample after heated. The measurement of moisture content done using ASTM D-3173.

Measurement of the value of volatile matter is done using ASTM D-3175. The volatile matter *VM (%)* is the amount of the substances that are lost when the briquettes sample are heated in a furnace at a temperature of 900 °C for seven minutes. Loss in weight was reported as volatile matter using this equation:

$$VM(\%) = [m_3 - m_4/m_2 - m_1] \times 100 \quad (3)$$

where: m_1 = weight of empty cup, m_2 = weight of empty cup + sample, m_3 = weight of empty cup + sample after taken from the stage (I), m_4 = weight of empty cup + sample after cooling.

The ash value measurement is done using ASTM D-3174. To calculate the ash content *AC (%)*, bio-briquettes are heated in the furnace at temperature 800 °C for 3 hours. Then the furnace is turned off, wait until the temperature below 400 °C, then the samples are removed from the furnace. Samples will be weighed if it is really cold and calculated using Equation4:

$$AC(\%) = [m_5 - m_1/m_2 - m_1] \times 100 \quad (4)$$

where: m_1 = weight of empty cup, m_2 = weight of empty cup + sample taken from stage (II), m_5 = weight of empty cup + ash left in the cup.

The fixed carbon, *FC (%)* is the result of a reduction of 100% sample with volatile matter, moisture content and ash content. The equation used to calculate the fixed carbon is:

$$FC(\%) = 100\% - MC(\%) - VM\,(\%) - AC(\%) \quad (5)$$

2.4.3 *Calorific Value*

Calorific Value is the amount of heat that develops from the mass (weight) in its complete combustion with oxygen in the calorimeter. The bomb calorimeter used is Leco – AC350. Sample preparation for a series of tests using ASTM D 5865, meaning briquettes provided are crushed and refined again and filtered through 60 mesh screen.

By using water as much as 2 liters and a sample to be tested weighed about 1 gram. Before placing the sample into the measuring instrument, oxygen is given with a pressure of 400 psi, then wire length is 8 cm long and fuse combustion heat of 4.1 Btu /cm. Changes in water temperature (ΔT) and the released of heat by burning briquettes will be recorded and calculated using installed program in a bomb calorimeter.

3. Results and Discussions

3.1. *Physical properties of bio-briquettes*

Average bulk density of bio-briquettes are shown in Table 1 below. The highest value of bulk density is sawdust of 0.89 g/cm^3 and the lowest is rice husks of 0.69 g/cm^3.

Table 1. Average bulk density of bio-briquettes

Bio-briquettes	Bulk density (ρ) *g/cm³*
Coconut Husks	0.76
Rice Husks	0.68
Coffee Husks	0.71
Sawdust	0.89

3.2. Proximate analysis of bio-briquettes
By using equation 2 to equation 5 the collected data can be presented in Figure 4 below. Figure 4 shows all the results of the test sample of all type of bio-briquettes.

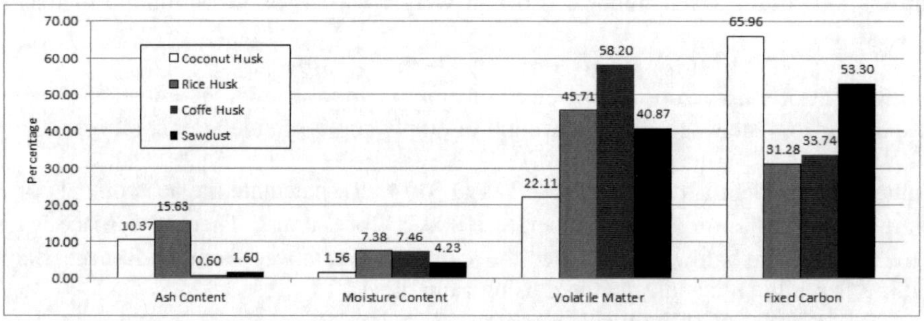

Figure 4. Proximate analysis test results of bio briquettes

Ash content of bio-briquette of coffee husks has lowest ash content of 0.60%, rice husks has a highest value of 15.63%. The lower the value of the ash content, the greater calorific value its has.
The moisture content of the coconut husks has the lowest value of 1.56%, rice husks has a highest value of 7.38%. The smaller the value of the water content in the bio-briquette is the better, if too much water, when used to fuel, the bio-briquettes will evaporate the water at first, and it should be avoided.

The volatile matter of bio-briquette from coconut husks has the lowest value of 22.11%, rice husks has a highest value of 58.20%.

The fix carbon of coconut husks bio-briquettes that have a high value of 65.96%. The higher the value of fixed carbon, the quality of bio-briquettes getting better. Because the amount of fixed carbon acts as a major generator of heat during combustion.

3.3. Calorific Value
Table 2 shows the test results calorific value bio-briquette from coconut husks has the highest calorific value is 5267 cal/g, but its lower compared to coal (6158 cal/g) and higher compared to woody charcoal (3158 cal/g). The higher the calorific value of the bio-briquette the higher quality it has.

Table 2. Calorific value of the four types of bio-briquettes compared to woody charcoal and coal

Bio-briquettes	Calorific Value *cal /g*
Coconut Husks	5267
Rice Husks	3350
Coffee Husks	4045
Sawdust	4247
Woody Charcoal*	3158
Coal (Carbonized)*	6158

*Source: * provided by Siti Jamilatun* [3]

Figure 5 showed that the increasing amount of fixed carbon percentage will increase the calorific value, that means the fixed carbon gives a rough estimate of heating value of the briquettes. In Figure 6, moisture content percentage shows increased at rice husks and coffee husks, hence the calorific value of these briquettes decreased. Moisture content is not only increased the heat loss, due to evaporation and superheating of vapor but also aids radiation heat transfer.

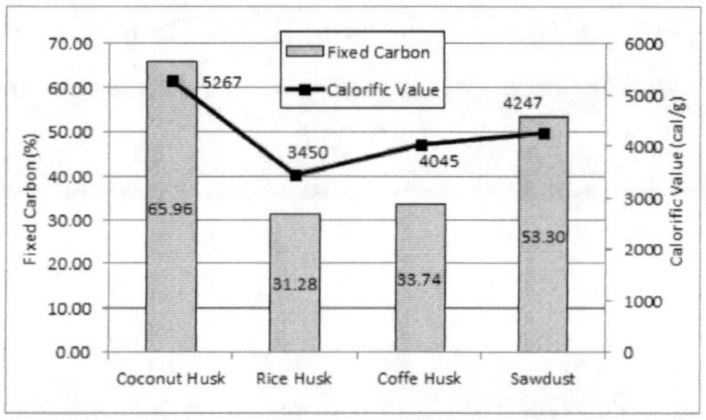

Figure 5. Fixed carbon - calorific value analysis

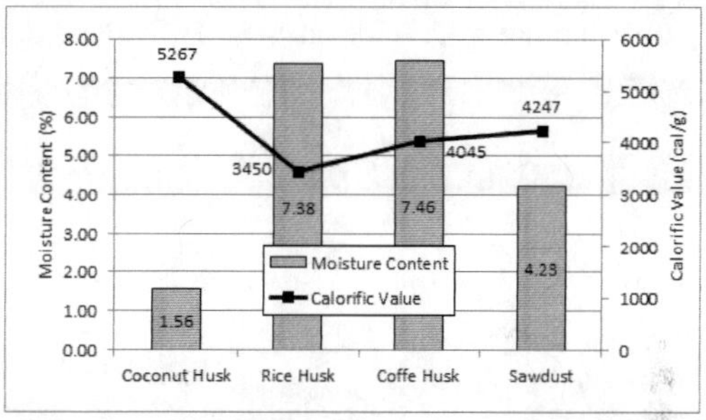

Figure 6. Moisture content calorific value analysis

Ash is an impurity that will not burn and caused reduces the burning capacity, besides it will effect in combustion efficiency. In Figure 7 showed the percentage of the ash content of rice husks is high and calorific value is low or vice versa, but it's not significant with coffee husks and sawdust. This may be because there are other impurity that affects such as volatile matter.

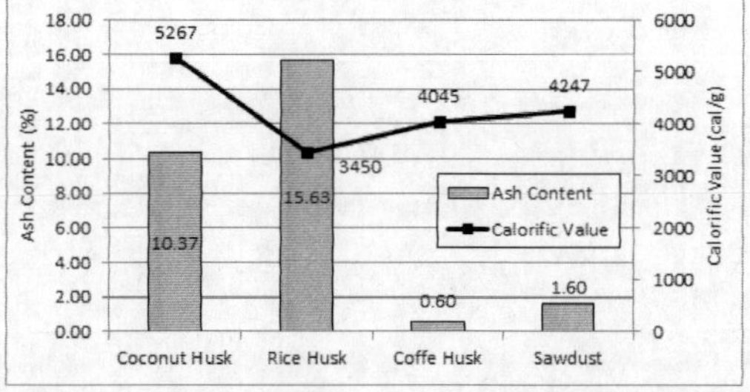

Figure 7. Ash content - calorific value analysis

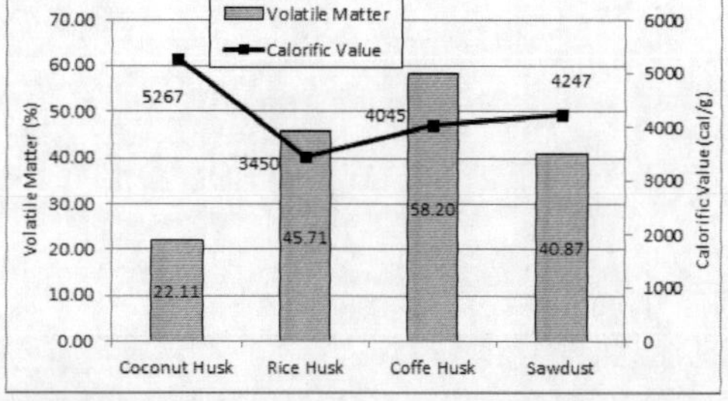

Figure 8. Volatile matter - calorific value analysis

Figure 8 showed that percentage of volatile matter in coffee husks and sawdust are significantly high. Volatile matter contains the composition of methane, hydrocarbons, hydrogen and carbon monoxide, and incombustible gases like carbon dioxide and nitrogen, that means volatile matter could demonstrate high ease of ignition bio-briquettes or vice versa, it depended on what composition is in.

4. Conclusions

High percentage of volatile matter doesn't mean will decrease the burning capacity. If its composition contains flammable gas, it will increase burning capacity by proportionately increases flame length, and helps in easier ignition of coal. But if its composition contains non flammable gas, it will decrease burning capacity.

Coconut husks bio-briquette has the highest calorific value, the volatile matter also low in percentage, percentage of water content and the ash value is at least. The calorific value of coconut husks higher than charcoal and lower than coal. With respect to use for fuel, it is preferable to utilize coconut husks bio-briquette than three other types of briquettes.

References

[1] UNEP 2013 *Technologies for Converting Waste Agricultural Biomass to Energy* (Osaka: United Nations Environmental Programme) p.2

[2] Chin O C and Siddiqui K M 2000 Characteristics of some biomass briquettes prepared under modest die pressures *Biomass and Bioenergy* **18** 223-228

[3] Siti Jamilatun 2008 Sifat-Sifat Penyalaan dan Pembakaran Briket Biomassa, Briket Batubara dan Arang Kayu *J. Rekayasa Proses* **2** 40

IOP Conf. Series: Earth and Environmental Science **65** (2017) 012013 doi:10.1088/1755-1315/65/1/012013

Determination of Indonesian palm-oil-based bioenergy sustainability indicators using fuzzy inference system

Y Arkeman[1,2], R A Rizkyanti [1], E Hambali[1,2]

[1] Department of Agroindustrial Technology, Bogor Agricultural University, Bogor, Indonesia

[2] Surfactant and Bioenergy Research Center, Bogor, Indonesia

Email : yandra.arkeman@gmail.com

Abstract. Development of Indonesian palm-oil-based bioenergy faces an international challenge regarding to sustainability issue, indicated by the establishment of standards on sustainable bioenergy. Currently, Indonesia has sustainability standards limited to palm-oil cultivation, while other standards are lacking appropriateness for Indonesian palm-oil-based bioenergy sustainability regarding to real condition in Indonesia. Thus, Indonesia requires sustainability indicators for Indonesian palm-oil-based bioenergy to gain recognition and easiness in marketing it. Determination of sustainability indicators was accomplished through three stages, which were preliminary analysis, indicator assessment (using fuzzy inference system), and system validation. Global Bioenergy partnership (GBEP) was used as the standard for the assessment because of its general for use, internationally accepted, and it contained balanced proportion between environment, economic, and social aspects. Result showed that the number of sustainability indicators using FIS method are 21 indicators. The system developed has an accuracy of 85%.

1. Introduction

Bioenergy is a potential alternative energy source from biomass. Reducing the dependency on fossil fuel and being eco-friendlier are some of the advantages from the utilization of bioenergy. Palm oil is the most potential resource to produce bioenergy in Indonesia because of its availability; the area of palm plantation in Indonesia is estimated at 11,3 million hectares with palm oil productivity reaching 31,3 million ton/year[1]. Various bioenergy products such as biodiesel, bio-oil, bioavtur, biopellet, biobriquettes and biogas can be produced from crude palm oil (CPO) and also the biomass namely shell, fiber, empty bunches and liquid waste. Thus, palm oil is a very potential raw material source for bioenergy.

One of bioenergy product which has a big potential to be exported is biodiesel. It is shown by biodiesel production capacity in 2014 was 3,961 million kl while the rate of domestic utilization was only 1,845 million kl[2]. Consequently, there was still production excess which could be exported, contributing to the increase of national income.

The development of palm-oil-based bioenergy in Indonesia faces an international challenge regarding to sustainability issue, environmental, economic and social impact. Deforestation, reduction of

Content from this work may be used under the terms of the Creative Commons Attribution 3.0 licence. Any further distribution of this work must maintain attribution to the author(s) and the title of the work, journal citation and DOI.
Published under licence by IOP Publishing Ltd

biodiversity and carbon stocks were considered as the impacts of palm oil plantation[3]. In addition, there are some other impacts that could occur in social and economic aspects such as decrement of land availability, rise of land price and conflicts between indigenous people or between indigenous people and the company[4].

There were some initiations in the making of standard which contain the criteria, indicators and the principle of the bioenergy sustainability. These initiations are the national, regional, international and multi-stakeholder initiations. Indonesia also has a sustainability standard for palm oil plantation named Indonesian Sustainable Palm Oil Certification System (ISPO). However, ISPO only contained indicators and principle of plantation sustainability or cultivation of palm oil. It has yet to include the indicators of palm-oil-based bioenergy production, whereas not all the existing indicators from other countries are appropriate to Indonesian palm-oil-based bioenergy sustainability indicator. Thus, Indonesia needs appropriate sustainability indicators for its palm-oil-based bioenergy.

The determination of the indicators has been carried out qualitatively using focus group discussion (FGD). This method has some advantages, for example varying data could be collected in shorter period time because the interaction between participants is more practical, flexible, and cost-efficient. Nevertheless, FGD has some disadvantages due to the complexity of the data collected so that it requires higher level- and longer duration of analysis[5]. Fuzzy inference system (FIS) is a method which maps an input to an output regarding to the determined factors. It has a mathematical framework which is capable of solving complex problems[6]. Therefore, FIS is proper to be used for measuring the appropriateness of Indonesian palm-oil-based sustainability indicators.

2. Metodology

This research was accomplished through three stages, namely preliminary research, appropriateness assessment and system validation. The preliminary research was conducted by reviewing bioenergy sustainability standards from various country and categorizing them by some aspects. In addition, we also determined the standards which contained the sustainability indicators that would be assessed their appropriateness with Indonesian palm-oil-based bioenergy. The determined standard was the one that internationally accepted, general for use, and it contained balanced proportion between environment, economic, and social aspects.

The assessment for the appropriateness of indicators was conducted using Mamdani fuzzy inference system (FIS) model towards the determined standard which was obtained after the preliminary research. The Mamdani FIS is a model which is able to be implemented into the input containing human linguistic variable. The output is in a fuzzy set[7]. The required steps to generate the output using Mamdani FIS model are determination of fuzzy set, application of implication function, composition of rules and defuzzification[8].

After that, a data analysis was conducted towards the output of the system generated by Matlab, so the appropriateness value of each indicator could be determined as well as the number of sustainability indicators which are appropriate to Indonesian palm-oil-based bioenergy. A validation process of the system was conducted in order to examine the accuracy of the system. This process was conducted by inputting a value into the system and comparing the output with the opinion from the expert. In this research, the system was using 81 cases. The accuracy calculation was performed with Equation 1 where X denotes number of case having the same result between system output and expert's opinion and Y denotes number of total cases.

$$Accuracy = \frac{X}{Y} x\ 100\ \% \tag{1}$$

3. Result and Discussion

3.1 Review of Global Bioenergy Sustainability Indicators

Sustainability becomes one of the important issue in the world. The development of various things is always linked to sustainability as it impacts towards the environment, society and economy. There were initiations that generate standards containing principles, criteria, and indicators about bioenergy sustainability. Those initiations are national, international, and multi-stakeholder initiation. Some of global standard regarding to bioenergy sustainability are Indonesian Sustainable Palm Oil Certification System (ISPO), International Sustainability & Carbon Certification (ISCC), Renewable Transport Fuel Obligation (RTFO), Global Bioenergy Partnership (GBEP), Roundtable on Sustainable Palm Oil (RSPO), and Roundtable on Sustainable Biomaterials (RSB). Each standard generally consists of the minimum requirements for biodiversity protection, reduction of gas house emission and other issues regarding to water, land, or social aspects[9].

After the analysis was done for each of the standard, a review of a global bioenergy sustainability standard was performed by categorizing them into five groups, namely environment, social, economic, legality (documentation, legislations), and implementation (production activity, standard operational procedure or SOP, transparency). This review can be seen in table 1. Each of the standards has a tendency in particular group. Standards that have a tendency to the environment indicators are ISCC, RTFO, RSPO and RSB. ISPO contains more about legality, whereas GBEP has a balanced proportion between environment, economic and social.

Tabel 1. Review of global bioenergy sustainability standards

Sustainability standard		Scope	Number of indicator in principles or criteria					Number of indicators
			En	S	E	L	Im	
	ISPO (*Indonesian Sustainable Palm Oil*)[10]	Plantation, plantation that produces palm oil	42	30	2	46	13	133
	ISCC (*International Sustaianbility and Carbon Certification*)[11]	Bio- and renewable energy based supply chain and raw material	56	34	-	2	3	95
Department for Transport	RTFO (*Renewable Transport Fuel Obligation*)[12]	Reduction of GHG emission from fuel used on-road and off-road	24	18	-	-	-	42
	GBEP (*Global Bioenergy Partnership*)[13]	The relation between production activity and the use of palm-oil based bioenergy in a sustainable development context	8	8	8	-	-	24
	RSPO (*Rountable Sustainable of Palm Oil*)	Plantation, industries producing palm-oil-based fuel, producer of palm-oil derivate products	62	51	2	14	9	138
	RSB (*Rountable on Sustainable Biofuel*)	Production, utilization and transportation of renewable energy based on biomass	19	13	-	1	3	36

Notes : environment (En), social (S), economic (E), legality (L), implementation (Im)

Based on the review, we found that GBEP is the suitable standards for undergoing appropriateness assessment for Indonesian palm-oil-based bioenergy. Several factors that support GBEP to be the most suitable standards for Indonesian are its international recognition and general use, having a balanced proportion between environment, economic, and social aspects as well as its inclusion of the overall bioenergy production activity from the cultivation of palm oil to the use of bioenergy. GBEP has 24 indicators which consists of 8 environmental indicators, 8 economic indicators and 8 social indicators. Sustainability indicators of GBEP can be seen in table 2.

Table 2. Sustainability indicators of GBEP

Environment	*Social*	*Economic*
1. *Life–cycle GHG Emissions*	9. *Allocation and tenure of land for new bioenergy production*	17. *Productivity*
2. *Soil quality*	10. *Price and supply of a national food basket*	18. *Net energy balance*
3. *Harvest levels of wood resources*	11. *Change in income*	19. *Gross value added*
4. *Emissions of non-GHG air pollutants, including air toxics*	12. *Jobs in the bioenergy sector*	20. *Change in consumption of fossil fuels and traditional use of biomass*
5. *Water use and efficiency*	13. *Change in unpaid time spent by women and children collecting biomass*	21. *Training and re-qualification of the workforce*
6. *Water quality*	14. *Bioenergy used to expand access to modern energy services*	22. *Energy diversity*
7. *Biological diversity in the landscape*	15. *Change in mortality and burden of disease attributable to indoor smoke*	23. *Infrastructure and logistics for distribution of bioenergy*
8. *Land use and land-use change related to bioenergy feedstock production*	16. *Incidence of occupational injury, illness, and fatalities*	24. *Capacity and flexibility of use of bioenergy*

Source : *The Global Bioenergy Patnership Sustainability Indicators for Bioenergy First Edition*[13]

3.2 Design and Documentation of the System

Mamdani FIS model used to assess the appropriateness of Indonesian palm-oil-based bioenergy sustainability indicators. The defuzzification method that used was Mean of Maximum (MOM). The steps to generate the output using Mamdani FIS model are fuzzification, application of implication, composition of rules, and defuzzification [8].

3.2.1 Fuzzification

Fuzzification is process to determine the fuzzy set that contains membership functions and domains [14]. Before the fuzzification, we need to determine the input and output variables of the system. The input variable is the assessment criteria of the appropriateness of Indonesian palm-oil-based bioenergy. The criteria are the relevant (targeted of its usefulness), user utility, reliability (provable in a scientific ways), and measurability (data is available, easy to collect, and always updated)[15]. Meanwhile, the output variables is the appropriateness of palm-oil-based bioenergy indicators. The relationship between the input and output variables can be seen at figure 2.

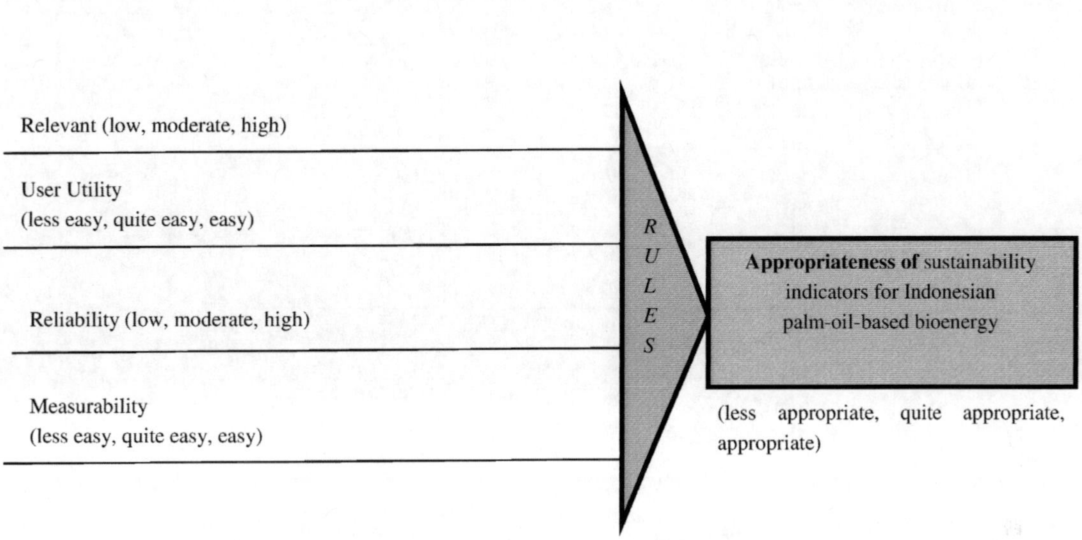

Figure 1. Dependency diagram of the input and output system

The input variables had three fuzzy sets, namely low, moderate, and high or less easy, quite easy, and easy. It had trapezoidal and triangular types of membership function curve. The output variables had three fuzzy sets, namely less appropriate, quite appropriate, and appropriate. It also had trapezoidal and triangular types of membership function curve. The sets of input and output variables can be seen in table 3.

Table 3. The set of input and output variables system

Function	Name of variables	Notattion	Fuzzy set		Type of currve	Domain	Range
			Name	Notation			
Input	Relevant	RL	Low	L	Trapezium	[0,4]	
			Moderate	M	Triangle	[3,7]	[0,10]
			High	H	Trapezium	[6,10]	
	User Utility	UT	Less easy	LE	Trapezium	[0,4]	
			Quite easy	QE	Triangle	[3,7]	[0,10]
			Easy	E	Trapezium	[6,10]	
	Reliability	RB	Low	L	Trapezium	[0,4]	
			Moderate	M	Triangle	[3,7]	[0,10]
			High	H	Trapezium	[6,10]	
	Measurability	MU	Less easy	LE	Trapezium	[0,4]	
			Quite easy	QE	Triangle	[3,7]	[0,10]
			Easy	E	Trapezium	[6,10]	
Output	Appropriateness	AP	Less appropriate	LA	Trapezium	[0,4]	
			Quite appropriate	QA	Triangle	[3,7]	[0,10]
			Appropriate	A	Trapezium	[6,10]	

This system only used two types of membership function, namely trapezium and triangle. The membership function of the system can be seen in figure 2.

Fuzification of relevant input

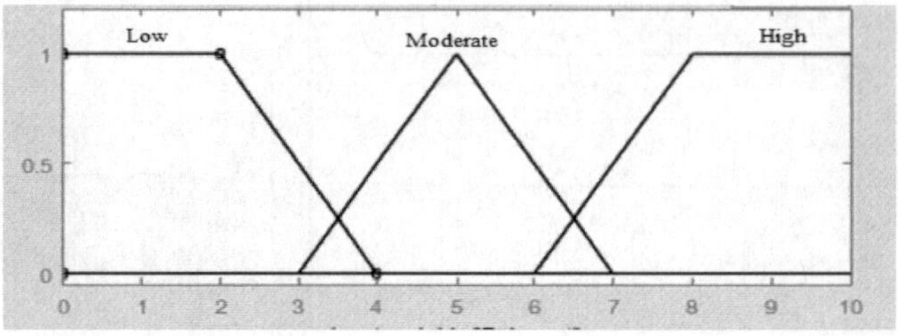

Figure 2. Membership function for relevant input

Membership function for each fuzzy set of relevant input can be written as in the equation below:

$$\mu_L[x] = \begin{cases} 0; & x \geq 4 \\ 1; & 0 \leq x \leq 2 \\ (4-x)/(4-2); & 2 < x < 4 \end{cases} \quad (2)$$

$$\mu_M[x] = \begin{cases} 0; & x \leq 3 \text{ or } x \geq 7 \\ (x-3)/(5-3); & 3 < x < 5 \\ 1; & x = 5 \\ (7-x)/(7-5); & 5 < x < 7 \end{cases} \quad (3)$$

$$\mu_H[x] = \begin{cases} 0; & x \leq 6 \\ 1; & 8 \leq x \leq 10 \\ (x-6)/(8-6); & 6 < x < 8 \end{cases} \quad (4)$$

Fuzification of user utility input

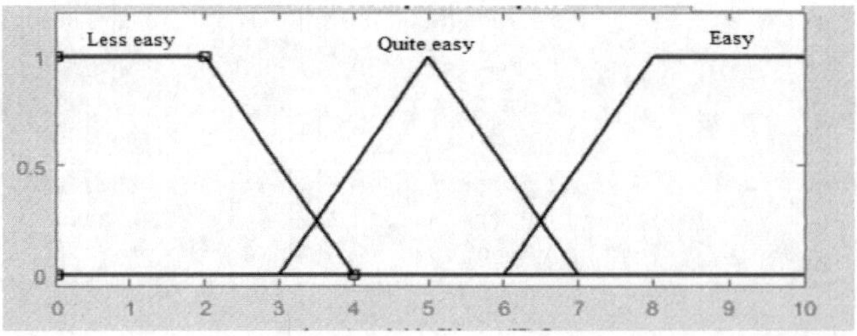

Figure 3. Membership function for user utility input

Membership function for each fuzzy set of user utility input can be written as in the equation below:

$$\mu_{LE}[x] = \begin{cases} 0; & x \geq 4 \\ 1; & 0 \leq x \leq 2 \\ (4-x)/(4-2); & 2 < x < 4 \end{cases} \quad (5)$$

$$\mu_{QE}[x] = \begin{cases} 0; & x \leq 3 \text{ or } x \geq 7 \\ (x-3)/(5-3); & 3 < x < 5 \\ 1; & x = 5 \\ (7-x)/(7-5); & 5 < x < 7 \end{cases} \quad (6)$$

$$\mu_{E}[x] = \begin{cases} 0; & x \leq 6 \\ 1; & 8 \leq x \leq 10 \\ (x-6)/(8-6); & 6 < x < 8 \end{cases} \quad (7)$$

Fuzification of reliability input

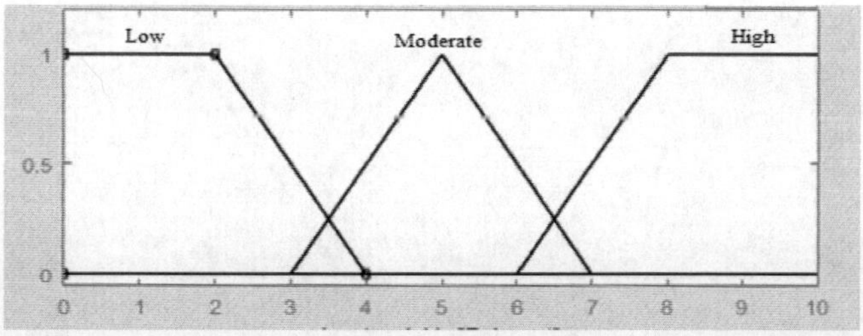

Figure 4. Membership function for reliability input

Membership function for each fuzzy set of reliability input can be written as in the equation below:

$$\mu_L[x] = \begin{cases} 0; & x \geq 4 \\ 1; & 0 \leq x \leq 2 \\ (4-x)/(4-2); & 2 < x < 4 \end{cases} \quad (8)$$

$$\mu_M[x] = \begin{cases} 0; & x \leq 3 \text{ or } x \geq 7 \\ (x-3)/(5-3); & 3 < x < 5 \\ 1; & x = 5 \\ (7-x)/(7-5); & 5 < x < 7 \end{cases} \quad (9)$$

$$\mu_H[x] = \begin{cases} 0; & x \leq 6 \\ 1; & 8 \leq x \leq 10 \\ (x-6)/(8-6); & 6 < x < 8 \end{cases} \quad (10)$$

Fuzification of measurability input

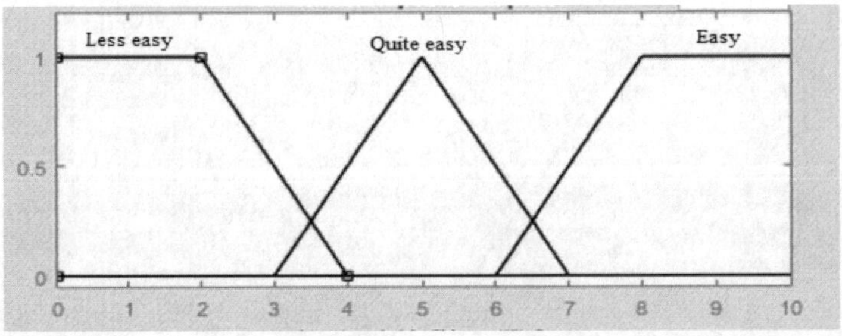

Figure 5. Membership function for measurability input

Membership function for each fuzzy set of measurability input can be written as in the equation below:

$$\mu_{LE}[x] = \begin{cases} 0; & x \geq 4 \\ 1; & 0 \leq x \leq 2 \\ (4-x)/(4-2); & 2 < x < 4 \end{cases} \quad (11)$$

$$\mu_{QE}[x] = \begin{cases} 0; & x \leq 3 \text{ or } x \geq 7 \\ (x-3)/(5-3); & 3 < x < 5 \\ 1; & x = 5 \\ (7-x)/(7-5); & 5 < x < 7 \end{cases} \quad (12)$$

$$\mu_{E}[x] = \begin{cases} 0; & x \leq 6 \\ 1; & 8 \leq x \leq 10 \\ (x-6)/(8-6); & 6 < x < 8 \end{cases} \quad (13)$$

Fuzification of appropriateness output

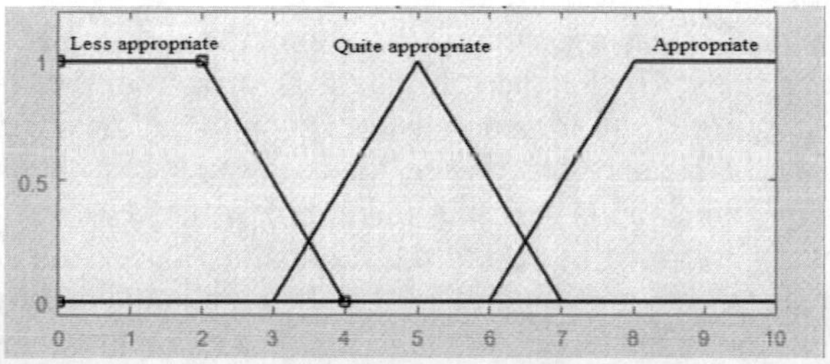

Figure 6. Membership function for appropriateness output

Membership function for each fuzzy set of the appropriateness output can be written as in the equation below:

$$\mu LA[x] = \begin{cases} 0; & x \geq 4 \\ 1; & 0 \leq x \leq 2 \\ (4-x)/(4-2); & 2 < x < 4 \end{cases} \quad (14)$$

$$\mu QA[x] = \begin{cases} 0; & x \leq 3 \text{ or } x \geq 7 \\ (x-3)/(5-3); & 3 < x < 5 \\ 1; & x = 5 \\ (7-x)/(7-5); & 5 < x < 7 \end{cases} \quad (15)$$

$$\mu A[x] = \begin{cases} 0; & x \leq 6 \\ 1; & 8 \leq x \leq 10 \\ (x-6)/(8-6); & 6 < x < 8 \end{cases} \quad (16)$$

3.2.2 Application of implication

The application of implication was used min implication. The determination of implication area was performed by generating rules. The number of rules generated from the four fuzzy sets are 81 rules. Before application of implication, it's important to determine the membership degree for each input value. The determination of membership degree for each input can obtained by inputting the input value into membership function. After the membership degree was obtained, application of implication function was performed by taking the minimum value of membership degree from each rule. An example of the application of implication can be seen in the description below. This example was performed for indicator numbered 16 in the 41th rule with input values are 5.47 (RL), 5.6 (UT), 5.8 (RB), 5.87 (MU).

$$\alpha R41 = min\, \mu_M RL[5.47] \wedge \mu_{QE} UT[5.6] \wedge \mu_M RB[5.8] \wedge \mu_{QE} MU[5.87]$$
$$= \text{min } (0.765; 0.7; 0.6; 0.565$$
$$= 0.565$$

The implication area of the system in Matlab can be seen in figure 7 at the red box outline area.

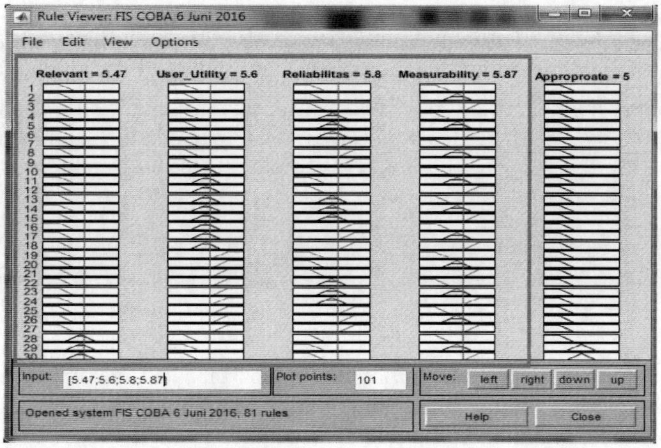

Figure 7. Implication area of system for indicator 16 in Matlab

3.2.3 Composition of rules

Composition of rules are aggregation of some rules. In this system, composition of rules was performed by taking the maximum membership degree between rules using OR operator. In general, composition between rules can be written as in equation (17) where $\mu_{sf}[x_i]$ denotes the value of fuzzy membership and $\mu_{kf}[x_i]$ denotes the value of fuzzy consequent membership.

$$\mu_{sf}[x_i] \leftarrow \max\left(\mu_{sf}[x_i], \mu_{kf}[x_i]\right) \tag{17}$$

The example of rules composition that performed towards the indicator numbered 16 can be seen in the red marked area in figure 8.

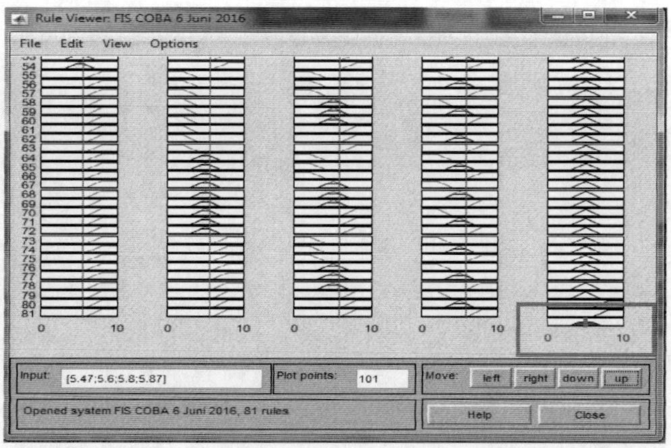

Figure 8. Composition of rules for indicator 16 in Matlab

3.2.4. Defuzification

Defuzification is a process that convert the output system into fuzzy set domain which system had. Defuzification was performed using Mean of Maximum (MOM) model that take average domains of maximum membership degree. For example, the indicator numbered 16 has maximum membership degree value at 0.565 in the 41th rule. After that, we determined which fuzzy set domain that had maximum membership degree at 0.565 by inputting it into membership function so that obtained two domains at 4.23 and 5.87. The determination process to obtain fuzzy set domain are in the description below.

$$\begin{aligned} 0{,}565 &= (x-3)/2 \\ 1{,}13 &= x-3 \\ x &= 4{,}23 \end{aligned} \qquad \begin{aligned} 0{,}565 &= (7-x)/2 \\ 1{,}13 &= 7-x \\ x &= 5{,}87 \end{aligned}$$

Defuzification was performed by averaging domain values that had maximum membership degree. The resulted output value for indicator numbered 16 from Matlab was 5 then translated into fuzzy set so that the output value means quite appropriate. The process of defuzification with MOM are in the description below.

$$Z = \frac{4{,}23 + 5{,}87}{2} = \frac{10}{2} = 5$$

3.3 *Indonesian palm-oil-based bioenergy sustainability indicators*

The determination process was performed using Mamdani fuzzy inference system (FIS) model. This process was performed by conducting appropriateness assessment for each indicator by the expert from Indonesian government (professional), practitioner (industrialist) and academic. This assessment was performed by filling the questionnaire. The results showed that there were 21 sustainability indicators which are appropriate for Indonesian palm-oil-based bioenergy. Those indicators consisted of 8 environmental indicators, 5 social indicators and 8 economic indicators. The appropriateness value for each indicator can be seen in table 4.

Table 4. Results of data processing in the Mamdani FIS system

No	Relevant		User Utility		Reliability		Measurability		Appropriateness	
1	9.53	H	8.4	E	8.07	H	7.87	E	8.95	A
2	9.4	H	8.27	E	8.2	H	7.73	E	8.9	A
3	8.07	H	7.93	E	7.73	H	7.8	E	8.9	A
4	8.6	H	8	E	7.13	H	7.4	E	8.6	A
5	9.07	H	8.6	E	8.73	H	7.6	E	8.8	A
6	8.73	H	8.13	E	8	H	7.27	E	8.65	A
7	8.13	H	6.87	E	6.67	H	5.87	QE	8.35	A
8	8.4	H	7.47	E	7.33	H	7.2	E	8.6	A
9	8.8	H	7.67	E	7.67	H	7.4	E	8.7	A
10	7.8	H	7.53	E	7.6	H	7.47	E	8.75	A
11	8.93	H	8.53	E	7.93	H	8.07	E	9	A
12	8.73	H	8.13	E	7.8	H	7.73	E	8.9	A
13	6.53	H	6.6	E	6.27	H	6.13	QE	5	QA
14	8.93	H	8.27	E	8	H	7.33	QE	8.7	A
15	5.87	M	6	QE	6	M	5.93	QE	5	QA
16	5.47	M	5.6	QE	5.8	M	5.87	QE	5	QA
17	8.8	H	8.4	E	8.4	H	7.6	E	8.8	A
18	8.33	H	7.33	E	7.93	H	7.27	E	8.65	A
19	8.2	H	7.6	E	7.47	H	7.27	E	8.65	A
20	8.73	H	8.13	E	7.93	H	7.67	E	8.85	A
21	7.87	H	7.6	E	7.47	H	7.27	E	8.65	A
22	8.87	H	7.87	E	8	H	7.27	E	8.65	A

No	Relevant		User Utility		Reliability		Measurability		Appropriateness	
23	8.53	H	7.93	E	7.8	H	6.8	E	8.4	A
24	8.2	H	8.07	E	7.53	H	6.87	E	8.45	A

Indicator 13, 15, and 16 which were not selected for Indonesian palm-oil-based bioenergy because based on its generated value by Mamdani FIS model, they were not appropriate with the condition of Indoensia and in the case of palm-oil-based bioenergy. The appropriate indicators that generated by Mamdani FIS model can be seen in table 5.

Table 5. Bioenergy sustainability indicators based on Mamdani FIS method

Environment	Social	Economic
1. Life–cycle GHG Emissions	9. Allocation and tenure of land for new bioenergy production	14. Productivity
2. Soil quality	10. Price and supply of a national food basket	15. Net energy balance
3. Harvest levels of wood resources	11. Change in income	16. Gross value added
4. Emissions of non-GHG air pollutants, including air toxics	12. Jobs in the bioenergy sector	17. Change in consumption of fossil fuels and traditional use of biomass
5. Water use and efficiency	13. Bioenergy used to expand access to modern energy services	18. Training and re-qualification of the workforce
6. Water quality		19. Energy diversity
7. Biological diversity in the landscape		20. Infrastructure and logistics for distribution of bioenergy
8. Land use and land-use change related to biodiesel feedstock production		21. Capacity and flexibility of use of bioenergy

Based on this model, there were 21 indicators which were appropriate for Indonesian palm-oil-based bioenergy industry. After that, a validation process was conducted by inputting a value into the system and comparing the output with expert's opinion. There were 81 cases that were tested into this system and the accuracy was 85 %. The system validation can be seen in table 6.

Table 6. System validation of the Mamdani FIS model

No	Relevant		User Utility		Reliability		Measurability		Appropriateness (output of system)		Appropriateness (expert's opinion)
1	2	L	4	LE	3.5	L	3	LE	LA	1.75	LA
2	0	L	1.5	LE	4	L	6.5	QE	LA	1.75	LA
3	1.25	L	3	LE	2.25	L	8	E	LA	1.5	LA
4	4	L	2.6	LE	7	M	4	LE	QA	5	LA
5	1.5	L	4	LE	6.5	M	7	QE	LA	1.75	LA
6	3.5	L	3.9	LE	5	M	8.7	E	LA	3.25	LA
7	0	L	3	LE	9.25	H	2.45	LE	LA	1.5	LA
8	2.5	L	2.9	LE	10	H	6.25	QE	QA	1.6	LA
9	4	L	0	LE	8	H	8.5	E	LA	5	LA
10	2.25	L	7	QE	2	L	2	LE	LA	1.5	LA
...											
80	10	H	10	E	9.25	H	7	QE	A	8.5	A
81	7	H	10	E	10	H	9	E	A	8.5	A

4. Conclusion

Sustainability becomes one of the important issue in the world. The development of various things is always linked to sustainability as it impacts towards the environment, society, and economy. Many countries have made standards which contain principles, criteria and indicators about bioenergy. The review of global sustainability indicators related to bioenergy revealed that International Sustainability & Carbon Certification (ISCC) has 95 indicators that focus on environment, Renewable Transport Fuel Obligation (RTFO) has 42 indicators that focus on the environment, Global Bioenergy Partnership (GBEP) has 24 indicators that focus on environment, economic, and social, Roundtable on Sustainable Palm Oil (RSPO) has 138 indicators that focus on environment, and Roundtable on Sustainable Biomaterials (RSB) has 36 indicators that focus on environment.

The Indonesian palm-oil-based sustainability indicators were obtained by assessing the appropriateness of the indicators using Mamdani FIS method towards *Global Bioenergy Partnership* (GBEP) standards for each indicator. The number of sustainability indicators for Indonesian palm-oil-based bioenergy with Mamdani FIS model was 21 indicators, consisting 8 environmental indicators, 5 social indicators, and 8 economic indicators. The system had an accuracy of 85 %.

5. Recommendation

The system can be further developed by adding other inputs that support the appropriateness assessment criteria for Indonesian palm-oil-based bioenergy sustainability. Moreover, the rules combination can be changed to get the best combination rules and also by adding the number of fuzzy sets so that the accuracy of the system can be improved.

References

[1] [DJP] Direktorat Jendral Perkebunan Kementrian Pertanian 2016 *Statistik Perkebunan Indonesia Komoditas Kelapa Sawit 2013-2015* (Indonesia: Kementrian Pertanian) p 3

[2] [EBTKE] Direktorat Jenderal Energi Baru. Terbarukan dan Konservasi Energi 2015 *Statistik EBTKE 2015* (Jakarta : EBTKE Kementrian Energi dan Sumber Daya Mineral) p 13

[3] Lee J S H, Abood S, Ghazoul J, Barus B, Obidzinski K and Koh L P 2014 Environmental impacts of large-scale oil palm enterprises exceed that of smallholdings in Indonesia *Conservation Letters* **7** 25–33

[4] Obidzinski K, Andriani R, Komaruddin H, and Andrianto A 2012 Environmental and social impacts of oil palm plantations and their implications for biofuel production in Indonesia *Ecology and Society* **17** 25

[5] Afiyanti Y 2008 *Focus Group Discussion* (Diskusi Kelompok Terfokus) sebagai metode pengumpulan data penelitian kualitatif. *Jurnal Keperawatan Indonesia* **12** 58-62

[6] Aly S and Vrana I 2005 Fuzzy expert marketing mix model *Journal of Agric.Econ.-Czech* **51** 69-79

[7] Hasibuan Y M, Nilamsari K, and Beni I 2014 Pengendalian kecepatan kendaraan roda empat dengan menggunakan fuzzy inference system metode mamdani *Buletin Ilmiah Math. Stat. dan Terapannya (Bimaster)* **3** 39-46

[8] Kusumadewi S and Purnomo H 2004 *Logika Fuzzy untuk Pendukung Keputusan*.Yogyakarta (Indonesia: Graha Ilmu) p 39

[9] Fritsche UR and Leire I 2014 Sustainability criteria and indicators for the bio-based economy in Erope: state of discussion and way forward *Journal Energy* **7** 6825-6836

[10] Menteri Pertanian Republik Indonesia 2015 Sistem sertifikasi kelapa sawit berkelanjutan Inndonesia (Indonesian sustainable palam oil certification system/ISPO) *Peraturan Menteri Pertanian Republik Indonesia Nomor 11/Permentan/OT.140/3/2015*

[11] Internasional Sustainability & Carbon Certification 2015 ISCC202 Sustainibility re-quirements for the productions of biomass ISCC PLUS Version 2.3 16 November 2015

[12] Departement for Transport 2014 RTFO Guidance Part One: Proces Guidanc Version 7.0 April 2014 (London : Departement for Transport)

[13] FAO 2011 *The Global Bioenergy Partnership Sustainability Indicators for Bioenergy First Edition* (Italy : Global Energy Partnertship)

[14] Navianti DR, I Gusti NRU, and Farida AW 2012 Penerapan Fuzzy Inference System pada Prediksi di Curah Hujan di Surabaya Utara *Jurnal Sains dan Seni ITS* **1** 23-28

[15] [OECD] Organisation for Economic Co-operation dan Development 2003 OECD Environmental Indicators: Development, Measurement, and Use *Reference Paper* (Paris : OECD)

IOP Conf. Series: Earth and Environmental Science **65** (2017) 012014 doi:10.1088/1755-1315/65/1/012014

Development of bio-fuel from palm frond via fast pyrolysis

M D Solikhah[1], A A Raksodewanto[1], A Kismanto[1], F Karuana[1], Y Heryana[1], Riza[1] and F T Pratiwi[1]

[1]Agency for the Assessment and Application of Technology, Build. 480 PUSPIPTEK, South Tangerang, Indonesia.

Email : maharani.dewi@bppt.go.id

Abstract. In order to fulfill the fuel demand in the future, Indonesia has to find a sustainable alternative for its energy. Energy source in the form of biomass is a promising alternative since its availability is abundance in this tropical country. Biomass can be converted into liquid fuel via fast pyrolysis by contacting the solid biomass into hot medium in the absence of oxygen. Hot sand is the common heat carrier for fast pyrolysis purposes but it is very abrasive and required high pyrolysis temperature (450-600 ^oC). This paper will discuss on the equipment design and experiment of fast pyrolysis of palm frond using high boiling point thermal oil as heat carrier. Experiments show that by using thermal oil as heat carrier, bio-oil can be produced at lower pyrolysis temperature of 350 ^oC, compared to the one using hot sand as heating carrier. The yield of bio-oil production is 36.4 % of biomass feeding. The water content of bio-oil is 52.77 % mass while heating value is 10.25 MJ/kg.

1. Introduction

Indonesia is currently facing a big problem in fulfilling fuel demand. The increasing of fuel demand has reached almost 1.5 million barrels per day at 2015 while production is only nearly 0.8 million barrels per day and decreasing year by year [1]. The gap between demand and production is filled with fuel import that causes pressure on national economics. Therefore, it is required to obtain alternative fuel to find a sustainable alternative for its energy.

As a country that blessed with sunlight and produce a lot of biomass, converting biomass into fuel could become the best way. Biomass may be converted into liquid fuel via pyrolysis. Pyrolysis is a thermo-chemical decomposition process in which organic material such as biomass is converted into a carbon-rich solid (charcoal), liquid, and gaseous products by heating in the absence of oxygen [2]. The amount of each product can be adjusted by adjusting the temperature and residence time of the process [3]. Slow pyrolysis that occurs for hours up to days produce more char than fast pyrolysis. Fast pyrolysis process can give much liquid from the pyrolysis processes of the biomass feedstock and only occurred in order of seconds at higher temperature. Therefore, fast pyrolysis gives advantages on

Content from this work may be used under the terms of the Creative Commons Attribution 3.0 licence. Any further distribution of this work must maintain attribution to the author(s) and the title of the work, journal citation and DOI.
Published under licence by IOP Publishing Ltd

producing liquid fuel. On the other hand, Gable and Brown [4] mentioned that bio-oil production on fast pyrolysis did not stop at 2 s and still produce until 42.4 s. These studies will be useful for the design of fast pyrolysis reactor.

Fast pyrolysis has been studied using various kind of reactor such as fluidized bed reactor, transported bed & circulating fluidized bed, rotating cone, and auger or screw [5]. Commercial scale pyrolysis reactor has been developed by Biomass Technology Group (BTG) of Netherland that using "rotating cone reactor" where biomass was heated by hot sands as heat carrier so that ablation and vaporization occurred [6]. Another commercial technology owned by Ensyn called "Rapid Thermal Processing" where biomass heated using hot sands in a fluidized bed reactor. Biomass will be vaporized and flowed up together with hot sands and char [7]. The vapor was separated from sand and char then condensed. Both technologies operated at temperature of 450 – 600 °C and can produced bio-oil of 55 – 75 % mass of biomass feeding. However, the hot sands caused abrasion on the equipment.

Other pyrolysis method has been studied by Jess, et. al. [8] that developed a fast pyrolysis process that using vacuum gas oil as a heat carrier. Jess *et al.* [8] produced bio-oil at lower temperature than Ensyn or BTG *i.e.* 320 – 400 °C with 60% mass of bio-oil production. This process is promising but vacuum gas oil is difficult to be found at open market. On the other hand, the similar heat transfer characteristics could be found on selected heat transfer oil that has high boiling point suitable for the operation temperature of the pyrolysis.

Pyrolisys process can utilize any kind of biomass. In Indonesia, the biggest source of biomass is oil palm plantation and industry since Indonesia has become the biggest producer of palm oil in the world. At 2014 Indonesia had more than 10 million hectares of oil palm plantation and almost 30 million tons of crude palm oil [9] that may produce biomass such as palm frond, palm trunk, fiber, shell, and empty fruit bunch. Palm fiber and palm shell has been widely utilized for own use power generation at palm oil mill while palm trunk only produced during replanting after 25 – 30 years. Empty fruit bunch's production may reach 4.4 tons/(ha·year) and still not widely utilized but it has high moisture content, very tough to be chopped and contain high mineral content that may affect the reaction [10]. On the other hand, palm frond has low moisture content, easy to be crushed, and has a low utilizations. The production of palm frond from pruning can reach 10.4 tons/(ha·year) [11]. Therefore, palm frond has a good future for biomass utilization.

From above discussions, it was found that the use of liquid heat carrier could produce bio-oil at lower temperature at productivity that competitive with the one using hot sand as heat carrier. So, this paper will discuss on the equipment design and experiment of fast pyrolysis of palm frond using high boiling point thermal oil as heat carrier.

2. Method

2.1. Material

Biomass used in this experiment was palm frond that collected from palm tree in PUSPIPTEK, South Tangerang, Indonesia. The leaves were removed from the frond prior to chopping and powdering until size of 16 – 30 mesh. The size comparison is presented at Figure 1. The water content of the frond powder was measured. Thermal oil heater used in this experiment was Mobiletherm 605 (Mobil Oil Co. Ltd., Surrey, UK).

(a) (b) (c)

Figure 1. (a) Fresh palm frond (b) Palm frond after chopping (c) Palm frond after powdering

2.2. The Reactor

A continuous pyrolysis reactor was designed and equipped with electric heater, inclined screw feeder for torrefaction, thermal oil recovery (char filter and screw conveyor), and condenser for bio-oil collection. Condenser was using water as cooling media. All heaters were controlled using temperature controller. All parts of reactor were isolated using high temperature rock-wool insulation of 10 cm. The residence time of the reactor was design for 30 – 60 s. Figure 2 shows the 3D design and the actual reactor used in this experiment.

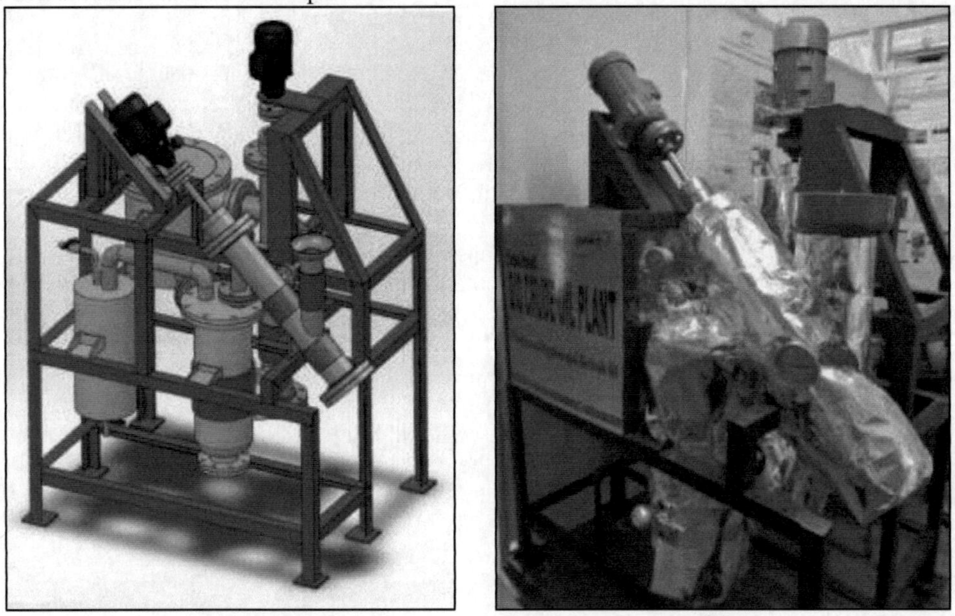

(a) (b)

Figure 2. (a) The 3D design and (b) the actual Pyrolysis Reactor

2.3. Fast Pyrolysis Experiment

Thermal oil of 21 kg was heated until 350 °C in the reactor. Torrefaction section was heated until 200 °C prior to reaction experiment to remove the water contained in the biomass. After all temperature was reached, 5.5 kg of biomass fed into torrefaction section at flow rate of 150-200 grams per minute. Biomass was heated in torrefaction section and flowed into reactor trough screw feeder to perform fast pyrolysis reaction. When biomass contacted with thermal oil inside the reactor, vapor was developed. The vapor flowed out from the top of the reactor and condensed at the condenser then collected as liquid at the output of condenser. This liquid was weighted and collected as liquid product. This reaction also produced char. To avoid char accumulation in the reactor, the char was removed from reactor using screw conveyor and transported into char filter. After char removal, heat transfer oil was circulated back into reactor. This circulation also enhanced heat transfer performance in the reaction, either from reactor wall into heat transfer oil as well as from heat transfer oil into biomass. The schematic diagram of the reactor is presented in Figure 3. The mass input and output from the experiment was measured to analyze the mass balance of the system. Furthermore the yield of bio-oil production was calculated as mass of bio-oil divided by mass of palm frond used in the experiment.

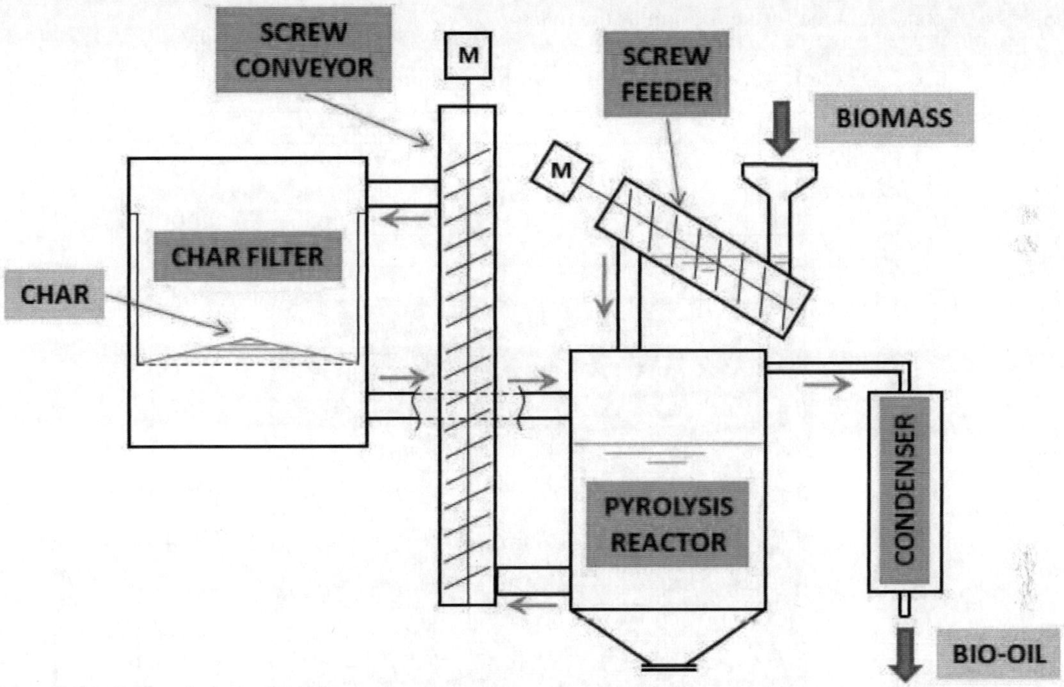

Figure 3. Schematic Diagram of the Fast Pyrolysis Reactor

2.4. Product characterizations

Liquid product collected from condenser formed in two phases: upper phase and bottom phase. The upper phase was in light color and water insoluble, called as sample A. The bottom phase was in dark color and water soluble, called sample B. Sample A and B was separated using funnel separator prior to analysis. Both samples were analyzed by GC/MS to determine the composition. A Shimadzu GC/MS QP 2010 (Shimadzu Corp, Tokyo, Japan) equipped with column of DB1 (30 m x 0.32 mm x 1 μm capillary column) was used on the analysis. A 0.8 μL sample was injected on the column that started at 40 °C and hold for 4 minute, after that, ramped at 5 °C/minute until 250 °C and then hold for 10 minute. Water content was analyzed on sample B using Karl Fisher KF200 (Mitsubishi Chemical Analytech Co.,Ltd., Kanagawa, Japan). Heating value was measured on sample B using Automatic Calorimeter AC500 (Leco Corp., Saint Joseph, USA). The pH analysis was performed on sample B using 848 Titrino plus (Metrohm AG, Switzerland).

3. Results and Discussions

3.1. Reactor Performance

Before filled into the reactor, the water content of frond powder was analyzed. The water content of palm frond was 11.54 % mass. The reactor was working well and can reach 350 ^{o}C after 1 h from ambient temperature. The liquid product produced from 5.5 kg palm frond powder was 5.38 kg (sample A + sample B). The mass balance of the reaction is presented at Table 1. This table shows that there is 7.9 % mass loss from the system. From observation during the experiments, some vapor has been loss from between flanges that may contribute to the mass loss. Therefore, prevention of leakage should be of concern in the further design of the reactor.

Table 1. Mass Balance on Bio-oil Production from Palm Frond

Input, kg		Output, kg	
Biomass	5.5	Liquid Product	5.4
Heating oil	21.0	Char	8.0
		Heating oil recovered	11.0
Total	26.5	Total	24.4

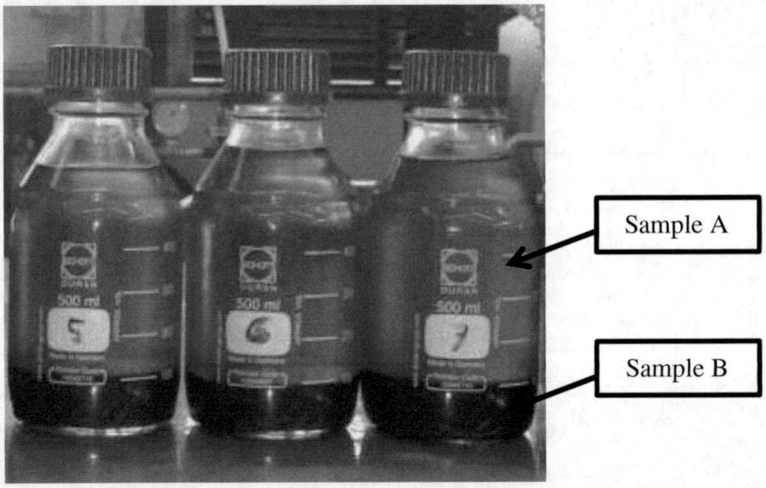

Figure 4. The appearance of liquid product from the reactor

3.2. Bio-oil Characterizations

Figure 4 shows the appearance of liquid product from the reactor. Liquid product was separate into two phases: water insoluble phase (sample A) and water soluble phase (sample B). GC/MS analysis was conducted on sample A, sample B, and heating oil for comparison. Figure 5 and figure 6 show the chromatogram of heating oil and sample A. From these figures, it is observed that both chromatograms are similar and most of sample A are heating oil. Therefore, it is concluded that sample A is not bio-oil.

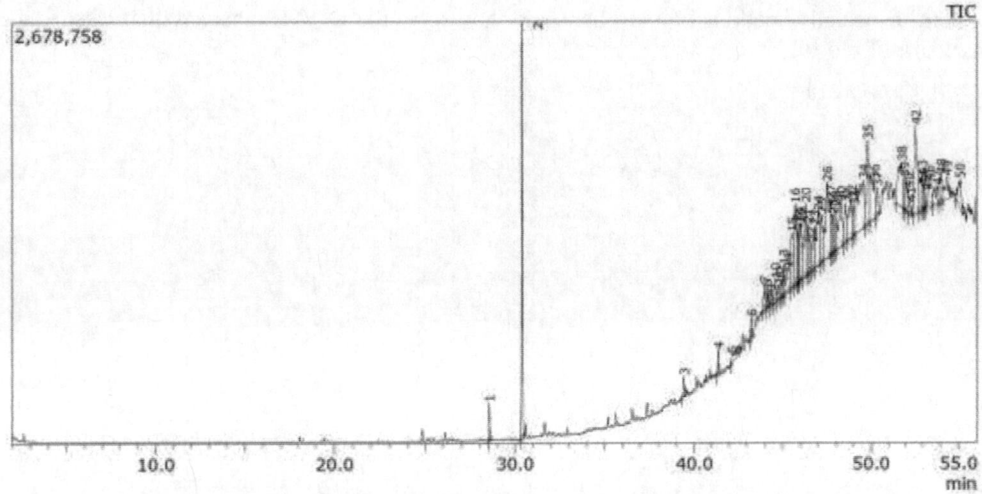

Figure 5. Chromatogram of Heating Oil (Mobileterm)

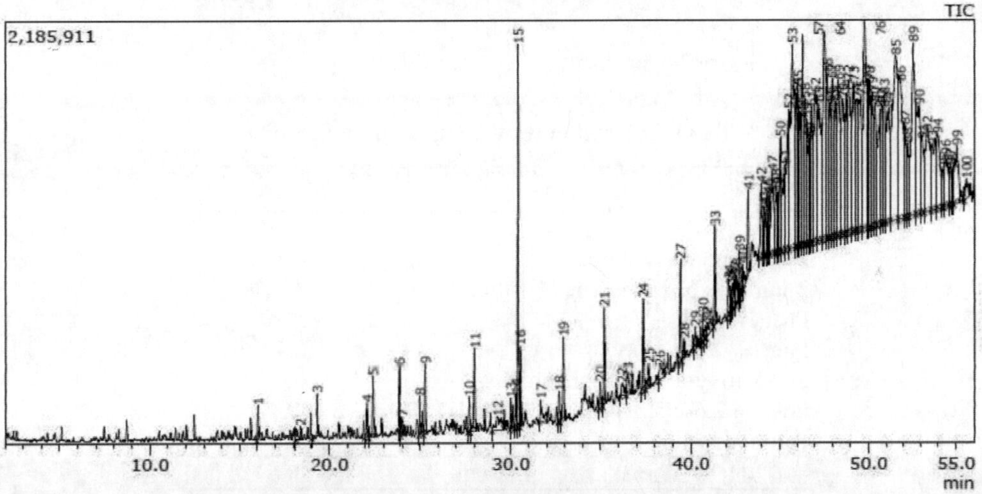

Figure 6. Chromatogram of Sample A (Water Insoluble Phase)

According to Bridgewater [5], pyrolysis oil typically is a dark brown, free-flowing liquid, and composed of a complex mixture of oxygenated compounds. By using this definition, sample B is the liquid that referred to as "bio-oil". After separation from sample A, 2 kg of bio-oil was obtained from the liquid product. Therefore, the yield of bio-oil production was 36.4 % of biomass feeding. This result is still between the ranges of 16.6 – 43.5 % that mentioned by Abnisa *et al* [11] .

The result of GC/MS analysis of bio-oil was presented at Figure 7. This figure shows chromatogram that is different from Figure 5 and Figure 6. The dominant component of bio-oil is presented on Table 2. This table shows that carboxylic acids, ketones, aldehydes, and phenol were indicated and most of the functional groups contained oxygen. The composition of bio-oil is in line with the result made by previous studies [12,13]. Presence of the above oxygenated components makes this phase becomes water soluble. The pH and water content of bio-oil was 2.2 and 52.77 % mass, respectively, that is in accordance to the work of Abnisa *et al.* [11]. Heating value of bio-oil was 10.25 MJ/kg that is lower than the result from Abnisa *et al.* [11]. This result may be caused by the high content of oxygenated components especially acetic acid and phenol as showed in the GC/MS analysis that can reduce the heating value of bio-oil.

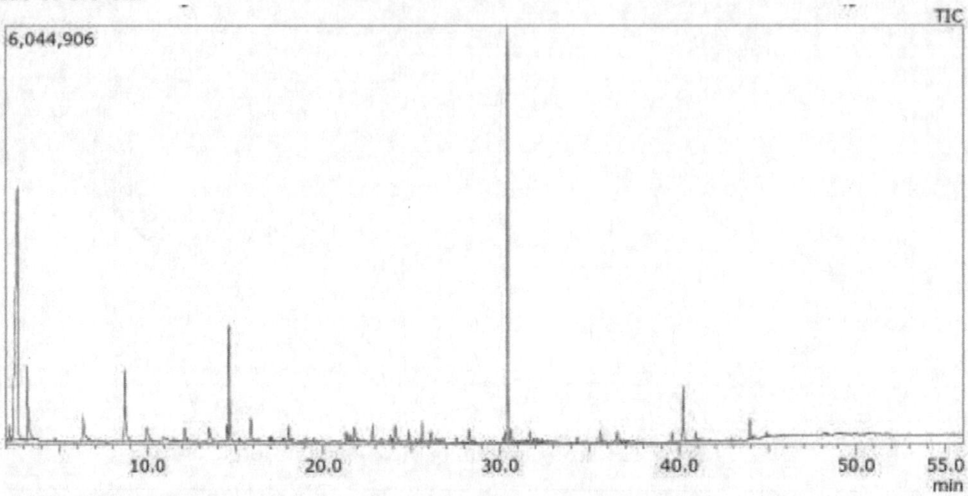

Figure 7. Chromatogram of bio-oil from palm frond

Table 2. The dominant component in Bio-oil

Components	% Area
Acetic Acid	30.20
2-Propanone (Acetone)	4.14
2-Furancarboxaldehyde (Furfural)	4.85
Phenol	5.70
Phenol, 2-methoxy-	0.75
2-Methoxy-4-methylphenol	0.46
Phenol, 2,6-bis(1,1-dimethylethyl)-	19.48
Palmitic Acid	3.09
Total	**68.67**
Others	**31.33**

4. Conclusions

The equipment design and experiment of fast pyrolysis of palm frond using high boiling point thermal oil as heat carrier has been conducted. The conversion of palm frond into bio-fuel via fast pyrolysis using heat transfer oil as heat carrier has been elucidated. This experiment shows that by using thermal oil as heat carrier, bio-oil can be produced at lower pyrolysis temperature of 350 °C, compared to the one using hot sand as heating carrier. Bio-oil has been characterized having heating value of 10.25 MJ/kg, water content of 52.77 %, and pH of 2.2. GC/MS analysis of bio-oil shows that it consisted of carboxylic acid, aldehydes, ketones, and phenols that caused bio-oil soluble in water. The most dominant component was acetic acid and phenol. The reactor can produce bio-oil at yield 36.4 % mass. However, 7.9 % of mass loss has been observed. Therefore it is suggested to improve the reactor by eliminate the vapor loss at flange connection and install a chiller condenser.

Acknowledgements
This research project is funded by "Badan Pengelola Dana Perkebunan Sawit" No. PRJ-30/DPKS/2015.

Reference

[1] Pertamina 2013 Greendiesel Business Development *Conf.* of Lokakarya Biodiesel III, October 25, 2013, Jakarta-Indonesia
[2] Demirbas A and Arin G 2002 An overview of biomass pyrolysis. *Energy Sources* **24** 471–82.
[3] Peryoga Y, Solikhah M D and Raksodewanto A 2014 Production cost assessment of palm empty fruit bunch conversion to bio-oil via fast pyrolysis *Int. J. Adv. Sci. Eng. Inf. Technol.* **4** 394-400
[4] Gable P and Brown R C 2016 Effect of biomass heating time on bio-oil yields in a free fall fast pyrolysis reactor *Fuel* **166** 361–366
[5] Bridgwater A V 2012 Review of fast pyrolysis of biomass and product upgrading *Biomass Bioenergy* **38** 68-94
[6] BTG 2016 Empyro Project Accessed September 5th 2016 (https://www.btg-btl.com/en/company/projects/empyro)
[7] Freel B 2011 *Rapid Thermal Conversion of Biomass* US Patent no.7,905,990, Ensyn, March 15, 2011
[8] Jess A, Kaufman D, and Daugaard D E 2012 *Biomass pyrolysis in Refinery Feedstock*, United States Patent No. 8,287,723 B2, Phillips 66 Company, October 16, 2012
[9] Badan Pusat Statistik 2015 *Indonesian Oil Palm Statistics* (Indonesia: Badan Pusat Statistik)
[10] Abdullah N, Sulaiman F and Gerhauser H 2011 Characterisation of oil palm empty fruit bunches for fuel application *J. Phys. Sci.* **22** 1–24.
[11] Abnisa F, Arami-Niya A, wan Daud W M A, Sahu J N and Noor I M 2013 Utilization of oil palm tree residues to produce bio-oil and bio-char via pyrolysis *Energy Convers. Manag.* **76** 1073–82
[12] Cai W and Liu R 2016 Performance of a commercial-scale biomass fast pyrolysis plant for bio-oil production *Fuel***18** 677-86
[13] Oh S J, Choi G G and Kim J S 2016 Characteristics of bio-oil from the pyrolysis of palm kernel shell in a newly developed two-stage pyrolyzer *Energy* **113** 108-15

Effect of kenaf short fiber loading on mechanical properties of biocomposites

J Andilolo[1], S Nikmatin[1,*], N Nugroho[2], H Alatas[1] and A S Wismogroho[3]

[1] Department of Physics, Bogor Agricultural University, Bogor, Indonesia
[2] Department of Forest Products, Bogor Agricultural University, Bogor, Indonesia
[3] Indonesian Institute of Sciences, South Tangerang, Indonesia

Email : mia.andilolo@gmail.com

Abstract. The research of biocomposite product with kenaf (*Hibiscus cannabinus*) short fiber as a filler and Acrylonitrile Butadiene Styrene (ABS) as the matrix had been done to understand the mechanical properties of this material. Kenaf short fiber was obtained from mechanical sieving after doing the mechanical milling. TAPPI method has been done to determine the chemical properties. In order to form a granular biocomposite a single screw extruder was performed with a variation of particle loading 10 and 15%. The original of acrylonitrile butadiene styrene (ABS) has been used as matrix. The fabrication of speciment had been done by molding injection process. Mechanical properties test was done by ASTM standarization. The results showed the density of the fibers of 1.008 g/cm^3 with a fiber length of 897.07 μm and a diameter of 66.38 μm. Tensile strength of kenaf short fiber loading 10 and 15% was 23.522 ± 8.36 MPa and 20.739 ± 6.79 MPa, respectively. The tensile properties showed a decreasing trend as the fiber loading was increased. The values of impact strength were 68.657 ± 4.89 kJ m^{-2} and 82.090 ± 5.56 kJ m^{-2}, respectively and the hardness values were 96.60 ± 6.03 HR and 105.20 ± 13.17 HR, respectively. Kenaf fiber can be a good reinforcement candidate for high performance polymer bio-composites.

1. Introduction

In recent years, natural fiber composite has attracted attention in the sector of composite application. This composite provides many advantages compared with synthetic fibers such as low density, abundance, lower cost, renewable, and biodegradable [1,2,3]. The contents of natural fiber are generally composed of cellulose, hemicellulose, and lignin. One of fiber plants that contain high cellulose is kenaf [4]. Kenaf (*Hibiscus cannabinus*) is a species of plant that is easily cultivated in tropical region such as Indonesia. It has short growing season (3-4 months) and long stem of approximately 4 m. Kenaf fiber utilization in Indonesia is still limited as a material for rope and burlap sack but nowdays it has replaced by plastic thus the economic's value of kenaf fiber are decreasing. One way to increase the economic's value of kenaf fiber is its utilization as a filler material in a composite.

Utilization of kenaf fiber as filler composite material has been researched and applied in industry. Recently, there are some observation report in various condition to make an optimum kenaf fiber composite. For example, Edeerozey *et al.* [5] dan Nosbi *et al.* [6] reported some microstructural image of kenaf fiber to examine the surface morphology. Morphological microscopic analysis of the fiber surface is very important to describe the structural changes that occur in the treatment. Edeerozey *et*

Content from this work may be used under the terms of the Creative Commons Attribution 3.0 licence. Any further distribution of this work must maintain attribution to the author(s) and the title of the work, journal citation and DOI.
Published under licence by IOP Publishing Ltd

al. [5] shows that the surface morphology of untreated and treated kenaf fiber have some differences in the features of smoothness and roughness of the surface. Jonoobi *et al.* [7] characterized kenaf (*Hibiscus cannabinus*) nanofiber which was obtaine from the unbleached and bleached pulp isolation with a combination of chemical and mechanical treatment.

The kenaf nanofiber was characterized by using environmental scanning electron microscopy (ESEM) and transmission electron microscopy (TEM). Thermogravimetric analysis (TGA) showed that both types of pulp and nanofiber have a highes thermal stability than the raw kenaf. Fourier transform infrared (FTIR) spectroscopy also showed that the content of lignin and hemicellulose decreased during the pulping process while lignin was almost lost during bleaching. Liu *et al.* [8] observed the effect of processing methods and fibre length on biocomposite soy based kenaf fiber.

On the other hand, thermoplastic materials, Acrylonitrile butadiene styrene (ABS) is an important engineering copolymer widely used in industry due to superior mechanical, chemical resistance, ease of processing and recyclability [9]. ABS has wide application and because of that there is a great need to develop material using ABS polymers. Therefore, in this study ABS was used as a matrix polymer to compound with the kenaf fiber for the purpose of manufacturing the biocomposites. The addition of kenaf fiber on ABS polymer and reuse into biocomposite products will increase the economic value of the two materials. Fiber loading can affect mechanical properties of polymer composites. The opportunity to design composite with specific mechanical properties makes this subject become great interest to be observed. This research aims to study the effect of kenaf short fiber loading on the mechanical properties of biocomposites. Biocomposites with filler kenaf short fiber are expected to have superior quality due to its ability to protect the matrix of cracks in composite application.

2. Material and Method

2.1. Materials

Fresh stem kenaf was obtained from kenaf plantation in PT Global Agrotek Nusantara, Pekanbaru, Sumatera, Indonesia. Kenaf fibers with a length of 4 m was obtained by soaking dynamically in water for 2 weeks and then sun dried. Furthermore, kenaf fiber were cutted with a uniform length , ie ± 1 cm and oven dried (Tipe YNC-OV, YENACO, China) at 40 °C for 24 h until a constant weight. The dried kenaf was taken from oven then milling mechanically until totally smooth by using a milling tool (Model MDY-1000, FOMAC, China). After milling, the sample was sieved by using 20 mesh sieve. In this study, the sample used was a sample that passes 20 mesh (short fiber). The Original Acrylonitrile Butadine Styrene (ABS) polymer was imported from Malaysia.

2.2. Chemical Properties of Kenaf Fiber

Chemical composition of kenaf fiber was determined by using TAPPI test standart. Five measured chemical composition have been evaluated, such the cellulose, hemicellulose, holocellulose, lignin and extractive content.

2.3. Density Test

The basic method to calculate the kenaf fiber density is by Achimedes law, as :

$$\rho = \frac{m_1}{(m_1 - m_2)} \times \rho_{water} \qquad (1)$$

where ρ denotes material density (g/cm^3), m_1 was the initial mass during exist in air (g), m_2 is the final mass when treated in water (g), and ρ_{water} denotes water density (g/cm^3).

2.4. *Fiber Dimension and Morphology Structure of Biocomposite*

Dimension of kenaf fiber was calculated from images captured by a light microscope (Model BX51, Olympus, Japan) combined with the software Olympus DP2-BSW and DP25 Olympus Microscope Camera. The biocomposite morphology structure was observed under the light microscope.

2.5. *Preparation of Biocomposite*

Granular biocomposite was formed by the mixing of kenaf short fiber as a filler, ABS polymer as a matrix, maleic acid as coupling agent and primary antioxidant with the compositions that showed in Table 1.

Table 1. Composition of granular biocomposites

Sample	Composition (g)							
	Kenaf Short Fiber		Original ABS Polymer		Additive			
					Primary Antioxidant		Compatibilizer Maleic Acid	
KSF10%	10 %	2000	87 %	17400	1%	200	2%	400
KSF15%	15 %	3000	82 %	16400	1%	200	2%	400

All sample were fabricated with the single screw extrusion (Model HXSJ-125/125, Kai Xin, China) with a total mass of 20,000 g. The next stage was to create a standard American Society of Testing Material (ASTM) test pieces by molding injection (Model HC-250, Hwa Chin, China). As much as 5 kg samples of granular biocomposite were put into the hopper of molding injection then preheating in 60 °C. Heating barrels of molding injection was done in 5 zones (120 °C - 200 °C).

2.6. *Mechanical Test*

The resulting test piece was tested in mechanical properties based on ASTM standarization. Tensile test specimen were made in accordance with ASTM D-638 to measure the tensile properties. The samples were 164 mm long, 13.05 mm wide and 4.02 mm thick. Izod impact test specimens were prepared in accordance with ASTM D-256 to measure the impact strength. The specimens were 64 mm long, 13 mm deep, and 10 mm wide. A sharp file with angle of 45° was drawn across the centre of the saw cut at 90° to the sample axis to obtain a consistent starter crack. Hardness test specimen were prepared in accordance with ASTM D-785 to measure the Rockwell hardness number. The samples were 49 mm and diameters 6.03 mm thick.

3. Results and discussion

3.1. Density and chemical composition of kenaf fiber

Testing the density can be done with reference to the law of Archimedes where objects that are partially or wholly in the fluid always gets the upward force of weight of the fluid displaced. The tools used to perform this test is a technical balance and water as media controls. The density of the kenaf fiber is 1.008 gr/cm^3.

In Table 2, it is presented kind of carbohydrate component contents and the standard methods that we used. The plant cell wall is composed of cellulose, lignin, hemicellulose, holocellulose, and the extractives. Thus, the surface energy of the plant material must be some combination of the surface energies (γ) of these components (Liu and Rials, 1998). Therefore, quantitative amount of these components especially in the surface of fiber will influence the properties of the fiber biocomposites.

Table 2. Chemical composition of Kenaf fiber

Content	(%)
Cellulose	66.47
Lignin	2.39
Holocellulose	75.43
Hemicellulose	9.43
Extractive	2.11

3.2. Fiber dimension and morphology structure of biocomposite

The average of kenaf short fiber dimension was length 897.07 μm and diameter 66.38 μm with aspect ratio 13.50. Figure 1 showed the morphology structure from biocomposites, where the image can be seen that there are pores in biocomposites. This was due to the absence of additional alkali in the synthesis process of kenaf fiber up to that pores still contained in biocomposites.

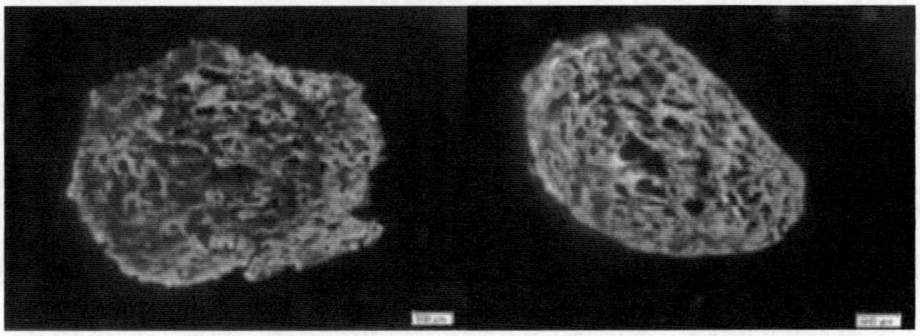

Figure 1. The observation of morphology structure of biocomposites used light microscope (a) Kenaf Short Fiber Loading 10% (b) Kenaf Short Fiber Loading 15%

3.3. Mechanical properties of biocomposite

Both of the matrix and the fiber properties are important in improving mechanical properties of the biocomposites. The chemical coupling method is also one of the important chemical methods, which improve the interfacial adhesion. In this method the fiber surface was treated with a compound which able to form a chemical bonds as bridge between fiber and matrix [10]

3.3.1 Tensile strength

Young's modulus is also called the modulus of elasticity (E). The key is to know the difference between elastic and plastic deformation. Young's modulus describes the relationship between stress and deformation of materials. Analytical model for the modulus of elasticity was derived from the Hooke Law and can be calculated by dividing the stress with strain. It is given as [11]:

$$E = \frac{\sigma}{\varepsilon} = \frac{F/A_0}{\Delta L/L_0} = \frac{FL_0}{A_0\Delta L} \qquad (2)$$

where E is modulus of elasticity (MPa). E is a direct measurement of the strength of the electron bonds between atoms/molecule. It indicates how much the force is required to stretch these bonds (without breaking then and without moving atoms / molecules around). From the definitions of stress and strain, we should see that stress corresponds to load and strain corresponds to extension. Variabel F and ΔL obtained from experimental data. Dimensions of the specimen is constant, L_0 (gauge length = 50 mm) and A_0 = w.h, where w is the width of the plate = 13,05 mm and h is the plate thickness = 4,02 mm.

The value of tensile strength, modulus of elasticity and ductility which can be seen in Table 3.

Table 3. The results of tensile test

Sample	Ultimate Tensile Strength (MPa)	Modulus of Elastisity (MPa)	Ductility (%)
KSF 10%	23.522 ± 8.36	439.94 ± 26.48	4.50 (brittle)
KSF 15%	20.739 ± 6.79	408.67 ± 18.32	3.50 (brittle)

Based on observations of test piece fracture after tensile testing known that the fracture was shaped pull out (due to withdrawal of fracture where it appears fibers) and broken fiber. Generally, the fiber composite was practically elastic influenced by the pull. Fiber-matrix interactions did not occur at the atomic level where the fiber harder and rigid compared to matrix so that its mechanical properties depend on the interface strength of the fiber and matrix.

The orientation also affected to tensile strength randomly oriented fiber had small reinforcement that there was no transfer stress. The more elasticity of sample that given more tensile force to. KSF 10 and 15% was brittle due to the deformation less than 5%.

Fiber loading of bio-composite also influence the value of tensile strength. As seen in Table 3 that the higher fiber loading given tendency to deacrease the value of tensile strength. This due to at the synthesis of kenaf fiber without addition alkali treatment so that in granular biocomposite there was pores which due to the tensile value decreases. In addition, natural fiber composites were anisotropic high maximum properties would be achieved when the entire fiber aligned in the direction of the axis of the fiber.

3.3.2 Impact strength

The value of impact strength was energy total that can be absorbed by materials result of sudden imposition which was the materials deformed. The Impact energy can be calculated using the equation below:

$$\Delta E = Wl\,(\cos\beta - \cos\alpha) \tag{3}$$

Where ΔE was total energy absorbed (kJ), W was weight of pendulum (N), α the initial angle (°) and β was the end angle (°). To determine the impact strength (I_s) so the total energy absorbed divided by cross-sectional area of specimen (A), which was described in the equation below :

$$I_s = \frac{\Delta E}{A} = \frac{Wl\,(\cos\beta - \cos\alpha)}{A} \tag{4}$$

I_s was the impact strength (kJ m^{-2})

Table 4. The results of impact strength

Sample	Impact Strength (kJ m^{-2})	Impact Energy (J)
KSF 10%	68.657 ± 4.89	0.223
KSF 15%	82.090 ± 5.56	0.255

Table 4 shows that kenaf short fiber loading in as balance as the impact strength. The higher of fiber loading, so the higher of impact strength.

3.3.3 Hardness

Effect of kenaf short fiber loading on hardness was tested. The result show that the number of Rockwell of biocomposite increase since the loading of kenaf short fiber which can be seen in Table 5. The increase of fiber loading properties means proper interaction between the fibers and matrix mutually helped in the transfer of load from latter to the former [12].

Tabel 5. The results of hardness test

Sample	Rockwell Hardness Number (HR)
KSF 10%	96.60 ± 6.03
KSF 15%	105.20 ± 13.17

4. Conclussions

The effect of fiber combination of Original ABS polymer as a matrix on mechanical properties like tensile strength, impact strength and hardness behavior were studied. The results of mechanical properties test of biocomposite was influenced by the kenaf short fiber loading. The tensile properties showed a decreased trend as the fiber loading was increased. As the kenaf short fiber loading increased in the biocomposites, the value of impact strength and Rockwell hardness number also increased.

Acknowledgements
This work was funded by BPPTN Research of Strategic application from the Directorate General of Strengthening Research and Development, Ministry of Research, Technology and Higher Education, Republic of Indonesia.

References

[1] Salleh Z, Hyie K M, Berhan M N, Taib Y M D, Latip E N A and Kalam A 2014 *Procedia Technology* **15** 856-861
[2] Nordin N A, Yussof F Md, Kasolang S, Salleh Z and Ahmad M A 2013 *Procedia Engineering* **68** 145-151
[3] Davoodi M M, Sapuan SM, Ahmad D, Ali A, Khalina A, and Jonoobi M 2010 *Materials and Design* **31** 4927-4932
[4] Mwaikambo LY 2006 *African Journal of Science and Technology* **7** 120 – 133
[5] Edeerozey A M M, Akil H Md, Azhar A B and Ariffin Z 2007 *Materials Letters* **61** 2023-2025
[6] Nosbi N, Akil H Md, Ishak Z A M and Bakar A A 2011 *BioResources* **62** 950-960
[7] Jonoobi M, Jalahuddin H, Alireza S, Manjusri M and Kristiina 2009 *BioResources* **4** 626-639
[8] Liu W, Drzal L T, Mohanty A K and Misra M 2007 *Composites : Part B* **38** 352
[9] Yang S, Castilleja J R, Barrera EV and Lozano K 2004 *Polymer Degradation and Stability* **83** 383-388
[10] Anuar H and Zuraida A 2011 Composites : Part B **42** 462-465
[11] Mattos H S C, Minak G, Gioacchino F D and Soldà A 2008 Materials and Design **30** 1674-1679
[12] Aji I S, Zainudin ES, Khalina A, Sapuan S M and Khairul M D 2011 Journal of Reinforced Plastics and Composites **30** 546-553

IOP Conf. Series: Earth and Environmental Science **65** (2017) 012016 doi:10.1088/1755-1315/65/1/012016

Empowerment model of biomass in west java

C Mulyana[1], N I Fitriani [1], A Saad[1] and Y Yuliah[1]

[1]Physics Department Faculty of Mathematical and Natural Science, Padjadjaran University, Bandung, Indonesia

E-mail: c.mulyana55@yahoo.com

Abstract. Scarcity of fossil energy accelerates the search of renewable energy sources as the substitution. In West Java, biomass has potential to be developed into bio-briquette because the resources are abundant. The objectives of this research are mapping the potency of biomass as bio-briquette in West Java, and making the model of the empowerment biomass potential involving five fundamental step which are raw material, pre-processing process, conversion mechanism, products, and end user. The main object of this model focused on 3 forms which are solid, liquid, and gas which was made by involving the community component as the owner biomass, district government, academics and researcher communities, related industries as users of biomass, and the central government as the policy holders and investors as a funder. In the model was described their respective roles and mutual relationship one with another so that the bio-briquette as a substitute of fossil fuels can be realized. Application of this model will provide the benefits in renewability energy sources, environmental, socio economical and energy security.

1. Introduction

The increase of energy consumption in many sector of industries and the shortage of fossil energy source drives some countries to make strategy mix energy in fulfilling the energy need. At the same times the world need of clean energy accelerates some countries to find new alternative energy source which is not polutif or green energy. The share of biomass in meeting current word's primary energy mix is at model level 10% [1], but the rising concern about global warming and sustainability, this share is very lkely to rise. Indonesia is agrarian countries, there are abundance of biomass energy source, one of them is in West Java. West Java has fertile soil, some region is cultivate with agriculture and plantation. Waste product from agriculture and plantation so far is used only for animal feeding and compost.

Biomass is formed from living species like plants and animal. Unlike fossil fuels, biomass does not take millions of years to develop. Altough only 5% (13.5 bilion metric tons) it can be potentially mobilized to produce energy about 26% of the world's energy consumption, which equivalent to 6 billion tons of oil [2]. Biomass is one of renewable energy source which has potency to be developed to substitue fossil energy. The main problem of using bulk biomass as energy source is the low energy

Content from this work may be used under the terms of the Creative Commons Attribution 3.0 licence. Any further distribution of this work must maintain attribution to the author(s) and the title of the work, journal citation and DOI.
Published under licence by IOP Publishing Ltd

IOP Conf. Series: Earth and Environmental Science **65** (2017) 012016 doi:10.1088/1755-1315/65/1/012016

density and the inconvenient form of biomass are major barriers to a rapid transition from fossil to biomass. Unlike gas or liquid, biomass cannot be handled, stored, or transported easily. This provides a major motivation for the conversion of solid biomass into liquid and gaseous fuel, which are more energy dense and can be handled and stored relative ease. There are three types primary fuels could be produced from biomass and are as liquid fuels, gaseous fuels, and solid fuel. For a long time indonesian uses biomass as energy and combustion, and it is the most utilization means of biomass in Indonesia. The other motivation for biomass conversion to comercial energy product because of three factors, Renewability benefits, Environmental benefits, and Socioeconomical benefits. Three potential soerces of reveneu for biomass conversion plant are as follows: a). Energy production through heat or electricity, b). Production of chemical or metallurgical feedstock. c), Production of solid fuel for cofiring or transport fuel as an alternative to diesel and gasoline.

In Indonesia, studies of biomass as energy source have been developed by reseacher in higher educational institution and research institution. The reseach are about converting biomass become solid fuel (bio briquittes carbonized and non carbonized), liquid fuel (ethanol, biodiesel, methanol, vegetable oil, and pyrolisis oil) and gaseous fuel (methane, carbon dioksida, hidrogen, and carbon monoxide). BPPT and LIPI are two prominent research institutions which have developed biomass reactor. Padjadjaran University (Unpad) also developed some reseach in bio briquettes: (1). The heat content of rice husk, corncob, coffe bean shell, hemp wood decortication's waste, (2). Physics properties of bio briquettes, (3). Chemical gass product of carbonized bio briquettes, (4). Prototype of manual, srew and hydrolic pressing machine has been developed, (5). Design and construct calorimeter bomb, (6). Measuring the performance of the excisting bio briquettes stove, (7). Modified and developed the excisting biobriquettes stove, (8). Prototype of carbonized has been developed. The reaserch in bio fuel and biogass also developed. Unpad also has net working with bio briquettes based machine product such as burner, catle, and steamer. Based on those facts, it is important to develope integrated and comprehensive study of biomass as national issues that biomass can substitute the fossil energy.

West Java is one of the Indonesia province which has area of agriculture and plantation about 1.394.900 Ha. After harvest, 40-60% of waste product can be used as biomass energy source. Ricehusk, rice straw, corncob, coffe bean shell are typical of biomass in West Java. In some province there are local spesific potency of biomass. The industrial plant of biomass like calianda, pine, hemp wood also grows in West Java. The sources of biomass are abundance. Converting biomass to bio energy product which can substitute fossil fuel need six related elements, there are: 1). Availability of biomass to suply the bio energy plan and market for the biomass derived product. 2). Society readiness to involve and participate in colecting the biomass and also ready to involve in production process. 3). There are skillfull human resource as expert and technology and expert to produce biomass energy product. 4). Financial structure including cost of money, goverment subsidy if any, and loan guarantee. 5). Capital and operating cost plant. 6). Study of environmental impact and aplicable regulation and the approval process [3].

There are two objective of this research, first is to make the map of biomass potency in West Java province with the biomass from rice husk, corncob, coffe bean shell with addition the spesific local biomass potency from each region. The result is converted energy unit each year. Second, the empowerment models of biomass conversion substitute fossil energy is developed by integrating the six elements above.

The map of biomass potency is very useful especially for West Java distric goverment. The potency data of biomass in each district can be used to mobilize the society to use biomass as energy source. The empowerment model of biomass conversion is important for West Java goverment. The model can be used as reference in making the integrated programe to develop biomass as comercial energy substituting the energy fosil. The model also can be apllied in other province in Indonesia. The other benefit of this model, it can generate income for society and self sustained energy.

2. Methodology

This research was started by searching the data of the availability of biomass to serve as product of energy either liquid, solid, or gas. The object of this research is West Java in connection with the diverse of biomass species, fertile soil, many farms, plantations, and forests. These data was taken from the local government, technology information, and direct survey to the village. These data include type of biomass, the amount of dry biomass in ton/years, the existing and potential land, rainfall, the properties of plant growth, the available transportation in the observe village. The next was analyzing the data to obtain minimum three seed biomass species. These data was expressed in kg of dry biomass per years, and then converted to kJ/kg, including fuel consumption. The result is potential distribution of biomass in every county and city in West Java. The types of product to be made were determined by considering the economic revenue. There are three main revenue sources of biomass conversion pruducts, energy revenue, revenue from chemical product, and revenue from transport fuel.

The next step was continued with determining biomass conversion process plant equipment and cost. It started with boimass colection system, pre processing, biomass conversion, gas cleaning/product treatment, and product utilization. After all the data of operating cost was available, the financial analysis of the plant could begin. The viability of project was measured in terms of several items, there are: 1). Cost of electricity or product. 2). Internal revenue requirement and return on investment. 3). Net present value. 4). Benefit cost ratio. 5). Pay back periode. And 6). Life-cycle cost. The goal of a financial analysis is to determine one or many of above indices to judge the economic viability of the project.

The next was measuring the owned carrying capacity of technology and capacity building which can be seen in how far the research of biomass conversion was done in Indonesia, who was involved, and how the product was marketed by community or low or moderate class industry.
Because the objective of this research is to involve the community so that they can take a role as buyer and business people, the study of readiness of the community was conducted. The social engineering was prepared to make the community as biomass collector, and to make a small-scale industry. The socialization for community was done to change the mindset, from fossil fuel user to biomass fuel user. In this program, the local and central government hold a role not only as facilitators, but also as provider of policy tools and loan fund provider so that the program can work well.

3. Results and Discussions

3.1. Biomass Potential Map in West Java

The model of community empowerment in West Java as energy commodity started with study of biomass availability in West Java. Selection of biomass species, which would be converted into energy, was done depending on the total of availability of biomass in certain district, availability of area, biomass heat content, and community characteristic that included in the biomass supply process. Biomass availability was design by the college by including the local government. Data of biomass availability was provided by local government (Agriculture Department and Forest Department), mass media with direct reviews to the field. Product of research potency is map of biomass potency in West Java, shown in Figure 1., depending on potency priority which converted into energy units. Data of biomass potency is very needed for local government when developing the biomass as energy source.

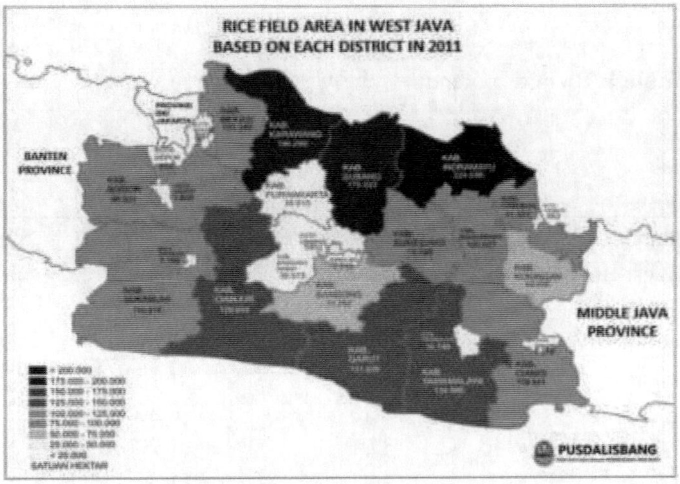

Figure 1. Planting area of rice paddy in West Java [5].

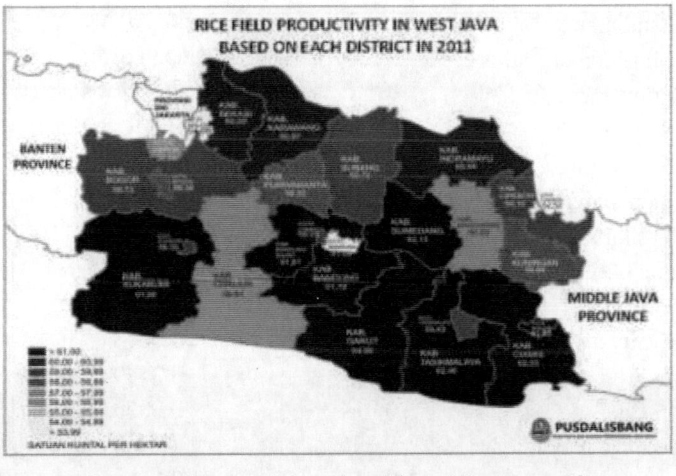

Figure 2. Productivity of paddy in West Java [6].

Figure 2 Shows map of productivity of paddy in West Java. The data from the map was used for calculating the availability of energy in West Java.

Table 1. Biomass potential from wasted paddy in West Java [6].

No	Area	Rice shell production (10^6 quintal/years)	Bio-briquette (10^6 quintal)	Energy (10^{11}) kkal
1	Dist. Bandung	1.75	0.702	2.59
2	Dist. West Bandung	0.87	0.350	1,29
3	Dist. Bekasi	2.43	0.974	3,60
4	Dist. Bogor	2.07	0.829	3,07
5	Dist. Ciamis	2.95	1.181	4,37
6	Dist. Cirebon	2.15	0.863	3,19
7	Dist. Cianjur	3.01	1.206	4,46
8	Dist. Garut	3.41	1.365	5,05
9	Dist. Indramayu	5.45	2.180	8,06
10	Dist. Karawang	4.78	1.914936	7,08
11	Dist. Kuningan	1.47	0.590	2,18
12	Dist. Majalengka	2.26	0.906	3,35
13	Dist. Purwakarta	0.81	0.325	1,20
14	Dist. Subang	4.27	1.710	6,32
15	Dist. Sukabumi	2.83	1.132	4,19
16	Dist. Sumedang	1.87	0.751	2,78
17	Dist. Tasikmalaya	3.37	1.348	4,98
18	Bandung City	0.028	0.0112	0.0415
19	Banjar City	0.17	0.0689	1.25
20	Bekasi City	0.019	0.0078	0.029
21	Bogor City	0.037	0.0149	0.055
22	Cimahi City	0.014	0.0056	0.020
23	Tasikmalaya City	0.336	0.134	0.497
24	Cirebon City	0.013	0.0556	0.0205
25	Depok City	0.020	0.0083	0.0307
26	Sukabumi City	0.088	0.035	0.131
			Total Energy	68.921

The process was done by calculating the potency of wetland area, production of waste in each district in ton per years, converted into total of bio-briquette which resulted in quintal. By knowing the heat content from rice shell, the potential energy can be produced in West Java.

Table 1. explains the potential of biomass from wasted paddy which converted into energy unit in each district in West Java. It assumed that the heat content of wasted paddy is 3700kkal/kg. The same thing was done for corn commodity. The data was collected from six district, District Majalengka, District Sumedang, District Garut, District Sukabumi, District Tasikmalaya, and District Bandung as shown in Table 2. It assumed that the heat content of corn is 4500kkal/kg. The total amount of energy resulted in wasted corn in West Java is 11.7 x 10^{10}kkal.

Table 2. Biomass potential from wasted corn in West Java.

No.	Area	Area (10^4 ha)	Wet waste (10^4 ton)	Dry waste (10^6 kg) estimated 30%	Bio-briquette (10^6 kg)	Energy (10^{10} kkal)
1	Majalengka	1.9	2.81	8.446	3.378	1.520
2	Sumedang	1.9	2.81	8.446	3.378	1.520
3	Dist. Garut	7.0	1.03	31.11	1.244	5.601
4	Sukabumi	1.51	2.24	6.748	2.699	1.214
5	Tasikmalaya	1.05	1.56	4.692	1.877	0.844
6	Dist. Bandung	1.31	1.94	5.842	2.336	1.051
					Total of Energy	11.75

In addition to potency of wasted paddy and corn, West Java has biomass potency from coffee skin. The data was obtained from Central Bureau of Statistic (BPS), and it was the total of data in West Java. According to the data, plantation area in West Java reaching 494,900 hectare, or 13.28% of total West Java area, which is 3.7 million hectare. The production of coffee in West Java has increased. In 2008, resulted 9,840 ton of coffee, as in 2014 reaches 12,943 ton [6]. In every 100kg processing of coffee can be produced 15.95kg (55%) coffee beans and 13.05kg (45%) skin dry logs. So the total of wasted skin of coffee reaches 5824.35ton. The next step to produce the bio-briquette is drying process of biomass. From the wet skin, it can be produced 22.31% of dry skin coffee. So the total is 1200.413ton. The heat content from coffee skin is 4427kkal/gram. The total energy of the wasted coffee skin is around 5.752 x 10^{12} kkal.

3.2. Biomass Empowerment Model as Energy Commodity

The biomass empowerment model as energy source in West Java was developed based on five processes which are raw material, pre-processing process, conversion mechanism, products, and end user as shown in Figure 3. There are five parts of process started by biomass selection which converted into energy. The results are the products that are ready to be marketed to the end users. There are specific treatments in every parts of process which involving the results of research, government roles, and public society and industry.

3.2.1. Raw Biomass

First step is collecting raw biomass. In this step the research of biomass potency in west Java had been done. Various types of plant / animal waste was researched and calculated in order to know both its quantity and its availability within one year. In this study we obtained various candidate of bio mass that would be collected. Furthermore, each type of biomass would be determined and converted to energy products which are divided into solid, liquid and gas and also would be calculated its heat content. The result was the datas of mass quantity and heat content of each raw biomass which had been collected before. In addition we also conducted a research which focused on the readiness of the society itself who involved in this process. This is necessary because the program is predicted to succeed if local society participate in. They were who have role in this stage are researchers from universities and research centers, a central statistics bureau which has the data that is needed, and the community that is around the location of biomass is located.

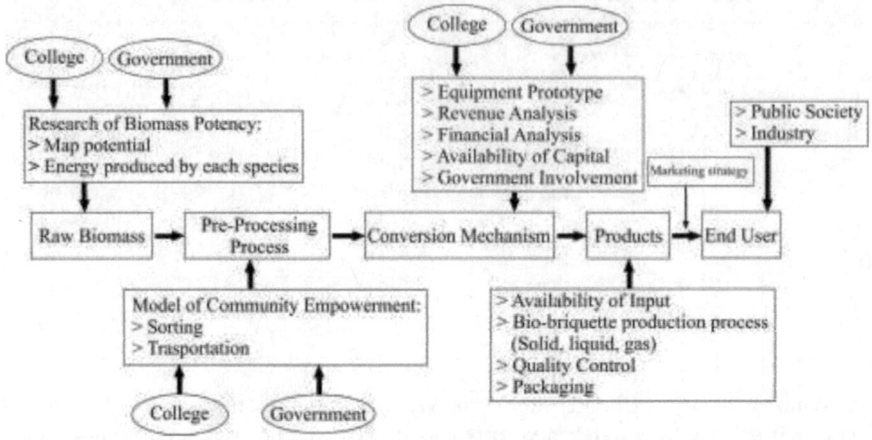

Figure 3. Diagram of biomass empowerment model as energy source.

3.2.2 Pre-processing Process
Pre-processing process was started by biomass collection which involving the community, classification process, sorting, handling, and drying the biomass. Pre-processing process done by local community in order to empower the community. It has two advantages, reduce the pre-processing cost and creates jobs for the community. To realize this process, it is needed the social engineering which designed by the college and full assistance from the government.

3.2.3 Conversion Mechanism
The biomass conversion process into energy other economical commodity is a part which has complex strategy because it involves many parties. The core of this process is setting the most appropriate technology in order to produce the product which has high economical value. There are two types conversion process. The first is thermochemical conversion including combustion, carbonization/torrefaction, pyrolysis, gasification, and liquefaction. The second is biochemical conversion including digestion (anaerobic and aerobic), fermentation, and enzymatic or acid hydrolysis. Researches from the colleges and researchers have a lot of results about this process. As an example the result of research in heat content of several biomass which is converted into bio-briquette, or the research in biogas and biodiesel from vegetables materials. The next is research in the form of prototype such as molding machine, carbonization machine, and small scale biodiesel reactor. The results of this research should be upgraded into the technology product which converted the biomass in large scale, so that can be used in this empowerment program. The involvement of private parties is also needed in order to manufacturing and procurement of tools. Each type of biomass conversion has own unique tools, although there are some product of biomass which used the equipment at the same time.

3.2.4 Cost Analysis
This process was started by calculating the procurement cost, procurement of biomass, human resources, and the availability of area. The owner of the capital is needed in this step. The next is finance analysis started by calculating the revenue from various biomass. The role of the government is so important because they should make appropriate regulations to support this program, such as subsidy, determining the selling price for the products, etc. The viability of any such project is measured in terms of following several terms: cost of electricity or product, internal revenue requirement and return on investment, net present value, benefit – cost ration, payback period, life cycle cost.

IOP Conf. Series: Earth and Environmental Science **65** (2017) 012016 doi:10.1088/1755-1315/65/1/012016

3.2.5 Production
There are three kinds of biomass product, solid fuel, liquid fuel, and gas fuel. The solid fuel consists of carbonization and non-carbonization, while the liquid fuel consists of bio-ethanol, bio-oil, etc. The gas fuel consists of methane, synthetic gas, and any other derivation products. The most important thing in this process is availability of the row material (low cost, easy to get, and sustainable), human resources, quality controls, selection of the appropriate technology, and market availability which will determine how much products will be produce.

3.2.6. End User
To get to the end users, this takes appropriate selling strategy, distribution channels, and community preparedness to for accepting the products through socializations and promotions. Biomass conversion products will be consumed by community and low and moderate industry.

4. Conclusions
The biomass potential map for biomass conversion (rice shell, corn, and coffee skin) into energy in West Java was successfully made. The energy total produced from wasted rice shell is 68.921 x 10^{11} kkal, 1.175 x 10^{11} kkal for wasted corn, and 57.52 x 10^{12} kkal for coffee skin. The total energy from all wasted biomass is 127.616 x 10^{11} kkal.

The integrated model of biomass empowerment in West Java was made by involving community, government, college and researchers, and industry. In this model, the biomass will be converted into solid fuel, liquid fuel, and gas fuel. If this model is realized, the potency of biomass in West Java will increase because the biomass used in the first conclusion it just involving three commodity of biomass.

References
[1] World Energy Council 2010 *2010-Survey Of Energy Resources* (London : World Energy Council) p 360

[2] IFP 2007 *Potential biomass mobilization for biofuel production worldwide In: Europe and in France,* Panorama 2007 (France : Innovation Energy Environment) <wwwifpfr/>

[3] Basu P 2013 *Biomass Gasification, Pyrolysis, and Torrefaction, Second Edition* (Amsterdam: Elsevier) pp 33 5

[4] Pusdalisbang 2011a *Luas Tanam Padi di Jawa Barat* Accessed October 2016 (http://pusdalisbangjabarprovgoid/pusdalisbang/images/datagrafikpeta/923LUAS%20TANAM%20PADI%20SAWAH%20DI%20JAWA%20BARAT%20MENURUT%20KABUPATEN%20DAN%20KOTA%20TAHUN%202011jpg)

[5] Pusdalisbang 2011b *Produktivitas Padi Sawah di Jawa Barat* Accessed October 2016 (http://pusdalisbangjabarprovgoid/pusdalisbang/images/datagrafikpeta/869PRODUKTIVITAS%20PADI%20SAWAH%20DI%20JAWA%20BARAT%20MENURUT%20KABUPATEN%20DAN%20KOTA%20TAHUN%202011jpg)

[6] Arief 2016 Luas kebun kopi Jawa Barat meningkat Accessed October 2016 (http://nasionalrepublika coid/berita/nasional/pemprov-jabar/16/01/03/o0cro3359-luas-kebun-kopi-di-jabar-terus-meningkat)

Enhancing the growth and yield of Ramie (*Boehmeria nivea* L.) by ramie biomass waste in liquid form and gibberellic acid

C. Suherman[1], A. Nuraini[1], A.P. Wulandari[2] and M. Kadapi[1]

[1]Faculty of Agriculture, Padjadjaran University, Bandung, Indonesia
[2]Faculty of Mathematics and Natural Sciences, Padjadjaran University, Bandung, Indonesia

Email : cucu.suherman@unpad.ac.id

Abstract. Ramie (Boehmeria nivea L.) is one of the most important sources of natural fibre, a sustainable biomass. The growth and yield of ramie are affected by mineral nutrients. In the present study, we usedfertilizers from waste of ramie biomass in liquid form (liquid organic fertilizer, LOF) and the other treatment is by gibberellic acid (GA3). This study was to obtain the effect of treatments on enhance the growth and yield of ramie. Hence, we measure the character that related to the important parameter for biomass product of ramie. Such plant height, stem diameter, dry plant weight, and ramie fresh stem weight of ramie clone Pujon 13. This research was conducted from January 2016 to March 2016 at Research Field Ciparanje, Faculty of Agriculture, Padjadjaran University, Jatinangor, Sumedang, West Java with an altitude of about ± 750 m above sea level. The type of Soil in this area is Inceptisolsoil order and thetype of rainfall according to Schmidt and Fergusson Classification is C type. The experiment used Randomized Block Design (RBD) which consisted of eight treatments (GA and LOF) and four replications. The concentration of GA from 0, 50, 100 and 150 ppm and for concentration of LOF is 40 mlL-1. We suggested the treatment of GA 150 ppm with 40 mlL-1 LOF was the best treatment on enhancing plant height and stem fresh weight of ramie clone Pujon 13.

1. Introduction

A sustainable biomass could be produced through agricultural product due to the agricultural crops is the cellulose resources. This could be a substitute material to produce the product based on petroleum resources to decrease the fossil fuel which the fossil fuel is non-renewable resources. The product from biomass is used in various field such textile, cosmetic, medical, automotive and electronic [1] and as bioethanol [2].

One of the important crops could be used in various field is ramie (*Boehmeria nivea* L.), this crop is a herbaceous perennial plant that belongs to the family Urticaceae, Generally ramie is can adapt and grown most of the tropics and subtropics regions such in China and some other Asian Countries including the Philippines and India [3]. As a crop mineral content of soil and fertilization practices in cultivation of this crop is important to reach high yield. Due to the rapid growth of ramie could rapidly reduce soil nutrients for normal growth and sustainable yield, therefore the supply of nutrition to for

Content from this work may be used under the terms of the Creative Commons Attribution 3.0 licence. Any further distribution of this work must maintain attribution to the author(s) and the title of the work, journal citation and DOI.
Published under licence by IOP Publishing Ltd

this crop to soil is very important [4,5]. Besides, the plant growth regulator (PGR) such gibberellic acid (GA) is very important in affected yield and fibre quality in ramie [3].

Information concerning the effects of nutrient management, particularly from waste of ramie biomass and the plant regulator (PGR) on the morphological traits and yield is nevertheless limited. In the present study, we accordingly aimed to assess the effects of the single or the combination use of liquid organic fertilizer (LOF) from ramie waste and GA3on the growth and yield of ramie.

2. Materials and Methods

This research was conducted from January 2016 to March 2016 at Research Field Ciparanje, Faculty of Agriculture, University of Padjadjaran, Jatinangor, Sumedang, West Java with an altitude of about ± 750 m above sea level. The type of Soil in this area is Inceptisol soil order and the type of rainfall according to Schmidt and Fergusson Classification is C type. The experiment used Randomized Block Design (RBD) which consisted of eight treatments (GA and LOF) and four replications. The concentration of GA from 0, 50, 100and 150 ppm and for concentration of LOF is 0 and 40 mlL-1. In each treatment consisting 3 plant of ramie of Pujon 13 clone, hence we used 96 plants in total. The treatments in this study are A = GA 0 ppm+ 0 mlL^{-1}LOF, B = GA 50 ppm + 0 mlL^{-1}LOF, C = GA 100 ppm + 0 mlL^{-1}LOF, D = GA150 ppm + 0 mlL^{-1}LOF, E = GA 0 ppm + 40 ml L^{-1}LOF, F = GA 50 ppm + 40 ml L^{-1}LOF, G = GA100ppm + 40 ml L^{-1}LOF, H = GA150 ppm + 40 ml L^{-1}LOF.

3. Results

According to the result of the analyzed of LOF (Table 1), the characteristic of LOF is alkali and contains the essential nutrients such a few nitrogen (N), phosphate (P) and potassium (K), (0,01;0.10;0,41 %). The plant height of ramie was affected by combination LOF and GA3, particularly in H treatment (GA150 ppm+ 40 ml L^{-1}LOF), in which the plant height of rami Pujon 13 clone showed the highest in 6, 10 and 12 week after sowing (Table 1). However, the statistically result showed that the diameter of stem was not different in all treatments. Although, it seems the value of stem diameter is increase in several treatments (Table 2).

In this study showed that increasing the fresh weight of ramie could be affected by GA3 as single treatment or combination of GA3 at the level D (150 ppm) and LOF at the level H (GA 150 ppm+ 40 ml L^{-1}) as seen at Table 3.

Table 1. Content and characteristics of ramie liquid of fertilizer (LoF)

No	Parameter	unit	value
1	pH	-	8,71
2	C-Organik	%	0,67
3	N-Total	%	0,01
4	P_2O_5	%	0,10
5	K_2O	%	0,41

Note: the liquid of fertilizer was analyzed by lab. Crop nutrition of Agriculture Faculty UNPAD.

Table 2. Plant height of rami Pujon 13 clone

Perlakuan	**Plant height (cm)**			
	6 week	8 week	10 week	12 week
A	49.63 c	79.21 a	115.92 b	125.92 c
B	68.49 ab	115.88 b	160.54 a	170.54 b
C	79.75 a	128.17 b	174.38 a	184.38 ab
D	83.75 a	132.08 b	180.21 a	190.79 ab
E	61.75 bc	94.13 a	135.42 b	136.08 c
F	82.88 a	120.00 b	160.21 a	171.63 ab
G	83.42 a	131.25 b	180.17 a	190.75 ab
H	85.67 a	132.38 b	180.83 a	201.13 a

Note : The numbers followed by the same letter at the same treatment were not significantly different at the Duncan test significance level of 5%

Table 3. Stem diameter of ramie Pujon 13 clone

Treatment	Stem diameter (mm)			
	6 week	8 week	10 week	12 week
A	22.15 a	27.95 a	33.88 a	37.68 a
B	24.33 a	30.78 a	35.98 a	40.63 a
C	25.35 a	32.23 a	37.58 a	42.33 a
D	25.40 a	32.63 a	38.15 a	43.00 a
E	22.55 a	28.25 a	33.63 a	37.88 a
F	25.13 a	31.53 a	37.55 a	42.25 a
G	26.03 a	32.85 a	38.48 a	42.68 a
H	26.95 a	33.58 a	47.58 a	43.35 a

Note: The numbers followed by the same letter at the same treatment were not significantly different at the Duncan test significance level of 5%

Table 4. Fresh weight of stem and Dry weight per plant of ramie Pujon 13 clone

Treatment	Stem fresh weight (g)	Plant dry weight (g)
A	270.00 b	225.0 a
B	323.75 b	226.3 a
C	278.75 b	192.7 a
D	330.00 ab	182.9 a
E	271.25 b	225.5 a
F	277.50 b	194.4 a
G	286.25 b	181.1 a
H	390.00 a	260.4 a

Note : The numbers followed by the same letter at the same treatment were not significantly different at the Duncan test significance level of 5%

4. Discussion

In our study, although LOF of ramie contains few amounts of essentials nutrient and low level of alkali suggests that the combination of treatment is effective to increase the plant height in ramie. According to Salisbury and Ross, the physiological effects typical application results GA3 is the lengthening of the stem, as a result of their activities in internodes cambium, so plants that are treated to be higher than normal plants [6]. Results of research conducted by Nasaruddin et al., showed that the liquid organic fertilizer concentration of 15 ml/L of water and 30 ml/L of water, increased height of cocoa seedlings [7]. However, in this study showed that the increasing of plant height was not accompanied with the widening of stem. This result suggests that each treatment and combination of treatment is not affect to stem diameter (Table 2). Our result was corresponded well with Dantas et al., in which their study showed the combination and the single effect of gibberellin and bio-stimulant to the growth of tamarind was affect to the stem elongation but not affect to the stem diameter [8]. In the level 150 ppm for GA3 is the best treatment in increasing the stem fresh weight. However, the increasing fresh weight at this level was no cause the increasing of plant dry weight.

Gibberellic acid (GA) is very important in affected yield and fibre quality in ramie [3]. According Mudyantini, GA can increase the amount of phloem, cellulose and lignin [9]. Cellulose and lignin is a determining factor for the quality of Ramie fiber. The results of the study Supriyadi et al., showed that the liquid organic fertilizer with a concentration of 150 ml/l of water, increased plant height, stem diameter, number of leaves, fresh weight of shoots and dry weight of shoots in nursery plants Jabon Red {Anthocephalus macrophyllus (Roxb.)} [10]. As well as the results of research conducted by Maruapey on the sugar cane plant cuttings, gibberellin with a concentration of 75 ppm can increase sugar cane plant height, leaf number, the number of segments of plants, stem diameter, number of roots of plants, and weight of cane biomass [11].

5. Conclussions

According to the result we conclude that the GA3at the level 150 ppm could increase the fresh weight of ramie and the combination of GA3 150 ppm and 40 ml.L-1 could increase the plant height. However, the stem diameter and dry weight per plant were not affected by single treatment or combination treatment.

The increasing of plant height will affect to the how long ramie fibers produced. We suggested the treatment of GA 150 ppm with 40 mlL-1 LOF was the best treatment on enhancing plant height and stem fresh weight of ramie clone Pujon 13.

References

[1] Siti A M, Abdul K H P S, Asniza M, Suhaily S S, Nur A A S and Jawaid M 2014 Agricultural biomass raw materials: the current state and future potentialities In Hakeem K R, Jawaid M and Rashid U ed. *Biomass and Bioenergy* (New York: Springer)

[2] Goh C S I, Tan K T, Lee K T, Bhatia S. 2010 Bio-ethanol from lignocellulose: Status, perspectives and challenges in Malaysia *Bioresour Technol.* **101** 34-41

[3] Liu F, X Liang, N Zhang, Y Huang and S Zhang 2001 Effect of growth regulators on yield and fibre quality in ramie (Boemherianivea L. Gaud.) China grass *Field Crops Res.* **69** 41–46

[4] Subandi M 2012 The effect of fertilizers on the growth and the yield of ramie (*Boehmeria nivea* L. Gaud) *Asian J. Agric. Rural Dev.* **2** 126–135.

[5] Huang C, Wei G, Luo Z, Xu J, Zhao S, Wang L, Jie Y 2014 Effects of nitrogen on ramie (Boehmeria nivea) hybrid and its parents grown under field conditions *J. Agric. Sci.* **6** 230–243.

[6] Salisbury F B, and Ross C W 1995 Plant Physiology Bandung: Institut Teknologi Bandung) **1**

[7] Nasaruddin dan Rosmawati 2011 Pengaruh pupuk organik cair (POC) hasil fermentasi daun gamal, batang pisang dan sabut kelapa terhadap pertumbuhan bibit kakao (http://isjd.pdii.lipi .go.id/index.php/) March 20th 2016.

[8] Dantas A C V L, Queiroz J M de O, Vieira E L, and Almeida V de O 2012 Effect of gibberellic acid and the biostimulant stimulate® on the initial growth of tamarind *Revista Brasileira de Fruticultura* **34** 8-14

[9] Mudyantini W 2008 Pertumbuhan kandungan selulosa, dan lignin pada rami (*boehmeria nivea* l. gaudich) dengan pemberian asam giberelat (GA) (http://biodiversitas.mipa.uns.ac.id/) September 10th 2015

[10] Supriadi D A and Setyorini D 2001 Bahan baku mutu pupuk organik (http://syekhfanismd.lecture.ub.ac.id/) September 21st 2015

[11] Marupaey A 2013 Efek berbagai konsentrasi zat pengatur tumbuh GA_3 terhadap dinamika pertumbuhan setek tebu (http://www. jurnal.lipi.go.id/data/1330473786/data/1377780020.doc) March 20th 2016

Equilibrium and kinetic modelling of cadmium (II) biosorption by Dried Biomass *Aphanothece* sp. from aqueous phase

Awalina [1,2], A Harimawan [1], GS Haryani [2], T Setiadi [1]

[1] Department of Chemical Engineering- Faculty of Industrial Technology, Bandung Institute of Technology, Jln. Ganesha No.10, Bandung, Indonesia
[2] Research Center for Limnology-The Indonesian Institute of Sciences, LIPI Cibinong Science Center, Cibinong-Bogor, Indonesia

E-mail : awalina@gmail.com

Abstract. The Biosorption of cadmium (II) ions on dried biomass of *Aphanothece sp.* which previously grown in a photobioreactor system with atmospheric carbon dioxide fed input, was studied in a batch system with respect to initial pH, biomass concentration, contact time, and temperature. The biomass exhibited the highest cadmium (II) uptake capacity at 30ºC, initial pH of 8.0±0.2 in 60 minute and initial cadmium (II) ion concentration of 7.76 mg/L. Maximum biosorption capacities were 16.47 mg/g, 54.95 mg/g and 119.05 mg/g at range of initial cadmium (II) 0.96-3.63 mg/L, 1.99-8.10 mg/L and 6.48-54.38 mg/L, respectively. Uptake kinetics follows the pseudo-second order model while equilibrium is best described by Langmuir isotherm model. Isotherms have been used to determine thermodynamic parameter process (free energy change, enthalpy change and entropy change). FTIR analysis of microalgae biomass revealed the presence of amino acids, carboxyl, hydroxyl, sulfhydryl and carbonyl groups, which are responsible for biosorption of metal ions. During repeated sorption/desorption cycles, the ratio of Cd (II) desorption to biosorption decreased from 81% (at first cycle) to only 27% (at the third cycle). Nevertheless, due to its higher biosorption capability than other adsorbent, *Aphanothece sp.* appears to be a good biosorbent for removing metal Cd (II) ions from aqueous phase.

1. Introduction

The existence of heavy metals in water and waste water has increasing due to the industrial waste water in the water bodies. Cadmium (Cd) is one of big toxic three heavy metals together with mercury (Hg) and lead (Pb) posing the greatest hazard to human health. Once Cd enters into aquatic ecosystem it can be accumulated through the trophic chain and produce toxic effect and *terratogenic* changes in aquatic ecosystem components including of human as top consumer in food chain [1]. Mainly, source of Cd which release into aquatic environment through waste streams come from process of electroplating, smelting, and some manufacturing (plastic, alloy, pigment, battery, mining, refining). In the other hand there are some strict regulation in term of limit of Cd content in wastewater and drinking water, which must be about 0.100 and 0.05 mg/L. To address that requirement, there are some

Content from this work may be used under the terms of the Creative Commons Attribution 3.0 licence. Any further distribution of this work must maintain attribution to the author(s) and the title of the work, journal citation and DOI.
Published under licence by IOP Publishing Ltd

conventional technology for Cd removal from wastewater, for example chemical precipitation, chemical oxidation or reduction, evaporation, adsorption and ion exchange has been developed. But, those processes seem ineffective or obviously expensive, moreover if concentration of metal in waste water occurs in order 0.1 to 100 mg/L. Therefore, it is necessary to develop more economically feasible technique to remove this metal from waste water [2].

Various biomass have been investigated for Cd uptake including of yeast, fungi, bacteria, agricultural waste, macro algae, microalgae and others. Recently, microalgae applications in waste water treatment for removing heavy metals (including of Cd) has been exploited. It has been known that microalgae have outstanding accumulation capacity of heavy metals even at low concentration. Particularly if microalgae originated and isolated from polluted waters. By using their biomass, it will allow to facilitate a development of inexpensive biosorbent for alternative metals removal technology. There are numerous potentials microalgae species with different capability to bind heavy metals ion, which each of capability depends on type of microalgae, type of heavy metals and physic-chemical conditions that may give significant effect on biosorption process. Therefore it is necessary to investigate a more detail regarding to specific microalgae capability to bind heavy metal in its biomass and many factors which affected on its process [2,3].

The aim of present study was to assess Cd biosorption Cd of wild strain freshwater microalgae *Aphanothece sp.* that had been previously isolated from a heavy polluted urban lake namely Situ Rawa Kalong–Depok, and then cultivated, harvested and processed into a biosorbent in laboratory of Bioprocess engineering-Dept. of Chemical engineering-ITB. Assessment involves of Fourrier Transform Infra-Red (FT-IR) spectrum characters; the influence of some operational condition parameter (pH, initial Cd concentration, biosorbent concentration, contact time, and temperature of solution) against biosorption efficiency. Biosorption uptake also has been quantified by means of two sorption models: Langmuir and Freundlich as well as two kinetic models: pseudo first order and pseudo second order. Furthermore, the effect of 1 M HCl as desorbing eluent on biosorption-desorption capacities and comparison due to its biosorption capability with other biosorbent were also considered.

2. Material and methods

2.1. Biosorbent preparation

Aphanothece sp was isolated from urban hyper eutrophic lake namely Situ Rawa Kalong-Depok. According to [3], this microalgae is belong to unicellular blue green algae (Cyanobacter), may form colonies with oval-shaped or oblong cells clustered densely and evenly within firm and abundant mucilage, although the mucous sheaths of individual cells are indistinct. The cells have granules but lack vacuoles. Its cells size are 1-4μm broad and 2-8 μm long. Cultivation in laboratory environment was conducted in a photobioreactor system which consisted of three bioreactors with each of them was 3 L in volume, illuminated with ~5700 to 6000 lux of daylight lamps, temperature range was 28-30 ºC and continuously aerated with 3L/minute of compressed air with atmospheric carbon dioxide input for 14 days. Harvesting was conducted by centrifugation in 6000 round per minutes (rpm) for 15 minutes in 25ºC resulting blue green wet pellets, which then dried in oven 60ºC for 7x24 hours. Dried biomass then crushed and sieved into particle size of 354 μm, stored in polyethilen bottle until biosorption experiments. Elemental analysis of % Carbon (C), Hydrogen (H), Nitrogen (N) and Sulfur (S) was conducted by using Labconco CHNS Analyzer. According to the result *Aphanothece sp* mainly involves 42.53%C, 6.43%H, 7.71%N and 0.51%S. Moisture content was 93.16-99.72%. The existence of functional groups which responsible for metal biosorption identification was confirmed by Fourier Transform Infra-Red (FTIR) analysis by using FTIR Prestige 21 Shimadzu (which recorded at 500 to 4500 cm^{-1}) prior to biosorption test.

2.2. Preparation and analysis of Cd solution

Cd stock solution (1000 mg/L) for biosorption was prepared in demineralized water with analytical grade salt of $CdCl_2.2.5H_2O$. The stock solution was diluted with demineralized water in order to prepare solutions of the desired concentrations. Biosorption studies were performed in 100 mL clear capped bottles. Pre-weighed dried biomass was added to each bottles and constantly agitated at 120 rpm at 30ºC, until the equilibrium was reached. At the end of biosorption, sample was filtered and collected in poly ethylene bottles. Cd content in filtrate was determined by an atomic absorption spectrophotometer model A-7000 (Shimadzu, Japan) at wavelength 226.502 nm. The absorption capacities were obtained by mass balance calculation (equation 1)

$$q_t = (C_0 - C_t)\frac{V}{w} \tag{1}$$

Where C_0 and C_t are the initial and concentration of solution at time t (mg/L), q_t is concentration of Cd on biosorbent (mg/g), V is the volume of Cd solution (L) and W is the mass of the biosorbent used (g). The efficiency of Cd biosorption was calculated as according to equation (2)

$$Biosorption\ (\%) = \left(\frac{C_0 - C_f}{C_0}\right) \times 100 \tag{2}$$

Where C_0 and C_f are the initial and final metal concentrations.
All the experiments were performed in a batch set up taking two replicates and average values were reported. Standard deviations were found to be within ±1.3%.

2.3. Biosorption studies

The biosorption of Cd (II) ions on dried microalgae biomass of *Aphanothece sp* was investigated in batch biosorption- equilibrium experiments. The effect of solution pH, biomass concentration, contact time and temperature on biosorption efficiency and initial Cd concentration on biosorption rate and capacity were studied according to [5,6]. The effect of pH was studied using a pH meter (Ohaus starter 3000, USA) in the range 3.0 to 11.0 and the desired pH of the solution was maintained by adding buffer solutions which respectively pH 3 (buffer KH-phtalate and 0.1 M HCl); pH 5 (buffer KH-phtalate and 0.1 M NaOH); pH 7 and pH 8 (buffer KH_2PO_4 and 0.1 M NaOH); pH 9 (buffer Na-tetraborate and 0.1 M HCl); and pH 11 (buffer Na-bicarbonate and 0.1 M NaOH). To determine the effect of biosorbent concentration, different biosorbent concentration 0.1 to 2.0 g/L were varied and suspended in Cd (II) solution of 7.9 mg/L. The sorption procedure was the same as described earlier.

To obtain biosorption isotherms and kinetics, the biosorbent (0.1 g/L) was suspended in Cd solutions (initial concentration range 0.96 to 53.38 mg/L), carried out in 30ºC. While thermodynamic study performed in 27, 30, 32, 37, 42, 47 and 50 ºC. The time required for reaching equilibrium condition was estimated by drawing samples at regular interval of time until equilibrium was reached.

2.4. Successive biosorption-desorption cycles

In order to determine the reusability of the biosorbent, consecutive biosorption-desorption cycles were repeated three times. HCl 1 M was evaluated for its capacity to desorb Cd (II) from metal loaded biosorbent. The biomass Cd loaded was placed in the HCl 1 M and agitated in shaker with 120 rpm agitation for 60 minutes at 30ºC. After each of biosorption-desorption cycle, the biomass was washed with demineralized water and reconditioned for biosorption in the succeeding cycle.

3. Result and discussion

3.1. FTIR spectrum of dried biomass of Aphanothece sp.

The transmittance spectrum of native or metal unloaded dried biomass of *Aphanothece sp.* with initial concentration of Cd 7.9 mg/L is shown in Figure 1. As seen on Figure 1, the broad and strong band at 3388.93 cm^{-1} is attributed to overlapping of -OH and N-H stretching vibration. While the band at 2926.01 cm^{-1} is assignment to the –CH stretch. The band peaks at 1651.07 and 1450.47 cm^{-1} recognized to asymmetric and symmetric stretching vibration of C=O groups. The bands 1242.16 and 1074 cm^{-1} assign to stretching of C-O groups on the biomass surface. Some bands in the fingerprint region attributed to sulfur esters S-OR and disulfide S-S. This FTIR spectra indicate the presence of ionizable functional groups (such like carboxyl, amino, amide and hydroxyl) which capable to interact with protons or ions. Those obtained data also suggested the presence of functional groups on the microalga cell surfaces and also on the biosorption mechanism which is dependent on functional groups especially carboxyl, hydroxyl, amines and amides.

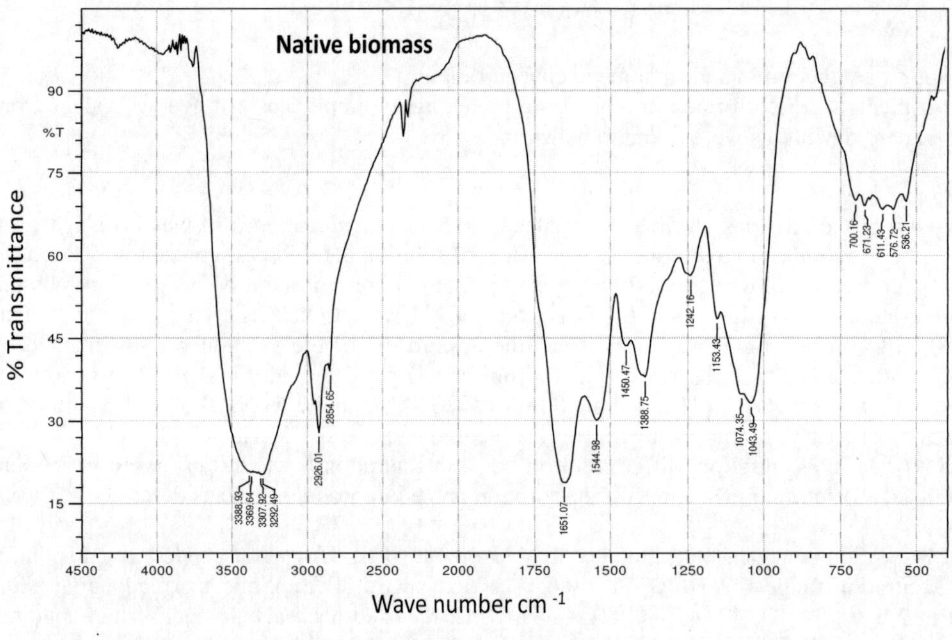

Figure 1. Fourier Transformed Infra-Red spectra of natural dried *Aphanothece sp* biomass.

3.2. *Biosorption of Cd (II) ion*

3.2.1. Influence of biosorbent concentration

The effect of biosorbent concentration on Cd (II) ions was examined using different biomass concentration in the range 0.1-2.0 g/L as presented on Figure 3. Results show that the biosorption efficiency is highly dependent on the decrease in biosorbent concentration. The biosorption efficiency of Cd ion in this study was attained at about biomass concentration 0.1 g/L and it sharply declined over higher biosorbent concentrations. This pattern could be explained as a consequence of partial aggregation of biomass at higher concentration, which results in a decrease in effective surface area for the biosorption. Similar trend also reported by [7] for the adsorption of Cd by dried biomass of red seaweed *Hypnea valentinae*. Therefore, the optimum biomass concentration was selected as 0.1 g/L for further experiments.

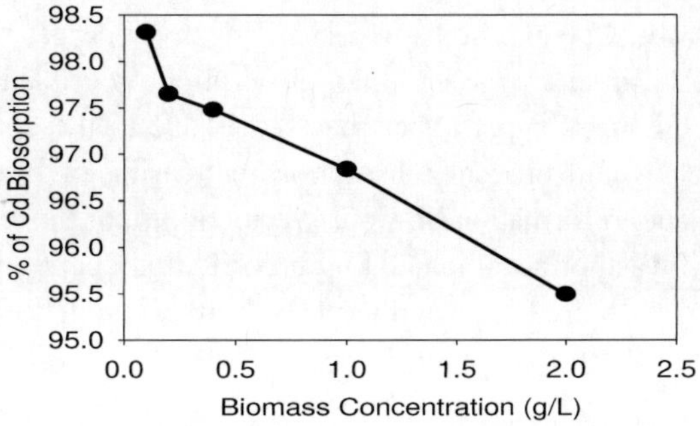

Figure 2. Effect of biosorbent concentration on the biosorption of Cd (II) ions on dried algal biomass *Aphanothece sp* from aqueous solutions: temperature 30ºC, initial Cd (II) concentration 7.9 mg/L, pH 8 (average values of two tests, error < 1.5%)

3.2.2. Effect of contact time

One of important parameter which affecting biosorption process is contact time. Figure 2 shows the effect of contact time on Cd biosorption on dried biomass of *Aphanothece sp* from aqueous solutions (at six different initial Cd concentration). The plot indicates that the remaining Cd ion becomes unchanged after 60 minutes. This represents the equilibrium time at which an equilibrium Cd ion is presumed to have been attained. These obtained data were further used to evaluate the kinetics of the biosorption process.

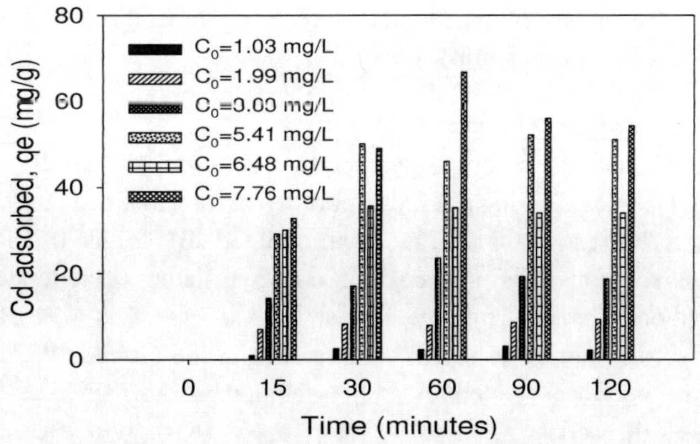

Figure 3. Effect of contact time on the biosorption of Cd (II) ions on dried algal biomass *Aphanothece sp* from aqueous solutions: temperature 30ºC, biomass concentration 0.1 g/L (average values of two tests, error < 1.5%)

3.2.3. Effect of pH

Based on earlier biosorption studies, metal biosorption is strongly affected by pH of solution. So, Cd (II) biosorption in this study is also a function of pH solution. As seen on Figure 4, biosorption efficiency were highest in pH 8 and steeply declined at pH 9. The pH value in solution influences both of cell surface metal binding sites and metal chemistry in water. At low pH, cell wall ligands were closely associated with hydronium ions (H_3O^-) and restricted the approach of metal cations as a results of the repulsive force. As the pH increased, more ligands such as carboxyl, imidazole, phosphate, hydroxyl and amino groups would be exposed and carried negative charges with subsequent attraction of metallic ions which have positive charge and then biosorption proceeded on to biosorbent surface [8]. The decrease of metal binding for pH upper than 8.0 is due to the complexation of Cd (II) by hydroxyl (OH^-) group which would prevent biosorption. Increasing of biosorption efficiency seem occurred at pH 11, but it is due to formation of more precipitation not biosorption. Since the highest biosorption efficiency which was not accompanied by precipitation occurred at pH 8, therefore all of the next biosorption experiments were conducted at pH 8. Further, modelling development was based on the experiment results which performed at pH 8.

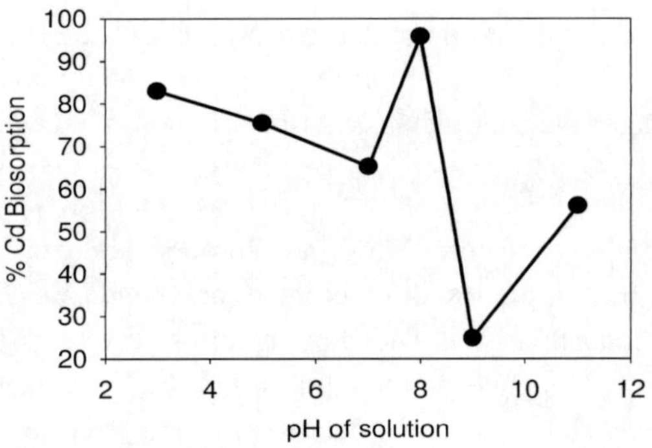

Figure 4. Effect of pH on the reduction of Cd (II) ions by dried algal biomass of *Aphanothece sp* from aqueous solutions: temperature 30ºC, initial Cd (II) concentration 1.3 mg/L (average values of two tests, error < 1.1%)

3.2.4. Effect of temperature

Figure 5 show the effect of temperature solution on Cd (II) biosorption by dried algal biomass *Aphanothece sp*. The highest biosorption efficiency attained at 30ºC while the lowest efficiency found at 47ºC. This result indicated biosorption of Cd (II) ions on dried algal biomass *Aphanothece sp* from aqueous solutions was proceed in exothermic nature. Temperature rising then lead to declining of biosorption efficiency (from 95.32% at 30ºC to 79.46% at 47ºC).This finding occurred may be caused by either the damage of active binding sites in the biomass [9] or the tendency to desorb metal ions from the interface of biomass to the solution as the increasing of temperature implemented in the process [10].

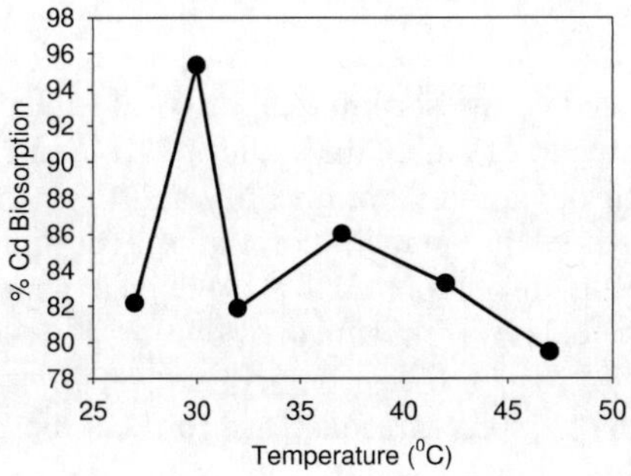

Figure 5. Effect of temperature on the reduction of Cd (II) ions by dried algal biomass *Aphanothece sp* from aqueous solutions: pH 8, initial Cd (II) concentration 7.9 mg/L and biomass concentration 0.1 g/L, contact time 60 minute (average values of two tests, error < 1.5%)

3.3. Isotherms modelling

Langmuir and Freundlich isotherms were used to quantitatively describe metal biosorption by the tested microalgae biomass. The Langmuir model is expressed by the linearized equation

$$\frac{C_e}{q_e} = \frac{C_e}{q_m} + \frac{1}{K_L q_m} \quad (3)$$

where q_e is the equilibrium metal ion concentration on the biosorbent (mg/g), C_e is the equilibrium metal ion concentration in the solution (mg/L), q_m is the monolayer biosorption capacity on biosorbent (mg/g), and K_L is the Langmuir biosorption constant (L/mg) relating to free energy of biosorption. The plots of $\frac{C_e}{q_e}$ versus C_e were drawn as shown in Figure 6 for three different initial concentration of Cd to calculate these constants whereas Table 1 present calculation results from the linearized equation form of the isotherms at all initial concentration over the whole concentration range studied and the extremely high correlation coefficients. These values of the correlation coefficients strongly support the fact that cadmium–dried biomass biosorption data closely follow the Langmuir model of sorption.

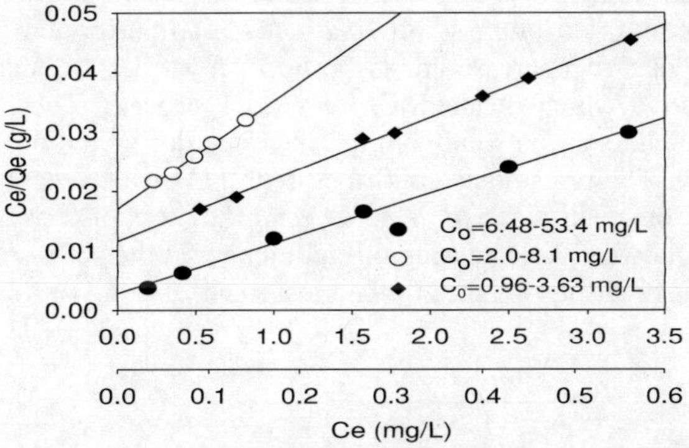

Figure 6. Langmuir isotherm equation fitting in linear form of Cd (II) on dried *Aphanothece sp.*

The linearized Freundlich model is represented by the equation:

$$\ln q_e = \ln K_F + \frac{1}{n}\ln C_e \quad (4)$$

where K_F and n are constant related to the sorption capacity and intensity. The plots of $\ln q_e$ versus $\ln C_e$ (figure not shown) were drawn to calculate the values of KF and 1/n which are given in Table 1. It was found that the plots exhibit deviation from linearity and the correlations coefficients indicate, the data are not well correlated to Freundlich correlation coefficients compared to Langmuir correlation coefficients. This result also reported by [9] which applied dried biomass of marine microalgae *Ulva lactuta* to remove Cd (II) from aqueous solution.

Table 1. Langmuir and Freundlich isotherm constants for the biosorption of Cd on dried *Aphanothece sp* biomass at different initial concentration of Cd in solution at pH 8, biomass concentration 0.1 g/L and temperature 30ºC with standard deviation ± 1.2 %

Range of Initial Cd	Langmuir constants			Freundlich constants		
conc. (mg/L)	q_{max} (mg/g)	K_L (L/mg)	R^2	K_F $(mg/g(L/g)^{1/n})$	n	R^2
0.96-3.63	16.47	5.23	0.998	17.07	2.190	0.949
1.99-8.10	54.95	1.06	0.993	27.030	1.520	0.759
6.48-53.38	119.05	3.00	0.997	101.93	1.760	0.683

3.4. Thermodynamic study

The free energy change (ΔG^0), enthalpy change (ΔH^0), and entropy change (ΔS^0) for this study were calculated using following equations:

$$\Delta G^0 = -\mathrm{RT}\ln K_d \quad (5)$$

Where R is universal gas constant (8.314 J/mol K), T is temperature (K) and K_d is $\frac{q_e}{C_e}$ represent the distribution coefficient (Rathinam et al., 2010). By considering the following equation, the ΔH^0 and ΔS^0 of biosorption were estimated from the slope and intercept of the plot ln K_d versus 1/T yields.

$$\ln K_d = \left(\frac{\Delta S^0}{R}\right) - \left(\frac{\Delta H^0}{RT}\right) \quad (6)$$

Table 2 show the calculation results from those calculations. As shown in Table 2, it can be said that biosorption of Cd (II) ions by dried biomass of *Aphanothece* sp performed in exothermic reaction since ΔH^0 was negative due to decrease in adsorption on successive increase in temperature. Meanwhile, ΔG^0 was negative which dictated a spontaneous process. The negative value of ΔS^0, reveals the decreased randomness at the solid-solution interface during the binding of Cd ions on the active site of the biosorbent. These results is confirmed with [11] which reported that biosorption of selenium by dried biomass of green algae *Maugeotia genuflexa* was also proceeded in spontaneous and exothermic reactions followed by decreasing of randomness at the solid-solution interface during the fixation of selenium ions on the active sites of the biosorbent.

Table 2. Thermodynamics parameters for the biosorption of Cd on dried *Aphanothece sp.* biomass at different temperature with standard deviation ± 1.0 %

Thermodynamics Parameter			
Temperature (K)	ΔG (kJ/mol)	ΔH (kJ/mol)	ΔS (kJ/mol K)
303	-12.80	-72.87	-0.20
310	-11.08		
315	-10.21		
320	-9.15		
323	-8.89		

3.5. *Kinetic modelling*

Conclusion on controlling mechanism of biosorption process commonly obtained from kinetic model fitting on experimental data. The high diversity of surface groups on algal cell wall such as carboxyl, phosphate, hydroxyl, amino, imidazole etc. suggested that there are many types of biosorbent-metal ions interactions. Two important kinetic models used in this study, namely the pseudo first order (equation 7) and pseudo second order (equation 8). Those linearized equations are written as follow

$$\ln(q_e - q_t) = \ln q_e - k_1 t \tag{7}$$

$$\frac{t}{q_t} = \frac{1}{k_2 q_e^2} + \left(\frac{1}{q_e}\right) t \tag{8}$$

Where q_t and q_e (mg/g) are the amounts of the metal ions adsorbed at t time and equilibrium, t is time (minute), k_1 is rate constant for pseudo first order (calculated from plotting $\ln(q_e - q_t)$against t) and k_2 is rate constant for pseudo second order (determined from plotting $\frac{t}{q_t}$ versus t). Those plotting results are presented on Table 3. It can be concluded from the R^2 and the similarity of q_e experiment to q_e calculated values, that the biosorption of Cd (II) ions on dried biomass of *Aphanothece sp* well followed the pseudo second order kinetic model. This pseudo second order kinetic model described kinetic behavior of biosorption process with chemical sorption being the controlling step [12].

Table 3. Biosorption rate constants q_e estimated and coefficient of correlation associated to the Lagergren Pseudo-first and second order for dried *Aphanothece sp* biomass at different initial Cd concentrations

Initial conc	$qe_{(exp)}$(mg/g)	Pseudo First Order rate constants			Pseudo Second Order rate constsnts		
(mg/L)		k_1 (min^{-1})	$qe_{(cal)}$(mg/g)	R^2	k_2 (g/mg.min)	$qe_{(cal)}$mg/g)	R^2
2.0	9.96	2.29E-02	9.45	0.969	5.40E-02	8.67	0.996
3.6	23.46	1.29E-02	23.57	0.949	1.91E-02	19.01	0.997
5.4	53.75	2.91E-01	42.84	0.962	4.95E-03	52.63	0.992
6.5	43.90	1.25E-02	42.84	0.952	6.84E-02	36.63	0.998
9.8	95.43	1.50E-02	88.2	0.863	5.38E-03	96.15	0.993
19.5	104.50	5.90E-02	129.32	0.945	6.36E-03	107.53	0.996
53.4	388.90	4.30E-02	320.86	0.937	7.18E-04	400.00	0.997

3.6. Comparison with other adsorbent

Table 4 compared maximum biosorption capacities obtained in this study with some other values reported by other researchers. The biosorption capacity for cadmium by using dried biomass of *Aphanothece sp* was higher in magnitude compared to other reported microalgae biomass as well as other types of biosorbent. As presented on Table 4, biosorption capacity of microalgae (*Chlorella vulgaris, Spirullina platensis, Oedogonium sp* and *Chlamydomonas reinhardtii*) were comparable with biosorption capacity of *Aphanothece sp.* Furthermore, biosorption capacity of *Aphanothece sp* compared to macroalge (*Gelidium sp*), peat and pine bark, is higher up to nine magnitudes. Meanwhile if compared to hazelnut shells (example of agricultural waste biomass), it was 19 times greater. According to [13] functional groups characters on surface of cell wall determines the extent of biosorbent affinity for metal ions bonding. Microalgae are more abundant with anionic functional groups than other type of biomass which presented in Table 4. Anionic functional groups such as carboxyl, phosphate, pyruvate, amide, extracellular polysaccharide, lipoproteins, etc are have great capability to bind metal ions by electrostatic forces and coordination bound (involving amino, carboxylic, sulfhydryl, phosphate, thiol and sulfate groups). Biosorption by dried biomass is better than precipitation since its ability to adjust of pH changes during sorption process. It also more advantageous than ion exchange and reverse osmosis in terms of its sensitivity to presence of suspended solids, organics, and other heavy metals. These finding demonstrate the eminence of dried biomass *Aphanothece sp* as biosorbent for removing Cd (II) ions from aqueous phase.

Table 4. Biosorption capacities for Cd (II) using different low cost biosorbent (at room temperature)

Biosorbent	Biosorption capacity (mg/g)	pH	References
Aphanothece sp	119.05	8.00	This study
Peat	22.5	5.00	[14]
Pine bark	28	7.50	[15]
Hazelnut shells	5.42	6.00	[16]
Chlorella vulgaris	85.3	4.00	[17]
Gelidium sp	18	5.30	[18]
Spirullina platensis	98.04	6.00	[19]
Chalmydomonas reinhardtii	42.60	6.00	[8]
Oedogonium sp.	88.20	5.00	[20]

3.7. *Biosorption-desorption cycle*

Biosorption-desorption experiments are useful in elucidation of the mechanism of sorption reaction and also to assess the regeneration capacity of the biosorbent for reuse in more economic manner. In this study, 1 M HCl desorbing solution was tested for Cd (II) desorption from the metal loaded biomass of dried Aphanothece sp (Table 5). The amount of desorption provides insight into the nature of biosorbent-Cd ions bonding and also on the ion exchange property of the biosorbent. Reusability of Aphanothece sp for Cd (II) removal from aqueous phase was three successive cycles of biosorption-desorption which carried out in batch system. Cadmium solution test was 7.2 mg/L was selected, since when concentration of metal below 100 mg/L, conventional technology to remove metal from aqueous phase will be inefficient. As seen, in three successive runs, the uptake capacity of biosorption was relatively unchanged. But in case of desorption, there was observed decreasing trend up to only 27% in third cycle from 81 % in first cycle. This result also confirmed [8] which reported declining trend in last desorption process around of 22% by using *Chlamydomonas reinhardtii* and acid desorbing eluent when removing Hg (II) ions from aqueous solution. Apparently, it caused by acids deteriorating effect on biomass and may dissolving certain polysaccharides which contain metal binding sites. It suggest HCl 1 M is too concentrated being used as desorbing eluent in this study.

Table 5. Successive cycle of biosorption and desorption of Cd (II) by dried *Aphanothece sp* biomass in batch system. Desorption by HCl 1 M, contact time 60 minutes. Data presented as means of two replicates with standard deviations 1.5 to 2.0 %.

	Uptake Capacity (mg/g)		
	Cycle 1	Cycle 2	Cycle 3
Biosorption (B)	59.93	63.51	58.23
Desorption (D)	48.71	44.77	15.62
Ratio (D/B)	0.81	0.70	0.27

4. Conclusion

The batch biosorption system in this study provides significant information regarding biosorption of cadmium on dried biomass of cyanobacteria *Aphanothece sp.* in terms of optimum pH and biomass concentration for maximum removal of Cd (II) ions from aqueous solution. This study indicated that dried biomass of *Aphanothece sp.* is an effective biosorbent for Cd (II) removal. The maximum Cd (II) biosorption capacity has been found to be 119. 05 mg/g of dry weight biomass at biosorbent concentration 0.1 g/L in 60 minute of contact time with initial Cd concentration 6.48-53.38 mg/L and optimum pH of 8.0. The Langmuir adsorption model and Pseudo second order kinetic models were fitted well to experimental data, it means that these adsorption and kinetics model are the best suited mathematical models for describing biosorption process of Cd (II) by dried biomass of *Aphanothece sp* from aqueous phase. HCl 1M as desorbing eluent is not affected biosorption capacity during 3 cycles of biosorption-desorption but it reduced uptake capacity of desorption process up to only 27% of capacity at third biosorption-desorption cycle. With the advantage of high metal biosorption capacity, the dried biomass of *Aphanothece sp.* has potential to be used as an efficient and economic biosorbent material for removing cadmium ions in aqueous phase.

Acknowledgements

Authors are thankful to Department of Chemical Engineering-FTI-ITB, Research Center for Limnology-LIPI, Research Center for Geotechnology-LIPI and Indonesian Ministry of Research and higher education for financial support and research facilities.

References

[1] Barwick M and Maher W 2003 Biotransference and biomagnification of selenium copper cadmium zinc arsenic and lead in a temperate seagrass ecosystem from Lake Macquarie Estuary NSW Australia *Marine Environmental Research* **56** 471-502

[2] Chojnacka K 2010 Biosorption and bioaccumulation – the prospects for practical applications *Environment International* **36** 299-307

[3] Brinza L, Dring M and Gavrilescu M 2007 Marine micro and macro algal species as biosorbents for heavy metals *Environmental Engineering and Management Journal* **6** 237-251

[4] Bellinger E and Sigee DC 2010 *Freshwater algae : identification and use as bioindicators* Chichester West Sussex UK JohnWiley and Sons Ltd

[5] Chojnacka K Chojnacki A and Górecka H 2005 Biosorption of Cr^{3+} Cd^{2+} and Cu^{2+} ions by blue–green algae *Spirulina sp*: kinetics equilibrium and the mechanism of the process *Chemosphere* **59** 75-84

[6] Cruz CC, Da Costa ACA, Henriques CA and Luna A S 2004 Kinetic modeling and equilibrium studies during cadmium biosorption by dead Sargassum sp biomass *Bioresource Technology* **91** 249-257

[7] Rathinam A, Maharshi B, Janardhanan SK, Jonnalagadda RR and Nair BU 2010 Biosorption of cadmium metal ion from simulated wastewaters using *Hypnea valentiae* biomass: A kinetic and thermodynamic study *Bioresource Technology* **101** 1466-1470

[8] Tüzün I, Bayramoğlu G, Yalçın E, Başaran G, Çelik G and Arıca MY 2005 Equilibrium and kinetic studies on biosorption of Hg(II) Cd(II) and Pb(II) ions onto microalgae *Chlamydomonas reinhardtii Journal of Environmental Management* **77** 85-92

[9] Sari A and Tuzen M 2008 Biosorption of Pb(II) and Cd(II) from Aqueous Solution using Green Alga (*Ulva Lactuta*) Biomass *Journal of Hazardous Materials* **152** 302-308

[10] Bulgariu D and Bulgariu L 2013 Sorption of Pb(II) onto a mixture of algae waste biomass and anion exchanger resin in a packed-bed column *Bioresource Technology* **129** 374-380

[11] Sari A, Uluozlü ÖD and Tüzen M 2011 Equilibrium thermodynamic and kinetic investigations on biosorption of arsenic from aqueous solution by algae (*Maugeotia genuflexa*) biomass *Chemical Engineering Journal* **167** 155-161

[12] Ho YS and *Mckay* G 2000 The Kinetics of Sorption of Divalent Metals onto *Sphagnum Moss Peat Water Res* **34** 735-742

[13] Kumar KS, Dahms H, Won E, Lee J, and Shin K 2015 Microalgae – A promising tool for heavy metal remediation *Ecotoxicology and Environmental Safety* **113** 329-352

[14] Gosset T, Transcart Jl and Thevenot Dr 1986 Batch metal removalby peat: kinetics and thermodynamics *Water Research* **20** 21-26

[15] Al-Asheeh S and Duvnjak Z 1998 Binary metal sorption by pine bark:study of equilibria and mechanisms *Separation Science and Technology* **33** 1303-1329

[16] Cimino G, Passerini A, and Toscano G 2000 Removal of toxic cations and Cr (VI) from aqueous solutions by hazelnut shell *Water Research* **34** 2955-2962

[17] Aksu Z 2001 Equilibrium and kinetic modelling of cadmium(II) biosorption by C vulgaris in a batch system: effect of temperature *Separation and Purification Technology* **21** 285-294

[18] Vilar VJP, Santos SCR, Martins RJE, Botelho CMS and Boaventura RAR 2008 Cadmium uptake by algal biomass in batch and continuous (CSTR and packed bed column) adsorbers *Biochemical Engineering Journal* **42** 276-289

[19] Rangsayatorn N, Upatham ES, Kruatrachue M, Pokethitiyook P and Lanza GR 2002 Phytoremediation potential of Spirulina (Arthrospira) platensis: biosorption and toxicity studies of cadmium *Environmental Pollution* **119** 45-53

[20] Gupta Vk and Rastogi A 2008 Equilibrium and kinetic modelling of cadmium(II) biosorption by nonliving algal biomass *Oedogonium sp* from aqueous phase *Journal of Hazardous Materials* **153** 759-766

Evaluation on community tree plantations as sustainable source for rural bioenergy in Indonesia

U J Siregar[1], B H Narendra[2], J Suryana [1], C A Siregar[2] and C Weston[3]

[1] Faculty of Forestry, Bogor Agricultural University, Indonesia
[2] Forest Research and Development Agency, Indonesia
[3] School of Ecosystem and Forest Sciences, University of Melbourne, Australia

Email : siregaruj@gmail.com

Abstract. Indonesia has forest plantation resources in rural areas far from the national electricity grid that have potential as feedstock for biomass based electricity generation. Although some fast growing tree plantations have been established for bioenergy, their sustainability has not been evaluated to date. This research aimed to evaluate the growth of several tree species, cultivated by rural communities in Jawa Island, for their sustainability as a source for bio-electricity. For each tree species the biomass was calculated from diameter and height measurements and an estimate made for potential electricity generation based on density of available biomass and calorific content. Species evaluated included *Acacia mangium*, *A. auriculiformis, A. crasicarpa*, *Anthocephalus cadamba*, *Calliandra calothirsus*, *Eucalyptus camaldulensis*, *Falcataria moluccana*, *Gmelina arborea*, *Leucaena leucochephala* and *Sesbania grandiflora*. Among these species *Falcataria moluccana* and *Anthocephalus cadamba* showed the best potential for bioenergy production, with up to 133.7 and 67.1 ton/ha biomass respectively, from which 160412 and 80481 Kwh of electricity respectively could be generated. Plantations of these species could potentially meet the estimated demand for biomass feedstock to produce bioenergy in many rural villages, suggesting that community plantations could sustainably provide much needed electricity.

1. Introduction

Indonesia is facing an electricity energy crisis with an 88.3% electrification ratio in 2015 [1], indicating that power plant development is unable to keep up with demand from a growing population. In the last few decades, the government and the private sector have built power plants that mainly use coal and fossil fuel. However in many remote areas, the need for electricity still cannot be fulfilled. Indonesia's millions of hectares of abandoned land could be planted with fast growing tree species that, with appropriate management, could serve as a feedstock for biomass-based electricity generated in remote areas. The use of biomass energy is very relevant to the Indonesian Government energy policy of development of renewable energy sources to assist in reducing fossil fuel energy use from 49% to 20% of total energy consumption.

Furthermore, the development of bioenergy plantations on abandoned land could reduce or eliminate the displacement of agricultural land for food production to biomass for energy production. Therefore fast growing tree plantations on land not suitable for agriculture would never sacrifice food security.

Content from this work may be used under the terms of the Creative Commons Attribution 3.0 licence. Any further distribution of this work must maintain attribution to the author(s) and the title of the work, journal citation and DOI.
Published under licence by IOP Publishing Ltd

When burned in a power plant, woody biomass produces less ash than the coal, saving on waste disposal [2]. Also, compared with fossil fuels, the use of biomass energy has lower CO_2 emissions, provides community income, and the residues can be used as fuel and fertilizer [3].

The case for biomass energy plantations in the tropics is promising, with several studies demonstrating their viability. For example, the study of [4] concluded that plantation trees are one of the largest sources of biomass in Thailand, with an annual energy potential of about 174–1600 PJ in 2010. This plantation biomass could supply up to 31.2% of estimated total energy consumption in Thailand at a biomass production cost of US$ 6–13.3 per ton, and with positive net present values for all the plantation options.

Some Indonesian tree species such as gamal, lamtoro, kihiyang, kaliandra, mete, red sengon, mindi, and gmelina are good sources of biomass energy. [5] using proxymate analysis according to the ASTM (*American Society for Testing Material)* standard, showed that stem of gamal, lamtoro, kihiyang, kaliandra, mete, and red sengon can be considered as good source of biomass energy, while mindi, gmelina had lower quality. [6] showed that tree species such as lamtoro, kihujan, turi, gamal, angsana, sengon, gmelina have sufficient caloric value to replace coal utilization in the cement industry, which made their development as a source of energy economically feasible.

Fast growing tree plantations are relatively low risk during cultivation and very popular among local communities to supply raw material for wood industries and provide a good economic return. The establishment of new areas of plantations is encouraged by the Indonesian Government due to their positive environmental impact, climate mitigation action as well as jobs provision [7]. Despite this rapid development most plantations are usually harvested for wood industry products and rarely as a source of bioenergy.

The development of fast growing tree species plantations as a source of bioenergy in Indonesia has not been well investigated, with most information on the potential for bioenergy from forest plantations available at experimental plot level. To assist with the further development of tree-based electricity generation, the objective of this study is to evaluate fast growing tree species in community forests as a potential source for bioelectricity.

2. Methodology

The study sites in West and East Jawa include a range community forests in lower (900 mm) to higher (over 4000 mm) annuall rainfall areas, each planted with fast growing tree species such as sengon, jabon and gmelina as cash crops either as agroforests or as monocultures. For each study site tree diameter at breast height (DBH), and tree height were measured and used with tree age to develop growth models and estimate biomass production. The theoretical electricity generation from the biomass (Table 1) was calculated using a formula developed by [8]. The opportunity for bioelectricity generation was then derived from the amount of land area available for planting with tree species often used as an energy source. The available lands included community forest and marginal land, and areas both in and out of designated forest.

Table 1. Calculation formula for conversion from biomass to electricity

Biomass : WP[a]	Calor (GJ/ton)[b]	Electricity (GWh)[b]	Efficiency 40%[c]
1.5: 1	1 ton WP = 19.8 GJ	1 GJ = 1 x 2.7^{-5} GWh	1 GJ = 1 x 1.1^{-4} GWh

Note : WP=wood pellet;
[a]Based on experiment;
[b][8];
[c][9]

Estimation of biomass from a plantation was done using allometric equations as shown in Table 2.

Table 2. Allometric equations for *A. mangium*, *P. falcataria* and other tree species

No	Tree Species	Allometric Equation	Source
1	*A. mangium*	$AGB = 0.199\ D^{2,148}$	[10]
2	*P. falcataria*	$AGB = 0.1126\ D^{2,3445}$	[11]
3	Other tree species	$AGB = \rho.V.BEF$ $V = 0.25\pi.(D/100)^2H.F$	[12]

Note: AGB =above ground biomass (kg);
D =diameter at breast height (cm);
ρ =specific weight (kg/m^3);
V =volume (m^3);
BEF =biomass expansion factor;
H =height (m);
F =tree form factor

3. Results and Discussions

3.1. Species utilized as bioenergy

The West Jawa (including Banten) sites in the Perum Perhutani area focused on the Government Forest Company districts of KPH Bandung Selatan, KPH Banten, KPH Garut, KPH Majalengka, KPH Tasikmalaya and KPH Sukabumi. District tree species, age and growth data are shown in Table 3.

Table 3. Tree stand characteristic for each species in West Jawa

District	Species	Age (yr)	Average height (m); Diameter (cm)	Planting space (m)	Tree growth (m3/ha/yr)	Biomass for energy (kg/ha/yr)
Bandung	*Pinus merkusii* (pinus)	65	50; 30	5x5	19.4 ± 1.3	2945
Banten and Tasikmalaya	*Paraserianthes falcataria* (sengon)	8	14; 20	5x5	22.9 ± 2.8	1183
Garut	*Calliandra calothyrsus* (kaliandra)	30	3; 18	1x1	32.7 ± 1.4	21909
Majalengka	*Acacia mangium* (acacia)	10	20; 35	6x6	23.3 ± 1.5	2175
Sukabumi	*Antocephalus cadamba* (jabon)	5	20; 30	5x5	25.4 ± 1.5	1246

In the wet conditions of Bandung, *Pinus* produced $19.4m^3$/ha/year, similar to estimates of [13], who each reported average increments of around 19.5 m^3/ha/year. If we assume that 40% of biomass is potentially used for energy wood, and that wood density is $550kg/m^3$, the pine can supply 7.76 m^3/ha/year at 46% water content, equal to 2945 kg/ha/year at 15% water content.

For a mature sengon tree, fresh trunks have a water content of 78% and can produce a total of 0.51 m^3 consisting of 0.30 m^3 of industrial wood per tree, and the remaining 0.21 m^3 can be used for energy wood. In other words, as much as 41 % volume of sengon trees can be used as energy wood. Based on

Table 3, if in one hectare there were 400 trees and the specific gravity is 340 kg/m^3, the sengon up to 6 years can potentially supply 9.4 m^3/ha/year or 1183 kg/ha/year of energy wood at 15% water content.

The West Java Kaliandra, planted in Garut in 1980, has been managed by the community for firewood by cutting and trimming the stems. Consequently, although the standing stock is quite old, the stem diameter is relatively small. From the cut stems, abundant shoots have grown, giving rise to short and dense tree form. Although the diameter at the base of the trunk is 15-20 cm, because of frequent trimming, the diameter of stems available for cutting is 2-4 cm; these branches were trimmed and weighed to evaluate biomass for bioenergy in this study. The average weight of kaliandra branches with twigs on each tree was 2.2 kg (oven-dry) and 2.5 kg (air-dry) showing that this part of the branch could potentially be used as raw material for wood pellets or firewood. Based on a Kaliandra wood density of 670 kg/m^3 and a plantation spacing of 1 x 1 m yielding 10,000 branches per ha every year, these branches with twigs could produce 32.7 m^3/ha/year, equivalent to 21909 kg/ha/year of biomass for bioenergy.

Observation on acacia trees indicated that the weight of the wet industrial wood assortment was 420 kg, while the total weight of the average acacia tree including the stem, branches and twigs was 830 kg. Total weight of wet biomass, which could be utilized for energy (including the tip of the stem, branches and twigs) is equal to 410 kg from each tree, or approximately 49.40% of the total weight of the tree biomass. Acacia fresh wood has a moisture content of about 42%, which made the potential for air-dried wood (15% moisture content) from all acacia trees for energy equal to 332.043 kg. If there are 278 trees/ha and the specific gravity is 700 kg/m^3, the acacia can potentially supply 11.5 m^3/ha/year or 2175 kg/ha/year of energy wood.

The weight for Jabon industrial wood was 245 kg, while total weight of the tree including stems, branches and twigs was 327 kg. Accordingly, wet weight of wood, which can be utilized for energy wood (including the tip of the stem, branches and twigs) is approximately 82 kg per tree, or about 25% of the total biomass. The moisture content of wet Jabon wood is 56% whereas for air-dried wood is 15% moisture. If there were 400 trees/ha and the specific gravity is 330 kg/m^3, jabon plantation can potentially supply 6.4 m^3/ha/year or 1246 kg/ha/year of energy wood.

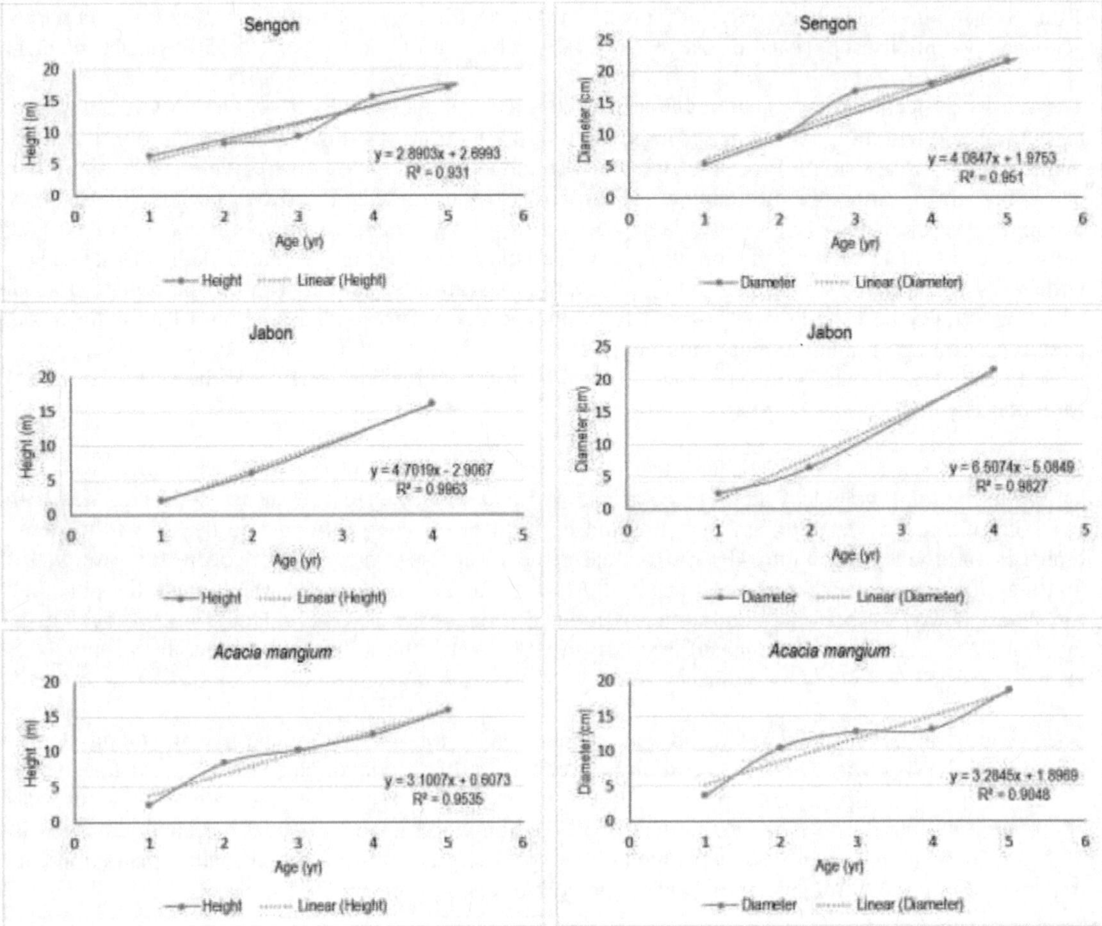

Figure 1. Growth model for height and diameter of sengon, jabon, and acacia plantations in West Jawa

In order to obtain estimation of tree growth in West Jawa, three fast growing tree species at different ages (sengon, jabon and acacia) were randomly selected in several community plantations for more detailed measurement of diameter and height relationship with age (regression equations given in Figure 1). In general all fast growing tree species studied increased growth rate to 5 years of age, although some species showed a decrease in growth, e.g. sengon showed decrease after only 2 years old and acacia after 3 years of age. Some unprecedented factors might affect those stands, such as early harvest by community for some selected trees, which left smaller trees in the field.

In East Jawa, the study focused on the four Districts of Malang, Situbondo, Kediri, and Bangkalan. In Malang, the study was located at Desa Mulyerejo, Kecamatan Ngantang. The location has 600 – 2800 m asl altitude and has Regosol/Entisol soil type with 2240-3850 mm/year annual rainfall. Potential species found in the location was sengon. In Kediri, the study was located at KPH Kediri, BKPH Pare that has 381 – 561 m asl altitude and Regosol soil type with 1886 mm/year annual rainfall. The main species in the location was also sengon. In Situbondo, the study was located at LMDH Sumber Jaya, Desa Sumberejo, Kec Banyuputih. This location has Latosol/Aluvial soil type with 994 – 1503 mm/year annual rainfall. Species found in the location was gmelina (*G. arborea*). The last area was located at Desa Kombangan, Kecamatan Geger. It has Litosol soil type with 2586 mm/year annual

IOP Conf. Series: Earth and Environmental Science **65** (2017) 012019 doi:10.1088/1755-1315/65/1/012019

rainfall. Species found in the location was kaliandra (*C. calothirsus*). Average of measurement result for each location is shown in Table 4.

Table 4. Average tree diameter and height in four Districts of East Jawa

Location	Species	Height (m)	Diameter (cm)	Age (Months)	Planting Space
Malang	*P. falcataria*	1.0	1.0	6	3 x 2 m
		8.3	8.4	30	3 x 2 m
		18.0	34.5	44	3 x 2 m
Kediri	*P. falcataria*	0.7	0.8	1	3 x 2 m
		8.1	8.8	2	3 x 2 m
		13.5	14.1	3	3 x 3 m
		17.3	21.4	4	3 x 3 m
		22.0	22.3	5	3 x 2 m
Situbondo	*G. arborea*	7.4	10.2	6	unregular
		10.0	17.0	7	3 x 2 m
		39.8	18.0	20	unregular
Bangkalan	*C. calothirsus*	6.1	2.2	8	1 x 1 m
		6.4	4.0	2	1 x 1 m

Based on this measurement result, regression equations were made as shown in Figure 2.

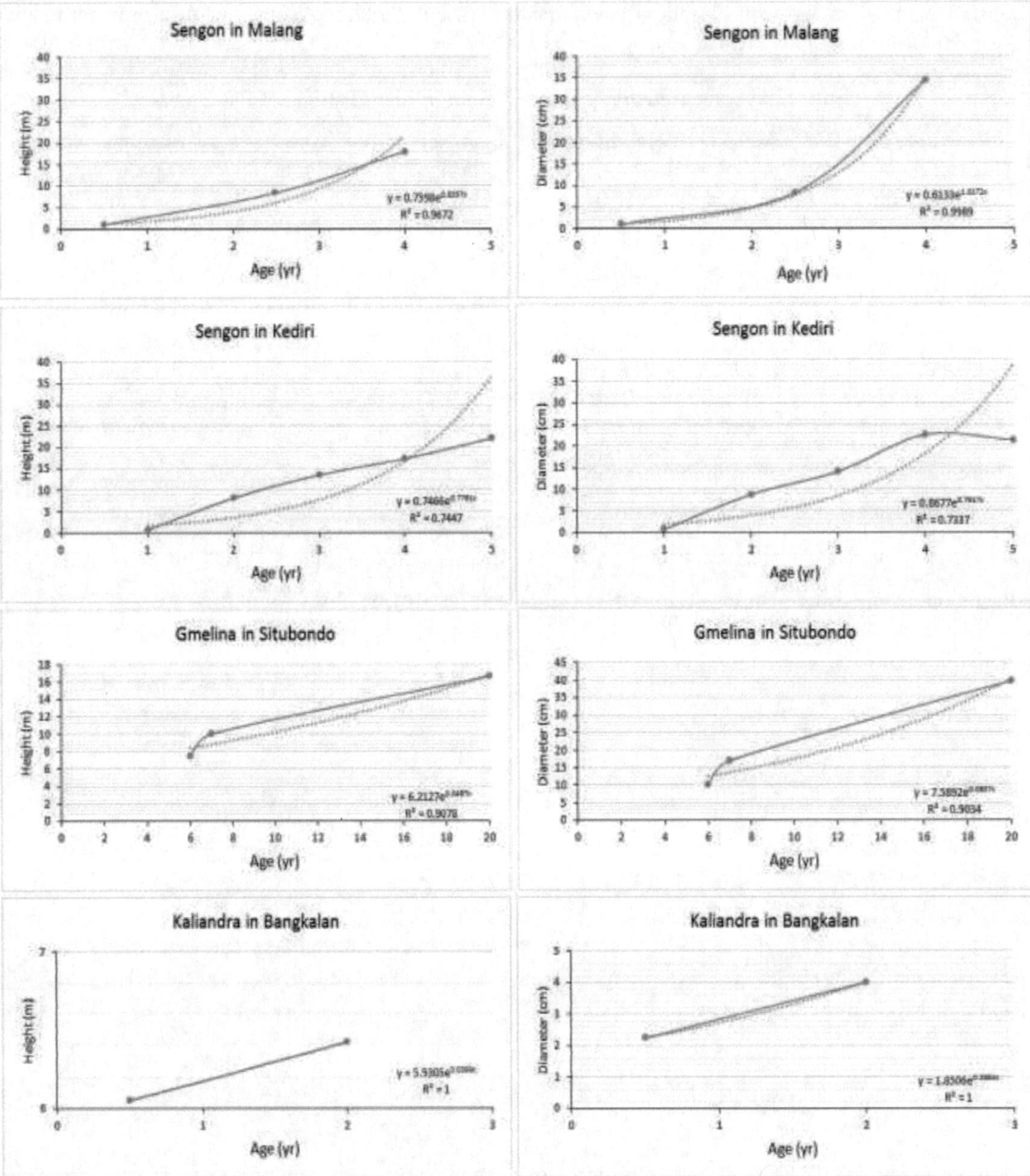

Figure 2. Growth model for height and diameter of sengon, gmelina, and kaliandra in East Jawa

3.2. Comparison on Biomass and Potentially Generated Energy

Based on height, diameter, and density measurement, the biomass value and potential wood pellet production rates were calculated, including electricity generation potential as shown in Table 5.

Table 5. Electricity characteristics resulted from some wood species

Location	Species	Age (yr)	Biomass (ton/ha)	Wood pellet (ton/ha)	Electric power (Kwh/ha/yr)	40% machine efficiency (Kwh)
West Jawa	Sengon	4	36.2	24.1	108467	43387
	Akasia	4	61.0	40.6	182878	73150
	Jabon	3	67.1	44.7	201203	80481
East Jawa	Sengon	4	133.7	89.1	401030	160412
	Jabon	3	56.4	37.6	169074	67630

Remark: electricity used by a house hold in a year = 450 watt x 24 hours x 30 days x 12 months

Sengon was found in all locations with the highest biomass occurring in East Jawa community forests. Utilization of one hectare of sengon as raw material for wood pellets can potentially supply electricity for 103 households. As a source of biomass, the growth of jabon species was better in the higher rainfall conditions of West Jawa compared to East Jawa.

Development of energy wood species is also initiated by community groups or the private sector. Table 6 shows the community forest area in each province that can be optimized for energy wood production. The Government should open the opportunity to utilize critical land for forest plantations to become a main source of energy. If it can be realized, the number of houses without electricity can be decreased. Up to 2013 there are 2.4 million house without electricity in West Jawa and 1.8 million in East Jawa at 2015 [14].

Table 6. Potential of energy wood development and resulted electricity

Location	Community forest (ha)[a]	Degraded land (ha)[b]	Total (ha)	Number of community Potentially get electricity	
				Household	Village[c]
West Jawa	973 860	1 387 711	2 361 571	48 884 155	21 853
East Jawa	659 414	2 126 619	2 786 033	114 946 792	54 451

[a][15];
[b][16];
[c]based on average household in each village [17]

Table 5 shows the number of households and villages that could receive electricity, based on total land area potentially available for plantations of energy wood species. West Jawa shows high potential to develop energy wood, although available land is not as high as in East Jawa. In West Jawa, potential electricity using woody biomass can support 21853 villages. East Jawa with higher available land for tree plantations, could potentially generate electricity from wood biomass to light 54451 villages. In West Jawa, jabon tree species potentially has the biggest role in supporting new electricity from woody biomass, while in East Jawa, sengon has the most potential.

4. Conclusions

Tree species found in West and East Jawa as energy sources were akasia (*Acacia mangium*), Jabon (*Anthocephalus cadamba*), Kaliandra (*Calliandra calothirsus*), *Gmelina arborea*, Sengon (*Falcataria moluccana* or *Paraserianthes falcataria*).

Jabon and sengon were two species which could potentially be adapted and developed as biomass energy plantations for electricity in West and East Jawa, due to their widespread occurrence and rapid

IOP Conf. Series: Earth and Environmental Science **65** (2017) 012019 doi:10.1088/1755-1315/65/1/012019

growth. Estimation for Jabon and sengon showed those species potentially resulting up to 133.7 and 67.1 ton/ha biomass respectively, as well as 160412 and 80481 Kwh of electricity respectively.

Some problems still need to be addressed, such as the price competition between wood for energy and industrial wood, village condition (no population vs availability of lands), electricity price by a National Electricity Company (PLN), and social engineering.

References

[1] Kementerian Energi dan Sumberdaya Mineral [KESDM] 2016 *Statistik Ketenagalistrikan 2015* Edisi no.29 (Jakarta: Ditjen Ketenagalistrikan)

[2] Electric Power Research Institute (EPRI) 1995 *Making Biopower Work for Utilities: A Rationale for Near-Term Investment in Integrated Biomass Power Systems Project #3407* (December)

[3] Nienow S, Mc Namara K T, Gillespie A R and Preckel P V 1999 A Model for the Economic Evaluation of Plantation Biomass Production for Co-firing with Coal in Electricity Production *Agricultural and Resource Economics Review* April 1999.

[4] Sajjakulnukit B and Verapong P 2003 Sustainable biomass production for energy in Thailand *Biomass and Bioenergy* **25** 557 – 570

[5] Mulyasari T 2013 Characteristics of some wood species as biomass energy raw materials *Undergraduate Theses* (Bogor: Faculty of Forestry, Bogor Agricultural University)

[6] Cahyono T D, Coto Z and Febrianto F 2008 Heat value analysis and economic feasibility of wood utilization as coal substitution in cement factory *Forum Pascasarjana* **31** 105-176

[7] Elauria J C, Castro M L Y and Racelis D A 2003 Sustainable biomass production for energy in the Philippines *Biomass and Bioenergy* **25** 531-540

[8] Oak Ridge National Laboratory [ORNL] 2008 *Conversion Factors for Bioenergy* (North Carolina: North Carolina State University)

[9] Smith W R 1982 *Energy From Forest Biomass.* (Washington: Academic Press)

[10] Krisnawati H, Adinugroho C W, Imanudin R 2013 *Pedoman Penggunaan Model Alometrik untuk Pendugaan Biomassa dan Stok Karbon di Indonesia* (Bogor: Pusat Penelitian dan Pengembangan Konservasi dan Rehabilitasi, Badan Penelitian dan Pengembangan Kehutanan – Kementerian Kehutanan)

[11] Siringoringo H H and Siregar C A 2006 Model persamaan allometrik biomassa total untuk estimasi akumulasi karbon pada tanaman Sengon *Jurnal Penelitian Hutan dan Konservasi Alam* **3** 541-553

[12] Krisnawati H, Adinugroho C W and Imanudin R 2012 *Model-Model Alometrik untuk Pendugaan Biomassa Pohon pada Berbagai Tipe Ekosistem Hutan di Indonesia* (Bogor: Pusat Penelitian dan Pengembangan Konservasi dan Rehabilitasi, Badan Penelitian dan Pengembangan Kehutanan – Kementerian Kehutanan)

[13] Hartoyo and Nurhayati 1976 Rendemen dan sifat arang dari beberapa jenis kayu Indonesia *Laporan Penelitian* no.62 (Bogor: Lembaga Penelitian Hasil Hutan)

[14] Kompas 2015 Ada 1,7 juta rumah tangga di jatim belum nikmati listrik *Harian Kompas* 6 Oktober 2015

[15] Dephut 2013 *Profil Kehutanan 33 Provinsi* (Jakarta: Pejabat Pengelola Informasi dan Dokumentasi (PPID))

[16] Kemenhut 2015 Penetapan peta dan data hutan dan lahan kritis nasional tahun 2013 *SK Dirjen BPDASPS No. SK.4/V-DAS/2015*

[17] BPS 2014 *Data sosial dan kependudukan* (Jakarta: Badan Pusat Statistik)

Extraction process of palm kernel cake as a source of mannan for feed additive on poultry diet

M Tafsin[1], N D Hanafi[1] and E Yusraini[2]

[1.] Animal Science Study Program, Faculty of Agriculture, University of Sumatera Utara, North Sumatera, Indonesia
[2.] Food Science and Technology Study Program, Faculty of Agriculture, University of Sumatera Utara, North Sumatera, Indonesia

Email : martafsin@yahoo.com

Abstract. Palm Kernel Cake (PKC) is a by-product of palm kernel oil extraction and found in large quantity in Indonesia. The inclusion of PKC on poultry diet are limited due to some nutritional problems such as anti-nutritional properties (mannan). On the other hand, mannan containing polysaccharides play in various biological functions particularly in enhancing the immune response and to control pathogen in poultry. The research objective to find out the extraction process of PKC and conducted at animal nutrition and Feed Science Laboratory, Agricultural Faculty, University of Sumatera Utara. Various extraction methode were used in this experiment, including fraction analysis used 7 number sieves, and followed by water and acetic acid extraction. The result indicated that PKC had different particle size according to sieve size and dominated by particle size 850 um. The analysis of sugar content indicated that each particle size had different characteristic on treatment by hot water extraction. The particle size 180-850 um had higher sugar content than coarse PKC (2000-3000 um). The total sugar content were recovered vary between 0.9-3,2% from PKC were extracted. Treatment grinding method followed by hot water extraction (100-120 °C, 1 h) increased total sugar content than previous treatments and reach 8% from PKC were extracted. Utilisation acetic acid decreased the total amount of total sugar from PKC were extracted. It is concluded that treatment by hot temperature (110-120 °C) for 1 h show highest yield to extract sugar from PKC.

1. Introduction

Palm kernel cake (PKC) is a byproduct from industrial processing and in Indonesia, PKC availability is high. Production of fresh fruit bunches of oil palm Approximately 12.5 to 27.5 tons / ha, and about 2 percent is palm kernel cake as by product [1]. Utilisation of PKC as potential feed reported on ruminant diet [2,3], chicken [4], even fish [5]. One limiting factor Utilisation of PKC especially for monogastric animal is high fiber content and its dominant component is a form of mannnose that reach 56.4% of total cell walls and dominated by β-mannan [6].

The content of the high mannan as well as the limiting factor may also be regarded as the potential to earn as a prebiotic feed additives that will improve the health of livestock. There was suppposted that there are similarities between the PKC with mannanoligosaccharides (MOS) that will improve the health and the immune system of poultry [7]. The research [8] using mannan derived from the hydrolysis of guar gum (*Cyamoposis tetragonolobus*) containing galactomannan with a molecular

Content from this work may be used under the terms of the Creative Commons Attribution 3.0 licence. Any further distribution of this work must maintain attribution to the author(s) and the title of the work, journal citation and DOI.
Published under licence by IOP Publishing Ltd

weight of about 20 000 Da. Another potential that can be used as a source of mannan is PKC. So far, the PKC is only used as one source of feed, but according to this potential can be increased into raw materials for feed additives.

On the other hand, microbes such as viruses or bacteria potentially harmful to animals that can be found in air, food or water. The bacteria that often contaminate chicken, either at the time of hatching, enlargement and postharvest is groups Salmonella sp. The bacteria will affect the health of livestock will also affect the food safety aspects. Several attempts have been made to overcome these problems such as vaccination, sanitation or the use of antibiotics. The effort is in addition to having many benefits also have limitations, for example to antibiotics now found several strains of bacteria that are resistant to antibiotics.

A new approach to prevent microbial infections found in recognition of the importance of the process of attachment of the digestive tract. It is known that type 1 fimbriae are sensitive to mannose plays role in the attachment of pathogenic microbes. Bacteria such as Salmonella, E. coli and Vibrio cholera have lectins on the cell surface that is specific to the mannose , thus mannose can inhibit the process of attachment harmful microbes in the digestive tract [9]. Another capability of MOS is able to stimulate the immune system and this effect also plays a role in the fight against bacteria Salmonella [10,11]. MOS can affect the immune system by stimulating the secretion of mannose-binding protein from binding liver capsule invading bacteria. Other studies show that MOS stimulates the immune system by enhancing the phagocytic activity of macrophages and increased levels of Ig (immunoglobulin). Furthermore the immune cells in the GALT (gut associated lymphoid tissue) to detect the presence of microbes due to their unique molecule called PAMP (pathogen-associated molecular pattern) which will further activate the immune system [12].

The objectives of this study is find out the source of mannan from PKC. So far, the PKC is only used as one source of feed, but with this potential can be increased into raw materials for feed additives. According to this background, we are interested in studying the production process and application of the PKC as as feed additives (feed additive) in poultry. So far, several commercial companies using materials from *Saccharomyces cerevisiae* as a raw material. Use of PKC can replace the material with a much cheaper price and high availability in Indonesia. The study on the extraction process of palm kernel cake is still limited, so that the research was conducted.

2. Materials and Methods

Combination treatment will be applied to obtain the most effective and efficient in extraction of PKC become source mannan that can be applied as feed additives on poultry diet.

2.1. Materials

The main ingredient needed is PKC derived from palm kernel oil processing plant owned by PTPN IV. Other materials mainly of chemicals obtained from a commercial company. The main equipment to be used is a water bath shaker, spectrofotometer, and grinder.

2.2. Method

Palm Kernel Cake obtained from the industry were separated using a sieve (screen) which aims to separate the rest of the crushed shells during the process of making palm kernel oil. Combination treatment will be applied to obtain the most effective and efficient in overhauling the structure of the material PKC become source of mannanoligosaccharides (MOS). The treatment is a combined treatment will be given in the form of extraction, the optimum temperature, and duration of incubation. The extraction process is done by using a solvent with a 1:10 ratio. Total PKC used as much as 4 grams for each experimental unit. Further stages performed using hot water extraction (hot water extraction; 121 ^{o}C, 15 minutes) that is using an autoclave. The process followed by centrifugation

(4200 rpm; 15 min; 10 °C). The next supernatants were collected and measured the total amount of sugar. The content of total sugar was measured using the reagent of concentrated sulfuric acid and phenol 5% was then measured using a spectrophotometer at a wavelength of 490 nm with D-glucose as standard [13].

3. Results and Discussion

3.1. Screening Process

The extraction stage started with screening of the PKC aimed at getting some fraction of its potential and gain the properties of the content of total sugar. Screening results are presented in table 1.

Table 1. Proportion of BIS based on particle size

Fraction	Sieve number	Particel size (μm)	Proportion (%)
1	5	4000	1.5
2	10	2000	25.5
3	20	850	32.5
4	30	600	15.0
5	40	425	4.0
6	50	300	6.5
7	80	180	15.0
Total			100.0

The images for each faction presented in the following figure:

Fraction no 1

Fraction no 2

Fraction no 3

Fraction no 4

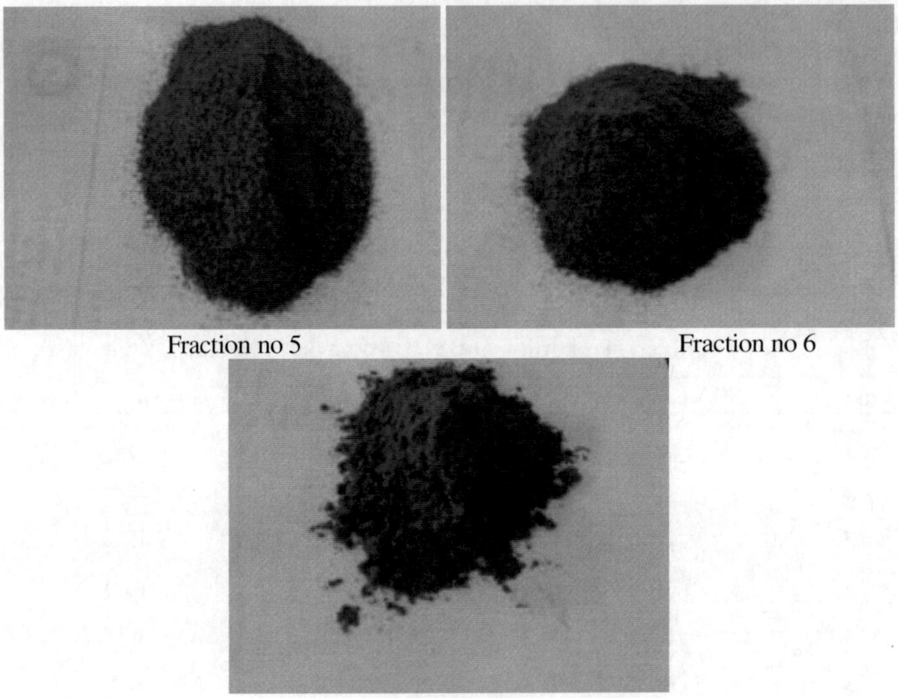

Fraction no 5 Fraction no 6

Fraction no 7

Figure 1. Appearance of palm kernel cake fractions based on particle size

The method used for the total amount of sugar is using phenol reagent and sulfuric acid. Observations were made using a spectrophotometer that is read at a wavelength of 490 nm. Observations on the content of total extractable sugar are presented in the following table:

Table 2. Total sugar content of each fraction PKC

Fraction PKC	Total sugar	
	g	%
1	0.1821	0.91
2	0.3762	1.88
3	0.5699	2.85
4	0.6493	3.25
5	0.5147	2.57
6	0.3420	1.71
7	0.5215	2.61

Table 2 showed that the total extractable sugar ranges between 0.91 - 3.25% of the PKC were extracted. The range is still relatively small, considering the type of solvent used is only using aquadest. The results of this stage was to get answers in fractions which obtained the best total sugars. Good results obtained in fraction 3 and so on. These results became the early references to perform an extraction process using materials that have been filtered by sieve size No. 10 or equivalent with a particle size of less than 850 um.

3.2. Effect of grinding, acetic acid and temperature on total extractable sugar
Improvements continue to do the extraction process to obtain a high yield of mannan contained in the PKC. Results of previous showed at around 0,9-3.2% of PKC. Further treatment provided through several stages: Grinding size using diskmill on materials that have been screened to a particle size of 850 um and the utlisation of acetic acid (1 and 2%) and a higher temperature (110; and 120 °C) with a longer time (1 h). The results are presented in the following figure:

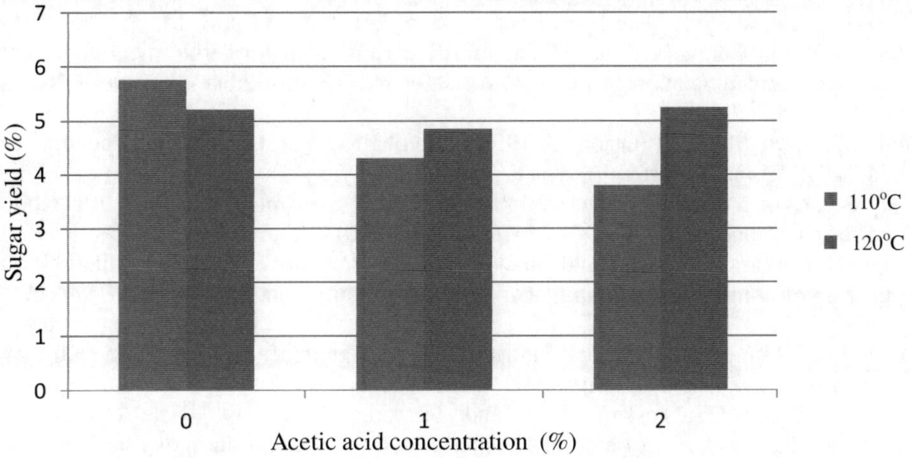

Figure 2. Effect of acetic acid and temperature on total extractable sugar

The figures above showed the total amount of sugar extracted from PKC. The use of high temperatures (110 and 120 °C) for one hour, combined with a particle size reduction PKC has raised a high yield of total sugar. Research results in previous stages showed the highest rate was 3.2%, whereas at this stage it reached 8.7%, and the yield is higher than reported [11]. Referring to the results of research which obtained the data content of total ADF 31.7% and total sugar component of 73% [6], so that the estimated number of potential sugar component of 23.1%, then the results of this study indicate that the total sugar were extracted reached 37%.

The utilisation of acetic acid to the level of 2% turned out to show the total amount of extractable sugar relative decline. It shows that the use of acid with a high temperature condition can destruct components of sugar extracted. further studies which we will do is the preparation to readings sugar component by using HPLC (high performance liquid chromatography).

4. Conclusion

Palm kernel cake extraction process can be done by using the PKC that has been filtered by the smaller size of 850 um. Grinding size reduction followed by hot water extraction may increase the total amount of sugar extracted approaching 8% of the total PKC. The utilisation of acetic acid to obtain extractable sugar of PKC does not exceed 1% concentration.

Acknowledgement
The author would like acknowledge to Badan Pengelola Dana Perkebunan Kelapa Sawit (BPDPKS) has provided research funding through a research grant program of palm oil in 2016.

IOP Conf. Series: Earth and Environmental Science **65** (2017) 012020 doi:10.1088/1755-1315/65/1/012020

References

[1] Sinurat A P 2001 Pemanfaatan bahan pakan lokal *Makalah pada dies natalis HIMASITER III* (Bogor: Fapet IPB)

[2] Elisabeth W and Ginting S P 2003 Pemanfaatan hasil samping industri kelapa sawit sebagai bahan pakan ternak sapi potong *Proc. Lokakarya Nasional Sistem Integrasi Kelapa-Sawit Sapi* 9-10 September 2003 Bengkulu Indonesia

[3] Mathius I W, Sitompul D, Manurung B P and Azmi 2003 Produk samping tanaman dan pengolahan buah kelapa sawit sebagai bahan dasar pakan komplit untuk sapi: suatu tinjauan. *Proc. Lokakarya Nasional Sistem Integrasi Kelapa-Sawit Sapi* 9-10 September 2003 Bengkulu Indonesia

[4] Sundu B and Dingle J 2005 Use of enzyme to improve the nutritional value of palm kernel meal and copra meal *Proc. Quensland Poult Sci Symp* Australia **11** pp 1-15.

[5] Keong N W and Chong K K 2002 The nutritive value of palm kernel meal and the effect of enzyme supplementation in practical diets for red hybrid tilapia (*Oreochromis sp.*) *Asian Fish Sci* **15** 167-176

[6] Daud M J, Jarvis M C and Rasidah A 1993 Fibre of PKC and its potential as poultry feed *Proc. 16th MSAP Annual Conference* Kuala Lumpur Malaysia

[7] Sundu B, Kumar A and Dingle J 2006 Palm kernel meal in broiler diets:effect on chicken performance and health *World's Poult Sci J.* **62** 316-325

[8] Ishihara N, Shu D C, Akachi S and Juneja L R 2000 Preventive effect of partially hydrolyzed guar gum on infection of *Salmonella enteridis* in young and laying hen *Poult Sci* **79** 689-697.

[9] [CFNP TAP] Center for Food and Nutrition Policy Technical Advisory Panel Review 2002 *Cell Wall Carbohydrates* : Livestock (Virginia CFNP)

[10] Spring P, Wenk C, Dawson K A and Newman K E 2000 The effects of dietary mannanoligosaccharides on cecal parameters and the concentration of enteric bacteria in the ceca of salmonella-challenge broiler chicks *Poult Sci* **79** 205-211

[11] Tafsin M 2007 kajian polisakarida mannan dari bungkil inti sawit sebagai pengendali *Salmonella thyphimurium* dan immunostimulan pada ayam Dissertation of Bogor Agricultural University Post Graduate Program Bogor Indonesia

[12] Shashidara R G and Devegowda G 2003 Effect of dietary mannan oligosaccharide on broiler breeder production traits and immunity *Poult Sci* **82** 1319-1325

[13] Dubois M, Giles K A, Hamilton J K, Reber P A and Smith F 1956 Colorimetric method for determination of sugars and related substances *Anal Chem* **28** 350-356

IOP Conf. Series: Earth and Environmental Science **65** (2017) 012021 doi:10.1088/1755-1315/65/1/012021

Fabrication and characterization of rice husk and coconut shell charcoal based bio-briquettes as alternative energy source

Y Yuliah[1], M Kartawidjaja[2], S Suryaningsih[3] and K Ulfi[4]

[1] Physics Departemen, Material Laboratory, Padjadjaran University, Jatinangor 45363, Indonesia
[2] Physics Departemen, Material Laboratory, Padjadjaran University, Jatinangor 45363, Indonesia
[3] Physics Departemen, Energy Laboratory, Padjadjaran University, Jatinangor 45363, Indonesia
[4] Physics Departemen, Energy Laboratory, Padjadjaran University, Jatinangor 45363, Indonesia

Email: y.yuliah@phys.unpad.ac.id

Abstract. Rice husk and coconut shell have been disposed or burned as waste. As biomass, both of materials are the potential sources of carbon which can be utilized as alternative energy sources. The energy content can be exploited more intensively when packaged in a brief and convenient. In this work, the mixtures of rice husks and coconut shells charcoal were prepared as briquettes. After going through the carbonization process, several measurements have been taken to find out the factors that determine the value of heat energy contains by each component of the charcoals. The basic ingredients briquettes prepared from rice husk and coconut shell charcoal with varying composition and addition of tapioca starch gradually as adhesive material to obtain briquettes in solid with the maximum heat energy content. After going through pressing and drying process, the briquettes with 50:50 percent of composition and the 6% addition of adhesive was found has the highest heat energy content, equal to 4966 cal/g.

1. Introduction

Fossil fuels are still the main energy source of the world today. In 2011, the need for fossil fuels totaled 10,668 million TOE (Tonnes Oil Equivalent) or 82% of the total demand and is expected to increase to 14,898 million TOE in 2035 [1]. Coal contributed 3,773 million TOE in 2011, will reach 44% in year 2035 and in 2020 is expected to take over the role of oil as the highest in the primary energy mix. Fossil fuels are un-renewably energy sources, will run out by the time if continuously used. While the use of renewable energy recorded 1,727 million TOE or 13% of world total usage energy [1]. Biomass is an alternative energy that can contribute to the primary energy mix as a replacement for coal when processed optimally. Biomass is material derived from plants that can be either directly or indirectly be utilized as fuel in huge so that it can be one of the alternative energy source of fuel oil (fossil).

Content from this work may be used under the terms of the Creative Commons Attribution 3.0 licence. Any further distribution of this work must maintain attribution to the author(s) and the title of the work, journal citation and DOI.
Published under licence by IOP Publishing Ltd

The potential of biomass energy in Indonesia is very large, that is 13,662 MWe, while the installed capacity is only 1,364 MWe2. One of the most prospective biomass materials in Indonesia is rice husk which is agricultural wastes that have been disposed as refuse [2]. Rice production in Indonesia in 2015 is estimated at 75.55 million tons of GKG (Gabah Kering Giling/dry rice unhusked) or increased by 4.70 million tonnes (6.64%) compared to 2014 records [3].

Processing rice husk into bio-briquettes is one way effective packaging rice husk. Rice husk charcoal briquettes can store heat energy average of 4384.043 cal/g [4]. Increased energy content of rice husk charcoal briquettes can be made by adding other kinds of biomass with high calorific value of energy. For example, the average heat energy briquettes rice husk increased to 4526.097cal/kg after being mixtured with wood charcoal. Improving the quality of briquettes can be done at the preparation stage of charcoal such as carbonization, grinding, formulating the optimum composition and the addition of appropriate adhesive material with appropriate levels as well [5,6].

In this research, rice husk charcoal briquettes were fabricated by adding a certain amount of coconut shell charcoal as a substitute for wood charcoal which has rared and is not economically profitable. Tapioca starch was used as an adhesive for quick burning and produces a high caloric. Characterization processes were performed to determine the calorific value and the factors affect the quality of the resulted biobriquettes.

Through the fabrication and characteristics test of briquettes made from rice husk charcoal is expected to be obtained the bio-briquettes production mechanisms and can contribute on how the utilization of waste rice husk as an alternative to fossil fuels.

2. Methods

2.1. Materials and Research Tools

The main materials used are rice husk, coconut and tapioca starch as depicted in the Figure 1. And the main tools consist of heater, softener, filter, briquettes mold and briquette compression tool.

(a) (b) (c) (d) (e)

Figure 1. The Base Materials Used: (a) Rice Husk, (b) Rice Husk Charcoal, (c) Coconut Shell, (d) Coconut Shell Charcoal, (e) Tapioca Starch.

2.2. Materials Preparation

The materials preparation processes were accomplished mostly to get the mixing of rice husk and coconut shell charcoals with definite compositions which were ready to be moulded. Prior made the charcoal, rice husk was dried for 10-15 hours, roasted until dark color and then cooled. The coconut shells charcoal was used from the market. The next process was down-sizeing, the rice husk charcoal smoothed by using a blender while coconut shells charcoal by means of ground. Separately both charcoals are filtered by using a 120 mesh filter.

Mixtures of rice husk: coconut shell charcoal were prepared with the composition-1 (50%:50%) and composition-2 (80%:20%). To each composition, adhesive of tapioca starch was added in vary levels (% mass) which were: A = 4%, B = 6%, C = 8% and D = 10%.

Table 1. Rice Husk : Coconut Shell Charcoal Briquettes Sample Identification

No.	Sampel ID	Charcoal Composition (%)		Adhesive Content (%)
		Rice Husk	Coconut Shell	
1.	1A	50	50	4
2.	1B	50	50	6
3.	1C	50	50	8
4.	1D	50	50	10
5.	2A	20	80	4
6.	2B	20	80	6
7.	2C	20	80	8
8.	2D	20	80	10

2.3. Briquettes Fabrication

The dough of each briquette was prepared from 80 g of the any charcoal composition and 60 ml of water and adhesive materials according to the its percentage level. The adhesive material was prepared in advance by mixing the tapioca with water in a pan and then heated on the stove until it coagulates. The charcoal then pour on the dough and stirring until blended. The doughs were then set up into a cylindrical mold with 2 cm of the diameter and then pressed using hydraulic compression tool with 3 ton of the power (21 swings). The last step was drying process to reduce the moisture content of the briquette so that make it easy to light when burned. All briquettes that have been printed were dried in an oven at a temperature of 55 ° C for one day.

2.4. Characterization

Characterization of briquettes conducted were included determination of water content lost in the heating process, the average density, the calorific energy of the briquettes resulted and proximate test [7].

Determination of calories energy value and proximate test were conducted at the Lab. Coal tekMIRA Research Center Bandung. The calories energy value was measured using a bomb calorimeter. Proksimate tests were conducted to determine standards quality of the briquettes produced include, fuel power, long flames and determine the standard of the sale value of the briquettes.
Test proximate tests performed consist of:

- Water Content Test
  The test results of water content can also be used to characterize other properties such as air-dry moisture lost.
- Volatile Matter Content Test (Fly Substance)
  This test intended to determine the content of volatile matter contained in briquettes produced. Value of volatile matter affects the perfection of combustion and flame produced which determine the marketability of briquettes.
- Ash Content Test
  Ash content test was performed to determine the levels of ash generated by the briquettes after burning.
- Fixed Carbon Determination

IOP Conf. Series: Earth and Environmental Science **65** (2017) 012021 doi:10.1088/1755-1315/65/1/012021

Intend to determine the levels of fixed/bonded carbon contained in the briquettes. Calculation of fixed carbon percentage was done in accordance with ASTM D 3172-13.

3. Results and Discussion

3.1. Briquette Fabrication

Briquettes have been made from mixtures of rice husk and coconut shell with two different compositions (50:50 and 80:20). For each composition, four adhesive content variations (mass % of tapioca): 4%, 6%, 8% and 10% were used. Each variant herein after are identified as Samples ID in the Table 1.

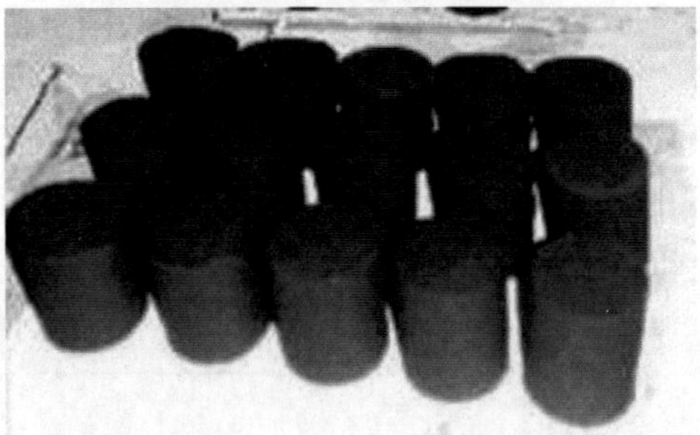

Figure 2. Rice Husk: Coconut Shell Charcoal Briquettes

Overall 8 variant cylindrical briquettes with 2 cm diameter of the surface are shaped same a like the example in Figure 2. The high of cylinders vary greatly depending on the composition and content of the adhesive. For samples with composition-1, the heights are ranges between 1.94 cm to 2.39 cm and for samples with compositions-2 are between 1.73 cm to 3.13 cm.

Table 2. The Water Content Lost and Dendity of Rice Husk:Coconut Shell Charcoal Briquettes

No.	Sampel	Charcoal Composition (%)		Adhesive Contain (%)	Water Content Lost (%)	Density (g/cm)
		Rice Husk	Coconut Shell			
1	1A	50	50	4	32.04	0.81
2	1B	50	50	6	32.09	0.79
3	1C	50	50	8	33.33	0.81
4	1D	50	50	10	32.37	0.82
5	2A	20	80	4	36.01	0.76
6	2B	20	80	6	37.70	0.68
7	2C	20	80	8	35.42	0.78
8	2D	20	80	10	28.56	0.84

3.2. Results of the Characterization

3.2.1. Properties of the Briquette

The physical properties of the resulted rice husk: coconut shell charcoal briquettes are shown in the Table 2. It appears that the level of water lost average increase linearly with starch content variation for each composition. With increasing levels of water lost, density briquettes are reduced. In some samples was obtained significant relationships that density with a high gluten content show a linear but not as a result with the water content. There are differences that are not linear on any variation of the levels of starch. It is because the nature of starch and charcoal those are not resistant to moisture so it is easy to absorb water from space [6].

Comparing the water content lost on the composition of charcoal briquettes with 50:50 and 80:20, it seem the composition 50:50 has higher water content. This is caused by the composition of coconut shell charcoal that is more than the composition of 80:20. Silica content is high enough in this composition supposedly resulting in high water levels due to the nature of silica itself is absorbing water. Silica content of rice husk was 16.98% while the silica content in the coconut shell charcoal is 52% [8].

The result of the average density calculation of the briquettes in Table 2 shows that for samples with starch content of 10% (sample D) has the highest density among other treatments for each type of briquettes composition, that the higher the levels of starch, the higher density briquettes. This happens because the briquettes pores will decrease due to the adhesive holding capacity increase highly [6]. The increase in density tends linear except at the sample B for each composition, the density appears to have declined.

If the density compared to each of the constituent composition of the briquettes, the briquettes with a 50:50 composition of greater density than the briquettes with a composition of 80:20. This is caused by the high density of coconut shell charcoal compared to the density of rice husk. Then the charcoal briquette compositions which more will make the briquettes have a high density as well.

3.2.2. Proximate Test Results.
The proximate test results which include determining of the water content, volatile matter, fixed carbon, ash content and calorific value of the average are shown in Table 3.

Table 3. Proximate Test Results and Calorific Value Average

Sampel	Water Content (% adb)	Volatile Matter (% adb)	Ash Content (% adb)	*Fixed Carbon* (% adb)	Calorific Energy Value (cal/g)
1A	7.92	21.70	20.75	49.63	4955
1B	7.63	23.12	20.21	49.04	4966
1C	8.52	23.40	20.16	47.92	4886
1D	8.52	24.46	19.92	47.10	4860
2A	7.52	21.88	30.44	40.16	4124
2B	7.59	22.91	29.66	39.84	4132
2C	7.59	24.63	28.97	38.81	4104
2D	6.27	25.61	29.24	38.83	4132

The water content of the briquettes were affected by the type of base material, type of adhesive and testing methods employed [6]. From the table above it appears that the greater the levels of starch, the higher the water content. Extra adhesive higher causes the water contained in the adhesive would go in and bound in the pores of charcoal, besides the addition of adhesive that the higher will affect the briquettes have a density higher the briquettes pores getting smaller and when drained of water trapped in the briquette pore hard to getaway.

The high water content was shown in the sample composition-1 with the starch content of 6% (sampel 1B) and 10% (sampel 1D). While the low water content in the sample composition-2 with a starch content of 10% (sampel 2D).

When viewed from the composition of the constituent materials, the composition-1 of rice husk and coconut shell charcoal has higher water content than the briquettes with the composition-2. This is thought to occur due to the silica content in the composition-1 is greater than the briquette with composition-2. The nature of the silica itself is hygroscopic, to absorb water very well, and therefore the amount of silica that is pretty much it causes the water absorption from the air. The adhesive or the water content of the material itself is still high as a result of carbonization less than perfect bound to the silica and causes the water content in the briquettes is getting higher. The cause of the silica content in the briquette composition-1 higher is because the levels of coconut shell charcoal is high, since the silica content in the coconut shell charcoal is higher than the rice husk. Silica content of rice husk is 16.98% while the silica content in the coconut shell charcoal is 52%.

Water levels resulting from this research meets the standards set by the government which, according to Minister of Energy and Mineral Resources of Indonesia Republk (ESDM RI) No. 047 In 2006, the maximum moisture content should be available for charcoal briquettes is 20%, and briquettes resulted in this research has a moisture content ranging from 6.27% - 8.52%.

Volatile matter or content of volatile matter is a substance that can be evaporated as a result of the decomposition of the compounds that are still present in the briquettes in addition to water, carbon bonded and ash [6]. From Table 3, comparing volatile matter content of the levels of starch show that the greater the levels of starch, the greater the levels of the volatile matter. This means that the addition of starch levels can make the content of volatile matter in briquettes increases. This is caused by the presence of volatile substances contained by an adhesive, such as CO, CO_2, H_2 and CH_4 [6]. The higher levels of starch, the amount of these substances were olso increase.

The most high volatile matter are owned by 10% starch levels, sample 2D that is equal to 25.61% and the lowest levels held by the sample 1A that is equal to 21.70%.

If we compare the levels of volatile matter for each composition, then the composition of rice husk and coconut shell charcoal 80:20 was higher in comparison with the composition of 50:50. When viewed from the silica content, composition briquettes 80:20 higher silica content than the composition of 50:50. Prof. Dr. Muhammad Zarlis, M.Sc found that the more the silica content in the briquette it causes more content of the volatile matter. This supposedly caused by a defective process of carbonization of rice husk. High and low a level of substance evaporates/volatile matter in briquettes caused by the perfection of the carbonization process and is also influenced by the time and temperature at the authoring process. The greater the temperature and time of composing the more substance evaporates wasted so those at the time of assay substance will obtained evaporate lower levels of substance.

Levels of volatile matter generated from this study do not meet the standards set by the government which, according to Minister of Energy and Mineral Resources of Indonesia Republik (ESDM RI) No.047 of 2006 levels of volatile matter that should be available for a maximum of charcoal briquettes is 15%, while briquettes research has a moisture content ranging 21.70% - 25.61%.

The next component of proksimate test is ash levels. The ash is metal oxides in a charcoal composed of minerals that can not evaporate in the carbonization process. The ash content is very influential to the quality of briquettes are made. The higher the ash content owned briquettes then the lower heat will be produced. From this research, it was found that the addition of starch concentration resulted in a decrease in ash content. According to research conducted by Maryanto et al in 2013 that the addition of starch levels cause increased levels of ash, because of the addition of starch levels will result in the addition of inorganic substances in briquettes, such as silica, Fe, MgO, and Fe2O3. So it can be said that the study does not prove the results Maryanto colleagues who say that the higher the levels of starch will result in increasingly high levels of ash produced.

When compared to the ash content of the briquettes with a composition of 50:50 and 80:20, then the composition of 80:20 has more ash content compared with the composition of 50:50. If the terms of the amount of silica contained in both types of briquettes, the silica content in the composition of 80:20 more than the composition of 50:50. According to research conducted Aditia Warman in 2005 and Prof. Dr. Muhammad Zarlis, M.Sc in 2009 stated that the more the silica content in a briquette, the ash produced will be more and more, because the residual silica materials including non-coumbustible. When compared to this study, the results of this research are not in accordance with the results of research conducted by two researchers.

When viewed from the water content, the water content of the briquettes with a composition of 50:50 compared to 80:20 lot more. This means that should the composition of the ash content 50:50 produce higher compared with the composition of 80:20.

Ash generated from this study do not meet the standards set by the government which, according to Minister of Energy and Mineral Resources of Indonesia Republk (ESDM RI) No. 047 2006 maximum ash content of charcoal briquettes that can be owned is <5%, while in this briquette research has a moisture content ranging from 19.92% - 30.44%.

By evaluation the fixed carbon content to the variation of starch levels it seem that the greater the level of starch, the smaller the bound carbon content is contained in the briquettes. This is due to the greater levels of starch, the higher the water content. Carbon content will be low if the levels of volatile matter and water content increase. One of the characteristics is, a good briquette have a high fixed carbon value, because it is what carbon compounds will produce heat when burned.

If the comparisons take between the briquettes with 80:20 rice husks and coconut shell charcoal composition with the composition 50:50, The 50:50 briquettes composition has a higher fixed carbon content. This is caused by the ash content and low levels of volatile matter. The existence of bonded carbon in charcoal briquette is affected by the ash content values and levels of substance evaporate. Levels of bound carbon are high when the value of ash content and levels of substance evaporates at charcoal briquettes are lower.

Fixed carbon levels resulting from this study did not meet the Indonesian National Standard (SNI) No. 1/6235/2000 namely briquette should have a fixed carbon content of at least 77%, while briquette research has a moisture content ranging from 38.81% - 49.63%.

Results of the determination of the calorific value of the average energy above shows that the addition of starch content causes decreased energy calorific value. This happens because the higher the water content contained in the briquettes with increased levels of starch, so that the heat generated in advance is used to evaporate water in the briquettes.

If the comparison between the type of sample to the composition of rice husk: coconut shell charcoal = 50:50 and 80:20 clearly seen on the graph that the type 1 (composition of rice husk: coconut shell charcoal = 50:50) has a value of more heat energy high compared with other types of briquettes 2 (composition of rice husk: coconut shell charcoal = 80:20). This is due to the silica content is pretty much on the composition of the briquettes 50:50. The high silica levels have led to higher water content for this silica absorbs water so that the briquettes are hygroscopic.

4. Conclusion

Bio-briquette models made from the mixtures of rice husk and coconut shell charcoal have been successfully created in this research. The highest calorific value of energy is possessed by the sample-1B with 4966 cal/g and the lowest is at sample-2C with 4104 cal/g.

Factors that might affect the content of heat energy value are moisture, volatile matter content, ash content, and the levels of carbon bonded/fixed carbon. Moisture, volatile matter content and ash content result in the higher will lower calorific energy value and vice versa. While higher levels of fixed carbon will result in increasing the calorific of briquettes.

Briquettes that have the optimal calorific / high value of energy is a sample in which the composition of rice husk: coconut shell charcoal is 50%:50%, a starch content of 6%, the water content of 7.63% adb, ash content of 20.21% adb, volatile matter content of 23.12% adb and level of fixed carbon 49.04%.

References

[1] Zed F *et al.* 2014 *The national energy council of the republic of indonesia, indonesian energy outlook 2014* (Jakarta: Ministry of Energy and Mineral Resources of Indonesia)
[2] Nur Shukri M 2014 Characteristics of waste rice for raw materials bioenergy *Indonesia Energy Outlook 2014 Energy Development Program in Support of Fuel Substitution* (Jakarta: Energy Resources Development Technology Center (PTPSE))
[3] Mukti and Hafidz 2015 Rice production highest in 10 years
[4] Soelaiman and Jalal R 2013 Characteristics comparison of various rice husk-based briquettes as renewable energy *thesis* Department of Physics, Faculty of Mathematics and Natural Sciences University of Jember
[5] Gandhi B and Aquino 2010 Effect of total mixed adhesives variation on corn cob charcoal briquette characteristics (paper) (Department of Chemistry, Faculty of Mathematics, University of Hasanuddin)
[6] Maryono *et al.* 2013. Preparation and Quality Analysis Coconut Shell Charcoal Briquette Judging from the content of Kanji (paper). Department of Chemistry, State University Negeri Makasar.
[7] Simorangkir A and Tiffani 2014 Proximate analysis, ultimate analysis and analysis on coal miscellaneous *skrispi* (Department of Mining Engineering Faculty of Mining Institute of Technology)
[8] Jamil and Sofhal 2015 Coconut shell charcoal (http://www.arangbatok.org/2014/08/komposisi-batok-kelapa.html) September 15th 2015

IOP Conf. Series: Earth and Environmental Science **65** (2017) 012022 doi:10.1088/1755-1315/65/1/012022

Formulation of insecticide profenofos using Surfactant Diethanolamide (DEA) based on palm olein

H S Dewi[1], M Rahayuningsih[2] and E Hambali[1]

[1]Surfactant and Bioenergy Research Center, LPPM – Bogor Agricultural University, Bogor, Indonesia
[2]Departement of Agricultural Industrial Technology, Faculty of Agricultural Technology, Bogor Agricultural University, Bogor, Indonesia

Email : nyonyapastimati@gmail.com

Abstract. Soybean is one of the major food commodities in Indonesia that the consumption is increasing each year, but this is not in line with the domestic soybean production capacity. One cause of the low production capacity is the armyworm attact. Generally, the armyworm attack controled by spread insecticide profenofos. Profenofos need to be dissolved, but profenofos couldn't dissolved in water. So that, it need the right formulation between the solvent and other ingredients which can supprotprofenofos performance. One of that ingredient is surfactant. This research used surfactant diethanolamide (DEA) based on palm olein. DEAfunction in insecticide formulation are as homogenizer, dispersant, sticker and spreader agent.The aims of this research are to obtain the best emultion insecticide product based on profenofos as the active ingredients and DEA as the surfactant, moreover it also to obtain information of the physico-chemical properties. The formulation test performed with compeletely randomized design (CRD) with two factors, first factor is DEA concentrationand the second factor is profenofos concentration. Data of physico-chemical properties test was analyzed by analysis of variance (ANOVA) and significant result tested by Duncant Multiple Range Test (DMRT).The result showed that, surfactant DEA could make good emultion between profenofos and sodium ethoxide as the solvent. The best treatment which obtain from formulation stage is concentrate with DEA 10% and profenofos 40%. Physico-chemical properties test result showed that droplet size is 1,76-2,07 µm, contact angle 11,575-24,218°, density 0,996-0,998 g/cm^3, surface tension 16,56-40,72 dyne/cm, viscosity 1,032-1,078 Cp and pH 6,87-8,22.

1. Introduction

Surfactant is a substance that has a surface active nature of molecules ampyphilic which has two groups with different properties in a single molecule is hydrophilic group and lypophilic. Surfactants have the ability to decrease the surface tension between the two phases that have different polarity,e.g liquid-liquids, solids-liquids, gas-liquid phases. Term of surface refer to interface that one of the two phases is gas [1]. Surfactant in low concentration has ability to ensconce between two interfaces of two medias that can't be dissolve, so that significantly change physical characteristics of the interface. Surface-active agents in surfactant has abilityto modifying the surface characteristics of a liquid or solid, to be coagulant, wetting agent, emulsifier, dispersant, adhesive, etc [2].

One of surfactant that potentially can be used to applied in insecticide to controlling the armyworm is dietanolamida (DEA). DEA will dispersing, homogenizing, leveling and sticking the active ingredient

Content from this work may be used under the terms of the Creative Commons Attribution 3.0 licence. Any further distribution of this work must maintain attribution to the author(s) and the title of the work, journal citation and DOI.
Published under licence by IOP Publishing Ltd

with other additives and the carrier media. Based on research conducted by a team of researchers SBRC in 2012, DEA has lowest surface tension value (20.97 dyne / cm) compared to other surfactants such as APG (21-22 dyne / cm), ethoxylate (23-25 dyne / cm), and lauryl betaine (31.17 dyne / cm) which is widely used in pesticides industry. Therefore, DEA potentially an increase insecticides effectiveness [3].

DEA can be produced from methyl esters and fatty acid reacted with diethanolamine. DEA on the market today produced from coconut oil andpalm kernel oil (PKO), which are quite expensive. However SBRC LPPM has been able to produce surfactant DEA with palm oleinthat cheaper then PKO. So that it needs a research to utilize DEA application in insecticide product.

Soybean is one of the major food commodities in Indonesia that the consumption is increasing each year, but this is not in line with the domestic soybean production capacity, so the government had to import soybeans each year to full fill the domestic need.Data from the Ministry of Agriculture showed soybean production in 2015 based on Forecast Figures (FF) Central Statistics Agency (CSA) is 998,870 tons of soybean dry seeds, whereas public consumption up to 2.54 million tons of soybean dry seeds. This soy deficit caused Indonesia to import soybeans to full fill domestic soybean need. One of the factors that causedthis lowsoybean production is armyworm attact. Armyworm (*Spodopteralitura)*causes loss of harvests up to 80% [4].Armywormspest controlling usually do by spraying profenofosinsecticide. The objectives of this research are to obtain the best emultion insecticide product based on profenofos as the active ingredients and DEA as the surfactant, moreover it also to obtain information of the physico-chemical properties.

2. Materials and Methods

2.1. *Instruments and Materials*

Vortex HeidolphReax top, and a rotor stator homogenizer Daihan HG-15D models, Tokebi homogenizer (22000 rpm) were tools that used in this research. Another tool used for sample analysis were pH meter (pH Meter Schott), density meter (Anton Paar DMA 4500M), potentiometer (Spinning Drop Tensiometer), viscometer (Rheometer Brookfield DV-III Ultra). The materials used in this researchwere surfactants diethanolamide (DEA) obtained from SBRC LPPM IPB, active ingredient was profenofosobtained from PT. Petrosida Gresik. Other materials wereNaOH and 96% ethanol.

2.2. *Solvent Preparation*

The insecticide concentrate was prepared by mix NaOH in etanol or natriumetoxcide to profenofos and followed by adding DEA, then they were homogenized for 5 minutes in 22000 rpm speed using Tokebi Homogenizer.

2.3. *Insecticide Formulation*

This stage aims to study the DEA and profenofos effect to the emultion physical character. Physico-chemical character that tested were contact angle (°), droplet size (μm), dencity, surface tension viscocity and pH. Completely Randomized Design (CRD) were used to done the insecticide formulation which is the first factor was DEA concentration (0, 10, 15 dan 20%) and second factor was profenofos concentration (40, 50, dan 60%).

3. Results and Discusion

3.1. *Solution Formulation Insecticides*

Generally, pesticides formulations can be classified into two major categories, they are liquid formulations and solid formulations. Liquid formulations are usually composed by active ingredients, solvents, and additives such as emulsifiers, leveling, adhesives and others. EC (emulsifiable

Concentrate), is one type of liquid formulation that prepared by dissolving the active ingredient in a particular solvent and by adding one or more surfactants or emulsifiers [5]. Insecticide solution in this study were made in EC formulation.

Most of insecticide productsin the market is produced by mixing active ingredient and additive (adjuvant). The active ingredients are toxic chemical compounds that kill the insects, but the presence of extra material is needed to increase theactive ingredient action mechanism. Additional materials are often used in insecticide formulations include solvents (solvent), surfactant (emulsifier), adhesive (sticker), wetting agents (wetting agent), colorant (coloring agent) and others [5].

Surfactant is an active compound tan can decrease the surface tension. Surfactants are amphipatic molecules which have a hydrophilic natures (polar), and hydrophobic nature (non-polar). This propertis caused surfactnat can be dissolved in a slition that has two defferent polarity degrees and hydrogen bonds such as water and oil. Surfactants are divided into four main groups and are widely used in almost all sectors of modern industry. The types of surfactants are anionic, cationic, nonionic and amphoteric surfactants [6]. The surfactant DEA that used in this study is a non-ionic surfactant.

Profenofos has the chemical name O- (4-bromo-2-chlorophenyl) O-ethyl S-propyl phosphorothioate [6]. Profenofos belonged toxic organophosphate stomach and contact poison [5].Refer to profenofos molecular structure, it has negative poles on Chlor and Brom, so generally profenofos has electronegativity. The big profenofos molecular structure caused it can't be dissolved in water and can't be evaporated in room temperature but it can recieve positive group from other compound. So that, it needs a right formulation to form good insecticdeemultion.Figure 1.Shows the profenofos chemical structure.

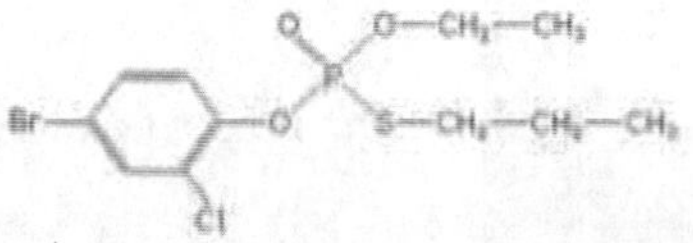

Figure 1. Profenofos Chemical Structure [6]

Rever to the negative charge, profenofos can be coated by kationic compound or a compound that has relatively positive charge, so in this research natrioumetixcide ($NaOC_2H_5$) have choosen. Natrium etoxcide has positive pole in the natrium and also in –CH_3 groups, so that natriumetoxcide forming Van Der Waals bond with profenofos. –Na grup that has positive charge will go to –Cl and –Br groups to forming Van Der Waals bond, it caused profenofos particle has a mantel with negative charge that formed by CH_3 group of natriumetoxcide. Thi mantel formed zeta potential that increase koloid stability. But this bond is not too strong, so if it diluted by water, it will detached and the emultion come apart. So that, it need to add a surfactant that can form a new mantel, in this reseach palm oleindietanolamide (DEA) was used.

Ethanol group of DEA has a relatively positive charge so that DEA will be heading -CH3 of sodium ethoxide, forming a new Van der Waals bond and a new outermantel,it makes the distance to the outer layer of profenofos particles getting further and the colloid becomes more stable because zeta potential that formed getting larger. If the emultion mixed into water, it will forming a water mantel from the bond of –CH3 group of DEA and water. Larger zeta potential makes more stable colloid. Figure 2.Showed DEA molecular structure.

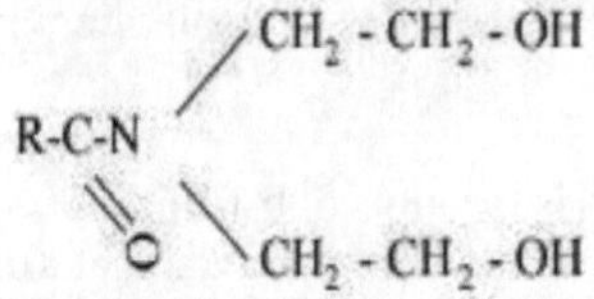

Figure 2. Dietanolamide chemical structure

Insecticide formulation in this research aims to study the DEA and profenofos effect to the emultion and also it'sphysico-chemical characters. The concentrate visual appearance that formed in this stage have transparent tawny color and specific profenofos scent that similar with very strong garlic scent. After 700 ppm of the concentrate dilluted in water, the solution become white color, still has specific profenofos scent but not too strong. Figur 3 and 4. Showed the insecticide concentrate and it's 700 ppm solution in water.

Figure 3. Profenofos insecticide concentrate solution

Figure 4.700 ppm profenofos insecticide solution in water

The physico-chemical test results showed droplet sizes range was 1.41-2.07 μm, the contact angle range was 11.56-24,22°, the density range was 0.996-0.998 g/cm3, the surface tension range was 15.833-40.717 dyne/cm, viscocity range was 1.032-1.115 cP and pH range was 6.328-8.207. Table 1. Showed the physico-chemical insecticide characteristic test result.

Table 1. The physico-chemical insecticide characteristic test result (0.07% profenofos in water)

Surfactant DEA (%)	Profenofos (%)		
	40	50	60
	Droplet size (μm)		
0	1,843 ± 0,334[b]	1,770 ± 0,127[b]	2,073 ± 0,046[b]
10	1,987 ± 0,287[c]	2,057 ± 0,165[b]	1,950 ± 0,190[b]
15	1,777 ± 0,053[b]	1,850 ± 0,275[b]	1,757 ± 0,061[b]
20	1,990 ± 0,045[b]	1,413 ± 0,074[a]	1,777 ± 0,045[b]
	Contact Angle		
0	22,830 ± 0,238[c]	14,398 ± 0,367[a]	24,218 ± 0,557[c]
10	18,477 ± 0,747[b]	13,631 ± 0,304[a]	23,543 ± 0,376[c]
15	13,730 ± 0,519[a]	12,137 ± 0,292[a]	11,575 ± 0,467[a]
20	11,556 ± 0,565[b]	16,413 ± 0,651[b]	21,070 ± 0,356[c]
	Density		
0	0,996 ± 0,00004[a]	0,996 ± 0,00004[a]	0,997 ± 0,00008[d]
10	0,997 ± 0,00004[b]	0,998 ± 0[e]	0,998 ± 0,00004[g]
15	0,997 ± 0,00009[c]	0,998 ± 0,00004[g]	0,998 ± 0,00004[f]
20	0,997 ± 0,00020[d]	0,997 ± 0,00023[d]	0,996 ± 0,00026[a]
	Surface Tension		
0	40,717 ± 1,276[e]	19,257 ± 1,138[b]	32,627 ± 0,677[d]
10	27,22 ± 0,682[c]	19,390 ± 1,074[b]	28,090 ± 3,275[c]
15	26,890 ± 1,553[c]	16,563 ± 0,959[a]	19,570 ± 1,740[b]
20	26,793 ± 2,704[c]	15,833 ± 0,927[a]	17,333 ± 0,684[a]
	Viscosity		
0	1,032 ± 0,003[a]	1,038 ± 0,003[a]	1,040 ± 0,013[a]
10	1,078 ± 0,008[b]	1,075 ± 0,005[b]	1,073 ± 0,012[b]
15	1,075 ± 0,013[b]	1,077 ± 0,008[b]	1,078 ± 0,002[b]
20	1,097 ± 0,007[c]	1,115 ± 0,015[c]	1,115 ± 0,015[d]
	pH		
0	8,207 ± 0,038[d]	6,873 ± 0,847[a]	7,787 ± 0,060[b]
10	8,153 ± 0,035[d]	7,887 ± 0,040[c]	7,310 ± 0,036[a]
15	8,118 ± 0,022[d]	7,757 ± 0,045[b]	7,210 ± 0,010[a]
20	7,863 ± 0,032[c]	7,860 ± 0,875[c]	6,328 ± 0,080[a]

Description: Figures in columns and rows followed by the same letter show no significant based DMRT at $\alpha = 0.05$

3.2. Droplet size (μm)

Droplet size in this study was measured in micro unit. In emultion system, the droplet size that have size 0.1 μm or 0.1-50 μm formed granules that dispersed well in other lequid. The smaller droplet size of emultion means that it has good character [7]. Droplet size in this study was measured in micro unit. In emultion system, the droplet size that have size 0.1 μm or 0.1-50 μm formed granules that dispersed well in other lequid. The smaller droplet size of aemultion means that it has good character. So that, droplet size obsevation was needed to ensure the insecticide emultion has droplet size in micro, it became indication that the insecticide emultion was already well formed and homogen.Droplet size analysis in this stage done by dilluted 700 ppm insecticide in aquadest.

Analysis of variance results showed that DEA has significant effect to the droplet size in convidence interval 95%. It means that DEA has function as mantel protector in the emultion system with dillution up to 700 ppm. The result also showed that profenofos has no significant effect and there were no interaction between DEA and profenofos in the insecticide solution although it has been dilluted up to 700 ppm.

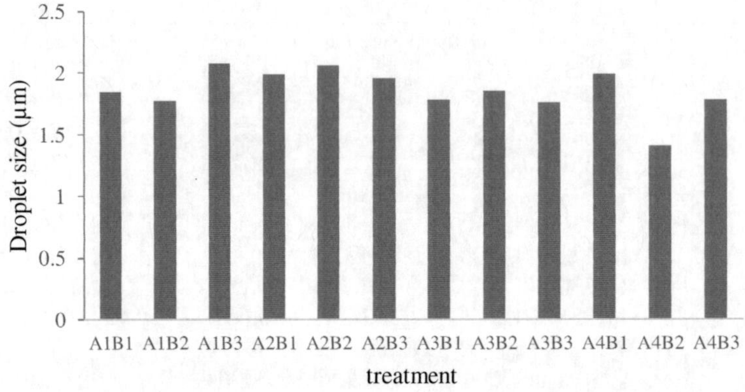

Figure 5. relationship of 0.07% profenofos insecticide solution in water to the droplet size

Information:
A1: The addition of DEA 0%
A2: The addition of DEA 10%
A3: The addition of DEA 15%
A4: The addition of DEA 20%
B1: The addition of Profenofos 40%
B2: The addition of Profenofos 50%
B3: The addition of Profenofos 60%

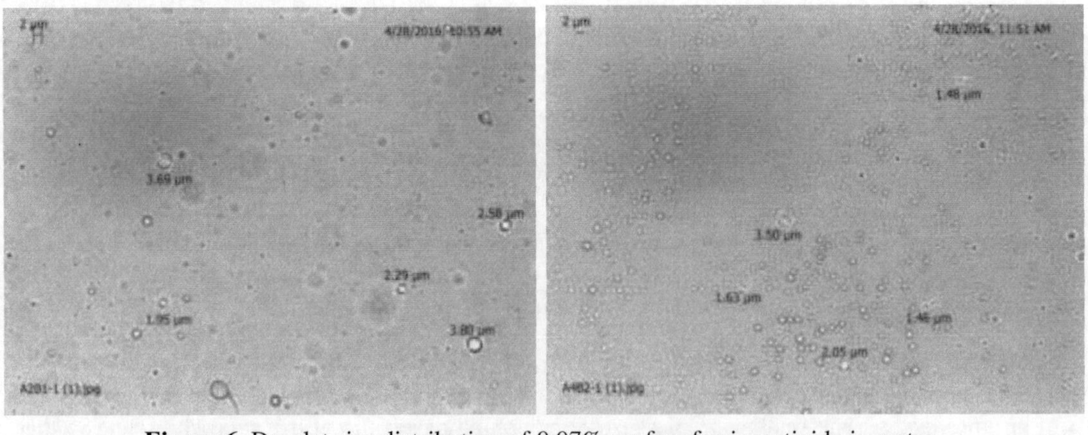

Figure 6. Droplet size distribution of 0.07% profenofos insecticide in water

Duncant Multiple Range Test (DMRT) result showed that DEA has significant effect at A2B1 treatment (20% DEA : 50% profenofos). It means that the bond of DEA-natriumetoxcide-profenofos formed stable emultion.

3.3. Contact angle (°)

Contact angle measured to see the insecticide solution ability to stick and spread on the soybean leaf. Insecticide products are expected to have as narrow as possible. It related to the insecticide ability to stick and spread on the object. 0° means there is no angle that formed so the droplet get well sticking on the object, 90° means that the droplet on the object can only stick but can't spread, and 180° means the droplet can't stick on the object and immediately slip [9]. Contact angle measured to see the insecticide solution ability to stick and spread on the soybean leaf. Insecticide products are expected to have as narrow as possible. It related to the insecticide ability to stick and spread on the object. 0° means there is no angle that formed so the droplet get well sticking on the object, 90° means that the droplet on the object can only stick but can't spread, and 180° means the droplet can't stick on the object and immediately slip. Contact angle of the insecticide in this research was 11.55°-24.2°, this range could be caused by the differences in natural absorption of the leaf.

ANOVA results showed that DEA, profenofos and they insteraction have significant effect to the 700 ppm insecticide in water. it could be caused by the defferences of the leaf to absorbing DEA and profenofos. Figure 7. Showed realtionshipbetween the insecticide with contact angle.

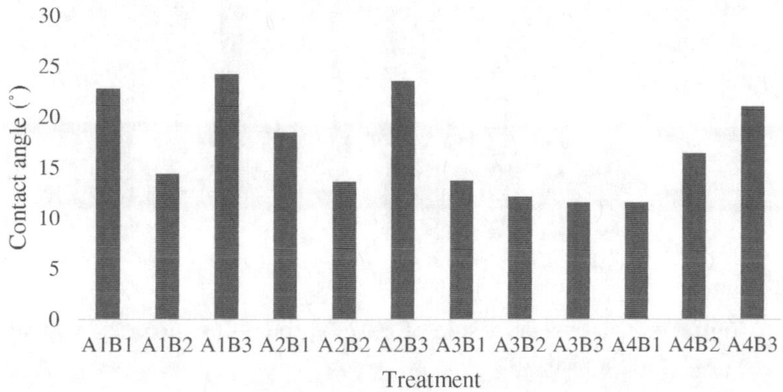

Figure 7. relationship of 0.07% profenofos insecticide solution in water to the contact angle

Information:
A1: The addition of DEA 0%
A2: The addition of DEA 10%
A3: The addition of DEA 15%
A4: The addition of DEA 20%
B1: The addition of Profenofos 40%
B2: The addition of Profenofos 50%
B3: The addition of Profenofos 60%

The significant effect that given by DEA to the insecticide in this research prove that DEA has function as dispersant, homogenizer, leveler and sticking agent in profenofos solution with natriumetoxcide as the solvent. This was in accordance with [8] who said that visual observation about droplet of a solution that containing surfactant could be spreaded out and pnetrated between fine hairson a leaf via it's capilarity. Whereas, the solution droplet without any surfactant could be spreaded out and could only sticked on a leaf.

3.4 Density (g / cm3)

Emultion system of the profenofos insecticide solution in this research was oil in water (o/w) that showed by the dispertion of oil phase into the dispersant that was water. Density is an important parameter to be analysed because density value related to surface tension value which is one of emultion stability factor.

Density defined as the ratio of the weight of a material to the total volume of the constituent materials. Density analysis performed by using density meter. The analysis shows that the density of the insecticide solution ranges from 0.996 to 0.998 g / cm3. The results of the data analysis by analysis of variance showed that the addition of surfactant DEA and profenofos have significant effect to the density of the solution, and there was an interaction between the DEA and profenofos in the insecticide solution 700 ppm in the water. Figure 8. Shows the relationship of 0.07% profenofos insecticide in water to the density.

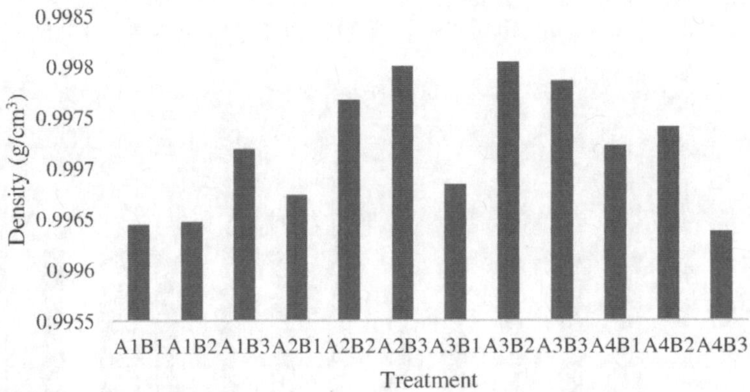

Figure 8. Curve relationship of 0.07% profenofos insecticide solution in water to the density

Information:
A1: The addition of DEA 0%
A2: The addition of DEA 10%
A3: The addition of DEA 15%
A4: The addition of DEA 20%
B1: The addition of Profenofos 40%
B2: The addition of Profenofos 50%
B3: The addition of Profenofos 60%

DMRT result showed that the DEA and profenofos have significant effect to density at 95% confidence interval, and there was an interaction in 700 ppm insecticide solution in the water. This results could be caused by DEA that has lessdensity than the profenofos density, solvent and water used as a diluent so that the addition of surfactant DEA can reduce the density of 700 ppm insecticide solution in water.

3.5 Surface Tension (dyne / cm)

The surface tension of a liquid is caused by the internal pressure of affinity molecules on the surface to the bottom surface of the liquid [9]. The surface tension is formulated as □□= F / d, □ is surface tension, F is the force of surface tension in N and d is the length or diameter with units of m so that the surface tension has units of N / m. Formula profenofos the DEA surfactant sodium ethoxide with an emulsion system so that the onset of surface tension caused by the particles in the emulsion and

IOP Conf. Series: Earth and Environmental Science **65** (2017) 012022 doi:10.1088/1755-1315/65/1/012022

solvent. Changes in surface tension is influenced by the content of particles and particle size of the emulsion surface. Curves formulations insecticide solution to the surface tension shown in Figure 9.

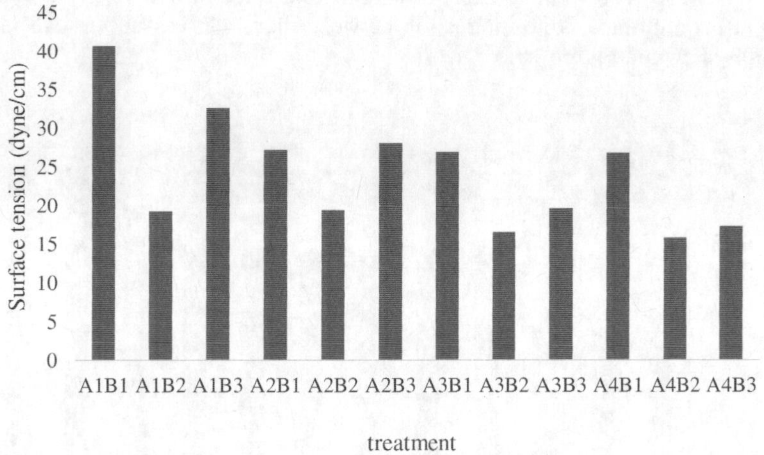

Figure 9. Curve relationship of 0.07% profenofos insecticide solution in water to the surface tension

Information:
A1: The addition of DEA 0%
A2: The addition of DEA 10%
A3: The addition of DEA 15%
A4: The addition of DEA 20%
B1: The addition of Profenofos 40%
B2: The addition of Profenofos 50%
B3: The addition of Profenofos 60%

Analysis results showed that the insecticide surface tension 700 ppm in water were 15.833-40.717 dyne/cm. ANOVA results showed that DEA and profenofos have significant effect to the surface tensiaon and there were interaction between them.

DMRT results showed that DEA and profenofos have significant effect to the surface tension in 95% interval confidence and there were interaction between them. Profenofos in natriumetoxide with DEA is an emultion system, so that surface tension formed by the particles in it's solvent. Surface tension change was influenced byparticles and particle size in the emultion surface. The significant effect could be caused by DEA that could decrease the density of insecticide formula, the smaller density makes distance between particles, so that the attractive force between particles become weaker and the surface tension got decrease.

3.6 Viscosity (cP)
Viscosity is one of the fluid properties are affected by the size and intermolecular forces. Viscosity indicates the viscosity level of a fluid. The higher viscosity means the higher the viscosity level of a fluid, which indicates changes in the structure and bonding between the molecules. The increase in viscosity due to increased concentrations of particles, as well as the flow pmaterials properties will depend on the viscosity and density of the liquid [7].
Armyworms controlling in the field done by spraying profenofos insecticide. One important vaktor that will affect the use of insecticide solution on sprey tool is viscosity, the higher viscosity of the

insecticide solution will be more difficult to be sprayed, the lower viscosity of the insecticide, it will be easier to be sprayed. Therefore, the viscosity of insecticides in this study need to be observed.
Data analysis results show that the viscosity of the insecticide were 1.032 to 1.115 cP. The results of the data analysis by ANOVA showed that DEA has significant effect to the viscosity, but profenofos has no significant effect and there is no interaction between them. Curves on viscosity formulations insecticide solution is shown in Figure 10.

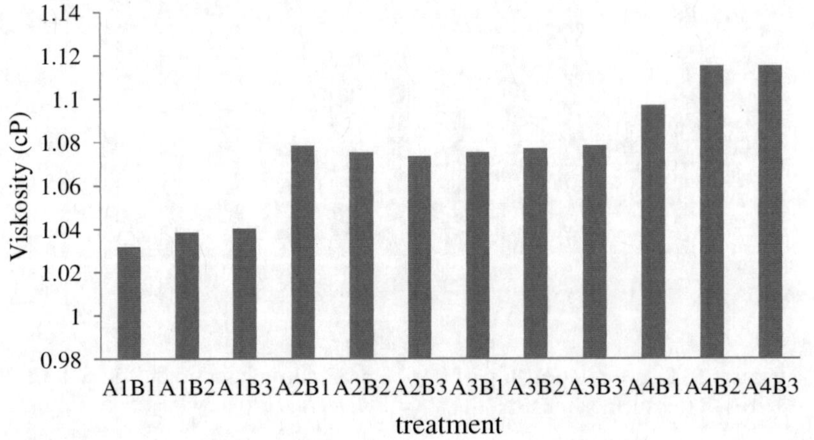

Figure 10. relationship of 0.07% profenofos insecticide solution in water to the viskosity

Information:
A1: The addition of DEA 0%
A2: The addition of DEA 10%
A3: The addition of DEA 15%
A4: The addition of DEA 20%
B1: The addition of Profenofos 40%
B2: The addition of Profenofos 50%
B3: The addition of Profenofos 60%

High viscosity due to high particle concentrations as well, so the flow properties of materials depend on the liquid viscosity and density. Liquid that easy to flow is the liquid that has low viscosity and materials that difficult to flow are the material with high viscosity.
The test further results by DMRT showed that the addition of surfactant DEA significantly different with the viscosity at 95% confidence interval, this can be caused by surfactants DEA that has a high viscosity compared with viscosity profenofos and solvent, so that the treatment the addition of surfactant DEA has a significant effect on the viscosity. This is also supported by previous data, the addition of surfactant DEA significantly affect the density, which is one factor of viscosity grades. In addition, the insecticide solution can be seen also in this study had a small droplet size, so it can be said emulsion insecticide solution has a small particle size, the smaller the particle size, the surface area for greater frictional forces so that the viscosity becomes.

3.7. The degree of acidity (pH)
Values of pH or acidity is used to express the degree of acidity or alkalinity of pH. pH also relates to the concentration of hydrogen ions, as a component of acidity and concentration of hydroxyl ions as a component of alkalinity (Rondinini et al.2001). At neutral pH conditions, the concentration of the two ions balanced but if the hydrogen ion concentration is greater than hydroxyl ions then tend to acidic

pH (lower). In general, the pH value of a substance in the range of grades 0-14. The range of pH values of 0-6 indicates an acidic substance, while the range pH 8-14 shows the alkaline material [7]. The results of data analysis showed that the pH of the resulting insecticide solution ranges from 6.328 to 8.207, but the data can be seen that the insecticide solution has tended toward alkaline pH, ie a pH greater than 7.

pH is the degree of acidity, expressed as -log [H +] and therefore in this study, the pH value of the data is transformed into a logarithm form in order to obtain the value of [H +] prior to analysis of variance and a further test. Curves formulations insecticide solution against pH shown in Figure 11.

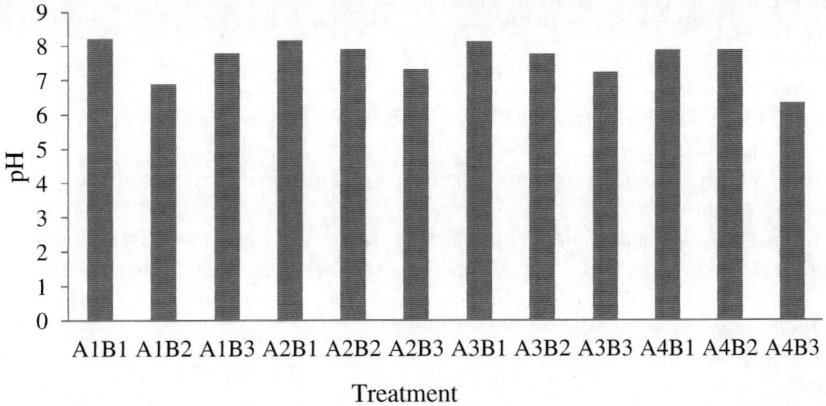

Figure 11. Curve relationship of 0.07% profenofos insecticide solution in water to the pH

Information:
A1: The addition of DEA 0%
A2: The addition of DEA 10%
A3: The addition of DEA 15%
A4: The addition of DEA 20%
B1: The addition of Profenofos 40%
B2: The addition of Profenofos 50%
B3: The addition of Profenofos 60%

The results of the data analysis by ANOVA showed that the addition of surfactant DEA and active ingredients profenofos gives significant effect on pH and there is interaction between the two. Further test results by DMRT showed that the addition of surfactant DEA and active ingredients profenofos significant effect on the pH of the 95% confidence interval, and there was an interaction between them. It can be caused by DEA surfactant that has alkaline properties that can raise konsektrasi [H^+] in insecticide solution so that the pH becomes alkaline.

4. Conclusion

The best insecticide solution obtained from this study is the insecticide solution with an active ingredient profenofos 40% in solvent and surfactant sodium ethoxide DEA 10% (A2B1). Chemical physical properties possessed by this insecticide solution that droplet sizes ranging from 1.41 to 2.07 μm, the contact angle ranged between 11.56 - 24,22°, the density ranges from 0.996 to 0.998 g / cm3, the surface tension ranges 15.833-40, 717 dyne / cm, 1.032 to 1.115 cP viscosity and pH ranged from 6.328 to 8.207.

Reference

[1] Sanatkaran N,Masalova I and Malkin AY 2014 Effect of surfactant on interfacial film and stability of highly concentrated emulsions stabilized by various binary surfactant mixtures *J Colloids and Surfaces A: Physicochem. Eng. Aspects* **461** 85 – 91.

[2] Bergenstahl B A and Claesson P M 1990 *Surface Forces in Emulsions*in: K. Larsson S E Firberg (Ed.) *Food Emulsi* (New York: Marcel-Dekker Inc.)

[3] Hambali E, Suryani A, Rivai M, Sutanto AI, Nisya F N and Nurkania A 2013 *Pengembangan Teknologi Proses Produksi Surfaktan Dietanolamida (DEA) dari Metil Ester Olein Sawit dan Aplikasinya untuk Personal Care Products* Konferensi Nasional Inovasidan Technopreneurship (Indonesia : Bogor Agricultural University)

[4] Marwotoand and Suharsono 2008 Strategi dan komponen teknologi pengendalian ulat grayak (Spodopteralitura Fabricius) pada tanaman kedelai Jurnal Penelitian dan Pengembangan Pertanian **4** 131-136.

[5] [Ditlintan] Direktorat Perlindungan Tanaman Pangan *Laporan Luas dan Serangan Hama dan Penyakit Tanaman Pangan di Indonesia* 2013 (Jakarta:Ditlintan)

[6] Knowles A 2008 recent developments of safer formulations of agrochemicals J. Environmentalist **28** 35 – 44

[7] De Man J M 1997 *Kimia Makanan* (Indonesia: Institute Technology of Bandung)

[8] Siegfired B D and Scharf M E 2001 *Mechanisms of organophospate resistence in insects* in Ishaaya I *Biochemical Sites of Insectiside Action and Resistence* (New York: Springer-Verlag) pp 269-87.

[9] [BPS] Badan Pusat Statistik 2015 *Produksi Padi, Jagung dan Kedelai berita Resmi Statistik* No. 62/07/ Th. XVIII

[10] Departemen Pertanian 2008 *Panduan pelaksanaan sekolah lapang pengelolaan tanaman terpadu (SL-PTT) kedelai* (Jakarta-Indonesia: Departemen Pertanian)

Granulation of coal fly ash by using different types of granule agents

H Agusta[1,2], F N Nisya[1], R N Iman[2], D B C Bilad[1]
[1] Surfactant and Bioenergy Research Center (SBRC), Bogor Agricultural University. Indonesia
[2] Department of Agronomy and Horticulture, Faculty of Agriculture, Bogor Agricultural University. Indonesia

Email : fifin.nisya@gmail.com

Abstract. The use of coal produces about 5% solid pollutant in the form of ash (fly ash and bottom ash). Of the total ash produced, about 10-20% is bottom ash and 80-90% is fly ash. This study was aimed at obtaining a type of adhesive which could be used as a fly granulation material for soil conditioner. The study was conducted at the pilot plant of Surfactant and Bioenergy Research Center (SBRC) LPPM IPB from April to August 2016. The fly ash used in this study was obtained from Kalimantan. A pan granulator was used in fly ash granule making process. Granule agent materials were diluted in the concentration of 5, 10, and 15%. Different types of granule agents, namely SBRC-M, SBRC-T, and SBRC-SC were used. The formed fly ash granules were then analyzed for their physical properties including particle density, fly ash granule pH, fly ash granule durability, and fly ash granule water holding capacity. Results showed that fly ash granules made from 15% of SBRC-M had the highest particle density (0.75 g/cm^3). Fly ash granules made with SBRC-M had higher pH (10) than those made by using SBRC-SC adhesive (9.3) and SBRC-T (9). SBRC-T was found as the granule agent material which produced fly ash granules with the highest durability levels on average. In this study, the use of SBRC-M granule agent resulted in higher water holding capacity (WHC) (40.62%) than did SBRC-SC (38.79%) and SBRC-T (36.85%). As a granule agent, compared to SBRC-SC and SBRC-T, SBRC-M could produce fly ash granules with highest particle density, highest pH, good durability, and best water holding capacity.

1. Introduction

The use of coal produces about 5% solid pollutant in the form of ash (fly ash and bottom ash). Of the total ash produced, about 10-20% is bottom ash and 80-90% is fly ash. Physically, fly ash is a material containing granules which are fine in size and gray in color. Fly ash is a heterogeneous mixture of amorphous and crystalline phases and generally considered as ferroaluminosilicate. It comprised of about 69 percent silt and clay size fractions [1]. Chemically, fly ash contains silica (SiO_2), alumina (Al_2O_3), ferric oxide (Fe_2O_3), and calcium oxide (CaO). It also contains some additional compounds and element including magnesium oxide (MgO), titanium oxide (TiO_2), alkaline (Na_2O and K_2O), sulfur trioxide (SO_3), phosphorus oxide (P_2O_5), and carbon. Fly ash was also found to have Ca and Mg contents, high pH value (11), and high cation exchange capacity [2].

Content from this work may be used under the terms of the Creative Commons Attribution 3.0 licence. Any further distribution of this work must maintain attribution to the author(s) and the title of the work, journal citation and DOI.
Published under licence by IOP Publishing Ltd

Fly ash has been much used as a soil conditioner material to solve agricultural problems in Greece [3], Thailand [4], India [5, 1, 6], USA [7], Japan [8,9], and other countries. Fly ash was proven to be able to repair the structure, bulk density, water holding capacity, conductivity, and particle size of soil [5]. Fly ash, which was alkaline, could increase soil pH. It was also found to increase plant production as it contained high K^+, Zn^{2+}, Ca^{2+}, Mg^{2+}, and Fe^{3+} ions which were important for plant growth.

These physical and chemical properties have made the potential of fly ash utilization as ameliorant/soil conditioner very high. In addition to improving soil pH, fly ash also provide nutrients required by plants. However, when it is used as a soil conditioner, there is a concern that fly ash may pollute the air and affect health when it is inhaled as its particles are very small in size and volatile. Therefore, fly ash needs to be granulated in order to make it uneasy to get inhaled by humans and easy to be transported. This study was aimed at obtaining an adhesive which could be used as a fly granulation material for soil conditioner.

2. Methodology

The study was done at the pilot plant of Surfactant and Bioenergy Research Center (SBRC) LPPM IPB from April to August 2016. Fly ash was obtained from Kalimantan. Granules were made by using a pan granulator. Granule agent material was diluted in a concentration of 5, 10, and 15%. During the granule production process, the pan granulator was kept rotating and granule agent spraying was done simultaneously. At least two personnel were needed, one as a stirrer and the other took care of the spraying. The granulation speed was set at 28 RPM. The granulation process was done for 15 minutes to produce granules with a diameter of 2-10 mm. The formed granule pellets were then taken out from pan granulator. The process proceeded to drying stage done by putting the granule pellets under direct sunlight and in an oven. The drying was done until the water content of the granules reached 9-15%. Good granule agents are easy to get, affordable, able to absorb water, and have good adhesive property and dispersion/release time which is correspondent to the rhythm of nutrient absorption by plants. In this study, granule agents were consisting of sugar by-products (SBRC-M), starch (SBRC-T), and soil cement (SBRC-ST).

The produced fly ash granules were analyzed for their physical properties, namely particle density, pH, dispersion time, durability, water holding capacity.

- Particle density: Particle density of fly ash granule was measured by putting 100 mL sample in a 100 mL beaker glass. The granules were weighed to get the particle density in 100 cm^3.
- pH : Five grams of fly ash granules were mixed with water in a ratio of 1:5. pH of the precipitate of the mixture was measured by using a pH meter.
- Durability: Five grams of fly ash granules were put in a beaker glass and stirred. Stirring was done by using a shaker set at the speed of 200 rpm for 10 minutes. Five replicates were allocated. When the shaking was finished, intact granules were taken out and weighed. The percentage of unbroken granules was determined.
- Water holding capacity: Water holding capacity was measured by referring to the method of [10]. The test was done by soaking the granules in water for an hour. Wet garanules were then weighed for their wet weight. The granules were then ovened at 60°C for 24 hours before they were weight for their dry weight. The test was done to determine the ability of soil conditioner granule to absorb water. Water holding capacity was calculated by using the following equation:
    - $$Water\ holding\ capacity = \frac{Wet\ weight - dry\ weight}{Wet\ weight}\ x\ 100\%$$

3. Results and discussion

3.1 Particle density

Particle density is the weight of fly ash granules per a volume unit. Mineral soil has a particle density of 2.6 g/cm^3 [11,12]. Particle density is stated in gram per volume unit (cm^3). Low value of its particle density established its potential for dust formation. The bulk density of fly ash was 1.01 mg/m3. Organic carbon content in the ash was 0.36 percent [1].

Results showed that the particle density of fly ash granules made from SBRC-M, DBRC-SC, and SBRC-T was 0.74, 0.68, and 0.67 g/cm^3, respectively. Organic soils in general have a particle density of 0.5-0.8 g/cm^3. The particle density of fly ash granules used in this study is listed in Table 1.

Table 1. Particle density fly ash granule (g/100 cm^3)

Granule agent (concentration)	*Particle density*	ST. Dev
SBRC-M 5% (M5)	0.73c	0.01
SBRC-M 10% (M10)	0.74b	0.01
SBRC-M 15% (M15)	0.75a	0.00
Average	**74.08**	
SBRC-SC 5% (SC5)	0.65f	0.01
SBRC-SC 10% (SC10)	0.70d	0.01
SBRC-SC 15% (SC15)	0.69d	0.01
Average	**67.91**	
SBRC-T 5% (T5)	0.65f	0.01
SBRC-T 10% (T10)	0.68e	0.01
SBRC-T 15% (T15)	0.68e	0.01
Average	**66.80**	

It was found that fly ash granules made from 15% of SBRC-M had the highest particle density (0.75 g/cm^3) while that made from SBRC-T 5% and SBRC-SC 5% shared the lowest particle density (0.65 g/cm^3). In general, higher adhesive concentration results in higher particle density. This can be used as a consideration in selecting adhesive material as particle density gives significant effect on granule packaging and transportation. The size of granule pellets affected the results of this study. Bigger granule size made more space in the beaker glass which then affected the total weight of the granules in the beaker glass.

Similar to soil, there were several factors including texture, organic material, structure, and bulk density affecting the particle density of fly ash granule [13]. Texture is a visual appearance of soil or granule resulted from qualitative composition of granule pellet size in a given granule mass. The texture of fly ash granule represents the composition of particles constituting the granule. High organic material content gives significant effect on particle density as found in coal fly ash granule used in this study. Fly ash obtained from coal-fired power plant was a very fine particle or a dust material. The size of fly ash of coal-fired power plant origin varies from the finest of 11-25 μm to the coarsest of 40-150 μm.

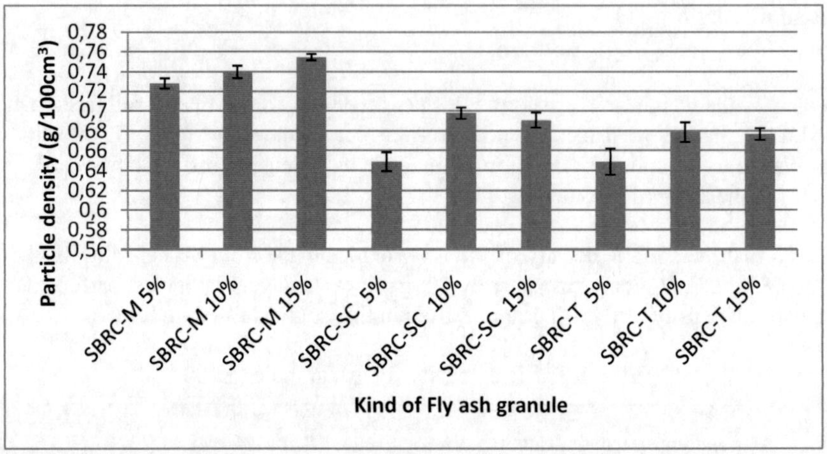

Figure 1. Particle density of fly ash granule made of some granule agents

3.2. pH

Plants require complete nutrients including macro, micro, and beneficial elements in proportional amount. Fertilizer formula containing these elements is beneficial in improving plant productivity and maintaining nutrient balance in the soil. Acidity level (pH) is an important factor affecting the process of nutrient absorption by plant roots. Soil with neutral acidity level is the most ideal condition as most nutrients are easily dissolved in water.

Soil pH level also shows the presence of elements which are toxic to plants. Aluminum (Al) is found a lot in acidic soil. Al is not only toxic to plants but it also binds phosphorus making the phosphorus not available for adsorption by plants. In addition, acidic soil contains excessive micro elements which could intoxicate the plants. In contrast, sodium (Na) and molybdenum (Mo) are found a lot in alkaline soil. Soil pH also affects microorganism development in soil. Organic material decomposer fungi and bacteria grow well in soil with pH 5.5-7. In such soil, microorganisms which are beneficial for plant roots also grow well.

Liming is commonly used to overcome acidity and high level of Al saturation in soil. Liming is done to increase soil pH from acidic to rather neutral or neutral and lower Al content. Lime has a pH of 8.5-10. Fly ash has a pH level of 11.7 and high content of K^{+}, Zn^{2+}, Ca^{2+}, Mg^{2+}, and Fe^{3+} ions which are important for plant growth. Dolomite commonly used by farmers only contains Ca^{2+} and Mg^{2+}. It has even been proven that fly ash was able to improve the structure, bulk density, water holding capacity, conductivity, and particle size of soil [1]. Therefore, fly ash is potential to be used as soil conditioner or ameliorant. The addition of fly ash increased the soil reaction (pH), the concentration of available P, exchangeable K, Na, calcium and magnesium [14].

In this study, the resulted fly ash granules had a pH of 9-10. It was found that types of granule agent significantly affected fly ash granule pH. Fly ash granules made with SBRC-M had higher pH (10) than those made by using SBRC-SC adhesive (9.3) and SBRC-T (9). This indicated that fly ash granules produced in this study could be used as materials to improve soil pH. pH levels of fly ash granules made by using different granule agents are depicted in Figure 2.

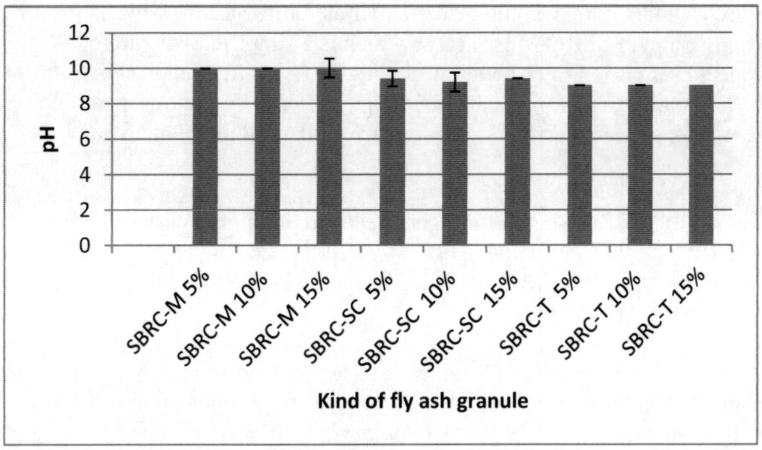

Figure 2. Graph of the effects of granule agent types on fly ash granule pH.

3.3. Durability

This test was aimed at assessing the durability of fly ash granules under a condition where there were certain friction and stress. Durability is a physical property indicating the granule's resistance to damages during transportation. The percentage of intact fly ash granules after steering process is listed in Table 2.

Table 2. Percentage of durable fly ash granules after stirring process

Material	Durability (%)	St. Dev
SBRC-M 5% (M5)	76.39	±8.29
SBRC-M 10% (M10)	93.91	±1.73
SBRC-M 15% (M15)	92.53	±3.53
Average	**87.61**	
SBRC-SC 5% (SC5)	86.74	±2.88
SBRC-SC 10% (SC10)	92.09	±2.74
SBRC-SC 15% (SC15)	90.69	±1.71
Average	**89.84**	
SBRC-T 5% (T5)	87.50	±6.21
SBRC-T 10% (T10)	93.99	±3.33
SBRC-T 15% (T15)	95.69	±1.48
Average	**92.39**	

It was shown in Table 2 that fly ash granules made by using SBRC-M had lower average durability (87.61%) than those made by using SBRC-SC (89.84%) and SBRC-T (92.39%). SBRC-T was found as the granule agent material which produced fly ash granules with the highest durability levels on average. The addition of fly ash can increases strength of soil [15].

The results of this study were found to be in line with the work of [16] who found that adhesive property of starch was better than that of sugar by product. This was caused by different chemical compositions of the adhesives. Starch contains sticky complex carbohydrate (amilopectin) while sugar by-product do not. The main component found in SBRC-M is sucrose which is a simple carbohydrate. While in SBRC-T, as it is sticky, amilopectin improves inter-material binding in fly ash granules which in turn improves the durability of fly ash granules to pressure and friction they receive during transportation. Sago and starch meal containing complex polysaccharides are commonly used as adhesive material.

3.4. Water holding capacity

In this study, the use of SBRC-M granule agent resulted in higher water holding capacity (WHC) (40.62%) than did SBRC-SC (38.79%) and SBRC-T (36.85%). The highest WHC (44.22%) was obtained fly ash granules made by using 10% SBRC-M granule agent. While the water holding capacity of fly ash without granulation was 56.9 percent [1].

Fly ash granules made by using SBRC-M were easy to adsorb and mix with water as the granules were immediately damaged when they were in touch with water. This phenomenon was more obvious with 10 and 15% SBRC-M which were easy to be washed out during the separation process of water and granules before the granules were weighed for their wet weight. This became an obstacle in the weighing process as water was still attached to the granules, making the wet weight of the granules was not easy to determine. Forced weighing would result in more granules being washed out in the water. In contrast, 5% SBRC-M were not immediately damaged when they were soaked in water. Therefore, no such obstacle as the one in 10 and 15% SBRC-M was found in the weighing process. Granules made by using SBRC-SC adhesive in various concentration did not absorb water well. However, no different water holding capacity was found on granules made by using all adhesive materials in all concentrations except in granules made by using 10% SBRC-M. SBRC-T in all concentration, except SBRC-T 5%, did not absorb water as high as SBRC-M 10%. Meanwhile, SBRC-T 5% and SBRC-T 15% could absorb water by 39.45% and 36.92%, respectively, but the two figures were not significantly different.

Water holding capacity is affected by the structure of the particles contained in granules. High water holding capacity of ash was due to its dominant silt and clay size fractions. Fly ash contained about 93 percent of silica and sesquioxides (Al_2O_3 and Fe_2O_3), and in the remaining potrion, Ca^{2+} was the dominany cation followed by Mg^{2+}, Na^+ and K^+[1]. And in the remaining portion, Ca^{2+} was the dominant cation The addition of fly ash granules to the soil will improve soil water holding capacity as fly ash granules have a capability to increase the number of soil micropores and macropores [17,18]. Macropores have pore diameter $\geq$ 0.1 mm while that of micropores is < 0.1 mm [19].

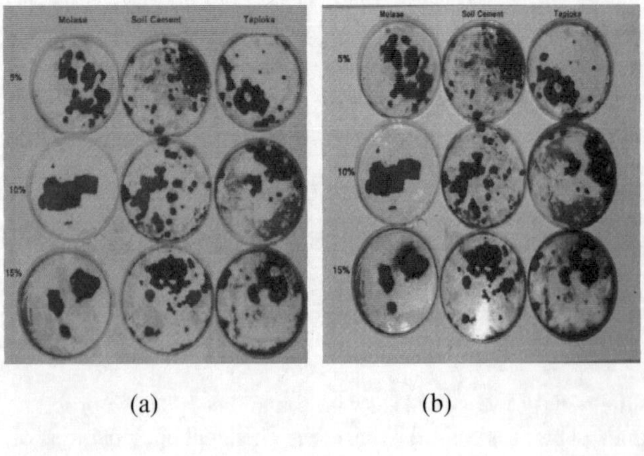

(a) (b)

Figure 3. Water holding process of fly ash granules made by using various types and concentration of adhesive materials, before drying (at) and after drying (b).

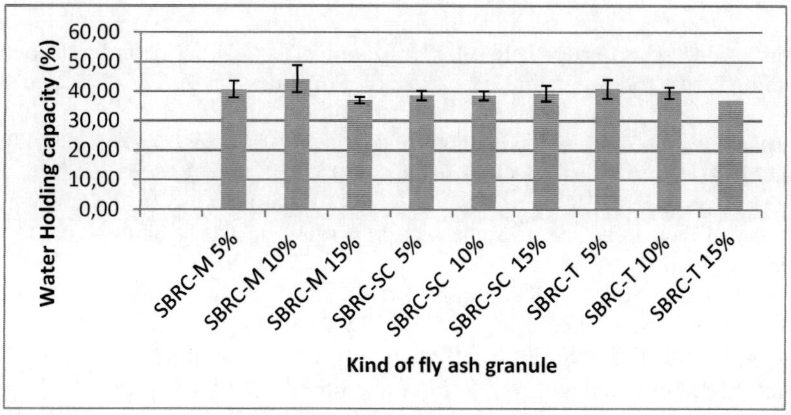

Figure 4. Water holding capacity (WHC) of fly ash granules in various types and concentration of granule agents

Table 3. Water holding capacity (WHC) of fly ash granules made by using various types and concentration of adhesive materials

Adhesive material	WHC (%)	St. dev
SBRC-M 5% (M5)	40.68ab	±2.91
SBRC-M 10% (M10)	44.22a	±2.93
SBRC-M 15% (M15)	36.95b	±4.62
Average	**40.62**	
SBRC-SC 5% (SC5)	38.56b	±1.10
SBRC-SC 10% (SC10)	38.46b	±1.59
SBRC-SC 15% (SC15)	39.35b	±1.49
Average	**38.79**	
SBRC-T 5% (T5)	40.75ab	±2.69
SBRC-T 10% (T10)	39.45b	±3.25
SBRC-T 15% (T15)	36.92b	±1.96
Average	**36.95**	

4. Conclusions
It was found that fly ash granules made from 15% of SBRC-M had the highest particle density (0.75 g/cm^3). Fly ash granules made with SBRC-M had higher pH (10) than those made by using SBRC-SC adhesive (9.3) and SBRC-T (9). SBRC-T was found as the granule agent material which produced fly ash granules with the highest durability levels on average. In this study, the use of SBRC-M granule agent resulted in higher water holding capacity (WHC) (40.62%) than did SBRC-SC (38.79%) and SBRC-T (36.85%). As a granule agent, compared to SBRC-SC and SBRC-T, SBRC-M could produce fly ash granules with highest particle density, highest pH, good durability, and best water holding capacity.

Acknowledgement
Research funding assistance from Ministry of Research, Technology, and Higher Education through the INSINAS 2016 program scheme was acknowledged.

References

[1] Aggarwal S, Singh G R, Yadav B R 2009 Utilization of Fly ash for Crop Production: Effect on the Growth of Wheat and Sorghum Crops and Soil Properties *Journal of Agricultural Physics* **9** 20-23

[2] Kunavanakrit W 1993 General properties of lignite fly ash. *In conference on The Potential of Lignite Fly Ash Utilization* (EGAT Thailand) p 2-15

[3] Tsadilas C D, Samaras V, Kazai P, Sgouras J 2002 Fly ash and sewage sludge application on an acid soil and their influence on some soil properties and wheat biomass production *12th ISCO Conference* (Beijing)

[4] Im-Erb R, Bamroongrugsa N, Kawashima K, Amano T, Kato S 2004 Utilisation of coal ash to improve acid soil. Songklanakarin *J. Sci. Technol* **26** 5

[5] Sharma S K and Kalra N 2006 Effect of fly ash incorporation on soil properties and productivity of crops: a review *J Sci Ind Res* **65** 383-390

[6] Parab N, Mishra S, Bhonde S R 2012. Prospects of bulk utilization of fly ash in agriculture for integrated nutrient management *Bulletin of the National Institute of Ecology* **23** 31-46

[7] Chou J S, Chou M M, Stucki J W, Warnock D, Chemler J A and Pepple M A 2005. Plant growth in sandy soil/compost mixture and commercial peat moss both amended with Illionis coal fly ash 2005 World of Coal Aash (WOCA) (Lexington, Kentucky, USA)

[8] Jayasinghe G Y, Tokashiki Y 2012 Influence of coal fly ash pellet aggregates on the growth and nutrient composition of *brassica campestris* and physicochemical properties of grey soils in Okinawa, Japan. *Journal of Plant Nutrition* **35** 3 453-470

[9] Jayasinghe G Y, Tokashiki Y, Kinjo K 2009 Recycling of coal fly ash and paper waste to improve low productive red soil in Okinawa, Japan *Clean – Soil, Air, Water* **37** 9 687–695

[10] Viji R, Rajesh PP 2011 Assesment of Water Holding Capacity of Major Soil Series of Lalgudi, Trichy, India. *J.Environ. Res. Develop* **7** 1

[11] Hardjowigeno 2003 Ilmu Tanah (Jakarta, Akedemika Pressindo) p 288

[12] Madjid 2010 Sifat dan Ciri Tanah (Fakultas Pertanian Institut Pertanian Bogor, Bogor)

[13] Hanafiah K A 2004 Dasar-dasar Ilmu Tanah Vol 2 *Dasar-dasar Ilmu Tanah Lanjutan* (Palembang)

[14] Inthasan J, Hirunburana N, Herrmann L and Stahr K 2002 Effect of fly ash application on soil properties, nutrient status and environment in Northern Thailand *17th WCSS Thailand* 1-6

[15] Budi G S, Cristanto A, Setiawan E 2003 Pengaruh fly ash terhadap sifat pengembangan tanah ekspansif *Civil Engineering Dimension* **5** 1 20–24

[16] Gunawan M S, Nawansih O, Nurainy F 2015 The application of The Use an Adhesive Material Towards Physical and Chemical Characteristic of Organic Fertilizers Granule That Feedstock of Compost The Rind of Cocoa

[17] Bot A and Benitez J 2005 The Importance of Soil Organic Matter (Food and Agricultural Organization of the United Nations, Rome)

[18] Pathan S M, Aylmore L A G, Colmer T D. 2003 Properties of Several Fly Ash Materials in Relation to Use as Soil Amendments. *J. Environ. Qual* **32** 687-693

[19] Shober, A.L. 2008. Soils and Fertilizers for Master Gardener: Soil Physical Properties. (http://edis.ifas.ufl.edu/pdffiles/MG/MG45100.pdf)

IOP Conf. Series: Earth and Environmental Science **65** (2017) 012024 doi:10.1088/1755-1315/65/1/012024

Indonesian CPO availability analysis to support food and energy security : a system dynamic approach

T Rahman[1], Y Arkeman [2], D Setyaningsih[2] and R Saparita[1]

[1]Development Centre for Appropriate Technology, Indonesian Institute of Sciences, Subang, Indonesia

[2]Department of Agroindustrial Technology, Faculty of Agricultural Technology, Bogor Agricultural University, Bogor, Indonesia

Email : tkr_rh@yahoo.com, tauf007@lipi.go.id.

Abstract. The development of biofuels could be a solution to overcome the energy problem. One of biofuel that has the potential to be developed, namely palm oil biodiesel that is also the raw material for food. As a provider of CPO raw materials, the production of palm biodiesel could trigger competitions, from biofuels demand growth and utilization of agricultural resources. Thus, it needs to be analyzed to determine the adequency of CPO supply to fulfill the need of food and policy recomendation which sets the development of palm oil biodiesel can be synergies with food need especially for the supply of raw material CPO. To obtain the optimal policy in the synergy between the raw material of CPO for food and energy is a need to establish some policy scenarios that allow to be applied and then chosen the best policy alternative of all scenarios. The purpose of this research were to : 1) analysis the availability of CPO to meet the needs of food and energy, 2) provide policy recommendation with regard biodiesel development of food security. The model made used system dynamic method. Several scenarios that used in the model are: 1) existing condition, 2) The scenario increase biodiesel production capacity and increase land productivity, 3) reduction scenario CPO export by 30%, 4) scenario use othe raw material for biodiesel by 20%. The simulation results showed the availability of CPO raw materials would answer all needs of both food and biodiesel when there was an increase in productivity, diversification of raw materials, and also a reduction in palm oil exports. It was needed an integrated policy from upstream to downstream along with the consistency of implementation. Policy suggestions that could be considered were increased productivity through agricultural intensification, enforcement disincentive policies of CPO to exports, and development of non-CPO biodiesel raw materials and development of renewable energy.

1. Introduction

Energy is one of the critical needs in mankindlife. Energy need increases in line with the increase of population, technological advances, and human activities. Majority of energy use in Indonesia currently is still using fossil fuel sources. Indonesia's oil production reached its peak in 1977 which amounted to 1,685 thousand barrels per day and continued to decline down to 909 thousand barrels per day in 2006, or decreased at a rate of 1.83% per year. On the other hand petroleum consumption

Content from this work may be used under the terms of the Creative Commons Attribution 3.0 licence. Any further distribution of this work must maintain attribution to the author(s) and the title of the work, journal citation and DOI.
Published under licence by IOP Publishing Ltd

continues to increase at a rate of 5.04% per year. This situation made Indonesia originally as an exporter turned into an importer since 2000 with an increasing level of deficit[1].

One of the sources of energy that can be an alternative is biofuel. Biofuel is fuel from biological sources. Biofuel currently becomes an option to be used as an energy source instead of petroleum. This is because biofuel is the easiest fuel to convert into energy fuel and electricity[2]. Another advantages of the use of biofuels are renewable, environmentally friendly, biodegradable, to be able eliminating the greenhouse effect, and the assurement of raw material continuity[3].

The selection of biofuel as an alternative energy is not without risk, as the source of biofuel raw materials are plants that require land, time, seed and maintenance. In fact, most of biofuel raw materials are also a need for human food such as sugar cane, cassava, corn, palm oil, and soybean[4]. On the one hand Indonesia must develop biofuels, but on the other hand Indonesia must continue to ensure food security and production sustainability of some kinds of food commodities.

Food is a basic need that is essential for human beings that need to be met before other needs in sustaining life and living. Food security in UU No. 7/ 1996 is defined as the fulfillment of the conditions of adequate food, both in quantity and quality, safe, equitable, and affordable. If competition between food and energy needs occurs through the development of biofuels, the food security should take priority first. Therefore, the development of biofuels should be consistent with the management of Indonesian food.

The availability of food and energy in sufficient quantities is very vital for every country. Both are consumed either to meet the main needs of beings living or as well as an input factor in the production process. The development of biofuels based on food crops can encourage competition with efforts to comply the food. Bioenergy development requires land for planting area and the availability of water. Therefore, the development of bioenergy raises the potential threat on food security as a result of the conversion and the competition between energy and food needs to the people[5,6,7].

One of the food raw materials that can be used as a biofuel raw material is crude palm oil (CPO). Utilization of domestic CPO all this time is used as an derivative industrial raw material of CPO which consists of eighteen kinds of products divided into the food and non-food industries. The increase in fuel prices on the domestic CPO demand will provide a broad impact in the palm oil industry. CPO demand for energy will compete with CPO demand for food. That is one of the factors causes food prices and cooking oil in Indonesia soared out of control. Its impacts are the appearing of shortage and decreased supply[8]. Influence of domestic policy which sets the production of biofuels has a positive impact that biofuels product has the competitiveness in the market. To increase the production and biofuels consumption required obligation of use biofuel policy, tax exemption and subsidies[9].

Palm oil is one commodity that has strategic value because it is one of nine basic needs of Indonesia. High demand for palm oil both inside and outside the country is an indication of the importance oil palm in the nation's economy. Palm cooking oil needs continue to increase from year to year as the increase of population, development of mills and food industry, and rising public consumption will be cooking oil to cook. The system dynamics approach is a good method to understand the behaviour of a complex system over time[10]. Some researchers developed system dynamic model in bioenergy development like to analyze the effects of government policy for stimulating biodiesel industry[11]. The purpose of this research were to : 1) analysis the availability of CPO to meet the needs of food and energy, 2) provide policy recommendation with regard biodiesel development of food security.

2. Methodology

2.1. Framework

The population growth increases every year will lead to the increase of demand on food, energy, and land for residence and industry. Limited reserves of fuel in Indonesia caused the government began to look at the use of biofuels as an energy source. Increased food and energy requirements need to be in line with an increase in supply or production of raw materials for food and bioenergy. Sources of potential bioenergy raw materials for development is also a source of raw materials for food such as palm oil. On the other hand the amount of agricultural land is also limited. Thus bioenergy development can potentially lead to food insecurity as a result of competition of raw materials and land acquisition for the food fulfillment for community. The causal relationship among involved variable groups can be seen in figure 1.

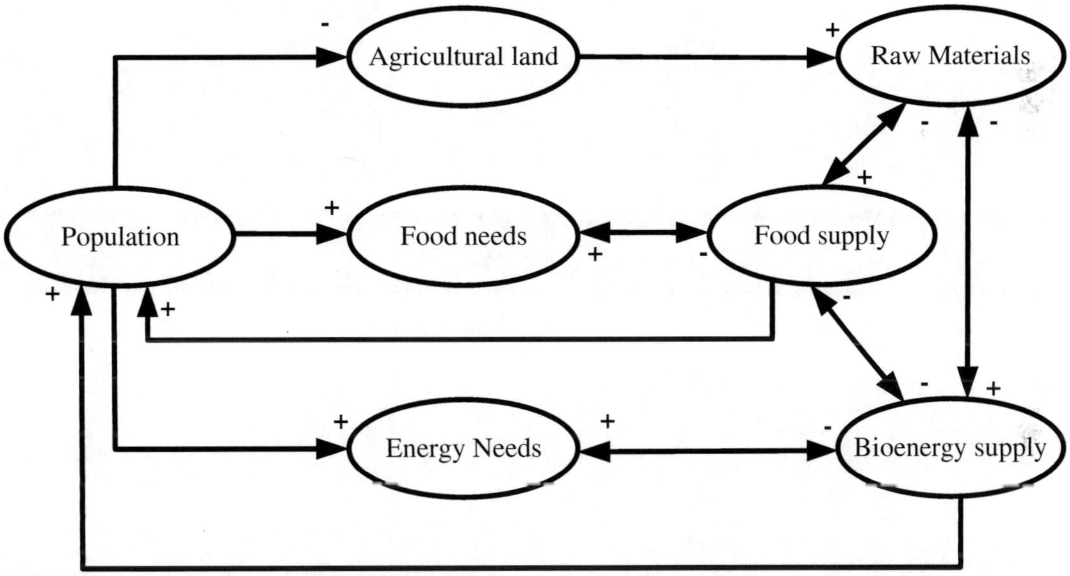

Figure 1. Causal relationship among involved variable groups

2.2. Types and sources of data

The data used in this study were primary and secondary data. Primary data was collected through interviews and discussions with respondents related to the research, which consisted of industry associations, the government (Ministry of Energy and Mineral Resources, Ministry of Agriculture, Ministry of Trade, and Ministry of Industry), and research institute. Meanwhile secondary data was time series data. Secondary data in the study were collected from various sources such as statistics of Indonesia report and also documents and research papers from various agencies associated with the research. Time period simulation that has been to be performed is 2015 – 2030.

2.3. Data analysis methods

System is the collection of elements or sub-system that interact with each other, work together to achieve a certain goal[12]. System dynamic is related to the behavior of system that changes according to the time, with the aim to explain and understand the feedback affect behavior of system, then designs the structure of the information feedback and control policies appropriate, through simulation and optimization of the system by using qualitative and quantitative methods. Feedback form a particular pattern of the dynamic model is one of the core of a system dynamic[13]. This methodology is focused on making policy and how that policy determines the behavior of the problem issues that can be modeled by the system dynamic[14].

Stage in this dynamic system approach begin with understanding the system, identification of problems, conceptualization system, the formulation of model, simulation and model validation, policy analysis and implementation models[10]. This process of dynamic system approach can be seen in figure 2.

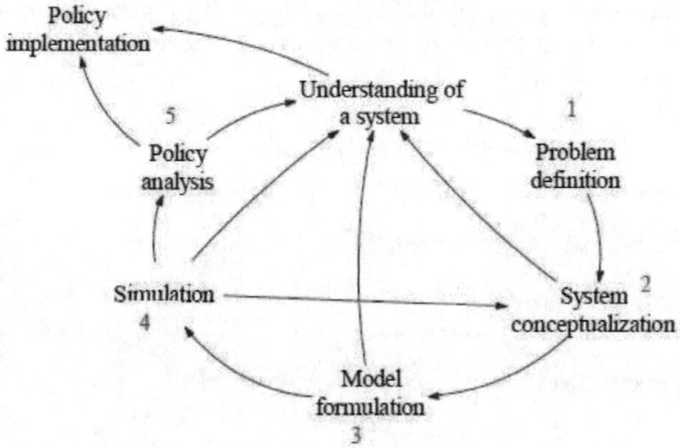

Figure 2. Stage of system dynamic approach

Basic equation of mathematical system dynamic is:

$$\frac{d}{dt} x(t) = f \; (x, p)$$

where “x” is a vector *stock,* “p” is parameter, and “f” is a nonlinear factor function [14]

2.4. Validation

Model validation was carried out to assess whether the model was set up to represent the system in the real world so that it could be concluded convincingly[15]. Error measurement method used in this study was RMSPE (Root Mean Square Percent Error) or an error percentage of square root average, which was calculated by the formula:

$$\text{RMSPE} = \sqrt{\frac{1}{n}\sum_{t=1}^{n}\frac{(S_t - A_t)^2}{A_t}}$$

By A_t as an actual value, S_t as the computer simulation value, n as a data amount reflected the length (time/data) collected.

IOP Conf. Series: Earth and Environmental Science **65** (2017) 012024 doi:10.1088/1755-1315/65/1/012024

3. Results and discussion

3.1 Model identification
Model identification is a chain of relationships between needs of system actors with existing problems to meet those needs. The linkage is illustrated in the causal loop diagram in figure 3. Causal loop diagram is the disclosure of the incident causal relationship (causal relationship) into the language of a particular image. The picture language is an arrow interlocking, thus forming a simple diagram , where the upper arrow reveals the causes and arrowhead reveals as a result[16].

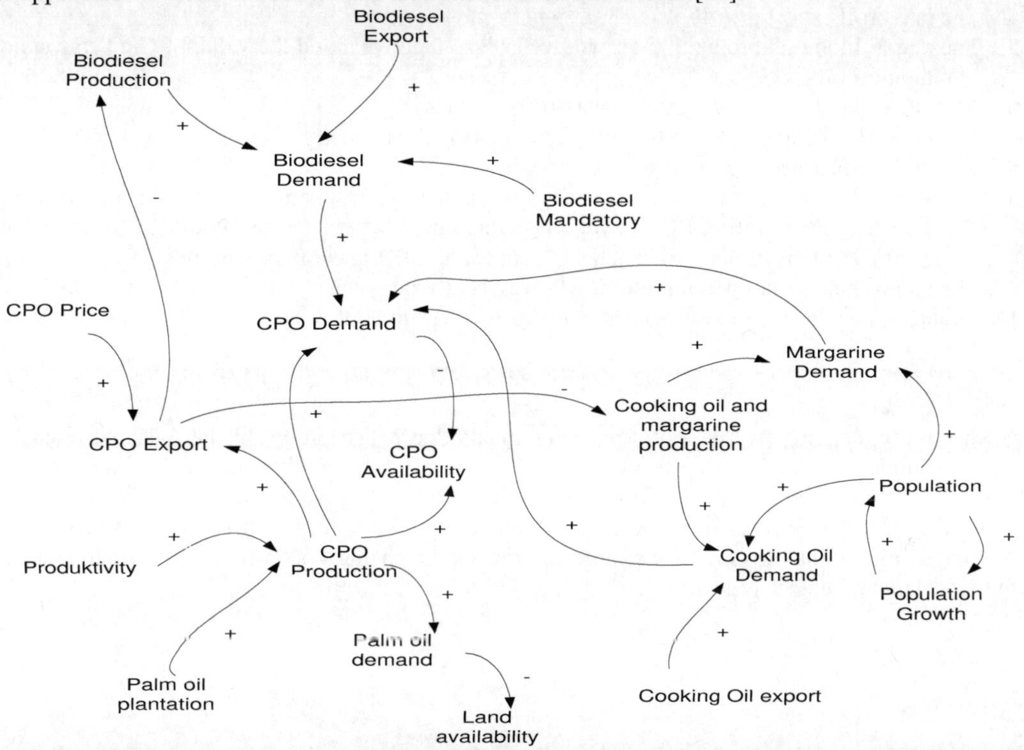

Figure 3. Causal loop diagram

The simulation result is expected to generate projections CPO sufficient material to meet the need of food (cooking oil and margarine) and biodiesel need (mandatory target).

3.2 Model development
Development of bioenergy development policy modelin taking food security was composed of four sub-systems, namely: (1) land and plantation subsystem, (2) CPO raw material subsystem, (3) food need subsystem, and (4) biodiesel need subsystem.

Based on the causal diagram that had been prepared, then was prepared a flowchart (stock flow diagram) to describe the model in more detail due to the influence of each time and the relationship between variables. Thus there would be variable indicated accumulation results in a system called a level, and variable as system activity and influenced level called rate. Aftersetting up the model,then was devised a mathematical formulation of the model could be simulated.

3.3 Model Assumptions

Some of the assumptions used in this modeling were:

1. The need for CPO were calculated based on food raw material needs, CPO exports, and biodiesel raw material demand. Food needs were calculated based on the needs of cooking oil and margarine.
2. CPO provision was only calculated based on the number of CPO production in the country. CPO import was negligible.
3. The rate of export CPO growth was10.85% per year.
4. The rate ofoil export growth was 13.02% per year.
5. The rate of biodiesel production growth was 24.43% per year and the value of CPO conversion into biodiesel was 97%.
6. The rate of biodiesel export growth was 10% per year.
7. The rate of cooking oil production growth rate was 12.71% per year.
8. The rate of solar needs growth was 32.70%.
9. The needsof margarine and cooking oil per capita were respectively 0.065 kg per year and 14.21 kg per year, with CPO conversion values into margarine and cooking oil were 6.67 tonnes of CPO/tons of margarine and 1.37 tonnes of CPO/tons of cooking oil.
10. The rate of Indonesia's population growth was 1.49% per year.
11. Simulation analysis period was limited to period 2015 to 2030.

3.4 Model validation

To get the correct conclusion based on the requirements that had been established was necessary to validate the model that had been made[17]. The test result showed that RMSPE value for palm plantation land area variable was 6,3763% and for CPO production value was 7,5875 (figure 4 and figure 5). This value was lower than the required maximum value limit of 10%. Deviation limits of both variables showed that this model was able to simulate the changes occured actually on the field so that the model could be declared valid.

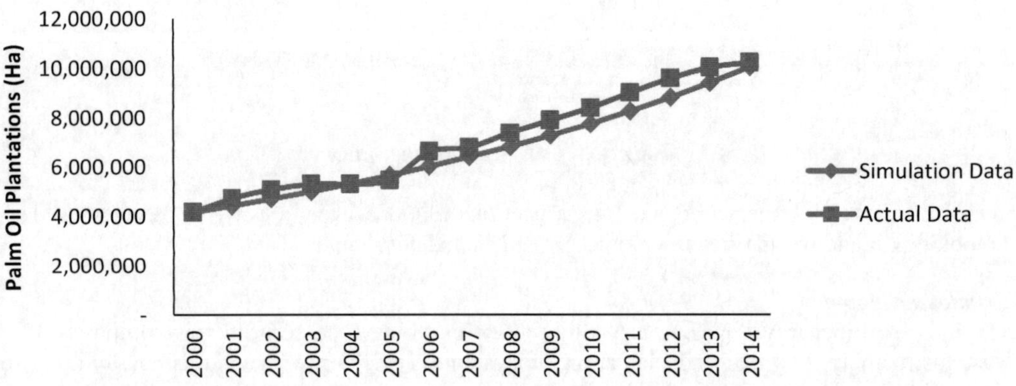

Figure 4. Actual and simulation results data graph for palm oil plantation area

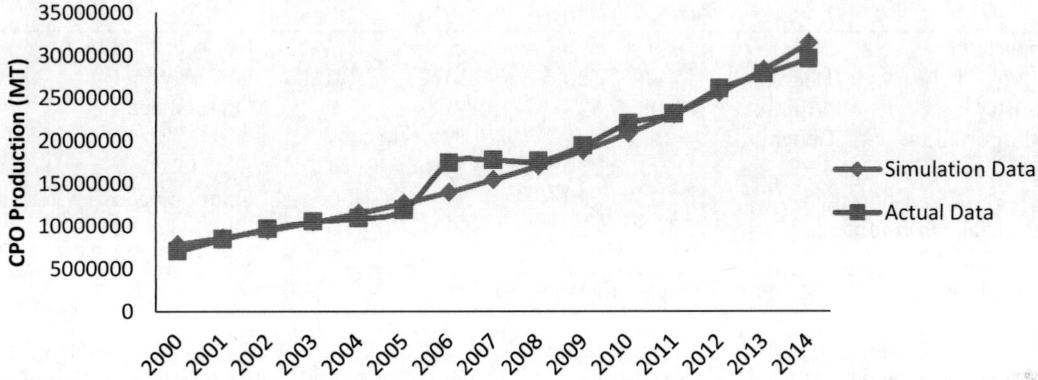

Figure 5. Actual and simulation results data graph for CPO production

3.5 Policy scenario and simulation results

In order to formulate policy of CPO raw material supply for food and energy was needed various simulations to determine the appropriate alternative policy options. Several scenarios used in this study were:

1. Basic scenario. In this scenario the system condition was kept as the condition of the present day.
2. The increase of biodiesel production capacity and land productivity scenario. At this scenario was assumed an increase in biodiesel production capacity so that biodiesel production capacity could meet the achievment of biodiesel mandatory target. Besides, land productivity also was increased gradually until it reached 4 tons/ha.
3. Reductionof 30% in CPO export scenario. At this scenario was used the previous scenario and added with a reduction in CPO exports amounted to 30% of the previous that was allocated to meet domestic demand.
4. The 20% use of other raw material for biodiesel scenario. On this scenario, the previous scenario was added with the assumption that the non-palm oil raw material biodiesel contribution was 20% to meet the mandatory target.

In 2015, the Ministry of Energy and Mineral Resources issued Regulation of Minister of Energy and Mineral Resources No. 12/2015 as a refinement of Regulation of Minister of Energy and Mineral Resources No. 32/2008 concerning the provision, utilization, and administration of biofuel as the other fuel. One of its content isto establish the obligation of a biodiesel minimum use as a fuel mixture for use in household, PSO and non-PSO transportation, industrial and commercial, and power plant sectors in stages starting in 2015 until 2025 as shown in Table 1.

Table 1. The obligation of minimal utilization of biodiesel to the total need

Sector	April 2015	January 2016	January 2020	January 2025	Description
Household	-	-	-	-	Not specified nowadays
Micro industry, fisheries industry, farming, Transportation and General Services (PSO)	15 %	20 %	30 %	30 %	On total needs
Non-PSO Transportation	15 %	20 %	30 %	30 %	On total needs
Industrial dan commercial	15 %	20 %	30 %	30 %	On total needs
Power plants	25 %	30 %	30 %	30 %	On total needs

Source: Regulation of Minister of Energy and Mineral Resources No. 12/2015

Every year oil palm plantation area always increase with a total land area in 2013 was estimated to reached approximately 10.010.824 hectares. Indonesia's palm oil production also continues to increase every year. In 1990 Indonesia's palm oil production reached 2.412.612 tons and in 2013 increased to 27.746.125 tonnes[18].

Biodiesel industry in Indonesia is basically a new industry (which began around 2006) that are vulnerable to external shocks. To reach the mature stage of industry, is needed long process and time, and also often require the government intervention. Some of problems faced by biodiesel businesses/producers in Indonesia are price fluctuation of CPO raw materials, fuel subsidy policy, certainty of domestic market, production capacity, and availability of land [19]. According to the Ministry of Energy and Mineral Resources, nowadays there are about 25 companies that have licenses to produce biodiesel with a total of installed capacity is 5,67 million kiloliters. All of biodiesel industries in Indonesia use palm oil as a raw material

As the population increases, also the industries and motor vehicles increases in Indonesia, the fuel needs will also increase. The existence of biofuel mandatory policy will automatically increase the domestic biodiesel needs. Thus, Indonesia should increase biodiesel production to meet the mandatory target.

Cooking oil has a strategic role in the national economy because it includes as one of the nine basic material needs that are important to the community. With the increasing number of Indonesian population, the consumption of cooking oil is likely to increase from year to year. Based on data [20], the amount of cooking oil consumption of Indonesian community in 2008 amounted to 9.6 kg/capita/year, and in 2011 increased to 10.7 kg/capita/year.

The simulation results of system dynamic model on basic scenario as seen in figure 6 and figure 7 showed that CPO production and food needs was continued to increase during simulation period. This increase occurred because the oil palm plantation area was expected to continue increasing every year due to demand increase for CPO raw material, both for domestic consumption and export.

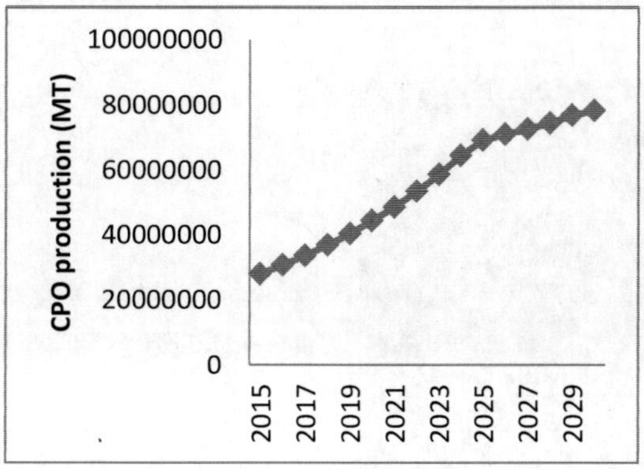

Figure 6. Simulation results for CPO production

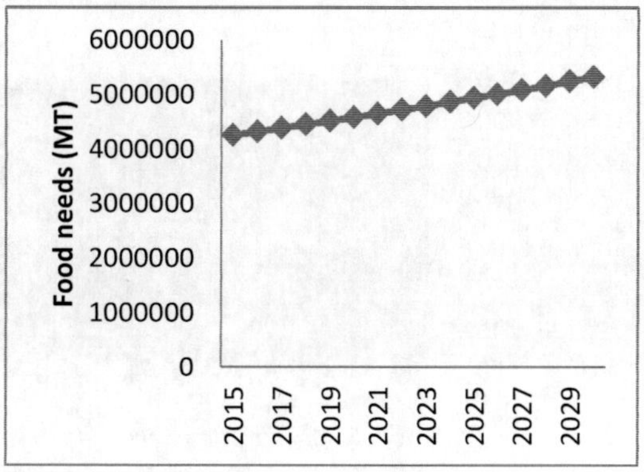

Figure 7. Simulation results for food needs (cooking oil & margarine)

In 2014 Indonesia's CPO production was 29 million tons and was expected that in 2030 CPO production will increase almost 3x more for about 80 million tonnes. As the growth of Indonesia's population,it is estimated that food need to meet domestic need is also expected to continue the increase every year. The simulation result (figure 6 and figure 7) showed that in 2030 will be required CPO raw material of 5.323.274 tons only to meet domestic food consumption (cooking oil and margarine).

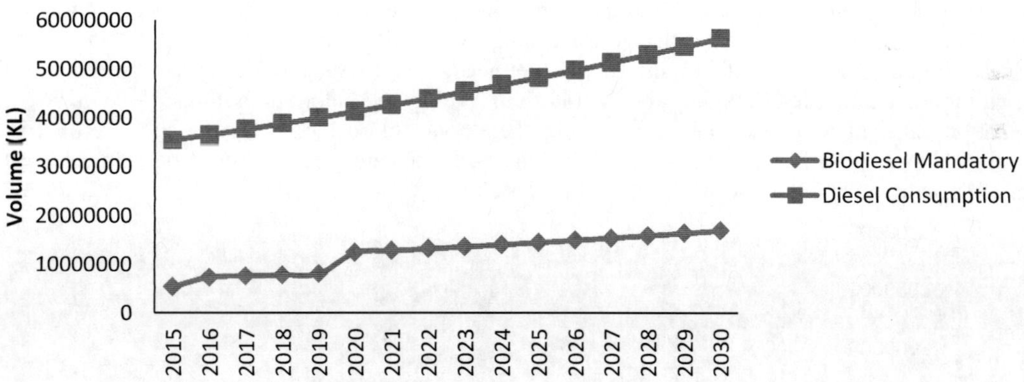

Figure 8. Simulation results for biodiesel mandatory and diesel consumption

The simulation result for energy need, particularly diesel, was also showed an increase during the simulation (figure 8). In 2014 diesel need reached 35.382.510 kiloliters and will increase to 56.305.728 kiloliters in 2030. Assuming that the mandatory target of biodiesel will be successfully implemented, the biodiesel need prediction in 2030 will be 16.545.761 kiloliters. On the other hand, production capacity of biodiesel industry currently was only about 6 million kiloliters. Thus the biodiesel need in the future will not be met with current production capacity, so that is necessary to increase production capacity. In proportion, it showed that the percentage of CPO demand for raw materials of biodiesel in order to meet mandatory target was much larger than CPO need portion to meet domestic food need. CPO demand portionfor domestic biodiesel raw material in 2030 was expected to reach approximately

75%, while the portion of CPO need to meet domestic food need was estimated only about 25% (figure 9).

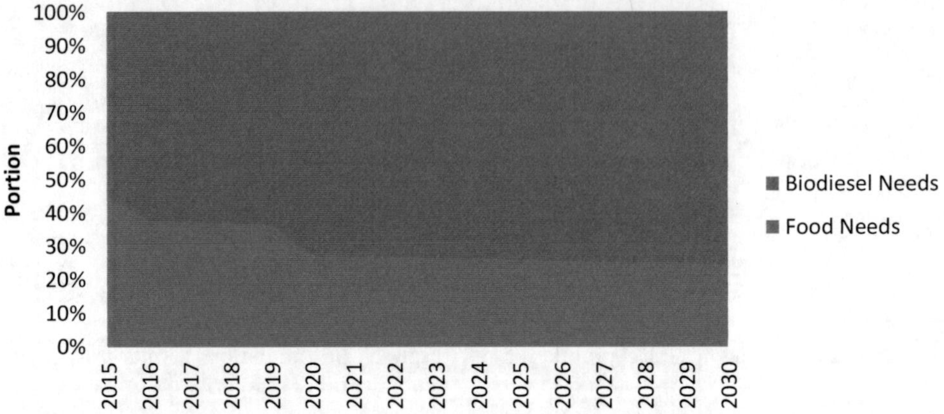

Figure 9. Simulation results for food needs and biodiesel needs for domestic

Figure 10 shows a CPO availability which is the difference between CPO production, CPO need for food, biodiesel, oleochemical and export. The simulation results of basic scenario and scenario 1 and 2 showed that at the beginning, CPO production was still able to meet their food needs, CPO export, and biodiesel need until 2025. After 2025, the availability of CPO showed a tendency to decrease with the increasing deficit rate and was estimated that CPO raw material supply would only meet all needs until 2029. It was because the estimate that in 2025 land availability for oil palm development would be reduced because of competition with residence needs, industry, and the development of other agricultural commodities. Besides, the moratorium policy on Indonesia's forest to converse into plantations also affected the rate of oil palm plantations. In scenario 3, simulation result of this scenario showed that during time simulation, biodiesel mandatory target could be achieved by not causing a shortage in CPO raw material for food needs.

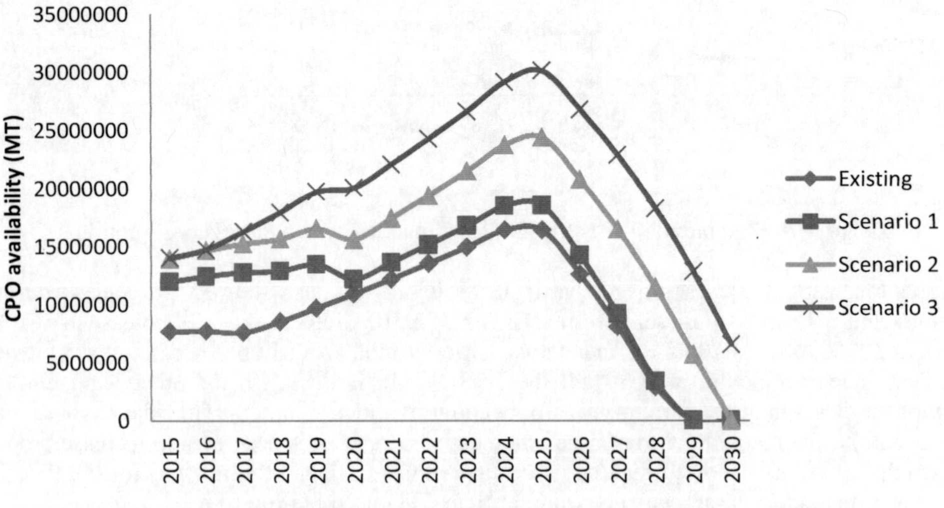

Figure 10. CPO availability simulation results

IOP Conf. Series: Earth and Environmental Science **65** (2017) 012024 doi:10.1088/1755-1315/65/1/012024

Figure 11 and figure 12 showed the total portion of CPO for food and energy needs, both for domestic consumption and export, to meet the needs of derivative products. The simulation results of the basic scenarion and scenario 1,2, and 3showed that total of CPO need for food was far greater than total of CPO need for biodiesel. CPO portion of food ranged between 40-60%, while the CPO portion of biodiesel ranged between 10-30%, and the remaining was used for CPO exports in raw form. CPO for biodiesel portion tendedto increase until 2020 due to a mandatory increase in mixture of biodiesel and increased dieselfrom 15% to 30%. After 2020, because there was no percentage increase of biodiesel mixture with diesel, so that CPO portion from biodiesel tended to decrease. Similarly, the CPO for food portions tended to decrease until 2020 and then increased after 2020.

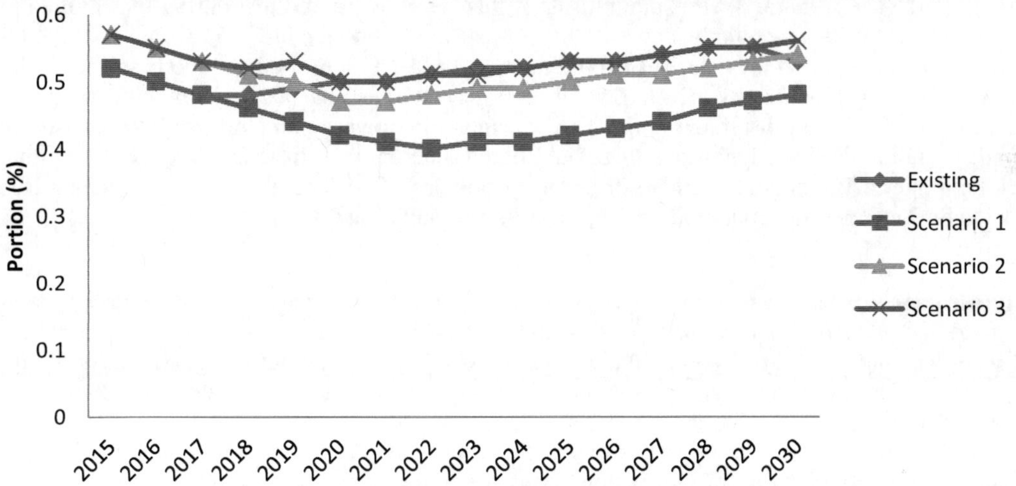

Figure 11. Simulation results of CPO portion for food

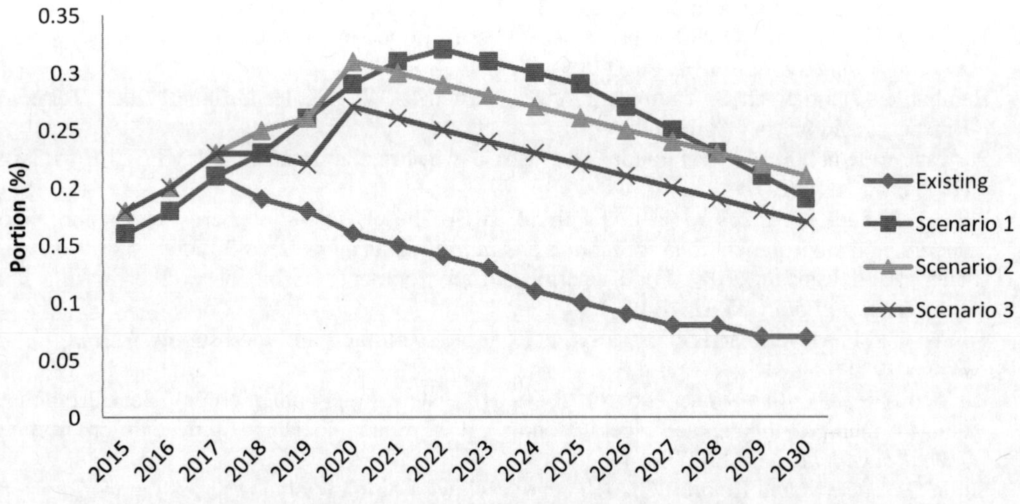

Figure 12. Simulation results of CPO portion for energy

4. Conclusions

Based on simulation results, it can be seen that the created model had been able to describe the increase in demand for CPO raw materials due to biodiesel mandatory target that might affect the availability of CPO raw materials to meet food needs. Simulation result shows that oil plam plantation area will continue to increase to certain extent, thus CPO production increases throughout the year. The policy of mandatory biodiesel can increase CPO demand for domestic food (cooking oil and margarine). CPO surplus simulation result for scenario 1 and 2 shows that in 2015, there is sufficient material CPO to fulfill all needs for food, energy and also export, even there is a surplus with surplus increasing tendency until 2025. After 2025, the availability of CPO showed a tendency to decrease with the increasing deficit rate and was estimated that CPO raw material supply would only meet all needs until 2029. In scenario 3, simulation result of this scenario showed that during time simulation, biodiesel mandatory target could be achieved by not causing a shortage in CPO raw material for food needs. Availability of CPO raw materials would meet all needs, both of food and biodiesel, when there was an increase in productivity, diversification of raw materials, and also a reduction in palm oil exports. It was needed an integrated policy from upstream to downstream along with the consistency of implementation. Policy suggestions that could be considered were increased productivity through agricultural intensification, enforcement disincentive policies of CPO to exports, and development of non-CPO biodiesel raw materials and development of renewable energy.

Acknowledgements

We would like to thank to The Ministry of Research and Technology that has provided research funds. This work was supported by Head of Surfactant and Bioenergy Research Center, Bogor Agricultural University and Head of Center for Appropriate Technology Development are gratefully acknowledged for their facilities supports. The authors would also like to thanks to Hendri Wijaya, Dhani Satria Wibawa, Laurinciana Sampebatu, Hermawan Prasetya and Prof. Ani Suryani for discussion, experimental assistance and useful suggestion during the course of this project.

References

[1] Susila W R and Munadi E 2008 Dampak pengembangan biodiesel berbasis CPO terhadap kemiskinan di Indonesia Jurnal Informatika Pertanian. **17** 1173 – 1194

[2] Thornley P and Cooper D 2008 The Effectiveness of policy instruments in promoting bioenergy *Journal Biomass and Bioenergy* **32** 903 – 913

[3] Hambali E, Mujdalipah S, Tambunan A H, Pattiwiri A W and Hendroko R 2007 Teknologi Bioenergi (Indonesia : Agromedia)

[4] Purwantoro R N 2008 Sekilas Pandang Industri Sawit. Majalah Usahawan LMFEUI Edisi No. 4 (Indonesia : LMFEUI)

[5] Popp J, Lakner Z, Rákos M H and Fári M 2014 The effect of bioenergy expansion: Food, energy, and environment *Renewable and Sustainable Energy Reviews* **32** 559 – 578.

[6] Ciaian P and Kancs A 2011 Food, energy and environment : is bioenergy the missing link? *Journal Food Policy* **33** 326–348

[7] Timilsina G F, Mevel S and Shrestha A 2011 Oil price, biofuels and food supply *Journal Energy Policy* **39** 8098 – 8105

[8] Widodo K H, Abdullah A and Arbita K P D 2010 Sistem supply chain crude palm oil Indonesia dengan mempertimbangkan aspek economical revenue, social welfare dan environment *Jurnal Teknik Industri* **12** 47 – 54

[9] Sorda G, Banse M and Kemfert C 2010 An overview of biofuel policies across the world *Energy Policy* **38** 6977-6988

[10] Turcksin Laurence and Macharis Cathy 2010 Biodiesel In Belgium: A System-Dynamics Model, Report for Task 4.2**,** Study performed under the authority of Belgian Federal Science Policy (Belgian: Belgian Science Policy Office)

[11] Akhmad Hidayatno, Aziiz Sutrisno, Yuri M Zagloel and Widodo WP 2011 System dynamics sustainablity model of palm oil based biodiesel production chain in Indonesia International *Journal of Engineering & Technology* **11** 1 - 6
[12] Wasson C S 2006 System Analysis, Design, and Development : Concepts, Principles, And Practices (USA : John Wiley & Sons, Inc)
[13] Coyle R G 1995 Management System Dynamics (London : Chapman & Hall)
[14] Richardson G P and Pugh A L 1986 Introduction to System Dynamics Modelling with Dynamo (Cambridge, England : The MIT Press)
[15] Sterman, J D 2000 Business Dynamics. System Thinking and Modeling for a Complex World (USA : McGraw-Hill)
[16] Muhammadi E, Aminullah and Soesilo B 2001 Analisis Sistem Dinamik: Lingkungan Hidup, Sosial, Ekonomi, Manajemen. (Jakarta : UMJ Press)
[17] Hartrisari 2007 Sistem Dinamik. Konsep Sistem dan Pemodelan Untuk Industri dan Lingkungan (Indonesia : Seameo Biotrop)
[18] Directorate General Of Estate Crops 2013 The Crop Estate Statistics Of Indonesia 2012 – 2014 (Jakarta : Ministry Of Agriculture)
[19] Indonesia Biofuels Producer Association (APROBI) 2014 Biodiesel supply capability in support of mandatory biodiesel and the problem. National Seminar Biodiesel Mandatory Policy.) (Bogor : Palm Oil Agribusiness Startegic Policy Institute (PASPI)
[20] Statistics Indonesia 2014 Indonesian Oil Palm Statistics (Indonesia : BPS)

IOP Conf. Series: Earth and Environmental Science **65** (2017) 012025 doi:10.1088/1755-1315/65/1/012025

Isolation and characterization of marine bacteria from macroalgae *Gracilaria salicornia* and *Gelidium latifolium* on agarolitic activity for bioethanol production

M Kawaroe[1 2], I Pratiwi [2] and A Sunudin[2]

1Surfactant and Bioenergy Research Centre, Bogor Agricultural University, Baranang Siang Campus, Bogor 16143, Indonesia
2Department of Marine Science and Technology, Faculty of Fisheries and Marine Science, Bogor Agricultural University, Darmaga Campus, Bogor 16680, Indonesia

Email : mujizatk@gmail.com

Abstract. *Gracilaria salicornia* and *Gelidium latifolium* have high content of agar and potential to be use as raw material for bioethanol. In bioethanol production, one of the processes level is enzyme hydrolysis. Various microorganisms, one of which is bacteria, can carry out the enzyme hydrolysis. Bacteria that degrade the cell walls of macroalgae and produce an agarase enzyme called agarolytic bacteria. The purpose of this study was to isolate bacteria from macroalgae *G. salicornia* and *G. latifolium*, which has the highest agarase enzyme activities, and to obtain agarase enzyme characteristic for bioethanol production. There are two isolates bacteria resulted from *G. salicornia* that are N1 and N3 and there are two isolates from *G. latifolium* that are BSUC2 and BSUC4. The result of agarase enzyme qualitative test showed that isolates bacteria from *G. latifolium* were greater than *G. salicornia*. The highest agarolitic index of bacteria from *G. salicornia* produced by isolate N3 was 2.32 mm and isolate N3 was 2.27 mm. Bacteria from *G. latifolium* produced by isolate BSUC4 was 4.28 mm and isolate BSUC2 was 4.18 mm, respectively. Agarase enzyme activities from isolates N1 and N3 were optimum working at pH 7 and temperature 30 °C, while from isolates BSUC4 was optimum at pH 7 and temperature 50 °C. This is indicated that the four bacteria are appropriate to hydrolyze macro alga for bioethanol production.

1. Introduction

Biomass is a potential raw material that can be used to produce bioethanol and one of biomass is macroalgae. Macroalgae has several advantages, ie. Low lignin content, do not compete with terrestrial biomass, not a major food source, have high sugar content, and in Indonesia, cultivation productivity is quite high [1].

G. salicornia and *G. latifolium* are macroalgae producer of gelatin that potential to be use as raw material for bioethanol. Both macroalgae are suitable for cultivation because of several factors ei. easily obtain and contain high levels of agar that can be hydrolyzed into oligosaccharides.

Content from this work may be used under the terms of the Creative Commons Attribution 3.0 licence. Any further distribution of this work must maintain attribution to the author(s) and the title of the work, journal citation and DOI.
Published under licence by IOP Publishing Ltd

The content of agar in *G. salicornia* in Indonesia generally range between 32-48% and the carbohydrate content of *G. latifolium* reach 67.85-76.15% [2].

Research on bioethanol from terrestrial biomass has long been done, but the use of macroalgae (*G. salicornia.*) as raw material for bioethanol production is still not a done deal. One study evaluating the crude production of bioethanol from *G. latifolium* and *G. verrucosa* done by Kawaroe et al [3], which produces crude bioethanol in the fifth and sixth day. Difficulties experienced is due to several factors that greatly affect the bioethanol production process at every stage that must be passed. Stages of the process is the process of hydrolysis (by acid and enzym) and fermentation. Factors that influence the hydrolysis process including the carbohydrate content of the raw material used, the conditions of pH and temperature, while the fermentation process is a type of microorganism, sugar resulting from the hydrolysis process, time, pH, and temperature.

Bacteria that degrade the cell walls of macroalgae and produce agarase enzyme called agarolitic bacteria and enzymes produced called agarase enzyme. The existence and characteristics of agarolitik bacteria can be determined by agarase activity test in qualitative and quantitative. Qualitative agarase activity performed by dripping Lugol reagent. This reagent can indicate the activity of the bacteria in producing agarase above culture vessels [4]. Agarase enzyme quantity is generally measured by spectrophotometry value of the increased concentration of reducing sugars by the method of Nelson or DNS method that uses D-galactose as standard. The enzyme activity (U/mL) was defined as the amount of enzyme required to release 1 mol of D-galactose per minute [5]. Agarase activity test is also necessary to know the characteristics of the bacteria to produce bioethanol optimum. It is the background that the potential of bacteria agarolitik of macroalgae for bioethanol needs to be developed further.

This study aims to investigate the characteristics of agarase enzyme activity of macroalgae *G. salicornia* and *G. latifolium* to produce bioethanol.

2. Materials and Method

This study was conducted on May 2015-March 2016 at Surfactant and Bioenergy Research Center (SBRC) and the Microbiology Laboratory of the Department of Marine Sciences and Technology, IPB.

2.1. Rejuvenation of bacteria

Bacteria cultivated by applying bacteria into nutrient agar. The bacteria were incubated in an incubator for 1 x 24 hours at 30°C. Colonies that have been purified subsequently inoculated into an agar medium slant NA in-use as stock.

2.2. Qualitative Test of Agarase Enzyme

Qualitative test performed with Lugol staining method. Bacterial isolates dripped onto an agar medium YPA. The bacteria were incubated for 3 days at 30 °C [3], then agarolitic activity tested by adding Lugol dye as many as 2 mL and allowed to stand for 15 minutes. Diameter of clear zone and the diameter of colonies formed were measured. Measuring the diameter of clear zone and the diameter of the colony did in two repetitions. Potential agarolitic bacteria in producing agarase enzyme can be seen qualitatively by calculating the index of agarolitic. Agarolitic index equation (1) is the ratio between the diameter of the clear zone and colony. The larger the index agarolitic generated, the greater the enzyme produced by the bacterial isolates. Agarolitic Index is an index of the ability of bacteria to degrade the media or produce agarase enzyme.

$$IA = \frac{dzb - dk}{dk} \quad (1)$$

Where:
IA: Agarolitic index (mm)
Dzb: Clear zone diameter (mm)
Dk: Colony diameter (mm)

2.3. Determination of Optimum Time Enzyme Production Agarase
The timing of the inoculum be done by culturing two loops of isolates in 5 mL of nutrient broth and incubated for 620 minutes. The cultures were incubated at room temperature in a shaker with agitation speed of 85 rpm. Sampling was performed during 15 hours of incubation with the sampling time span of 1 hour and measured values Optical Density (OD) at a wavelength of 600 nm.

A total of 5 mL nutrient broth containing cell cultures was inoculated in 500 mL production medium and incubated in a water bath. Inoculum pouring time seen from the exponential growth of bacteria at growth curve. Sampling was done every 6 hours for 42 hours during incubation. 1 mL of the reaction and 3 mL DNS reagent was centrifuged for 15 minutes, then T stirred using vortex and heated in boiling water for 5 minutes. Agarase activity is expressed in International Units U/mL. One unit is the amount of enzyme required to break one mol agarase be reducing sugars per minute in test conditions. Glucose levels resulting from hydrolysis by enzymes agarase based on the absorbance at λ 550 nm. Then, agarase activity was calculated based on the equation of the agarase enzyme activity in the modification (2) [3].

$$\text{Agarase enzyme activity (U/mL)} = \frac{glukosa\ (mg/L)\times 1000}{BM\times V\times t} \times fp \qquad (2)$$

Where
BM: Glucose molecule weight (0.18 mg/μmoll); *V*: Enzyme volume (1 mL); *T*: incubation time (30 minute)

2.4. Crude enzyme production Agarase
Production of agarase enzyme was done according to the procedures and incubation time of the highest agarase activity that was known from agarase activity curves. Media production were incubated at 30°C in a water bath with agitation speed of 100 rpm, then harvested during the agarase enzyme production time highs that have been obtained previously. During the incubation period, sampling was done for the calculation of reducing sugars do as much as 1 mL with the addition of 3 mL. Cell culture on production media that containing extracellular agarase enzyme was centrifuged at 2500 rpm for 15 min at 4°C to separate the enzyme solution with bacterial pellets. Supernatant resulted from centrifugation then stored at 10°C as the crude extract enzyme.

2.5. Agarase Enzyme characteristic
Adding 0.2 mL of enzyme reacted with 1.8 mL of the substrate-tested effect of pH on enzyme activity. Substrates made by a mix of 1.8 g from 0.5% bacto agar into a buffer with different pH levels at 3,4,5,6,7,8 and 9. Each enzyme was incubated at 30° C for 30 minutes. Agarase enzyme activity was measured according to the test procedures before.

The influence of temperature on the enzyme activity carried out by reacting 0.2 mL to 1.8 mL enzyme substrate wherein the substrate is made by mixing 1.8 g bacto that the optimum pH buffer. Enzymes that have been mixed with the substrate and then incubated at a temperature level of between 20-80 ° C with an interval of 10 ° C for 30 minutes of incubation time. Agarase enzyme activity was measured according to the test procedures before.

IOP Conf. Series: Earth and Environmental Science **65** (2017) 012025 doi:10.1088/1755-1315/65/1/012025

3. Results and Discussion

3.1. Agarase Activity

Clear zone formed demonstrated that the enzyme activity of agarase extraseluler issued by N1 and N3 isolates from *G. salicornia* and BSUC2 and BSUC4 of *G. latifolium* (Table 1).

Table 1 Agarolitic index

Isolate	Agarolitic Index (mm)
N1[1]	2.27
N2[1]	2.32
BSUC2[2]	4.18
BSUC4[2]	4.28

[1]Bacteri isolated from *G. salicornia*
[2]Bacteria isolated from *G. latifolium* [6]

Potential of agarolitic bacteria in producing agarase enzyme can be seen qualitatively by calculating their agarolitic index. Table 1 showed the agarase enzyme activity of bacterial isolates from *G. latifolium* greater than from *G. salicornia*. The big agarolitic index formed from isolates BSUC4 (*G. latifolium)* was 4.28 mm.

3.2. Optimum time of Agarase Enzyme Production

The peak time of growth can be used as a determination of the best time for pouring the inoculum media to the production media for the optimum number of cells that are expected to generate curves in corresponding enzyme activity. Isolates N1 and N3 had the best time for pouring inoculum at the same time, i.e. at the 13th hour. While, isolates BSUC2 and BSUC4 of *G. latifolium* optimum at the 14th hour and 8^{hr} all through the 30 minutes [3]. Once known, the next step is to determine the best enzyme production time and can be seen from the curves of enzyme activity (Figure 1).

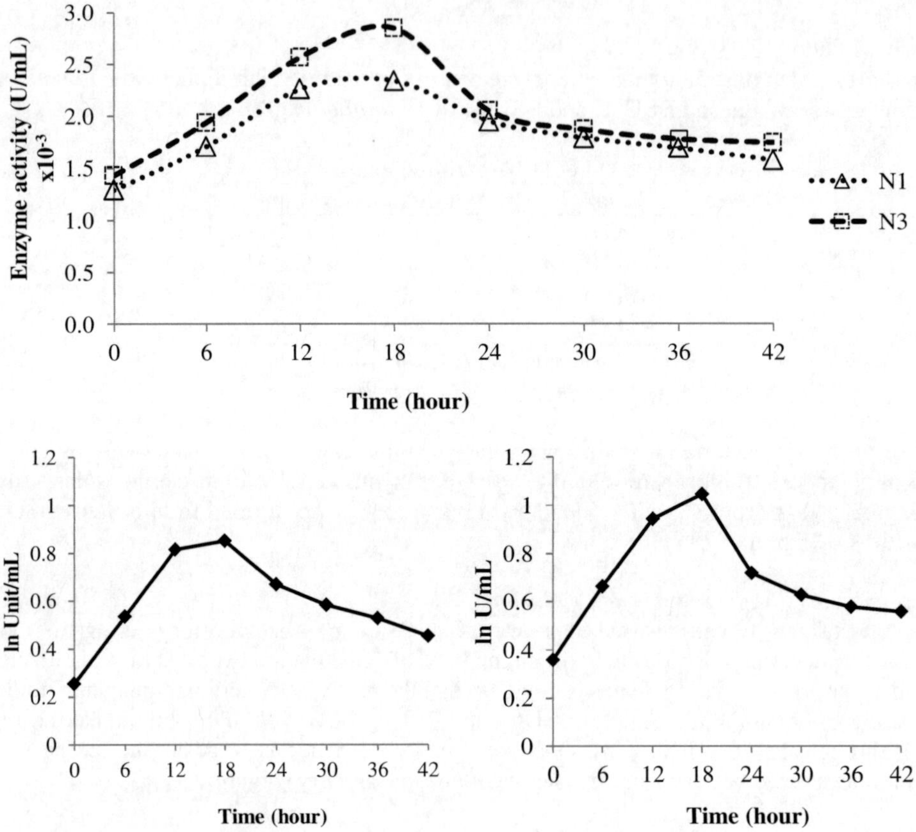

(b) (c)

Figure 1. (a) Agarase enzyme activity and rate of enzyme activity from (b) isolate N1 and (c) isolate N3.

The rate of agarase enzyme activity that isolated from bacterial can be written with the equations $N_t/N_0 = \exp(0.0009t)$ to isolate N1 and $N_t/N_0 = \exp(0.0006t)$ to isolate N3. Isolate N1 and N3 have had the highest enzyme activity time at the 18th hour. The high enzyme activity of isolate N1 was 2.34×10^{-3}U/mL and N3 was 2.84×10^{-3}U/mL. The enzyme activity of isolate BSUC4 also optimum at the 18th hour with the value of the enzyme activity that was 2.14×10^{-3} U/mL. The enzyme activity was not seen in BSUC2 isolate that cause these isolates did not include in further tested [6].

3.3. Characteristic of Agarase Enzyme

3.3.1. Optimum of pH

The enzyme activity is influenced by the pH where the pH or pH changes that do not fit will cause the area and catalytic enzyme conformational changes. In addition to changes in pH also causes the denaturation of enzymes causing loss of enzyme activity [7]. N1 isolates agarase enzyme activity optimum at pH 7 buffer with agarase activity value of 1.38 × 10-3 U/mL and the optimum pH N3 isolates were also obtained at pH 7 buffer with agarase activity value of 1.31 × 10-3 U/mL (Figure 2).

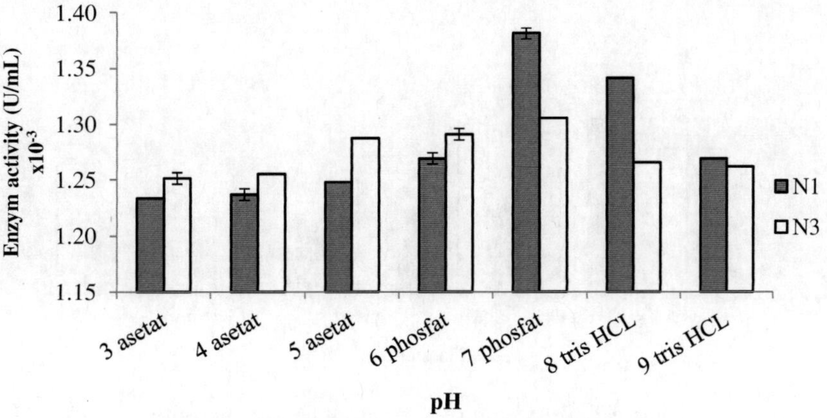

Figure 2. Effect of pH on agarase activity

Changes in pH effect on enzyme activity will happen if the conditions of low or high pH can also cause denaturation. The enzyme activity will drop drastically at a pH greater than 8.5 [6]. Energy reserves begin to run out and nutrients can lead to reduced enzyme activity decreased. Both isolates showed maximum yield at pH 7 phosphates, this shows that the pH is pH optimum for producing the agarase enzyme. These results do not differ much from the results of previous studies that the research carried out used the bacterium Bacillus subtilis and found that pH 8 is an optimum pH for growth [3] and agarase enzyme activity and research conducted by Kawaroe *et al.* [3] using bacterial isolates from *G. latifolium* shows the optimum results at pH 7 [3].

Characteristics of enzymes measured in this study were the effect of pH on the activity of the agarolitic enzyme with pH interval of 3-9. Yu et al**.** [7] mention that the pH is good for agarase enzyme activity at 3-8.5 [8] and Fu et al. [5] stated that the optimum pH at a range 6.5-7.5. Changes in pH can effect on enzyme activity if the conditions of pH is low, or high pH can also cause denaturation and will be clicking causes enzyme activity to decrease [8]. The enzyme activity will drop drastically at a pH greater than 8.5 [7]. Energy reserves begin to run out and nutrients can lead to reduced enzyme activity decreased. Both isolates showed that maximum yield at pH 7 phosphate. This indicates that this pH level is optimum for producing the agarase enzyme. These results do not differ much from the results of previous studies used the bacterium *Bacillus subtilis* [4] and found that pH 8 is an optimum pH for growth and agarase enzyme activity and research conducted by [6] using bacterial isolates from *G. latifolium* shows the optimum results at pH 7 [6].

3.3.2. Temperature

The enzyme activity will increase as the temperature rises until it reaches maximum activity and subsequently the enzyme activity decreases when the temperature is getting hotter due to denaturation of proteins [8]. The highest agarase enzyme activity generated by both isolate N1 and N3 obtained at 30^0C with a value of enzyme activity were 1.31×10^{-3} U/mL and 1.32×10^{-3} U/mL respectively (Figure 3).

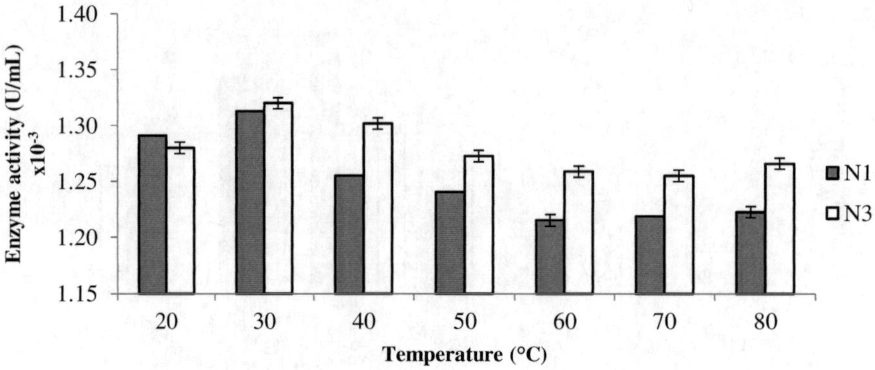

Figure 3. Temperature of Agarase Enzym Activity

The growth of bacterial cultures and production of enzymes optimum at temperature range of 20-50^{0}C in the basal medium [4]. The influence of higher temperatures can inhibit enzyme activity through a process of agarase enzyme protein when denaturation. Temperatures are getting warmer will decide the secondary bonding enzymes and may alter or eliminate the secondary and tertiary structure of the enzyme so that the enzyme activity will decline. The results showed that the optimum temperature suitable for enzyme production process was at a temperature of 30^{0}C. Research conducted by [6] with isolates from *G. latifolium* showed that the highest agarase activity occurred at 50^{0}C with enzyme activity value at 1.33 × 10^{-3} U/mL [8]. This suggests that the effect of temperature on the enzyme activity of agarase optimum at temperature range of 30-50^{0}C.

4. Conclusions

Agarase enzyme activity from bacterial that was isolated from *G. latifolium* is greater than isolated from *G. salicornia*. Characteristics of culture bacterial were indicated from the agarolitic index of isolates N1 (2.32 mm) and N3 (2.27mm) as well as from bacteria that isolated from *G. latifolium* that is BSUC2 at 4.18 mm and BSUC4 at 4.28 mm. Agarase enzyme activity from isolate N1 and N3 were optimum at pH 7 and a temperature of 30 ^{0}C, while the isolate BSUC4 optimum at pH 7 and a temperature of 50^{0}C. The whole of these characteristics indicated that there were bigt potential for bacterial culture of macroalgae *G. salicornia* and *G. latifolium* as raw material for bioethanol.

References

[1] Kawaroe M, Sunuddin, A, Hwangbo J and Shaumi A. 2015 Characteristics and selulotic activities of endophytic fungi in macroalgae (*Sargassum* sp., *Gracilaria* sp., *Gelidium* sp., and Caulerpa sp.) from seagrass habitat in Pari Island, Thousand Islands, Jakarta *Int. J. Sci: Basic Appl. Res.* **22** 149-160

[2] Wi G W, Kim H J, Mahadevan S A, Yang D J, Bae H J 2009 The potential value of the seaweed Ceylon moss (*Gelidium amansi*) as an alternative bioenergy resource *Bioresource Technology.* **100** 6658–6660

[3] Kawaroe M, Wulansari D, W. S, Hwangbo J and Santoso J. 2015 The optimum fermentation periode to produce bioethanol from Gelidium *latifolium* and *Gracilaria verrucosa J. Eng. and Tech.* **47** (6) 650-663

[4] Suzuki H, Sawai Y, Suzuki T, Kawai K 2003 Purification and characterization of an extracellular β-agarase from *Bacillus* sp. MK03 *J Biosci Bioeng.* **95** (4): 328-334

[5] Fu XT, Kim S M 2010 Agarase: review of major sources, catacories, purification method, enzyme characteristic and applications *Marine Drugs.* **8** 200-218

[6] Kawaroe M, Rusmana I and Nurafni 2014 Production of bioethanol from macroalgae gelidium sp. using agarase enzymes of marine bacteria *Int J. Environ. Bioenerg*. **9** 243-251.

[7] Yu F S, Hui C Y, Yen L 2009 The discovery of agarolytic bacterium with agarase gene containing plasmid, and sone enzymology characteristics *Int J Appl Sci Eng.* **7** (1) 25-41
[8] Fu W, Han B, Duan D, Liu W, and Wang C 2008 Purification and characterization of agaraes from marine bacterium *Vibrio* sp. F-6. *J Ind Microbiol Biotechnol.* **8** 915-922.

Isolation and characterization of triacylglycerol-accumulating microorganisms which can utilize wood polysaccharide

S A Susanto[1]

[1]Institute Teknologi Del, Jl. Sisingamangaraja, Sitoluama, Laguboti, Toba Samosir Regency, North Sumatera, Indonesia

Email: sonyadhisusanto@outlook.com

Abstract. Triacylglycerol is an ester which is made of glycerol and three fatty acids. This compound is an important feedstock for biodiesel production. In this study, several strains of oleaginous bacteria were isolated from environmental sample based on their ability to grow in mineral salts medium supplemented with wood-derived sugars such as cellulose, arabinose, xylose, mannose, and galactose. The lipid accumulating bacteria were selected based on fluorescent signal from hydrophobic inclusion in the cytoplasm after incubation in selective medium containing lipophilic dye 0.5 % (w/v) nile red. The lipid content was analyzed using thin layer chromatography (TLC) and gas chromatography-mass spectrometry (GC-MS). In this study, three bacterial isolates 2HPCS1R4, 1LPCS2R2, and 1LPCS2R14 were selected among several candidates. TLC analysis of hydrophobic substance from 1LPCS2R2 and 1LPCS2R14 showed two overlapped discrete bands corresponded to triacylglycerol reference band. While 2HPCS1R4 displayed a faint band located above the reference band. GC-MS analysis confirmed that the bands consisted of fatty acid methyl esters with alkyl length varied from C12 to C17. Kinetic study showed that the fastest growing strain was 1LPCS2R2 had the highest growth rates and when grown in glucose ($\mu = 0.29\ h^{-1}$) and xylose ($\mu = 0.16\ h^{-1}$). In conclusion, this study has identified of prospective bacterial isolates for commercial biodiesel production.

1. Introduction

Agriculture and plantation are the most important economic sector in Indonesia. Palm oil and pulp-paper industry contributes to almost US$ 31.27 billion from total of US$ 86.01 billion of Indonesian's export[1]. By the year of 2013, Indonesia had become the second largest palm oil producer after Malaysia with total production of 12 million tons of crude palm oil (CPO) accounted to 44% of total world CPO production[2]. The tremendous number of production is also followed by significant amount of waste. Most of the agriculture and plantation waste is lignocellulosic biomass. It is a recalcitrant substance served as structural backbone of plant cells and composed of cellulose (40%-50%) and hemicellulose (25%-30%) bound together by lignin (15% - 20%). Cellulose is a homogenous polymer consisted of several glucose units through β-1-4 glycosidic bond. Hemicellulose is a branched-and linear heteropolymer composed of mixture of 200 to 400 units of pentose (C5), such

Content from this work may be used under the terms of the Creative Commons Attribution 3.0 licence. Any further distribution of this work must maintain attribution to the author(s) and the title of the work, journal citation and DOI.
Published under licence by IOP Publishing Ltd

as xylose and arabinose, and hexose (C6), such as mannose, glucose, and galactose, sugars[3]. Lignin is an aromatic heteropolymer complex substance consisted of mostly three main monomers, guaiacyl (G) unit from coniferyl alcohol precursor, syringyl (S) unit from sinapyl alcohol precursor, and p-hydroxyphenyl (H) from p-coumaryl alcohol precursor. Each different plant species has different ratio of G, S, and H[4,5].

Due to the recalcitrant nature of the lignocellulosic biomass, pretreatment is required to degrade the overall structure of the biomass in order to release fermentable sugars for energy and carbon source for microbial-based biotechnological processes[6]. To date, there are several established pretreatment technologies which are described elsewhere[7]. One of them is acid hydrolysis that converts lignocellulosic biomass to polysaccharides such as cellulose, hemicellulose and monosaccharides such as arabinose, xylose, mannose, galactose and glucose and the composition varied based on the concentration of the acid and the enzyme[8]. These compounds can serve as cheap and abndant carbon source which is required for bulk low cost biodiesel production.

Biodiesel consists of monoalkyl esters of long-fatty acids with short-chain alcohols such as methanol or ethanol which constitute fatty acid methyl esters (FAMEs) or fatty acid ethyl esters (FAEEs) respectively. Though biodiesel has several advantages over petroleum diesel, commercialization of this environmental friendly energy is still limited (a detail discussion is available elsewhere[9]. Production costs, sustainability and geographic-dependent availability of feedstock are the main reason of the hurdle. Therefore the use of lignocellulosic biomass can reduce the production cost. Some of lignocellulosic biomass are available as underutilized waste in agriculture, plantation, and wood-related industry. Low cost bioconversion using oleaginous microorganisms, that can utilize the feedstock for biodiesel production, can give alternative solution for the issue[10].

In this report, explorative study to screen and characterize novel oleaginous microorganisms which were able to utilize lignocellulosic biomass was presented. Some environmental sample from various sources such as compost and sludge were collected and then the materials were incubated in screening medium supplemented with wood-derived sugars such as cellulose, xylose, arabinose, and glucose. Oleaginous microorganisms was screened using microscope fluorescence and validated by thin layer chromatography and gas chromatography – mass spectrometry.

2. Materials and Methode

2.1. Samples

Samples were collected from late stage compost soil located at composting facility in municipal garden, at Komplek Perumahan Dosen ITB, Jl. Kanayakan Bandung, Indonesia. The sewage sludge was collected from waste water and shrimps ponds in Perumahan Kompleks Pertamina Klayan, Gunung Jati, Cirebon, Indonesia. All samples used for this study were transferred to the laboratory and stored at 4 ^{o}C.

2.2. Screening of Cellulosic Biomass Utilizing Microorganisms

In order to isolate bacteria which are capable to utilize cellulosic biomass, a 2 ml resuspended sludge sample and 2 g of each compost soil were enriched in 200 ml of mineral salts medium (MSM), the high phosphate MSM (HPMSM)[11] and the low phosphate MSM (LPMSM)[12] and each was supplemented with 2% cellulose and starch as the sole carbon source. The aerobic incubation was then performed in temperature-controlled room (30°C) for 48 hours. The selection of microorganisms was conducted by transferring 500 μl of the liquid media containing selectively enriched microbes into agar plate HPMSM and LPMSM, each supplemented with 2% (w/v) of cellulose, arabinose, xylose, mannose, and galactose as the sole carbon source.

The plates were incubated at 30°C for 4 – 7 days until the bacterial colonies was grown. In order to obtain axenic cultures, the mixed culture bacterial colonies were restreaked into the new identical medium for three times. To store the selected bacteria, 2 ml of 10 % (w/v) sterile skim milk (Difco) was added to the bacteria grown in solid agar media, transferred to the cryo vial tube containing several sterile paper disc. The bacterial cells were freezed for at least 6 hours at -20°C then lyophilized with freeze dryer. The bacterial strains were then stored at -80°C.

2.3. *Screening of Hydrophobic Substance Accumulating Microorganisms*

PHAs granule which is intracellularly accumulated by certain microorganisms can be visualized using lipophilic stain, nile red, bound specifically with hydrophobic substances like lipid, wax esters, and PHAs. Modification of original method[13] was employed during this study. The detail of the protocol is as the following: Nile red (Biochemika) was dissolved with DMSO to give final concentration of 0.5% (w/v). The selected microorganisms was incubated in this solution for at least 5 minutes. Afterwards, the hydrophobic granule was observed under fluorescence microscope (Carl Zeiss) at a wavelength of 598 nm.

2.4. *Thin Layer Chromatography*

Thin layer chromatography (TLC) was performed to rapidly characterize the hydrophobic content of the sample from bacterial isolate acquired from the microscope fluorescence screening. The technique was modification of established methods which is described elsewhere[14]. The detail of the method is as followed. The bacterial cell candidate was firstly incubated in -20°C for at least 12 hours. Afterwards, the frozen cell was dried using freeze dryer for at least for 12 hours. The biomass was pulverized. Hydrophobic extraction was performed by weighing approximately 10 mg of the cell powder and mixed it with 1 ml of chloroform : methanol (2:1) in a microtube. Afterwards, the mixture was rigorously mixed with vibromixer for 1 minute at room temperature. In order to separate the extracted hydrophobic-rich substances from the rest of the cell biomass, the mixture was centrifuged for 10 minutes at 13,000 rpm. Afterwards, the extraction procedure was repeated with the rest of the cell biomass. The second supernatant was mixed with the first, afterwards it was evaporated until the volume achieve to approximately 0.2 ml. The sample was applied to the silica gel plate (Silica Gel 60, DC Kieselgel 60, Merck). Afterwards the plate was incubated in air-tight closed container containing mobile phase mixture composed of hexane : diethyl ether : acetic acid (90 : 10 : 1 v/v/v). The sample visualization was performed by exposing the plate with iodine vapour to characterize the unsaturated fatty acids[15]. To detect lipid fractions which did not contain double bond (saturated fatty acids), the sample was sprayed with 30% (v/v) aqueous sulfuric acid and heated in an oven to 200°C until the spot is visualized[14].

2.5. *Gas Chromatography Mass Spectrometry*

The lipid content of the bacterial cell which accumulating hydrophobic granule was analyzed using gas chromatography mass spectroscopy (GC-MS). The method was modification from a method which is described elsewhere[16]. The detail of method is described as followed. The selected isolates, contained large hydrophobic granule, were frozen at -20°C. Afterwards, the cells was lyophilized for a minimum of 24 hours. The cell was pulverized and 10 mg was transferred to a test tube. Alternatively, the hydrophobic content of the lyophilized cell was firstly separated with TLC, then the expected bands were scratched and 10 mg of the powder was transferred to test tubes. Afterwards methanolysis procedure was performed to convert the lipid sample to volatile derivative. To achieve this goal, 1 ml of a mixture of 85% (v/v) methanol and 15% (v/v) sulfuric acid was added, incubated at 100°C for 4 hours. Water was added to the reaction product to dissolve and separate the hydrophilic portion. A 3 µl organic phase was analysed in GC-MS instrument after splitless injection. The instrument was Series 6890 GC system equipped with a series 5973 Ei MSD mass-selective detector (Hewlett Packard). The column was BPX 35 capillary column (60 m x 250 µm, film thickness 250 µm (SGE Analytical Science). The carrier gas was helium, constant flow of 0.6 ml/min. The temperature injector and detector were 250°C and 240°C respectively. The following temperature program was applied: 120°C for 5 minutes, increased by 3°C/minutes to 180°C and by 10°C/minutes to 220°C. The data were compared with NIST-Mass Spectral Search Program available online at http://chemdata.nist.gov/mass-spc/ms-search/.

2.6. Growth Experiment

Growth experiments were conducted to determine growth curve and rate of the selected strains. Three selected bacterial strains were pre-enriched in 10 ml of nutrient agar medium (Oxoid) for 12 – 16 hours at constant temperature of 30°C, shaking incubator 144 rpm. Afterwards, the pre-enriched culture was inoculated to 200 ml of respective mineral salts medium supplemented with 2% (w/v) xylose or glucose in 1 l of klett flask. The cultivation was performed at 30°C, within 3-5 hours. The turbidity of the growth culture was measured using klett colorimeter (Manostat Corporation). The data was processed with Microsoft® Excel. The klett unit was converted to natural logarithm and plotted against incubation time in hours. In order to derive the growth rate, linear regression line was calculated through the log phase and the rate was deduced from the slope.

3. Result

3.1 Selection of Microorganisms

In this study, three distinct bacterial isolates were selected (table 1). All of them were able to use 2 % (w/v) cellulose as sole carbon source. Two of them were derived from low phosphate MSM while the other was selected from the high phosphate selective medium. All isolated bacterial cultures were able to live in the high salinity medium.

Table 1. Selected Bacterial Isolates and Their Properties

Strain Identifier	Source[1]	Salinity	MSM
1LPCS2R2	Compost	1%	Low phosphate
1LPCS2R14	Compost	1%	Low phosphate
2HPCS1R4	Sludge	2%	High phosphate

[1] The location where the bacterial sample was isolated. Compost was derived from composting facility, and sludge was derived from shrimp's pond

3.2 Hydrophobic Inclusion

Lipophilic dye, nile red, penetrated the plasma membrane of the selected cells and bound to the hydrophobic compound. Monochromatic light with a wavelength of 598 nm excited the non-polar substances bound nile red in the cells. The emission light was detected with microscope fluorescence and the results of the observation indicated variety of hydrophobic substances in the cells (figure 1). Bacterial strain 2HPCS1R4 contained two hydrophobic inclusions (figure 1B, red and blue arrow) in the cytoplasm. Whereas, the hydrophobic inclusion in 1LPCS2R2 was distributed around the plasma membrane (figure 1D, arrow). A small discrete inclusion was found at the edge of the cells (figure 1D). Similarly, 1LPCS2R14 showed faint distributed inclusion (figure 1F), and small discrete inclusion (figure 1F, blue arrow).

3.3 Analysfs of Hydrophobic Inclusion with Thin Layer Chromatography

Treatment of 10 mg of cell dry mass of 1LPCS2R14, 1LPCS2R2 (figure 2) and 2HPCS1R4 (figure 3) with organic solvent (chloroform: methanol) released the hydrophobic content of the cells. In the TLC plate, the mixture of hydrophobic compound was separated using silica gel plates using mobile phase composed of hexane/diethyl ether/acetic acid. 1LPCS2R2 contained triolein-like hydrophobic compound (figure 2, arrow 3) and oleic acid-like hydrophobic compound (figure 2, arrow 4) which were overlapped to each other. Whereas, 1LPCS2R14 gave faint triolein-like hydrophobic band (figure 2, arrow 5) and two discrete bands of oleic acid-like (figure 2, arrow 6). These respective bands lay slightly above the control bands. The 2HPCS1R4 contained different hydrophobic subtances which was comparably different with the previous two. The discrete band lay above the triolein-like hydrophobic band (figure 3, arrow 3).

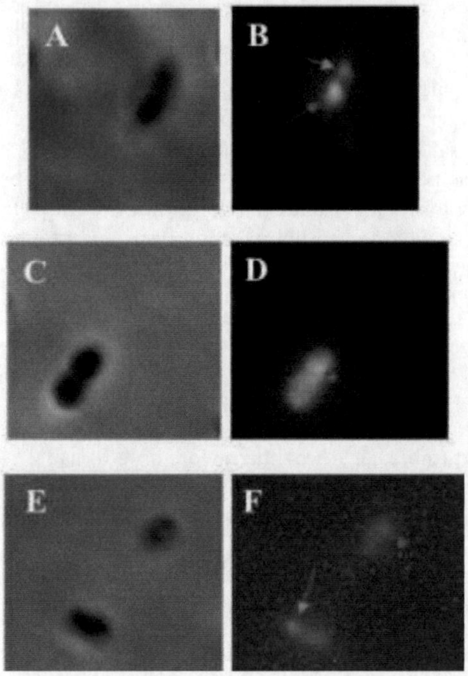

Figure 1. Fluorescence microscopy observation of three selected bacterial isolates (A) and (B) 2HPCS1R4, (C) and (D) 1LPCS2R2, (E) and (F) 1LPCS2R14. (A), (C) and (E) Light and (B), (D) and (F) fluorescence image of the respected bacterial strain: The white signal (red arrows) in the cytoplasm of bacterial cells was derived from interaction of a lipophilic dye (nile red) with hydrophobic substances. 2HPCS1R4 had two kinds of hydrophobic inclusion (B) (red and blue arrows). Whereas 1LPCS2R2 and 1LPCS2R14 had mainly membrane fluorescence (D), (F). Original magnification x 400.

3.4 Gas chromatography analysis

The elucidation of the identity of the hydrophobic compound in the TLC bands showed mixtures of different long chain fatty acid residues. The two discrete bands derived from bacterial strain 1LPCS2R2 (figure 2, arrow 3 and 4) revealed some mixtures of methyl ester ranged from myristic acid (tetradecanoic acid) methyl ester (C14) (figure 4, peak A), small amount of pentadecanoic acid methylester (C15) (figure 4, peak C, E), branched lauric acid (dodecanoic acid) methyl ester (figure 4, peak D), palmitic acid (hexadecanoic acid) methyl ester (figure 4, peak F, G, H).

GC-MS analysis of two discrete bands from 1LPCS2R14 (figure 2, arrow 5, 6) showed some mixtures of fatty acid methyl ester predominantly composed of lauric acid (dodecanoic acid) methyl ester (C12) (figure 5, peak C), myristic (tetradecanoic acid) methyl ester (C14) (figure 5, peak D, F), pentadecylic acid methyl ester (C15) (figure 5, peak E, G), palmitic acid (hexadecanoic acid) methyl ester (figure 5, peak H), and margaric acid (heptadecanoic acid) methyl ester (figure 5, peak I).

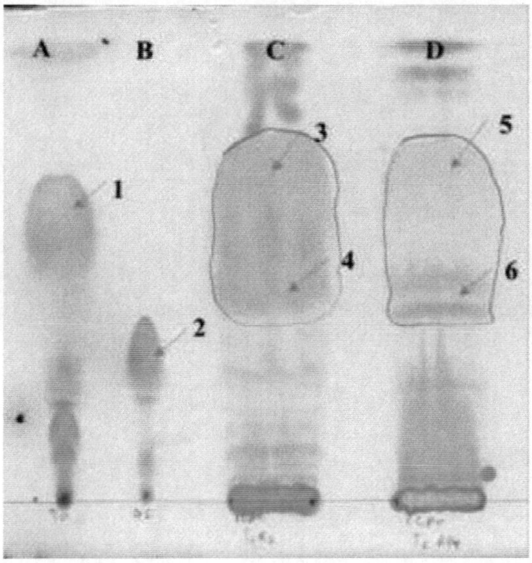

Figure 2. Thin layer chromatography (TLC) of hydrophobic content from selected bacterial isolates. (A) Triolein, (B) oleic acid. They were served as TLC standard of triglyceride (arrow 1) and free fatty acids (arrow 2) respectively. (C) TLC of hydrophobic content of 1LPCS2R2. Chloroform-methanol-exracted 10 mg of 1LPCS2R2 cell dry mass was analyzed with TLC. Two discrete bands (arrow 3, 4), which were overlapped each other, were observed after incubation of the TLC plate with iodine. (D) TLC of hydrophobic content of 1LPCS2R14. Chloroform-methanol-extracted 10 mg of 1LPCS2R14 cell dry mass was analyzed with TLC. Three obvious discrete bands (arrow 5, 6) were detected.

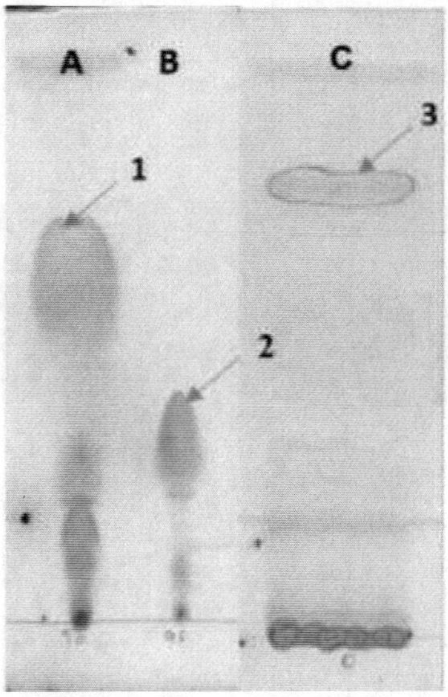

Figure 3. Thin layer chroma-tography (TLC) of hydrophobic content from 2HPCS1R4. (A) Triolein, (B) oleic acid. They were served as TLC standard of triglyceride (arrow 1) and free fatty acids (arrow 2) respectively. (C) TLC of hydrophobic content of 2HPCS1R4. Chloroform-methanol-extracted 10 mg of 2HPCS1R4 cell dry mass was analyzed with TLC. A discrete band (arrow 3) was detected just above the triolein standard.

Whereas, GC-MS analysis of hydrophobic substances in the cytoplasm of 2HPCS2R4 showed only few recognizable peaks (figure 6). They were palmitic acid (hexadecanoic acid) methyl ester (figure 6, peak A), oleyl alcohol (figure 6, peak B), and stearic acid (octadecanoic acid) methyl ester (figure 6, peak C).

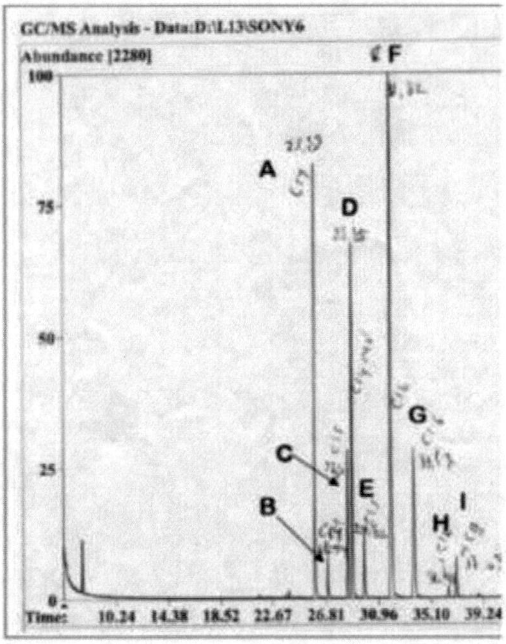

Figure 4. Gas chromatography – mass spectroscopy analysis of hydrophobic content of 1LPCS2R2. The hydrophobic content derived from discrete bands shown in figure 2, arrow 5 and 6. The chromatogram showed different length of fatty acids of methyl ester ranged from C14 to C16. Peak A (ret. time 25.89 min.) and peak B (ret. time 26.94 min.) were methyl tetradecanoate; peak C (ret. time 28.5 min.) and peak E (ret. time 29.86 min.) were methyl pentadecanoate; peak D (ret. time 28.85 min.) was 12-methyl- methyl tetradecanoate; peak F (ret. time 31.82 min.) and peak G (ret. time 33.67 min.) were methyl hexadecanoate: peak H (ret. time 36.45 min.) was 8 methyl- methyl decanoate.

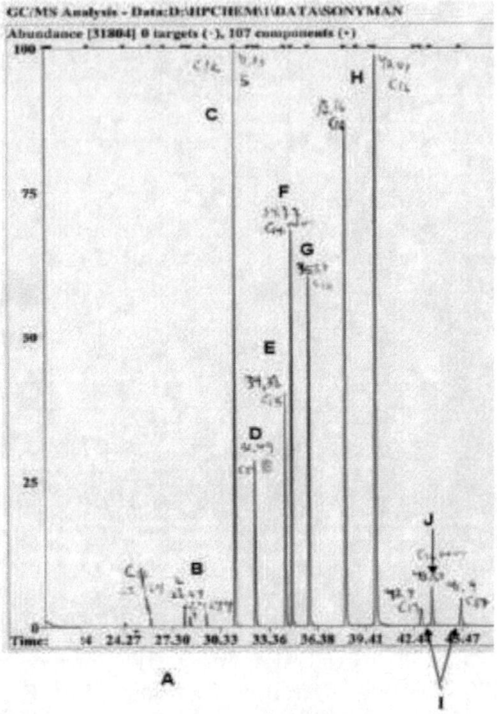

Figure 5. Gas chromatography – mass spectroscopy analysis of hydrophobic content of 1LPCS2R14. The hydrophobic content originated from TLC bands shown in figure 2, arrow 5. The chromatogram showed different length of fatty acids of methyl ester ranged from predominantly C12 to C17. Peak A (ret. time 25.97 min.) and C (ret. time 31.23 min.) were methyl dodecanoate; peak B (ret. time 28.04 min.) was methyl tridecanoate; peak D (ret. time 32,49 min.) was methyl tetradecanoate; peak E (ret. time 29.86 min.) was methyl pentadecanoate; peak F (ret. time 31.82 min.) and peak G (ret. time 33.67 min.) were methyl hexadecanoate; peak H (ret. time 36.45 min.) and peak J (ret. time) 37.08 min) were 8-methyl-methyl decanoate.

3.5 Growth Kinetic of Selected Bacterial Isolates

In order to evaluate whether the selected isolates are suitable for the industrial process, growth kinetic of selected bacterial isolates was investigated by incubating them in 2% (w/v) glucose and xylose. 1LPCS2R14 showed different growth patterns. When glucose was used as a sole carbon source (figure 7A), the log phase of 1LPCS2R14 commenced quite early, around 3rd to 23rd hours after inoculation. The calculated line of regression linear spanned the indicated times was y= 0.0469X + 3.8869, so that doubling time was 0.049 h^{-1}. 1LPCS2R14 grew in xylose with doubling rates of 0.0248 h^{-1}. The log phase spanned throughout the measuring time (figure 7B), however the linearity of the regression line from natural logarithm of klett unit vs incubation time was low (R^2 = 0.9016). 1LPCS2R4 incubated in 2% (w/v) glucose showed defined growth curve (figure 7C). The exponential phase was detected at

12th– 18th hours of incubation time. The rate of doubling time as appeared in the log phase was 0.29 h^{-1}. When the isolate was incubated in 2% (w/v) xylose (figure 7D), the doubling time was decreased to 0.17 h^{-1} and the log phase started at the early phase. No lag phase was observed in the growth curve. Similarly, log phase of 2HPCS1R14 (figure 7E), started at the early phase, (0 – 15th hours of incubation time). The growth rate of the isolates was 0.16 h^{-1}. However, when 2HPCS1R14 was grown in 2% (w/v) xylose (figure 7F), the doubling rate decreased dramatically to 0.089 h^{-1}.

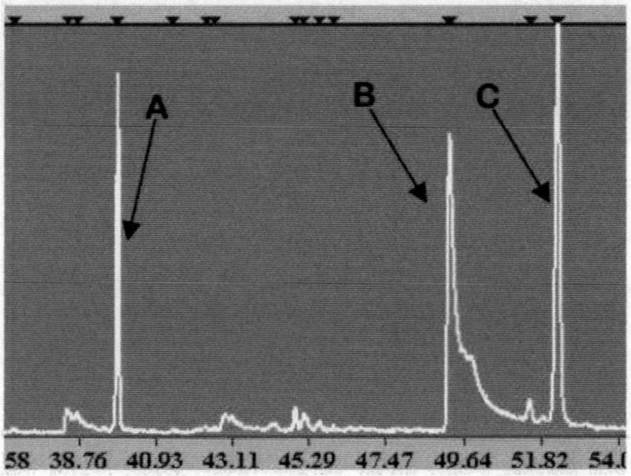

Figure 6. Gas chromatography – mass spectroscopy analysis of hydrophobic content from 2HPCS1R4. The gas chromatography analysis of discrete band (figure 3, arrow 3) showed three major distinct peaks, (A) hexadecanoic acid (C16), (B) oleyl alcohol (C8), and (C) octadecanoic acid (C18).

4. Discussion

This explorative research was aimed to screen and characterize potential microorganisms which were able to utilize wood-derived sugars as energy and carbon source to accumulate triglyceride as feedstock for biodiesel production. Some potential bacterial strains were isolated from sludge of municipal sewage, 2HPCS1R4, and two bacterial strains from late phase of compost soil, 1LPCS2R2 and 1LPCS2R14. All of them were potential bacterial strains derived from cellulose supplemented screening media.

Optically, 2HPCS1R4 showed the most abundant hydrophobic inclusion aggregate in comparison to the other two isolates (figure 1B, arrow blue and red). There are numerous types of hydrophobic inclusion known so far. *Rhodococus opacus* strain PD630 forms inclusion bodies which consists of several agglomerations of fat bodies triglycerides. They are fully filled the cytoplasm of the cells and between inclusion bodies were bordered only with wall membrane layer separating them[17]. Some *Actinobacter*-like *A. calcoaceticus* ADP 1 showed less dense hydrophobic inclusions[18]. They are separated each other and formed spherical disc shape similar to 2HPCS1R4 as shown in figure 1B, blue and red arrows. In some colonies *A. calcoaceticus* ADP 1 lipid body inclusion showed only single colony as shown in 1LPCS2R2 and 1LPCS2R14. However, in this study, the identity of the three selected bacteria was not elucidated. In the 1LPCS2R2 and 1LPCS2R14, the whole cytoplasm seemed to illuminate fluorescence light and it.

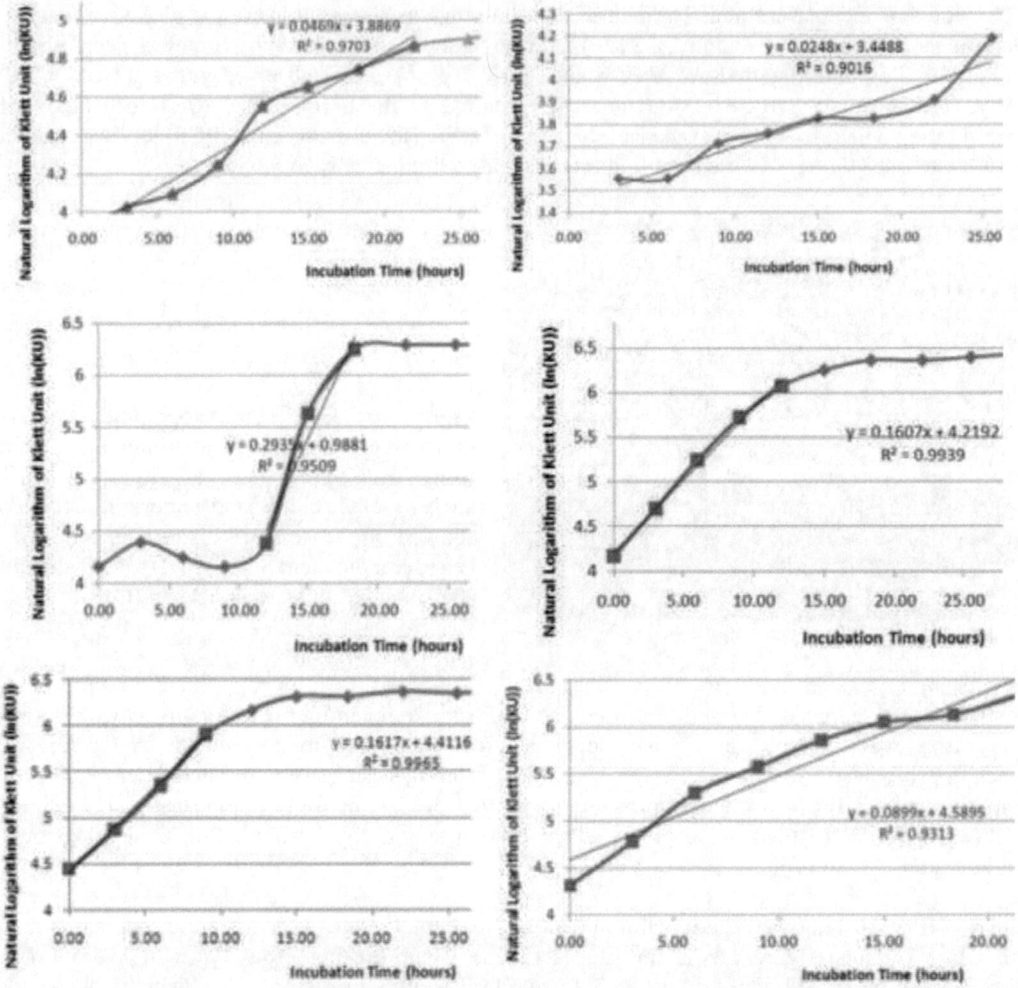

Figure 7. Growth Curve of Selected Bacterial Isolates. (A) 1LPCS2R14 grown in 2% (w/v) of glucose as sole carbon source. Graphical plot of natural logarithm of klett unit vs incubation time indicated early log phase spanned from 3rd hours to 23rd hours. (B) 1LPCS2R4 grown in 2% (w/v) xylose as sole carbon source. Log phase emerged irregularly throughout the growth phase. (C) 1LPCS2R2 incubated in 2% (w/v) glucose as sole carbon source. This bacteria when was grown in glucose adopted common growth phase. (D) 1LPCS2R2 incubated in 2% (w/v) xylose. Log phase was occurred already at the early phase of the bacterial growth (until 14th hours). (E) 2HPCS1R14 incubated in 2% (w/v) glucose. Log phase started at directly after inoculation and stopped at 10th hours of incubation time. (F) 2HPCS1R14 incubated in 2% (w/v) xylose. Log phase appeared throughout the measurement (until 22nd hours of incubation time). Afterwards the growth curve became steady.

Almost encased the signal from the lipid bodies. Spiekermann and coworkers have reported that nile red staining is not effective to detect hydrophobic inclusion in the gram-positive bacteria [13]. The gram analysis of 1LPCS2R2 and 1LPCS2R14 showed that these bacteria were gram-positive (data not shown), therefore the background signal was very high. Alternative method for identification of lipid substances in bacteria is sudan black staining[19]. However, this technique cannot be used for characterization of hydrophobic inclusion in viable bacterial cells. Because it requires replica that may increase the complication of the screening method which is unfavorable when the study has to deal with large number of bacterial strains. Nile red offers practical solution for large scale screening of hydrophobic-containing microorganisms. However, the limitation of using the dye for gram-positive bacteria can be problematic. Example of these bacteria derived from genus *Nocardia* and *Rhodococcus* which are known as lipid-accumulating strains[15]. Therefore establishment of alternative lipophilic dye that can be used to screen lipid-containing viable gram-positive bacterial cells is essential.

In this study, TLC was preferred as low cost technique to screen large number of bacteria derived from fluorescence microscopy initial selection for their ability to accumulate hydrophobic substances. The results showed that an unknown substance was found in 1LPCS2R2. The hydrophobic substances derived from the cells showed two overlapped discrete bands. These spots had comparable retardation factor with the reference TAG. The GC-MS analysis revealed that collection of various lipid with various alkyl length (C14 – C17) was observed. It was not clear whether the substance was TAG because it was subjected to methanolysis which convert the chemical nature of non-volatile lipid to volatile compound, fatty acid methyl ester.

This study evaluated also the prospect of the screened microorganisms for industrial application. Growth rate in the fermentation condition should be high to meet the requirement for industrial application. Therefore, kinetic study of the selected bacterial candidate was performed with glucose and xylose as sole carbon source. The selection was based on the fact that both sugars are the most available monosaccharide on the Earth. The result of the kinetic study with the isolated bacteria showed that 1LPCS2R2 was the fastest growth bacteria (0.29 h^{-1}). This isolates showed regular defined lag, log, and stationary phase while the other did not show such characteristic. However, the doubling rates of the isolate was reduced to half when the isolate was incubated with . The utilization of xylose requires the ability of the microorganisms to adopt xylose isomerase, Weimberg or XR/XDH pathways (xylose reductase and xylitol dehydrogenase enzymes)[20,21,22]. Some bacteria, such as E. coli, prefer to utilize glucose, using glycolytic pathway which end up in tricarboxylic acid cycle, than xylose[23]. This bacterium has carbon catabolite repression (CCR) that causes sequential consumption of glucose and xylose. The latter is a major component of hemicellulose with diverse proportion which is depended on type of the woods. Softwood comprises of maximum 10% (w/w) while hardwood 25% (w/w) dry weight of xylose[22].

In conclusion, the study demonstrated an effort to isolate new strains that utilized wood derived sugars which are majority component of lignocellulosic biomass and accumulated TAG which are feedstock for biodiesel production. Fluorescence microscopy revealed three isolates that accumulated considerable amount of hydrophobic compound. Two isolates, 1LPCS2R2 and 1LPCS2R14, seemed to show signals derived from hydrophobic-like membrane-bound substances which is typical for gram-positive bacteria. TLC study showed that two overlapping discrete bands were observed for 1LPCS2R14 and 1LPCS2R2 whereas 2HPCS1R4 showed only single band with the height just above the reference band of lipid. GC analysis confirmed the presence of the lipid that consisted of different C14-C16 fatty acid length. Growth kinetic analysis showed that 1LPCS2R2 was the most rapidly growing isolate among the three in two model carbon sources (glucose and xylose). The result of study proposed strain that can be used for biodiesel production.

References

[1] Kemenperin 2016 Perkembangan ekspor indonesia berdasarkan sektor *Ministry of Industry* (http://kemenperin.go.id/statistik/peran.php?ekspor=1) April 26th 2016

[2] FAO 2016 Production quantities by country (palm oil) (http://faostat3.fao.org/browse/Q/QC/E) March 1st 2016

[3] Gray K A, Zhao L and Emptage M 2006 Bioethanol *Current Opinion in Chemical Biology* **10** 141–146

[4] Bugg T D H *et al.* 2011 Pathways for degradation of lignin in bacteria and fungi *Natural Product Reports* **28** 1883

[5] Brethauer S and Studer M H 2015 Biochemical conversion processes of lignocellulosic biomass to fuels and chemicals – A Review. *CHIMIA International Journal for Chemistry*, **69** 572– 581

[6] Wyman C E 2003 Potential synergies and challenges in refining cellulosic biomass to fuels, chemicals, and power *Biotechnology Progress* **19** 254–262

[7] Jönsson L J and Martín C 2015 Pretreatment of lignocellulose: Formation of inhibitory by-products and strategies for minimizing their effects *Bioresource Technology* **199** 103–112

[8] Cara C *et al.* 2008 Conversion of olive tree biomass into fermentable sugars by dilute acid pretreatment and enzymatic saccharification *Bioresource Technology* **99** 1869–1876

[9] Krawczyk T 1996 Biodiesel – alternative fuel makes inroads but hurdles remain *Inform* **7** 801–829

[10] Yousuf A 2012 Biodiesel from lignocellulosic biomass - prospects and challenges *Waste Management* **32** 2061–2067

[11] Schlegel H G, Kaltwasser H and Gottschalk G 1961 Ein submersverfahren zur kultur wasserstoffoxydierender Bakterien: wachstumsphysiologische untersuchungen *Archiv für Mikrobiologie* **38** 209–222

[12] Nicholson C A and Fathepure B Z 2004 Biodegradation of benzene by halophilic and halotolerant bacteria under aerobic conditions *Applied and environmental microbiology* **70** 1222–5

[13] Spiekermann P *et al.* 1999 A sensitive, viable-colony staining method using Nile red for direct screening of bacteria that accumulate polyhydroxyalkanoic acids and other lipid storage compounds *Archives of microbiology* **171** 73–80

[14] Uthoff S and Stöveken T 2005 Thio wax ester biosynthesis utilizing the unspecific bifunctional wax ester synthase/acyl coenzyme A: diacylglycerol acyltransferase of Acinetobacter sp. strain ADP1 *Applied and ...*, **71** 790–796

[15] Alvarez H M, Kalscheuer R and Steinbuchel A 1997 Accumulation of storage lipids in species of Rhodococcus and Nocardia and effect of inhibitors and polyethylene glycol *I Forschungsbeitrage/Research Papers* 239–246.

[16] Ibrahim M H A and Steinbüchel A 2009 Poly(3-Hydroxybutyrate) production from glycerol by *Zobellella denitrificant* MW1 via high-cell-density fed-batch fermentations and simplified solvent extraction *J. Appl. Environ. Microbiol* **75** 6222-6231

[17] Alvarez H M *et al.* 1996 Formation of intracytoplasmic lipid inclusions by Rhodococcus opacus strain PD630. *Archives of microbiology* **165** 377–86

[18] Wältermann M *et al.* 2005 Mechanism of lipid-body formation in prokaryotes: How bacteria fatten up *Molecular Microbiology* **55** 750–763

[19] Hartman T L 1940 The use of sudan black B as a bacterial fat stain *Stain Technology* **9153** (June)

[20] Fraenkel D G and Vinopal R T 1973 Carbohydrate metabolism in bacteria *Annual Review of Microbiology* **27** 69–100

[21] Radek A *et al.* 2014 Engineering of Corynebacterium glutamicum for minimized carbon loss during utilization of d-xylose containing substrates *J. Biotechnol* **192** 156-160

[22] Sànchez Nogué V and Karhumaa K 2015 Xylose fermentation as a challenge for commercialization of lignocellulosic fuels and chemicals *Biotechnology Letters* **37** 761–772

[23] Desai T A and Rao C V 2010 Regulation of arabinose and xylose metabolism in *Escherichia coli Applied and environmental microbiology* **76** 1524–32

IOP Conf. Series: Earth and Environmental Science **65** (2017) 012027 doi:10.1088/1755-1315/65/1/012027

Kinetic study of nickel laterite reduction roasting by palm kernel shell charcoal

E Sugiarto[1,2], A D P Putera[2] and H T B M Petrus[1,2]

[1] Chemical Engineering Department, Faculty of Engineering, Universitas Gadjah Mada, Jl Grafika 2, Sleman, Daerah Istimewa Yogyakarta, Indonesia
[2] Center of Advanced Material and Mineral Processing, Universitas Gadjah Mada, Jl Grafika 2, Sleman, Daerah Istimewa Yogyakarta, Indonesia

Email : edi.sugiarto@mail.ugm.ac.id

Abstract. Demand to process nickel-bearing laterite ore increase as continuous depletion of high-grade nickel-bearing sulfide ore takes place. Due to its common nickel association with iron, processing nickel laterite ore into nickel pig iron (NPI) has been developed by some industries. However, to achieve satisfying nickel recoveries, the process needs massive high-grade metallurgical coke consumption. Concerning on the sustainability of coke supply and positive carbon emission, reduction of nickel laterite ore using biomass-based reductor was being studied.In this study, saprolitic nickel laterite ore was being reduced by palm kernel shell charcoal at several temperatures (800-1000 ^{o}C). Variation of biomass-laterite composition was also conducted to study the reduction mechanism. X-ray diffraction and gravimetry analysis were applied to justify the phenomenon and predict kinetic model of the reduction. Results of this study provide information that palm kernel shell charcoal has similar reducing result compared with the conventional method. Reduction, however, was carried out by carbon monoxide rather than solid carbon. Regarding kinetics, Ginstling-Brouhnstein kinetic model provides satisfying results to predict the reduction phenomenon.

1. Introduction

Nickel is a key mineral used to manufacture various metal-based product. Nickel however, mostly used in manufacture of stainless steel [1]. Nickel ore comes in two types of deposit, namely laterite and sulfide ores. Because of difficulty to extract nickel from laterite, about 58% of nickel demand is supplied by sulfide ores [2,3]. However, 78% of nickel deposit comes within laterite ores [2].

Indonesia has abundant deposits of laterite ores with 15% possession in the world[2]. Before 2013, this deposit successfully place Indonesia as first producer of nickel alongside Philiphines [4]. However, it is exported as raw ore. In 2013, Ministry of Energy and Natural Resources issued a regulation to process any mineral-bearing ores before exporting [5]. This issue forces various country to invest smelter in Indonesia, in which several had already been planned to built.

Content from this work may be used under the terms of the Creative Commons Attribution 3.0 licence. Any further distribution of this work must maintain attribution to the author(s) and the title of the work, journal citation and DOI.
Published under licence by IOP Publishing Ltd

Technology used to process nickel laterite involve massive metallurgical coke which definitely not environmentally friendly neither sustainable. The process uses carbon within the coke to reduce the oxide form of nickel and iron within laterites. Replacement of the coke using bioredutors is being developed by some researchers. In 2015, yield of reducing laterite using bioreductor has been studied [6]. Kinetics, however, has not been studied.

Indonesia owns established palm oil industries in which subsequently provide biomass. Based on this fact, kinetics of reducing nickel laterite by palm kernel shell charcoal is being studied. The objective of this study is to predict the phenomenon involved in the reduction of nickel laterite using palm kernel shell charcoal as well as provide kinetic model for the reduction.

2. Materials and Methods

2.1. Laterite Ore

Saprolitic laterite ore was obtained from Pomalaa, South East Sulawesi, Indonesia. XRF analysis shows that the sample has high Fe and Si content (Table 1). The sample also has high nickel and magnesium content, indicating that the ore belongs to saprolitic type.

Table 1. XRF Results of Nickel Laterite Sample

Si [wt%]	Fe [wt%]	Ni [wt%]	Mg [wt%]	Co [wt%]	Mn [wt%]	Cr [wt%]	Al [wt%]
13.8	32.4	3.9	6.69	0.18	0.6	1.21	1.49

2.2. Reductors

Both reductors was obtained from LIPI Lampung, Indonesia.. Proximate analyses of the reductors are given in Table 2.

Table 2. Proximate Analysis of Anthracite and Palm Kernel Shell Charcoal

Reductor	**Ash Content [wt%]**	**Moisture Content [wt%]**	**Volatile Matter [wt%]**	**Fixed Carbon [wt%]**
Anthracite	2.5	2.3	7.4	87.9
Palm Kernel Shell Charcoal (PKSC)	4.6	7.7	20.4	67.3

2.3. Sample Preparations

Both laterite and reductors were crushed and screened using standard ASTM sieve screen to obtain particle size of -100+120 mesh. Laterite and reductant were mixed together with reductant:laterite ratio of 1:4 while the PKSC was also mixed with ratio of 1:3. Mixture of 5 g was taken, added by 3 mL of distilled water, and then shaped into ball pellet. Pellet was dried at 90 ^{o}C for 2 h, and then continued at 110 ^{o}C for 3 h.

2.4. Reductions

Muffle furnace with heating rate of 20 ^{o}C/min was used to process the pelletized samples. Muffle was heated into the desired temperature and then 10 pellets within 5 ceramic crucibles were charged into the muffle. Each of ceramic crucibles represents reducing time of 5, 15, 30, 60, and 120 min. Once the reduction process reached the designated time, the representing crucible was taken out of the furnace and the samples within were cooled to room temperature inside enclosed compartment to prevent re-oxidation of samples.

IOP Conf. Series: Earth and Environmental Science **65** (2017) 012027 doi:10.1088/1755-1315/65/1/012027

2.5. Measurements
Pelletized samples were weighted before and after reduction at room temperature. Percentage of reaction is calculated as mass loss in the designated time divided by maximum mass loss. As for XRD, certain pellets with highest mass loss percentage were chosen to represent certain data points. The chosen pellets were then grounded to be analyzed by XRD.

3. Results and Discussion

3.1. Comparison of Reduction by Anthracite and Palm Kernel Shell Charcoal
Figure 1 and Figure 2 show the XRD patterns of saprolite reduced by anthracite coal and palm kernel shell charcoal at different reduction temperatures. It is shown in these figures that peaks formed were similar, which means that reduction by palm kernel shell charcoal offers similar yield compared to coke. Furthermore, magnetite intensity of laterite reduced by coconut shell charcoal was higher in both temperatures, which means that coconut shell charcoal may has better capability to reduce laterite than anthracite.

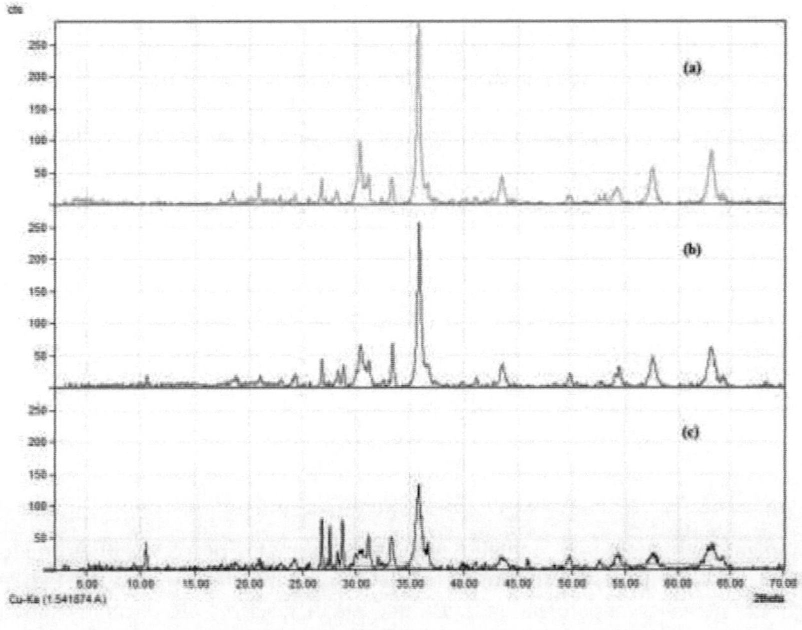

Figure 1. XRD pattern of saprolite reduced by anthracite coal at 1000 °C (a), 900 °C (b), and 800 °C for 2 h

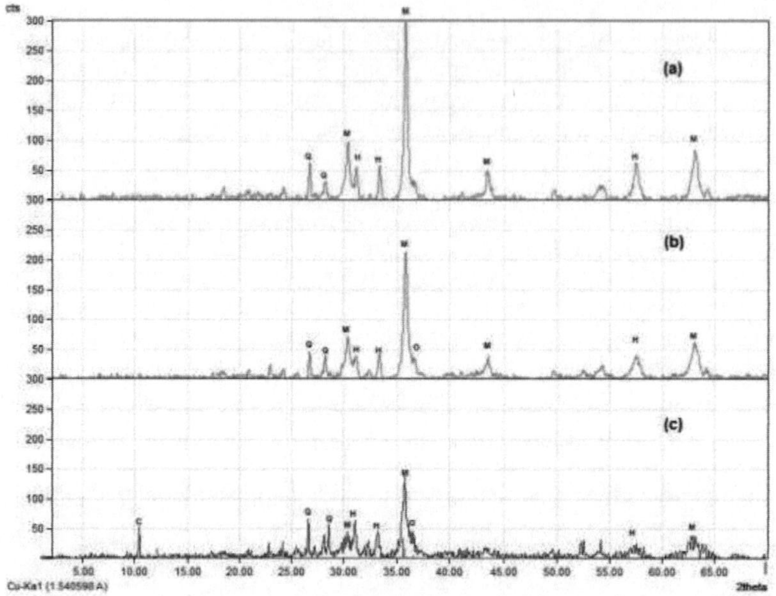

Figure 2. XRD pattern of saprolite reduced by palm kernel shell charcoal at 1000 °C (a), 900 °C (b), and 800 °C for 2 h

3.2. Reduction Mechanism

According to the previous study on coal-based reduction of nickel laterite, there are several reaction which occur in the pellet while it was heated [7]. The reaction is shown below, metal oxide is denoted as Me.O.

$$C + O_2 \rightarrow CO_2 \quad (1)$$
$$CO_2 \rightleftharpoons CO + 0.5\ O_2 \quad (2)$$
$$CO + Me.O \rightarrow CO_2 + Me.O_{-1} \quad (3)$$
$$C + Me.O \rightarrow CO + Me.O_{-1} \quad (4)$$

To study the dominance of reaction within pellet reduced by bioreductor, reduction is conducted at two different composition. With increasing the contact between carbon and metal oxides, the direct contact reaction of (4) should plays larger role and increase the rate of reaction. However, as shown by Figure 3 and 4, the reaction rate does not change significantly. This finding denotes that majority of reduction by bioreductor follows reaction (3), which is diffusion-based reaction. Because of this finding, Ginstling-Brouhnstein diffusion based models is used to predict the kinetics of nickel laterite reduced by palm kernel shell charcoal [8].

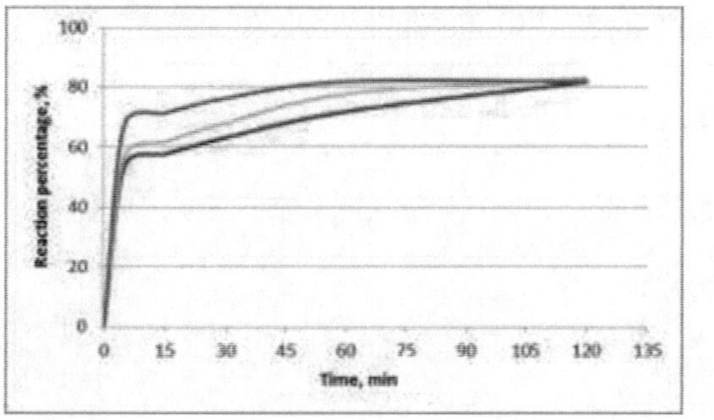

Figure 3. Graphical plotting of reaction percentage as a function of time with reduction at 800 °C (bottom), 900 °C (middle), and 1000 °C (top) with composition of 1 bioreductors to 4 laterites

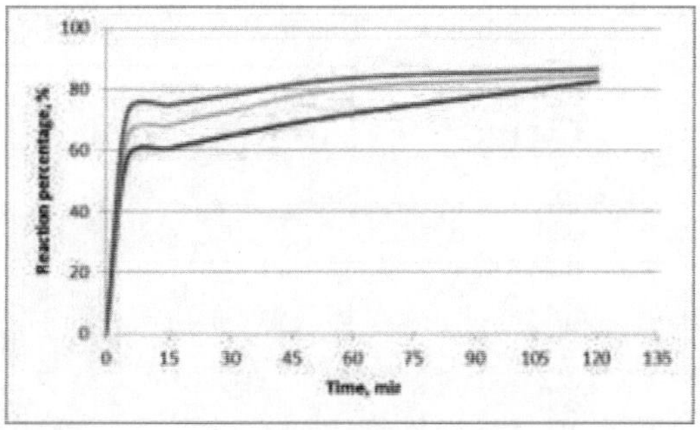

Figure 4. Graphical plotting of reaction percentage as a function of time with reduction at 800 °C (bottom), 900 °C (middle), and 1000 °C (top) with composition of 1 bioreductors to 3 laterite

3.3. Reduction Kinetics

Kinetics of reduction was evaluated by Ginstling-Brounshtein model [8]. In this kinetics model, slopes of Ω versus time were calculated to determine the kinetics. Data which function only to show that reduction has not been completed are not included in the model. The model shows that higher temperature will produce faster kinetics. Table 2. shows that higher temperature will produce higher diffusion kinetics.

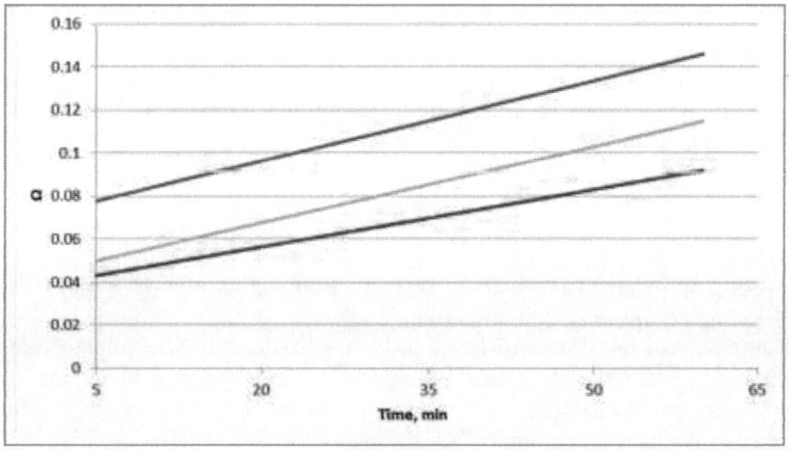

Figure 5. Graphical plotting of Ginstling-Brounhstein of reduction at 800 °C (bottom), 900 °C (middle), and 1000 °C (top) with particle size of 100-120 mesh and composition of 1 bioreductors to 4 laterites

$$\Omega = 1 - {}^{2}/_{3}\, r - (1-r)^{2/3} \qquad (5)$$

$$\Omega = k_d t \qquad (6)$$

Where, Ω is a result of ginstling correlation, while r represents reduction percentage and k_d denotes diffusion reaction rate constant in terms of time (t).

Table 3. Diffusion Rate Constant (k_d) from Ginstling-Brounhstein Regression

Reduction Temperature [°C]	Time [min]	Ω	k_d*10^4 [min^{-1}]
800	5	0.0153	8.8734
	15	0.0255	
	30	0.0314	
	60	0.0431	
900	5	0.0204	11.8125
	15	0.0304	
	30	0.0380	
	60	0.0596	
1000	5	0.0341	12.4342
	15	0.0421	
	30	0.0626	
	60	Not included	

4. Conclusions

Reduction carried out in this experiment denotes that reduction of saprolitic laterite ore by bioreductor is a diffusion-based reaction. Reduction proceeds faster with higher reducing temperature. However in this research, yield of nickel may be better comprehended if the reduced sample is further being analyzed. SEM-EDX is an option to comprehend the phase separation of nickel-bearing iron from the laterite body. Magnetic separation is also a good option to shows the quantitative nickel yield of reduction.

Acknowledgements

The authors wish to thank Badan Pengelola Dana Perkebunan Kelapa Sawit (BPDP-KS) for the support of this research, and also to Chemical Engineering Department, Faculty of Engineering Universitas Gadjah Mada that had supported this project.

References

[1] Johnson J, Reck B K, Wang T and Graedel T E 2008 The energy benefit of stainless steel recycling *Energy Policy* **36** 181–192
[2] Dalvi A D, Bacon W G and Osborne R C 2004 The past and the future of nickel laterites *PDAC 2004 International Convention* March 7-10 pp 1–27
[3] Mudd G M 2009 Nickel sulfide versus laterite: the hard sustainability challenge remains *Proc. 48 th Annual Conference of Metallurgists* (Canada : Canadian Metallurgical Society) pp 1–10
[4] US Geological Survey 2014 *Mineral Commodity Summaries 2014* Miner Commod Summ 196
[5] Minister of energy and mineral resources 2013 Ministrial Regulation No 20, 2013
[6] Chen G, Hwang W, Liu S, Shiau J and Chen I 2015 The Effect oF Bio-Coal on the on the Carbothermic Reduction of Laterite Ores *Mater. Trans.* **56** 550–555
[7] Li Y J, Sun Y S, Han Y X and Gao P 2013 Coal-based reduction mechanism of low-grade laterite ore *Trans. Nonferrous Metal Soc. of China* **23** 3428–3433
[8] Ginstling A M and Brounshtein B I 1950 Concerning the diffusion kinetics of reactions in spherical particles *J. Appl. Chem.* **23** 1327–1338

IOP Conf. Series: Earth and Environmental Science **65** (2017) 012028 doi:10.1088/1755-1315/65/1/012028

Molecular analysis on the utilization of oil palm empty fruit bunches fiber as reinforcement for acrylonitrile butadiene styrene biocomposites

B Hermawan[1], S Nikmatin[1], H Alatas[1], Sudaryanto[2] and S G Sukaryo[2]

[1] Department of Physics, Bogor Agricultural University, Bogor, Indonesia
[2] National Nuclear Energy Agency of Indonesia, South Tangerang, Indonesia

Email : eyosganteng@gmail.com

Abstract. Oil palm empty fruit bunches (OPEFB) was one of the solid waste produced by the palm oil factory and were totally plentiful in biomass. OPEFB fiber used as reinforcement of polymer matrix acrylonitrile butadiene styrene (ABS). The use of FTIR is to see that there is no changes in the molecules of the constituent biocomposite ABS and OPEFB. The reactivity of butadiene and styrene through the double bond- π conjugated system, contributed to the bond reaction with the maleic acid as compatibilizer witch is grafted to the system. It is concluded that the posible grafting reaction occurs by the addition of the MAH to the double bond of the butadiene and styrene. The hydroxyl group of cellulose can interact with this maleic acid to form a bond through the carboxyl group.

1. Introduction

Utilization of oil palm fruits as raw material to produce CPO leaves a problem associated with the processing of solid and liquid waste, one of which is oil palm empty fruit bunches (OPEFB). The development of environmentally-friendly materials derived from natural materials, especially waste with a touch of technology becomes very important and need special attention. OPEFB processed by physical method of heating and mechanical milling techniques without chemicals to form fiber[1]. OPEFB fiber is then added as filler ABS polymer matrix to be processed into biocomposite. Heat received by the composites reinforced with natural filler (rattan 5%) has a much higher melting point[2].

Molecular vibration analysis on biocomposite ABS will be carried out to see whether or not changes to a particular functional group, which will then be used to ensure the bond between ABS and OPEFB. Molecular vibrations can range from the simple coupled motion of the two atoms of a diatomic molecule to the much more complex motion of every atom in a large polyatomic molecule. In order to begin with a basic model, a molecule can be looked upon as a system of masses joined by bonds with properties like spring. Stretching or compressing a spring produces potential energy. When the spring is released, this energy is translated into a restoring forces which results in vibration about the

Content from this work may be used under the terms of the Creative Commons Attribution 3.0 licence. Any further distribution of this work must maintain attribution to the author(s) and the title of the work, journal citation and DOI.
Published under licence by IOP Publishing Ltd

equilibrium length of the spring. For a diatomic molecule, only one vibration that corresponds to the stretching and compression of the bond is possible. It is no longer obvious how many modes of vibration they have when we consider polyatomic molecules containing more than three atoms[3,4].

Molecular vibration spectrum is a unique nature and a molecular characteristics. Natural vibrations of molecules is a marker for the identification of molecules contained on the materials. Each molecule has a specific natural vibration depending on the type of molecules and molecular bonds. Vibration is presented through FTIR measurements in the form of peaks. Infra Red spectrum can be used as a fingerprint for identification by comparing the sample spectrum with "reference" spectrum. Characterize molecular structures, either "backbone" of the molecule or functional groups attached to the molecule will produce the absorption spectrum. This information can indicate backbone on the structure and specific functional groups of aromatic rings and show backbone on a straight chain or branched chain[5]. Therefore, we wanted to get information whether there is a reaction between components of a mixture of biocomposites through FTIR spectra.

2. Material and Method

OPEFB was taken from PKS-PTPN VIII Cikasungka , Bogor, Jawa Barat. The research was held at laboratory of physics department Bogor Agricultural University (IPB) and PT MUB Jaya Cibinong.

Materials used in this research is the oil palm empty fruit bunches obtained from palm oil mill PTPN VIII, Cikasungka Bogor, West Java, Indonesia. OPEFB is a complete waste which has been separated from the seeds. OPEFB were cleaned and cut into chips and dried for 2 days then put into oven at 100°C for 6 hours, m = 2 Kg. Into a mechanical milling (FOMAC, type FCT-Z2000, China) a dried chips milled about 10 minutes, v = 25000 rpm, m = 0,5 Kg. To obtained OPEFB micro fiber a score of times milling and sieving (100 mesh) were done.

The fabrication of biocomposites granules- MF20 performed by mixing the OPEFB micro fibers with the ABS polymers (2:8) using Single Screw Extruder (SSE), in the barrel samples blended with a gradient temperature of 195 ^{O}C – 215 ^{O}C – 220 ^{O}C – 220 ^{O}C – 220 ^{O}C – 225 ^{O}C – 225 ^{O}C – 225 ^{O}C.

OPEFB chips, OPEFB micro fibers, ABS polymers, and biocomposites granular then analyzed further related to functional groups of constituent materials using Spectrophotometer Fourier Transform Infra Red (ABB, model MB300, Canada). OPEFB micro fiber powder made into pellets with KBr method, while granular ABS and MF20 is made into a thin film (± 1 micron).

3. Results and Discussion

An infrared (IR) spectrum provides a rapid and simple method for obtaining preliminary information on the identity or structure of a molecule. The spectrum is a plot of the amount of IR radiation that passes through the sample (transmittance T) versus the wavelength, λ (or wavenumber, υ) of the radiation. The sample absorbs IR radiation at various wavelengths (or wavenumbers) and the positions of the absorptions give clues to the structure of the molecule, as indicated in figure 1. The spectra may also be exhibited in absorbance (A=log10 1/T) versus wavenumber units. Absorption maxima (or transmission minima) are referred to as peaks, bands, or absorptions[6].

IR spectra of a large number of compounds containing a particular functional group such as carbonyl, phenyl, amino, nitro, and so on are found to have certain features that appear at more or less the same frequency for every compound containing the group. Then, it is reasonable to associate these spectral features with the functional group, provided a sufficiently large number of different compounds containing the group have been studied.

Upon analysis of the spectrum in figure 1a we see that there is several peaks associated with the functional groups of cellulose at peak 1034 cm^{-1}, whereas the peak at 1730 cm^{-1} as a hint of functional group associated with lignin[4,7]. Figure 1b shows the peak at 2237 cm^{-1} was associated with acrylonitrile, while peak 910 cm^{-1} related to butadiene. The aromatic band at 1605 cm^{-1} is free from interference, and is utilized as marker of styrene[8].

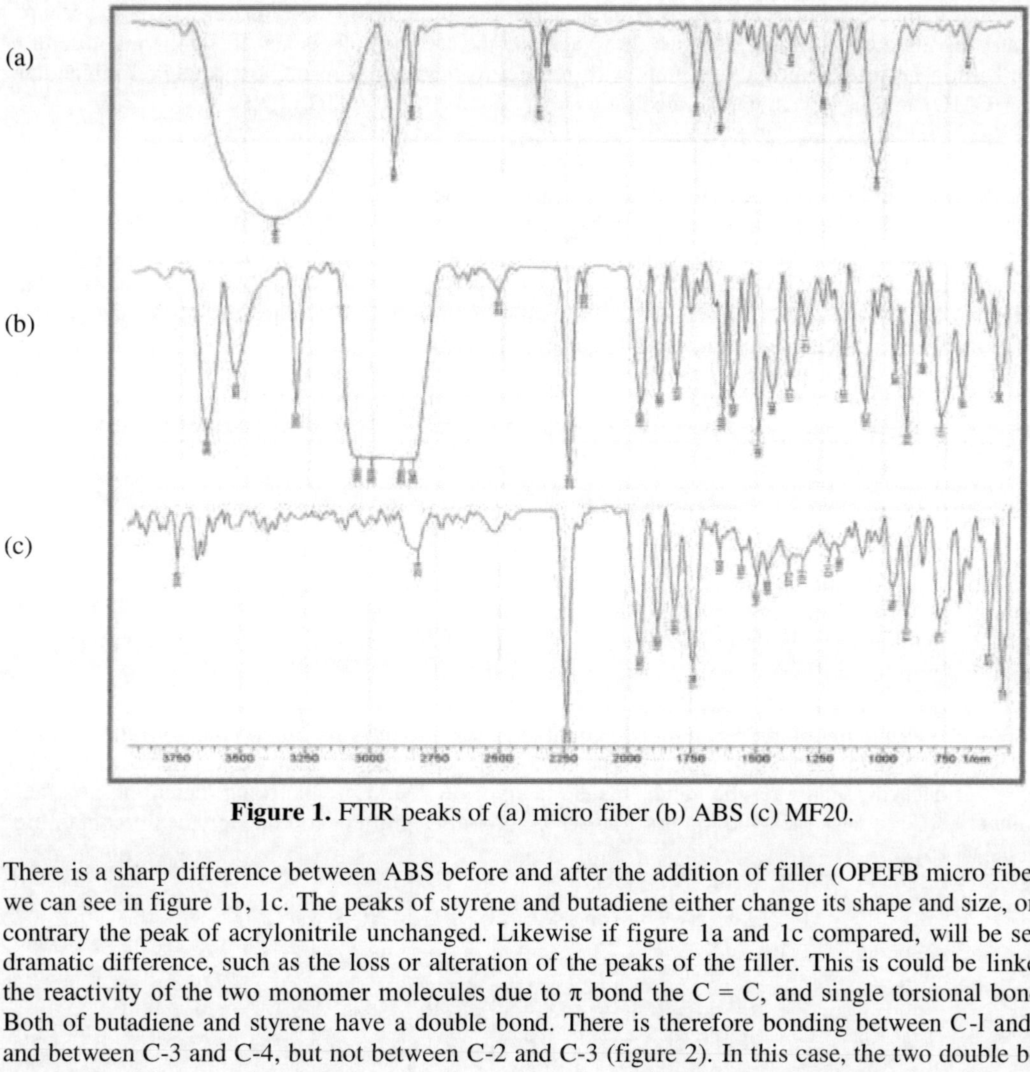

Figure 1. FTIR peaks of (a) micro fiber (b) ABS (c) MF20.

There is a sharp difference between ABS before and after the addition of filler (OPEFB micro fiber) as we can see in figure 1b, 1c. The peaks of styrene and butadiene either change its shape and size, on the contrary the peak of acrylonitrile unchanged. Likewise if figure 1a and 1c compared, will be seen a dramatic difference, such as the loss or alteration of the peaks of the filler. This is could be linked to the reactivity of the two monomer molecules due to π bond the C = C, and single torsional bond[9]. Both of butadiene and styrene have a double bond. There is therefore bonding between C-l and C-2 and between C-3 and C-4, but not between C-2 and C-3 (figure 2). In this case, the two double bonds are separated by one σ bond. Due to the close proximity of the two π bonds, they can conjugate with one another.

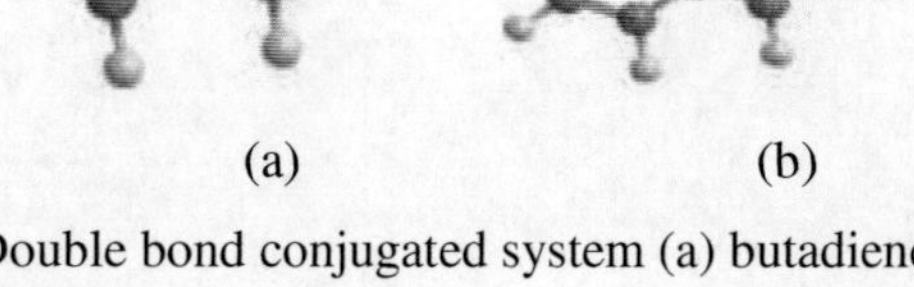

(a) (b)

Figure 2. Double bond conjugated system (a) butadiene (b) styrene

We shall be concerned mainly with organic molecules, for which we shall develop the Huckel molecular-orbital theory. A molecular orbital (MO) may be considered to be built up from a combination of atomic orbitals (AOs), and to extend over the whole molecule. Through the Huckel theory, two p orbitals are described as being conjugated with each other in making the π bond. Electrons are delocalized in conjugated systems. This generally lowers the overall energy of the molecule and increase the stability and are subject to the symmetry of the molecule, its point group. Conjugated bonds are stabilizing because they increase the number of resonance stabilized structure that can form. The most important orbitals with respect to reactivity are highest occupied molecular orbital (HOMO) and the lowest unoccupied molecular orbital (LUMO) [10,11,12,13].

We continue to separate completely the σ frame- work (using the 2s, 2px and 2py orbitals on carbon with the 1s orbitals on hydrogen) from the π system made up from the 2pz orbitals. Since this conjugated double bond consists of 4 2p orbitals of carbon to house 4 electron delocalized. After setting up the secular problem by appropriately constructing LCAO-MO (linear combination atomic orbital-molecular orbital) wave functions from 2pz AOs and subjecting these to the secular determinant, it is solved after making the following approximations (equation (1)). The linear combination of these orbitals takes the form of equation (1)[14].

$$x^4 - 3x^2 + 1 = 0$$
$$x^4 - 3x^2 + 1 = (x^2)^2 - 3x^2 + 1 = 0$$
$$x = \pm\left(\frac{1}{2}(1+\sqrt{5})\right), \pm\left(\frac{1}{2}(\sqrt{5}-1)\right)$$
$$x = \pm 1.618, \pm 0.618 \quad (1)$$

With energy put one by one into secular equations and evaluate determinant there are four level of molecular orbital energy each described in equation (2). therefore we construct the wave function as described in figure 3.

A common way of polymer blends is adding a compatibilizer to reduce the interfacial tension, increase the adhesion, and obtain better properties. The double bond in maleic anhydride is also extremely reactive. Methacrylic acid, acrylic acid, maleic anhydride (MAH), and long chain unsaturated monomers are among the monomers most commonly grafted onto polymers containing butadiene[15,16].

$$\psi_1 = 0.372p_1 + 0.602p_2 + 0.602p_3 + 0.372p_4$$
$$\psi_2 = 0.602p_1 + 0.372p_2 - 0.372p_3 - 0.602p_4$$
$$\psi_3 = 0.602p_1 - 0.372p_2 - 0.372p_3 + 0.602p_4$$
$$\psi_4 = 0.372p_1 - 0.602p_2 + 0.602p_3 - 0.372p_4 \quad (2)$$

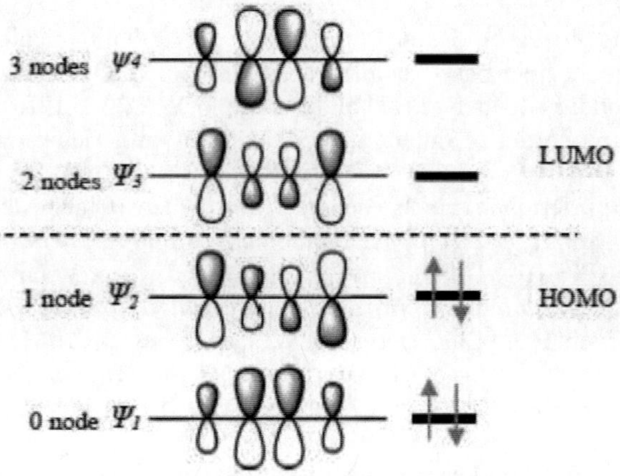

Figure 3. π molekular orbital of conjugated double bond.

Figure 4 shows the posible grafting reaction occurs by the addition of the MAH to the ABS, however we still can not convince the bond between MAH and cyclic of the styrene. The hydroxyl groups at C-2, C-3 and C-6 of cellulose are reactive and can react with anhydride. The intensity of the peak at 3379 cm^{-1} in the reduced obviously, indicating that more intermolecular hydrogen bond is destructed compared to the intramolecular hydrogen bond[17]. The reaction mechanisms for cellulose with MAH described in figure 5.

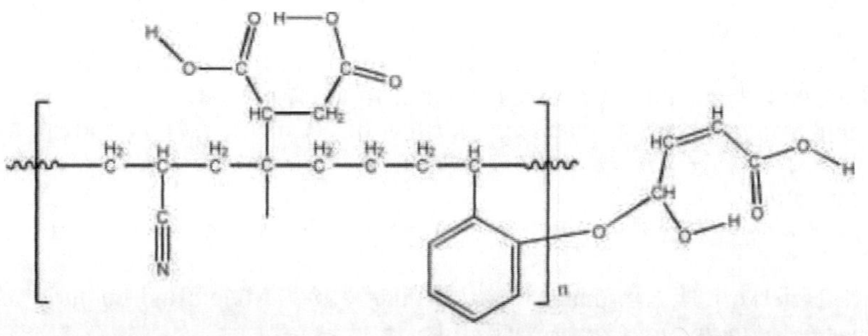

Figure 4. Posible reaction of ABS and maleic anhydride.

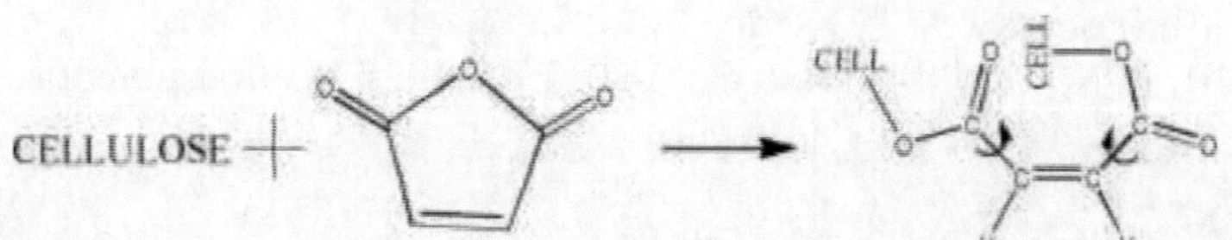

Figure 5. The mechanism reaction for cellulose with maleic anhydride.

4. Conclusions

The reactivity of butadiene and styrene through the double bond- π conjugated system, contributed to the bond reaction with the maleic acid as compatibilizer witch is grafted to the system. It is concluded that the posible grafting reaction occurs by the addition of the MAH to the double bond of the butadiene and styrene. The hydroxyl group of cellulose can interact with this maleic acid to form a bond through the carboxyl group.

References

[1] Iwamoto S, Nakagaito A N, Yano H and Nogi M 2005 Optically transparent composites reinforced with plant fiber-based nanofibers *Appl. Phys. A* **81** 1109-12

[2] Nikmatin S, Syafiuddin A, Kueh A B H and Purwanto Y A 2015 Effects of nanoparticle filler on thermo-physical properties of rattan powder-filled polypropylene composites Jurnal Teknologi **77** 181-7

[3] Hollas J M 2002 *Basic Atomic and Molecular Spectroscopy* (Chichester: Wiley)

[4] Stuart B H 2004 *Infrared Spectroscopy: Fundamentals and Applications* (Chicester: John Wiley & Sons)

[5] Coates J 2000 *Interpretation of infrared, a practical approach Encyclopedia of Analytical Chemistry* ed Meyers R A (Chicester: John Wiley & Sons) pp 10815-837

[6] Shurvell H F 2002 *Spectra-structure correlations in the Mid- and Far-infrared Handbook of Vibrational Spectroscopy*, vol 3, ed Chalmers J M and Griffiths P R (Chicester: Wiley) pp 1-34

[7] Kumar A, Negi Y S, Choudhary V and Bhardwaj N K 2014 Characterization of cellulose nanocrystals produced by acid-hydrolysis from sugarcane bagasse as agro-waste *Journal of materials physics and chemistry* **2** 1-8

[8] Nishikida K and Coates J *Infrared and Raman Analysis of Polymers Handbook of Plastics Analysis* ed Lobo H and Bonilla J V (New York: Marcel Dekker, inc) chapter 7

[9] Choi C H and Kertesz M 1997 Conformational information from vibrational spectra of styrene, trans-stilbene and cis-stilbene *J. Phys. Chem. A* **101** 3823-31

[10] Beez M, Bieri G, Bock H and Heilbronner E 1973 The ionization potentials of butadiene, hexatriene, and their methyl derivatives: Evidence for through space interaction between double bond π-orbitals and non-bonded pseudo- π orbitals of methyl group? *Helvetica Chimica Acta* **56** 1028-46

[11] Skaarup S and Boggs J E 1976 Contribution of resonance, hybridization, and nonbonded interactions to the structure of butadiene Tetrahedron **32** 1179-81

[12] Ladd M 1998 *Symmetry and group theory in chemistry* (Chicester: Horwood publishing)

[13] Fleming I 2009 *Molecular Orbitals and Organic Chemical Reactions*, Refference edition (Chicester: Wiley)

[14] Yates K 1978 *Huckel Molecular Orbital Theory* (New York: Academic press) chapter 2 pp 27-87

[15] Lohbeck K, Haferkorn H, Fuhrmann W and Fedtke N 2000 Maleic and Fumaric acids Ullmann's *Encyclopedia of Industrial Chemistry* **22** 145-55

[16] Qi R, Qian J and Zhou C 2002 Modification of acrylonitrile-butadiene-styrene terpolymer by grafting with maleic anhydride in the melt. I. preparation and characterization *Journal of Applied Polymer Science* **90** 1249-54

[17] Zhang Y, Li H, Li X, Gibril M E and Yu M 2014 Chemical modification of cellulose by in situ reactive extrusion in ionic liquid Carbohydrate Polymers **99** 126-131

IOP Conf. Series: Earth and Environmental Science **65** (2017) 012029 doi:10.1088/1755-1315/65/1/012029

Oil recovery test using bio surfactants of indigenous bacteria in variation concentration of carbon source

B Yudono[1], W Purwaningrum[2], S P Estuningsih[3], S Kaffah[4]

[1,2,4]Department of Chemistry, University of Sriwijaya, Palembang, Indomesia
[3]Department of Biology, University of Sriwijaya, Palembang, Indonesia

Email: yudonob@hotmail.com

Abstract. Recovery tests of crude oil by using bio surfactant of indigenous bacteria *Pseudomonas peli*, *Pseudomonas citronellolis*, *Burkholderia glumae* and *Bacillus firmus*. The bio surfactants were prepared with the variation concentrations of molasses carbon source; 0, 5, 10, 15, 20, and 25 %. The results showed that 10 g samples, which concentration 18.64% TPH could be dissolved in the bio surfactant 10%. Optimally in the molasses carbon source concentrations for each bacterium at 5, 10, 20 and 15 % with oil recovery as much as 31.92, 17.65, 22.32, and 14.38 % respectively. Oil components which extracted by bio surfactant were analyzed by using GLC (Gas Liquid Chromatography). The bio surfactants of *Pseudomonas peli* could dissolve oil fraction temperatures; 139.85; 144.69; 149.98; 1.55.03: 174.22 °C, *Pseudomonas citronellolis* could dissolve oil fraction temperatures; 139.13; 142.64;147.99; 155.03; 159.85; 164.50 °C, *Burkholderia glumae* could dissolve oil fraction temperatures 144.69; 149.98; 155.03; 159.85; 164.50 °C, and *Bacillus firmus* could dissolve oil fraction temperatures; 149.98; 155.03; 158.46; 164.50 °C.

1. Introduction

Petroleum is a natural resource that can be utilized to increase the sources of foreign exchange for the country. The needs of the world including Indonesia in petroleum from year to year increase. The role of oil until now has not replaceable whereas petroleum searches more difficult, as well as its exploitation increasingly expensive. Microbial Enhanced Oil Recovery Technology (MEOR) is an alternative method that is used to retrieve oil reserves are not drawn conventionally performed by utilizing microbes work processes contained in the reservoir which is dominated by bacteria [1].

This method utilizes the activity and bio product (gas, bio surfactant, biopolymers, acids, solvents, biomass) produced by microorganisms [2]. MEOR method utilize microbes to produce bio surfactant, where microbes grown on certain media and certain conditions. Bio surfactants can reduce the surface and interfacial tensions [3]. Bio surfactants consist of molecules of hydrophilic and hydrophobic. The existence of both these groups cause the bio surfactant can reduce the surface tension and interfacial

Content from this work may be used under the terms of the Creative Commons Attribution 3.0 licence. Any further distribution of this work must maintain attribution to the author(s) and the title of the work, journal citation and DOI.
Published under licence by IOP Publishing Ltd

tension and can form a stable emulsion that can make hydrocarbons soluble in water or otherwise water-soluble hydrocarbons [4]. MEOR method requires only slight modifications to the facilities on the ground, MEOR very suitable for oil reservoir with carbonate rocks, have low toxicity, biodegradability and biocompatibility specificity, selectivity, extreme temperature, pH and salinity [5].

Microbes characteristics which must be owned by a bacterium that is capable of processing MEOR; hydrocarbons (hydrocarbonoclastic), produce bio surfactant, produces gas, small size, barophilic, thermophiles, halophile, not pathogenic, and indigent [6]. Hydrocarbonoclastic bacteria are able to use carbohydrates such as glucose, fructose, and mannitol as a carbon source to support bio surfactant [7]. The concentrations of carbon source are focus in this study. This is because the concentration of carbon source is one that affects bacterial growth and employment in the bio surfactant. Based on research [8] studied the effect of carbon source (stearic acid, oil, and molasses) on the growth of *Pseudomonas peli*. The results showed that the best carbon source is molasses with a percentage of 16.01% at a concentration of 4%.

In this study, the recovery of petroleum tested using bio surfactant of bacteria indigent *Pseudomonas peli, citronellolis Pseudomonas, Burkholderia glumae* and *Bacillus firmus* by varying the concentration of carbon source molasses 0, 5, 10, 15, 20 and 25%.

2. Research Methodology

2.1. Tools and Materials Research

The tools used in This study is an autoclave, Erlenmeyer 1000 mL, 500 mL Erlenmeyer, Erlenmeyer 250 mL, test tubes, beakers 250 mL, measuring cups 100 mL, a set of GLC-Shimadzu, a set of tools soxhlet, analytical balance, microscopes, incubators, needle ose, petri dishes, filter paper, funnel, beaker glass, pipette, aerator, aerator stone, plastic hose, cotton, aluminum foil, Bunsen, paper labels, paper, mesh, camera, spatula, bowl, mortar, rotary shaker, hot plate, and a magnetic stirrer.

Materials used are bacteria isolates; *Pseudomonas peli, citronellolis Pseudomonas, Burkholderia glumae, Bacillus firmus* , distilled water, K_2HPO_4, peptone, $FeSO_4$, yeast extract, alcohol 70%, n- hexane technical, NA medium, molasses (taken from the sugar mill waste PT. Cinta Manis, Tanjung Raja, Ogan Ilir, South Sumatra), medium clay and sludge oil obtained from the Village Babat Toman, Banyuasin, South Sumatra [9].

2.2. Stages of Research

2.2.1. Sample preparation

Clay as much as 1 Kg dried in an oven with a temperature 110°C for 3 days. Clay is dried and then ground using a mortar until smooth and sifted so that the clay is still shaped blobs can be separated with clay that has been refined. Clay 700 g and 300 g as much petroleum sludge is then mixed and stirred until homogeneous.

2.2.2. Indigenous bacteria preparation

Each bacterium, namely; *Pseudomonas peli, citronellolis Pseudomonas, Burkholderia glumae, Bacillus firmus* inoculated with a zigzag motion in NA slant agar (Nutrient Agar) aseptically, then incubated at a temperature of 37°C for 24 hours and will be acquired stock cultures and working cultures were ready for use [10].

2.2.3. Preparation of medium zobell

Medium Zobell made with dissolve 5 g peptone, 1 g yeast extract, 0.012 g K2HPO4, and 0.01 g $FeSO_4$ in distilled water by volume solution of 1000 mL. The mixture is boiled over a hot plate and homogenized with a magnetic stirrer. Then the mixture sterilized by autoclaving at 121°C for 15 minutes [10].

2.2.4. Starter bacteria indigen preparation

Bacterial culture is taken as much as 5 ose, then subculture to put into each Erlenmeyer containing 100 ml of medium Zobell, then aerated for 24 hours, after 24 hours as much as 100 mL was added to the mixture until the total volume of 200 mL and the mixture is aerated again until the shortest generation time of bacteria. The shortest generation time for the bacteria *Pseudomonas peli* is 12 hours, the bacteria *Pseudomonas citronellolis* is 9 hours, the bacteria *Burkholderia glumae* which is 12 hours and the *bacteria Bacillus firmus* i.e. 15 hours [8].

2.2.5. Initial measurement TPH

A total of 10 g soil is inserted using the appropriate filter paper. The initial TPH concentration is measured by using soxhletation method, dried in an oven, cooled in a desiccator and weighed.
TPH percentage of petroleum is calculated using the formula:

$$\%TPH = \frac{w_1 - w_2}{w_3} x100\% \tag{1}$$

Where:
W_1 = Weight boiling flask and extract petroleum oil (g), W_2 = weight of the empty boiling flask (g), W_3 = weight of the sample (g) [11].

2.2.6. Production of crude bio surfactants

Varying concentrations of molasses 0, 5, 10, 15, 20, and 25%, mixed Zobell media, and then inoculated with indigenous *Pseudomonas peli, citronellolis Pseudomonas, Burkholderia glumae, Bacillus firmus* bacteria respectively. The cultures were incubated by the shortest generation time of each bacterial [8].

2.2.7. Test oil recovery

Erlenmeyer already containing 10 g sample (a mixture of soil and sludge) was added to 200 mL of crude bio surfactant of *Pseudomonas peli* at varying concentrations of molasses 0, 5, 10, 15, 20, and 25%, then aerated with an aerator for 10 days. The mixture is then filtered using a filter paper; the precipitate was washed using distilled water and dried in an oven at 100°C for 24 hours. The precipitate was dried and then weighted. The same treatment was also carried out using *Pseudomonas citronellolis, Burkholderia glumae*, and *Bacillus firmus* bacteria. The oil recovery calculated from the difference TPH contents between initial and final samples.

2.2.8. GC analysis

The samples were diluted with n-hexane then injected 2.5 μL into the GC-Shimadzu instruments. Injector temperature is 290°C, temperature hold 40°C for 5 minutes, and then the temperature is increased every minute to the temperature rise to 280°C with temperature step 5°C per minute and hold for 5 minutes. The mobile phase used is Helium [12].

3. Results and Discussion

3.1. Sample

The samples mixed petroleum oil and soil as initial TPH concentration of samples. The measurements were done by solvent soxhlet extraction using n-hexane solvent with three repetitions; the average concentration of TPH was 18.64% [13].

3.2. Oil Recovery Test

Crude bio surfactant of each bacterium was applied to the soil samples with various concentrations molasses as carbon resources. Variations molasses carbon source provided on crude bio surfactant was 0, 5, 10, 15, 20 and 25 % for each of the bacteria and the mixture aerated for 10 days. The results of the acquisition and application of soil containing petroleum with crude bio surfactant TPH also be measured, as well as the measurement of the initial TPH. TPH measurement data used to calculate the recovery of petroleum for each bacterium. Petroleum recovery results using the bio surfactant of each bacterium can be seen in the figure below:

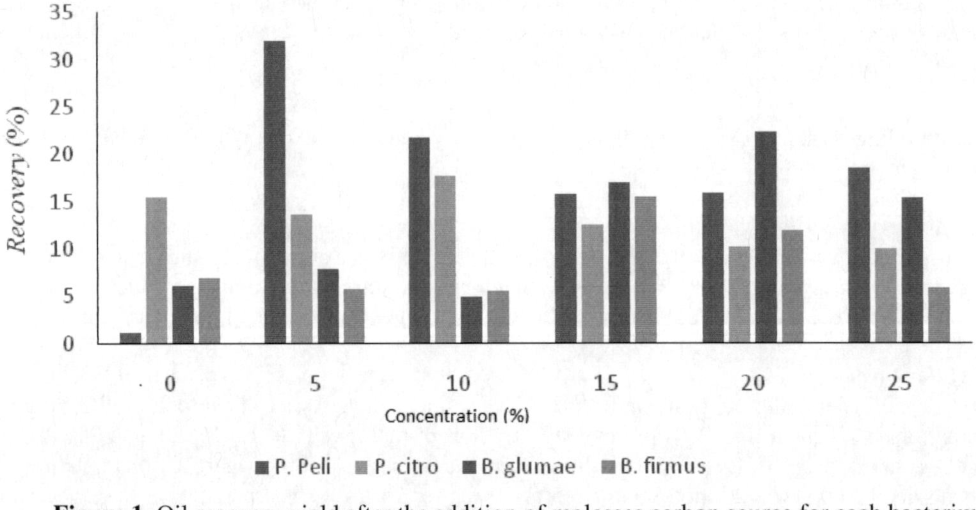

Figure 1. Oil recovery yield after the addition of molasses carbon source for each bacterium

Based on that data, four of these bacteria produce bio surfactant different, where the bacteria *Pseudomonas peli* bacteria produce oil recovery highest molasses carbon source 5%. This indicates that the *Pseudomonas peli* bacteria could effective recovery of petroleum. In this study, the number of medium Zobell added is constant, so it does not affect the performance of the bacteria because bacteria performance only influenced by the addition of molasses carbon source. The ability of the bacteria can be seen from the clear zone on an agar medium [9]. Each of these bacteria produces a clear zone different extent. Clear zone formed quite well above 20 mm, the wider clear zone produced, the greater the potential for bacteria to produce bio surfactant. In former the study showed that the clear zone formed on the bacteria *Pseudomonas peli* amounted to 25.79 mm. Based on the extensive clear zone are formed and the bacteria *Psudomonas peli* is a potentially good bacteria to produce bio surfactant than other bacteria. Bio surfactant activity is closely associated with the ability of bacteria to consume the carbon source [14]. The use of carbon sources too little will make the substrate does not have sufficient contact with the cells, while the source of carbon is too much will make the cell has enough room to interact with the carbon source, so the production process bio surfactant not be optimal [15]. In this study, *Pseudomonas peli* grown on molasses carbon source at a concentration of 5% has the greatest recovery value.

3.3. Chromatogram Analysis Results using GLC

GLC analysis provides information on the number of components of the oil before and after using crude bio surfactant with molasses carbon source concentration variations of each bacterium. Furthermore, the work of bacteria then identified by comparing identical compound fractions of each bacterium and is equipped with an abundance of compounds percentage between samples before and after treatment. The determination of percent identical compounds and those abundances can explain the carbon fraction degraded or dissolved. The chromatograms before and after treatments, and the component hydrocarbons percent abundance either dissolved or degraded through histogram described as follows:

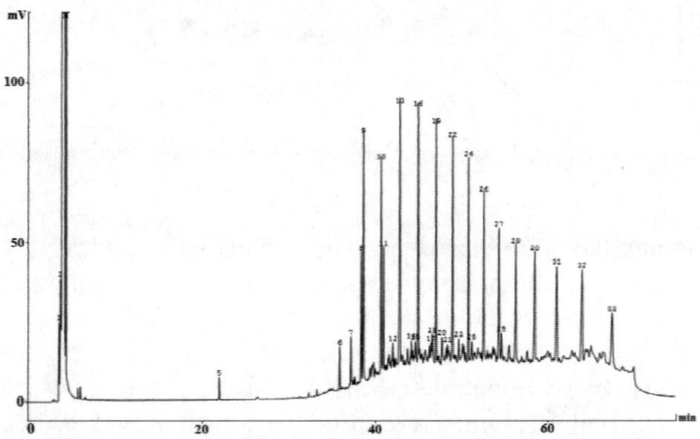

(a)

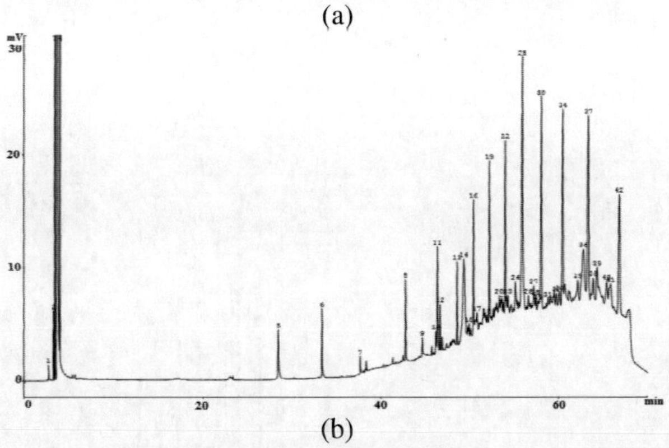

(b)

Figure 2. The chromatogram of initial condition sample (a), the chromatogram of sample after dissolved with bio surfactant of *P. peli* bacteria with carbon source concentrations of 5% molasses (b)

In the chromatogram of sample before treatment showed 33 peaks. Based on the total retention time for 67 minutes, the peaks of the chromatogram occurred at retention times 38.56; 40.65; 42.88; 44.99; 47.01; 48.94; 50.80; 52.58. After sample treated bio surfactant of *Pseudomonas peli* bacteria with the best concentration of carbon source molasses 5% have additional peaks from 33 to 42. Based on the total retention time for 67 minutes, there are six higher peaks in the chromatogram with retention time of 52.20; 53.96; 55.85 57.98; 60.45; 63.38 while the other had retention times of chromatogram with lower peaks

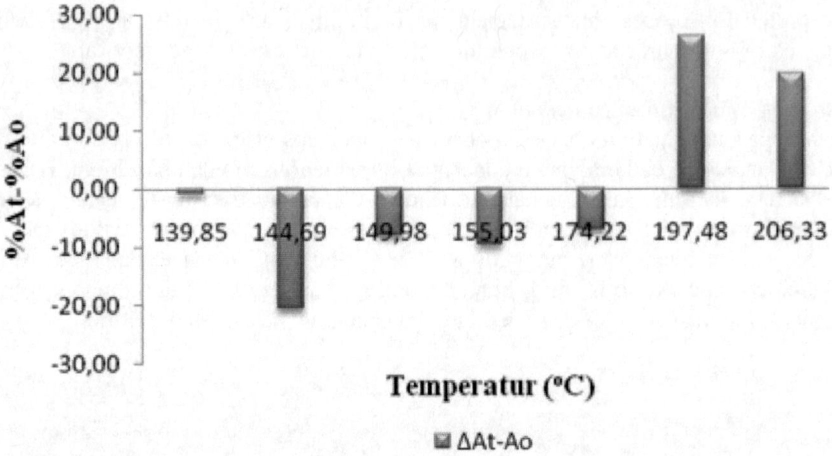

Figure 3. Histogram of dissolved hydrocarbons before and after the addition of crude bio surfactant *P. Peli* bacteria

Based on the above histogram, the difference in per cent calculated by subtracting the percent abundance peak after the addition of crude bio surfactant (At) of the *Pseudomonas peli* bacteria reduced percent of peak abundance prior to the addition of crude bio surfactant (A_0). The results were negative at baseline subtraction histogram shows that hydrocarbons short chain C atoms dissolved. Bio surfactant of *Pseudomonas peli* bacteria can dissolve petroleum compounds at temperatures fractions; 139.85; 144.69; 149.98; 155.03; 174.22°C. The highest temperatures dissolve compounds of hydrocarbons at 174.22°C.

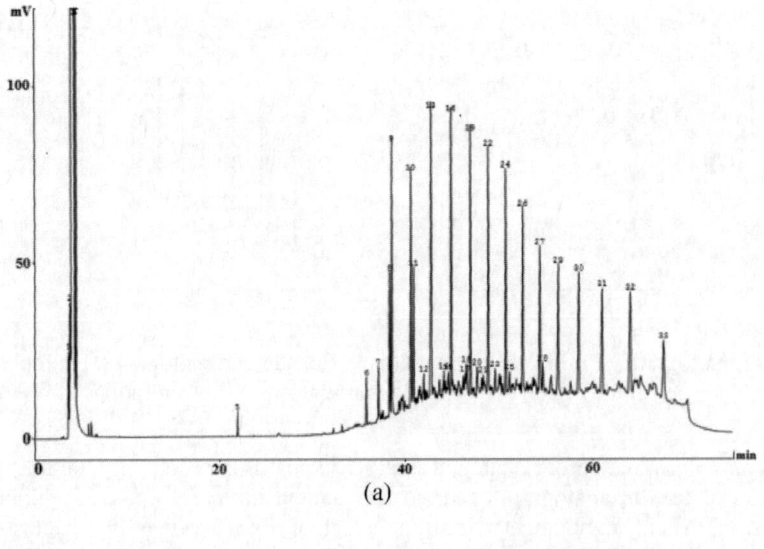

(a)

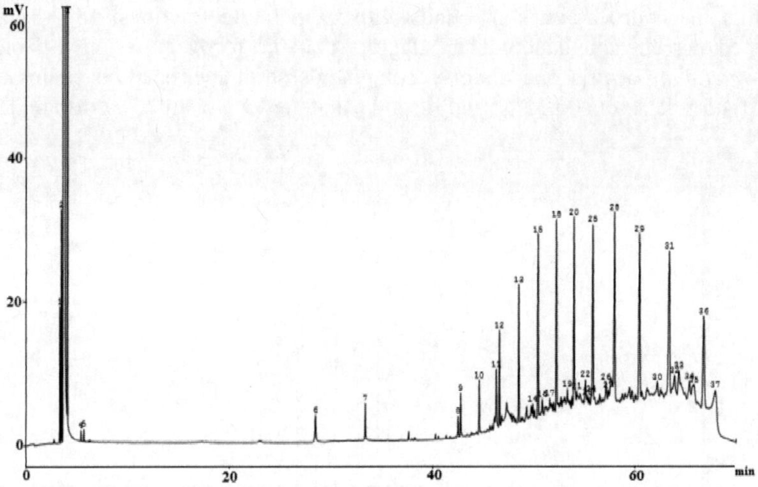

(b)

Figure 4. The chromatogram of initial condition sample (a), the chromatogram of sample after dissolved with bio surfactant of *P. citronellolis* bacteria with carbon source concentrations of 10% molasses (b)

Figure 4 shows the chromatogram of extracted sample by bio surfactant of *Pseudomonas citronellolis* bacteria with the best concentration of carbon source molasses 10% compared with the chromatogram constituents of petroleum before treatment have additional peaks from 33 to 37. Based on the total retention time for 68 minutes, there are eight highest peaks in the chromatogram with retention time; 48.54; 50.41; 52.20; 53.96; 55.85; 57.98; 60.44; 63.38 while the other had a retention time of chromatogram with lower peaks.

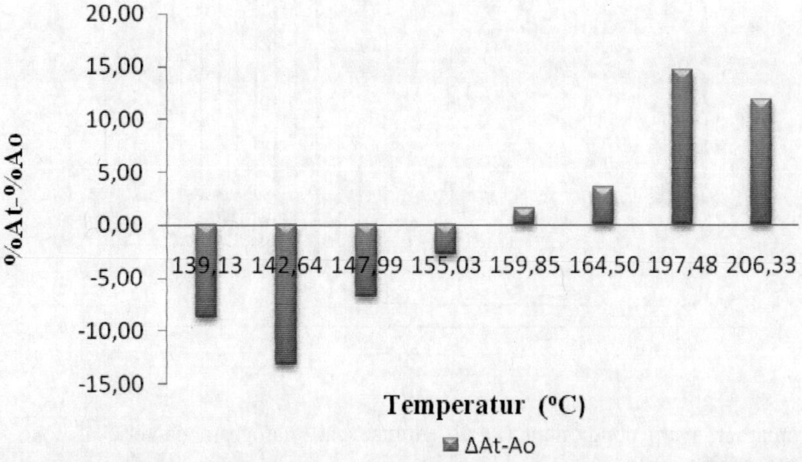

Figure 5. Histogram dissolved hydrocarbons before and after the addition of crude bio surfactant of *P. citronellolis* bacteria

Figure 5 shows that the hydrocarbon compounds with temperature fractions; 139.13; 142.64; 147.99; 155.03 °C decreased percent abundance. This suggests that the recovery of oil use bio surfactant of *Pseudomonas citronellolis* bacteria can dissolve compounds short hydrocarbon chains at temperatures fraction; 159.85; 164.50; 197.48; 206.33°C, and it shows that the long chain C atoms are degraded to form new compounds.

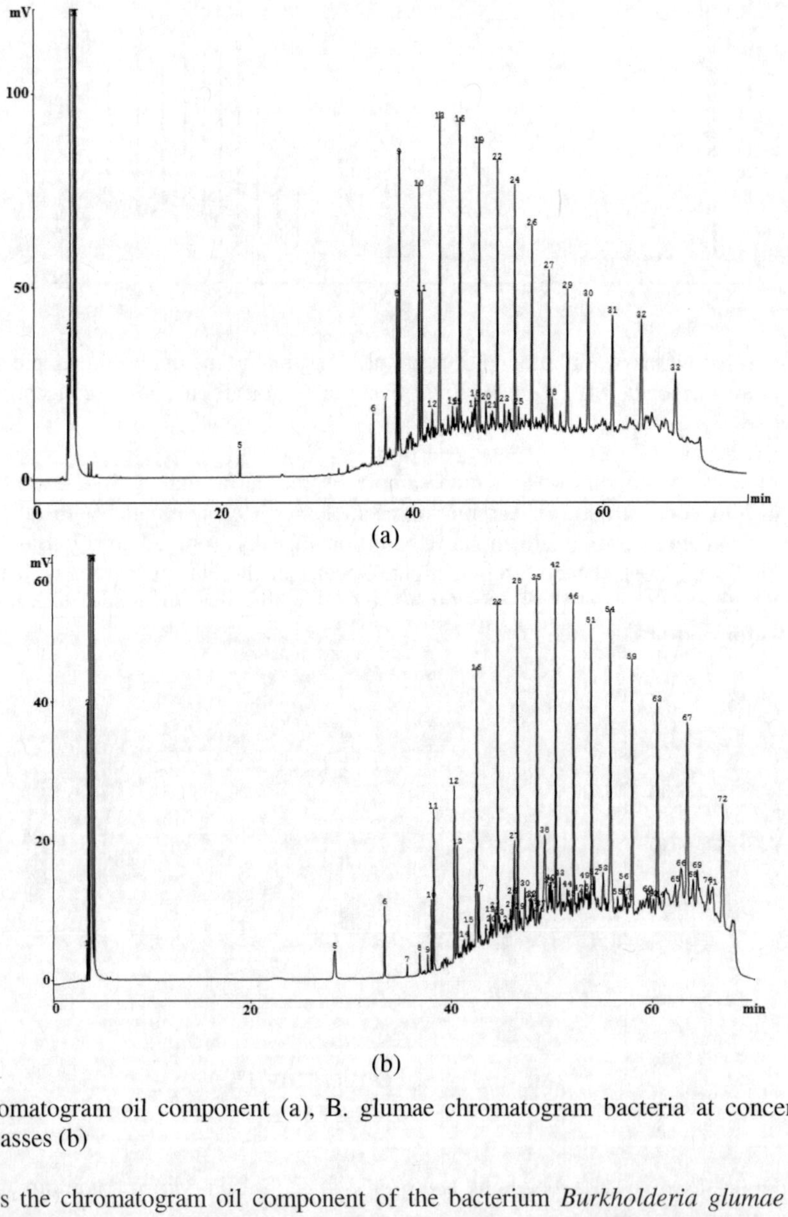

Figure 6. Chromatogram oil component (a), B. glumae chromatogram bacteria at concentrations 10% molasses (b)

Figure 6 shows the chromatogram oil component of the bacterium *Burkholderia glumae* with the best concentration of carbon source molasses 20% compared with the chromatogram constituents of petroleum before treatment experienced a peak increase of 33 to 72 peak chromatogram peak. Based on the total retention time for 67 minutes, there were 11 chromatogram peaks higher than other peaks occur at retention times of 42.51; 44.63; 46.65; 48.59; 50.45; 52.23; 53.99; 55.88; 58.01; 60.49; 63.45.

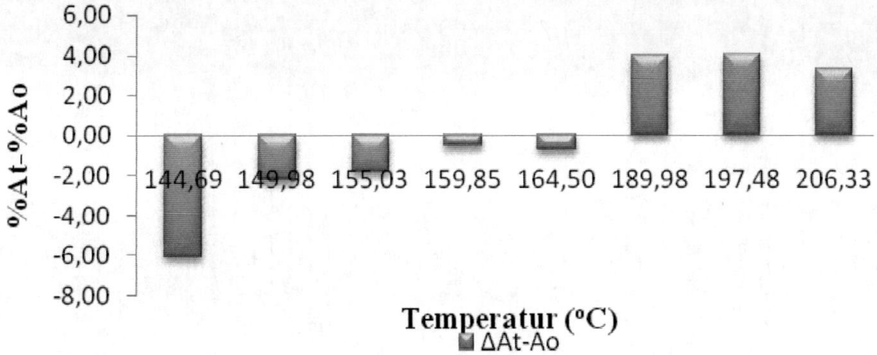

Figure 7. Histogram of dissolved hydrocarbons before and after the addition of crude bio surfactant of B. *Glumae* bacteria

Based on the above histogram, wide abundance peaks in early TPH before treatment with after the addition of crude bio surfactant of bacteria *Burkholderia glumae* is able to dissolve compounds C atoms shorter chain hydrocarbons at temperatures of 144.69; 149.98; 155.03; 159.85; 164.50 °C. In this bacterium maximum temperature to dissolve hydrocarbons short chain C atom is 164.50 °C

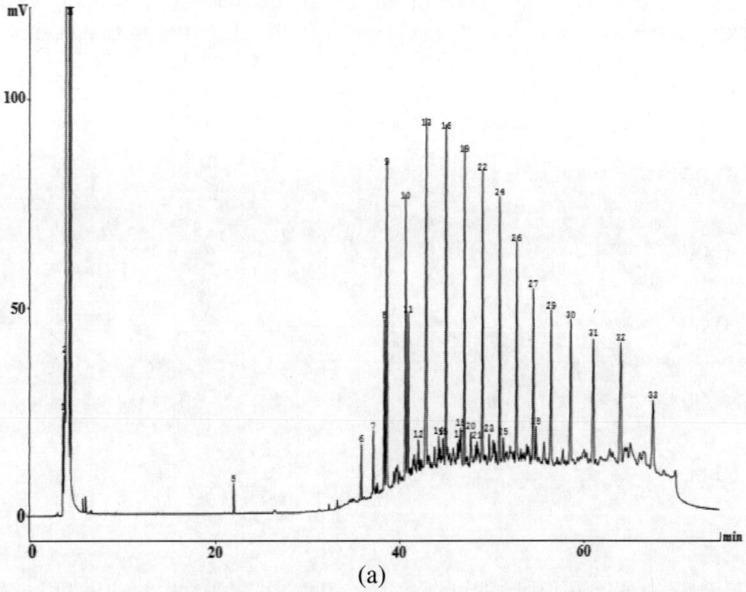

(a)

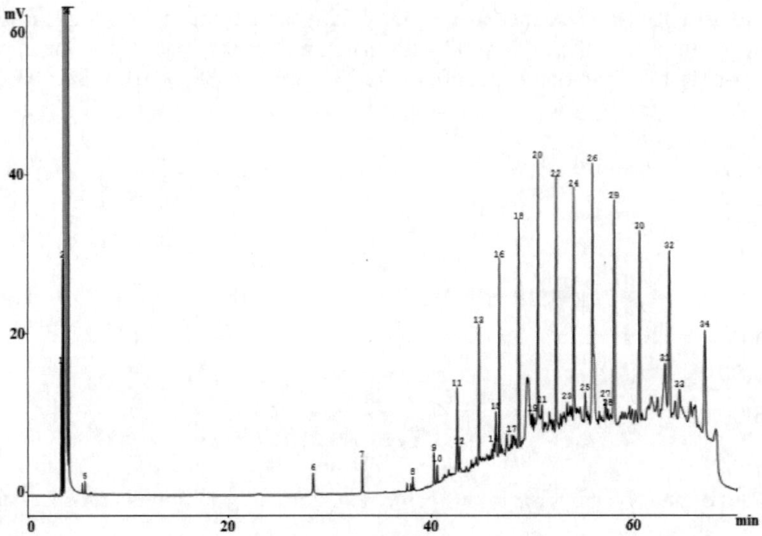

(b)

Figure 8. Chromatogram petroleum components, *Bacillus firmus* bacteria at concentrations of 15% molasses

Figure 8 shows the chromatogram oil component of the bacterium *Bacillus firmus* with the best concentration of carbon source molasses 15% compared with the chromatogram constituents of petroleum before treatment with bio surfactant of *Bacillus firmus,* the number peaks increase 33 to 34. Based on the total retention time for 67 minutes, there were 9 chromatogram peaks higher than other peaks occur at retention times; 46.63; 48.57; 50.43; 52.22; 53.98; 55.87; 58.01; 60.45; 63.42.

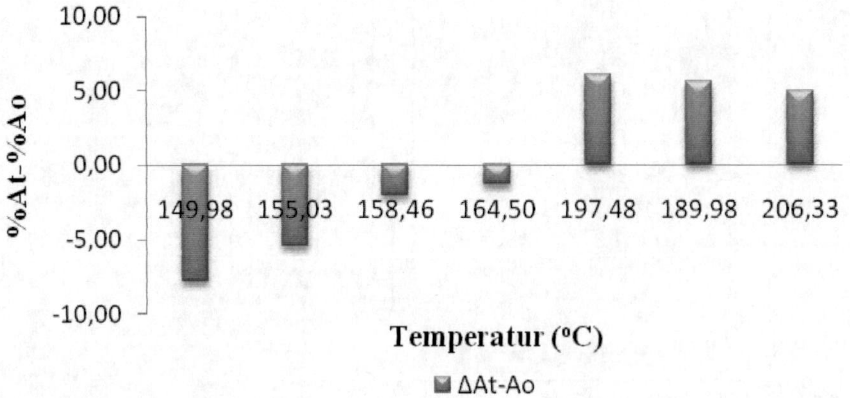

Figure 9. Histogram of dissolved hydrocarbons before and after the addition of crude bio surfactant of *Bacillus firmus* bacteria

Figure 9 shows that the hydrocarbon compounds with temperature fractions; 149.98; 155.03; 158.46; 164.50 °C decreased abundance percent or negative. This indicates that the recovery of petroleum using bio surfactant from *Bacillus firmus* bacteria can dissolve hydrocarbons short chain C atoms. The highest temperature fraction dissolving the hydrocarbon short C chain is at 164.50°C. The fifth chromatogram shows that petroleum is a compound that is composed of many components of the hydrocarbon. Comparison of chromatograms of petroleum and recovery results with bio surfactant chromatogram of each bacterium has a change. Changes chromatogram peaks can occur because the process of degradation by bacteria. The results may indicate a degradation peaks are missing, emerging, increasing peak heights or lowering peak height. Lowering peaks height shows that the bacteria are able to degrade hydrocarbons perfectly. The bacteria degrade and break down long carbon chains into radicals, so that the amount of carbon captured GLC showed the presence of additional amounts of carbon and increasing the number of peak of the initial components [16]. Bacteria have the largest percentage recovery are different at each concentration of molasses. *Pseudomonas peli* produces the greatest recovery, namely the addition of molasses concentration as much as 5% in the amount of 31.92%, the bacteria *Pseudomonas citronellolis* produces the greatest recovery in the addition of molasses concentrations as much as 10% in the amount of 17.65%. Meanwhile, the *Burkholderia glumae* bacteria on the addition of molasses concentration of 20% with a recovery yield of 22.32%, and the bacterium *Bacillus firmus* on the addition of molasses concentration of 15% with a recovery yield of 14.38%. From the result of the recovery in each bacterium is a quantitative analysis, while for qualitative analysis is to analyze hydrocarbons dissolved and remains as a residue which is shown in the graph percent abundance. The fraction of carbon chains contained in the chromatogram can be identified based on the methodology temperature program [16].

4. Conclusions

Based on the research results can be concluded that:

- The optimal concentrations of molasses carbon source to produce bio surfactant effective for the recovery of oil is 5% molasses concentration to the *Pseudomonas peli* bacteria, molasses concentration of 10% for the *Pseudomonas citronellolis* bacteria, molasses concentration of 20% for the *Burkholderia glumae* bacteria, and molasses concentration of 15% for *Bacillus firmus* bacteria.
- The *Pseudomonas peli* bacteria with optimal concentrations of 5% molasses are the highest % recovery with a value of 31.92%.
- Based on the results of GLC analysis, the four bacteria are able to dissolve hydrocarbon compounds with C atoms of short chain hydrocarbons.

References

[1] Juli N and Virmuda B 2001 Penelitian awal terhadap delapan isolat bakteri reservoar dalam mengembangkan volume minyak bumi secara monokultur *Prosiding Simposium Nasional* IATMI 2001 3-5 Oktober 2001 Yogyakarta

[2] Baker C and Herson D 1994 *Bioremediation* (New York : Mc-Graw Hill, Inc.)

[3] Karanth N G K, Deo P G, and Veenanadig N K 2005 Microbial Production of Biosurfactants and Their Importance (Indian Institute of Science) pp 1-18

[4] Gudina E J, Jorge F B A, Ligia R R, Joao A P C, and Jose A T 2012 Isolation and Study of Microorganis from Oil Samples for Application in Microbial Enhanced Oil Recovery. *International Biodeteriovation and Biodegradation* **68** 56-64

[5] Desai J D and Banat I M 1997 Production of a Bio surfactant by *Pseudomonas fluorescens*. *Microbiol. Mol. Biol.* Rev. **61** 47–64

[6] Laini R E, Napoleon A, and Munawar. Isolasi Bakteri Termofilik Penghasil Biosurfaktan yang Ber-potensi sebagai Agen MEOR (Microbial Enhanched Oil Re-covery) dari Sumur Minyak di Sungai Angit *Jurnal Penelitian Sains* **17** 1-5

[7] Zajic J E, Grignard H, Gerson F D 1977 Emulsifying and Suface Active Agents from

Corybacterium hydrocarbonclatus.Biotech and Bioeng **19** 1285-1301

[8] Fajri C 2014 Test biosurfactants production by Pseudomonas peli on Different Carbon Sources as Agent Anti crust (scale inhibitor) to the Pipe Distribution of Petroleum (Indonesia : Dept.of Chemistry University of Sriwijaya South Sumatera).

[9] Yudono B 2013 eksplorasi bakteri hidrokarbonoklastik dari rhizosfer di lahan tambang minyak rakyat, kecamatan babat toman, sumatera selatan *Jurnal Penelitian Sains* **13**

[10] Sharpley J M 1996 *Elementary Petroleum Microbiology* (Texas : Gulf Publishing Company) pp 37 –109.

[11] Munawaroh M 2014 Uji Recovery minyak dengan biosurfaktan dari bakteri yang toleran terhadap konsentrasi garam (Indonesia : Departemen of Chemistry, Mathematic and Natural Science Faculty of Sriwijaya University)

[12] LECO Coorperation 2012 *Analysis of Light Crude Oil Using Gas Chromatography High Resolution Time of Flight Mass Spectrometry*. (Michigan USA : Saint Joseph.).

[13] Agustin S 2014 *Biodegradasi minyak bumi dengan menggunakan bakteri indigen dari sumur tua desa babat toman musi banyuasin (MUBA)* (Indonesia : Dept.of Chemistry University of Sriwijaya South Sumatera)

[14] Fiecter A 1992 Bio surfactant moving towards industrial application *Tibtech.* **10** 208-216

[15] Fatimah 2007 Bio surfactants Production from *Pseudomonas sp* In Different Substrates *Berk. Penel Hayati* **12** 181-185

[16] Yudono B 1994 *An Investigation Into The Prenature Cracking Asphalitic Pavement in Hot and Climated Thesis School of Chemistry* (England :University of Bristol)

On the use of sodium lignosulphonate for enhanced oil recovery

M M Azis[1], H Rachmadi[1], J Wintoko[1], A T Yuliansyah[1], W Hasokowati[1], S Purwono[1], Rochmadi[1] and B Murachman[1]

[1]Chemical Engineering Department, Faculty of Engineering, Gadjah Mada University, Jln. Grafika 2, Yogyakarta, Indonesia

Emai : muhammad.azis@ugm.ac.id

Abstract. There has been large interest to utilize oil reservoirs in Indonesia by using Enhanced Oil Recovery (EOR) processes. Injection of surfactant as a part of chemical injection technique in EOR is known to aid the mobility and reduction in surface tension. One potential surfactant for EOR application is Sodium Lignosulphonate (SLS) which can be made from various sources particularly empty fruit bunch of oil palm and black liquor from kraft pulp production. Here, we will discuss a number of methods for SLS production which includes lignin isolation techniques and sulphonation reaction. The use of SLS alone as EOR surfactant, however, is often not feasible as the Interfacial Tension (IFT) value of SLS is typically above the order of 10^{-3} dyne/cm which is mandated for EOR application. Hence, brief discussion on SLS formulation screening is provided which illustrates an extensive labwork experience during the SLS development in our lab.

1. Introduction

With a growing concern to improve national oil production, enhanced oil recovery (EOR) technology appears to be a promising technology in upstream operation and thus will play an important role in the near future. In addition to primary and secondary techniques, tertiary recovery of EOR is expected to be crucial with the aid of chemicals, thermal, miscible flooding and microbial flooding (Figure 1). Thermal flooding may include hot water and steam stimulation and in situ combustion. For chemical injection, polymers, surfactants and alkaline are common chemicals to aid improvement in oil production. Miscible flooding such as injection mixture of hydrocarbon solvents, carbon dioxide, nitrogen as well as biological (microbial) EOR are two promising techniques that is rapidly growing especially in research scale [1]. Despite of large interest in tertiary EOR, it is generally acknowledged that the knowhow and field experience of tertiary EOR is still limited in Indonesia.

Content from this work may be used under the terms of the Creative Commons Attribution 3.0 licence. Any further distribution of this work must maintain attribution to the author(s) and the title of the work, journal citation and DOI.
Published under licence by IOP Publishing Ltd

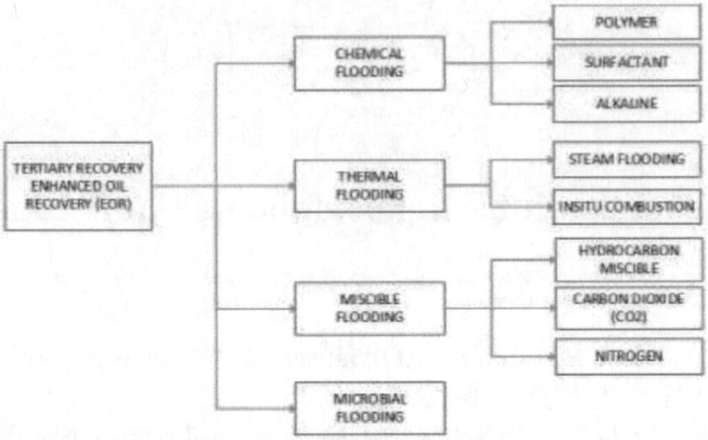

Figure 1. Tertiary recovery techniques for Enhanced Oil Recovery (EOR)

Screening of polymer or surfactant for EOR has been widely studied in literature [1]. Surfactant is a surface tension-lowering active compound which has bipolar structure. The head part favors water (hydrophilic), while the tail part behaves like an oil and thus water-insoluble (hydrophobic). As a result, surfactants tend to be at the interface between different phases degree of polarity such as water and oil. Surfactant serves to reduce surface tension, interfacial tension, improve the stability of the dispersed particles and control the types of emulsion formation [2].

Based on the charge on the ion, surfactant can be divided into four, namely anionic surfactants, cationic surfactants, zwitterionic surfactants and nonionic surfactants. [3]. For chemical injection, petroleum based sulfonate is a common surfactant for EOR. In order to decouple the price of surfactant and oil, there is a need to find an alternative which is economical, produced from abundant resources and reliable for the EOR application. One potential surfactant based on lignin is Sodium Lignosulfonates (SLS).

SLS has been known as plasticizer, flocculation agent and dispersant. As a surfactant, SLS has been able to lower the interfacial tension of two liquid phases and thus may be beneficial to increase the oil recovery from oil wells [4]. SLS is an anionic surfactants because it contains a negative charge on the hydrophilic part. The presence of sulfonate (SO_3^-) and its salts ($NaSO_3^-$) as well as the head of a hydrocarbon group as a tail causes SLS to be water soluble. The structure of SLS is depicted in Figure 2.

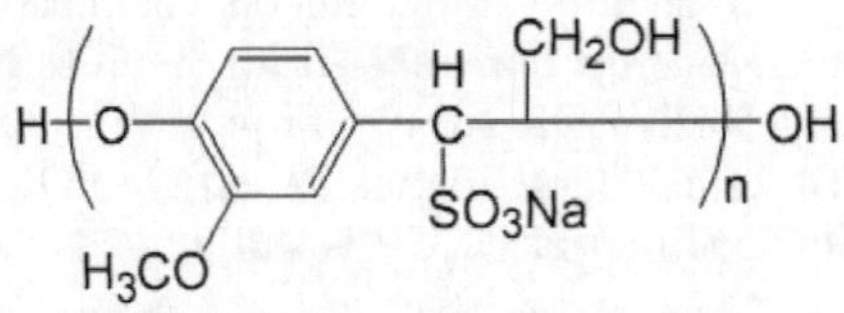

Figure 2. The structure of SLS

This paper aims to provide a mini review of SLS production from lignin, sulphonation reaction of lignin and SLS formulation to be used as EOR surfactant.

2. Sodium Lignosulphonate

2.1. Lignin isolation from kraft pulp black liquor

Lignin is derived from the Latin word "lignum" meaning wood. Lignin is a main constituent component of the vascular plants. The content of lignin varies by the type of wood. Softwood has a lignin content of about 24-33%, while hardwood has lignin content of 19-28% [5]. A number of softwood that can be found in Indonesia are hibiscus wood, wood bark, pine, wood rosin, pine and spruce bonsai. Indonesia also has a number of hardwood such as meranti wood, mahogany, teak, amber and wood suren.

Lignin can also be obtained from paper mill i.e. from the pulping process. Pulping is the process of dissolving the lignin with certain chemical compounds and separates them from the cellulose compounds. For large scale paper mill, there are two methods namely sulfite pulping and kraft pulping. Sulfite pulping uses sulfite compounds, thus one may obtain sulfonate compounds that dissolve in water. Kraft pulping method uses caustic and sodium sulfide compounds to degrade lignin into smaller molecules that dissolve in the black liquor.

Currently, black liquor is known as side product in paper mill and it has been used as a fuel in the boiler pulping. There is an interest to utilize a small portion of black liquor for an added value chemicals rather than just as a fuel. For this purpose, production of SLS from black liquor may provide an attractive route for black liquor utilization especially for EOR application.

Isolation of lignin is the initial stage in SLS surfactant production from black liquor. This process aims to extract lignin by precipitation method by reducing the pH. Typically, precipitation of lignin is conducted by adding certain amount of inorganic acid compound such as sulfuric acid and hydrochloric acid. Addition of acid aims to trigger acid precipitation to occur as a gel. According to Lim [6], the optimal pH for lignin precipitatation occurs at pH = 2. The precipitated gel produced in isolation lignin, however, complicates the filtration process. To tackle this problem, chloroform is typically added as an organic compound that is insoluble in water. With the presence of chloroform, hydrogen bond will form between acid molecule in lignin and chloroform and thus replaces the water molecules that originally bound to the lignin acid. Replacement of water by halogenated organic compounds will increase the specific gravity and increasing the lignin polymer hydrophobicity. With the increase of specific gravity, lignin deposition will occur together with the organic compounds. Due to the increase of hydrophobicity, the deposited lignin can be easily separated from water (Lim, 1994). The drawback of lignin isolation method using an inorganic acid is not environmentally friendly as large amount of used acid needs to be disposed. Therefore, an alternative method using CO_2 has been reported in literature by Luong *et al.* [7]. Addition of CO_2 plays role to decrease the pH and addition of Alum ($KAl(SO_4)_2.12H_2O$) resulted on large precipitation of lignin at pH of 4. With this process, lignin isolation is faster, efficient and more environmentally friendly.

2.2. Lignin isolation from empty fruit bunch (EFB) of oil palm

There has been large interest to utilize empty fruit bunch of palm oil as raw material for SLS production. Palm oil business is one of the largest business sector in Indonesia where together with Malaysia, Indonesia is the largest producer of palm oil in the world. Utilization of solid waste from palm oil processing for SLS production may be beneficial particularly from economic point of view. The composition of palm oil solid waste as well as their content is presented in Table 1.

Table 1. The composition of different part of a palm oil tree

No	Solid Waste	Composition (wt.%)					
		Cellulose	Hemi Cellulose	Lignin	Wax	Ash	Others
1.	Fiber	19,0	15,2	30,5	9,1	7,0	19,2
2.	Shell	14,7	16,4	53,6	2,3	2,3	10,7
3.	Empty fruit bunch (EFB)	35,8	21,9	17,9	4,0	3,0	17,4
4.	Palm frond	31	17,1	22,9	2	2,8	24,2
5.	Trunk	39,9	21,2	22,6	3,1	1,9	11,3

Synthesize of SLS from empty fruit bunch (EFB) of oil palm has been the focus of SLS research at UGM. Isolation of lignin from EFB is the first step in SLS synthesize. Figure 3 shows the process of lignin isolation from EFB.

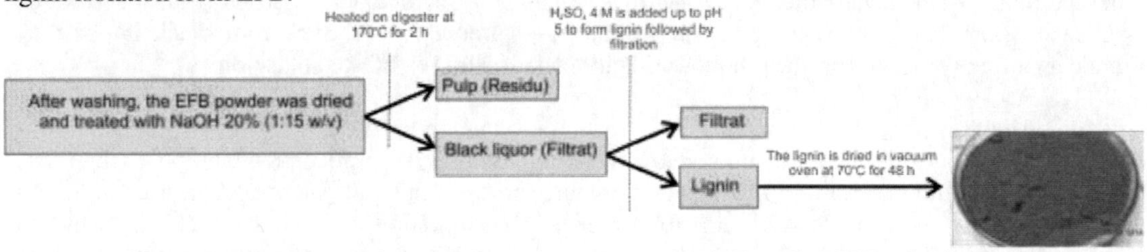

Figure 3. Lignin isolation technique from Empty Fruit Bunch (EFB) of Oil Palm

Initially, EFB powder was reacted with NaOH to degrade the lignin in an autoclave (digester). Subsequently, the black liquor was then filtered with addition of H_2SO_4 to obtain solid lignin. The solid lignin was then dried in an oven to obtain lignin powder.

2.3. Lignin sulphonation

SLS can be synthesized from lignin by sulfonation reaction. Sulfonation reaction aims to insert sulfonate groups and thus changes the physical properties of lignin to be more polar as lignosulfonate salt[8]. There are several methods to carry out lignin sulphonation as has been reported by Inwood [4] in Table 2.

Table 2. Various methods for lignin sulphonation

Starting Material		Reagent	References
Phenolated lignin	sulfuric	1. 1 M NaOH + 37% formalin → 1M HCl 2. 40% $NaHSO_3$	[9]
Phenolated lignin	sulfuric	0,5 M NaOH + $CH_2(OH)SO_3Na$	[9]
Phenolated lignin	sulfuric	1. HCHO + 3M NaOH 2. C_2Cl_4 + $ClSO_3H$	[10]
Alkali lignin	-	1. HCHO 2. Na_2SO_3	[11]
Lignin	-	SO_3^{2-} + CH_2O	[12]

In general, it can be inferred from the tabulated data that concentrated sulfuric acid, sodium bisulfate, sodium sulfate, chlorosulfuric acid and sulfur trioxide are capable of sulfonating lignin with a high efficiency. However, each reagents may have advantages and weaknesses such as corrosion problem. Lignin sulphonation in our lab was conducted with the aid of $NaHSO_3$ according to the following reaction:

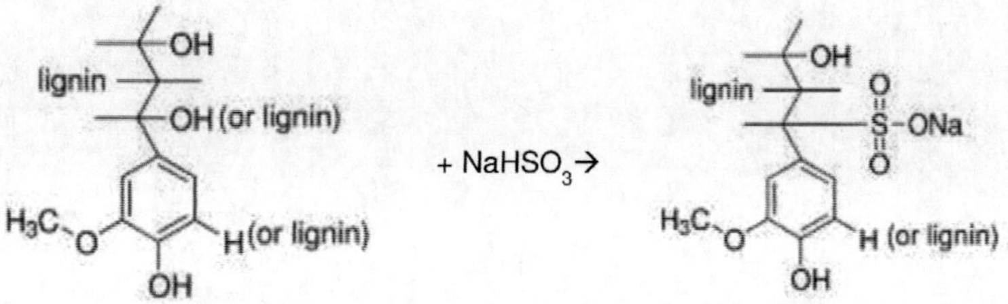

A brief schematic method to carry out lignin sulphonation is presented in Figure 4. The process involves reaction and distillation step.

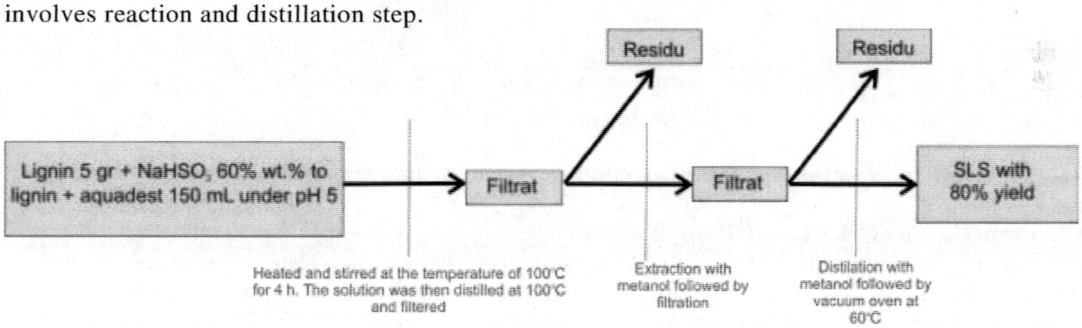

Figure 4. Schematic diagram for lignin sulphonation with $NaHSO_3$

FTIR analyses were conducted to evaluate the insertion of SO_3 group into lignin. Figure 5 shows the FTIR spectrophotograms of lignin and SLS using Shimadzu IR Prestige-21. In general, the spectra of lignin and SLS are relatively similar. However, the spectrophotogram of SLS shows more intensities to the presence of S-O vibration at 623 cm^{-1}, vibration of SO_3 group at 1115 cm^{-1} and vibration of symmetrical stretch of S-O at 1037 cm^{-1}. Hence, it can be concluded that insertion of SO_3 group into lignin structure took place during sulphonation of lignin with $NaHSO_3$.

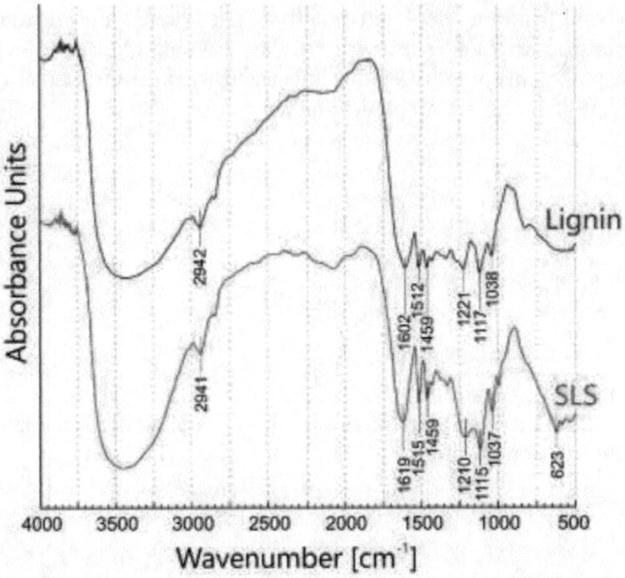

Figure 5. FTIR spectrophotograms of lignin and SLS

3. Formulation of SLS for EOR application

There are requirements that has to be met by any surfactant for EOR application. These requirements are shown in Table 3. The most crucial part for SLS development for EOR surfactant was to find the appropriate formula in order to meet the requirements as mandated by SKK Migas. It is generally accepted that the EOR surfactant is developed based on field specific based on the field characteristic such as reservoir condition, brine etc. The up-scaling process from lab scale to field test is often far from trivial due to the field condition (practical issues) and the large amount of surfactant that is needed for the injection.

Table 3. The requirements of EOR Surfactant by SKK Migas

IFT	10^{-3}-10^{-4} dyne/cm
Phase behavior	Type III (Middle phase) or Type II (bottom phase)
Compatibility	Clear > 7 days
Thermal stability	IFT in the range of 10^{-3} dyne/cm for at least 90 days
Filtration ratio	< 1.2
Adsorption	Average < 0.4 mg/g
pH	6-8
Incremental recovery	≥ 10%
Price	ca. 5 USD/kg

One of the challenge was to obtain IFT value within the range of 10^{-3}-10^{-4} dyne/cm. In our lab, measurement of IFT was typically conducted on TX-500 Spinning Drop Tensiometer. The result of IFT measurement of the SLS that we produced from black liquor gave 9.15 .10^{-1} dyne/cm. As a result, formulation of SLS is needed by mixing it with long chain alcohol and PFAD (palm fatty acid distillate). We have been able to obtain a surfactant formula with the alcohol contain of less than 10% (wt.%) and gave the IFT value around 10^{-3} dyne/cm. However, other aspect which is also crucial was to find a good thermal stability for 90 days test. Our test showed that the IFT value fluctuated and it might go beyond the 10^{-3} dyne/cm value during the 90 days test. As a result, finding the right formulae of EOR surfactant is probably the most intensive work in EOR development as large screening process is needed with the aid of ternary diagram. This issue is still under investigation and better formulation as well as better stability is continuously studied in our lab.

4. Conclusions

With the growing concern to increase petroleum production, the use of tertiary EOR will play an important role in the future. Therefore, it is important to develop local surfactant which fulfils all the requirement for EOR surfactant. SLS appears as one of promising surfactant for EOR and may be produced from abundant sources such as black liquor and empty fruit bunch of palm oil. Here, we have presented the method for SLS production from those sources. Eventually, our experience shows that SLS may have a promising future with appropriate formulation and economically competitive with other surfactants.

Acknowledgements

We would like to acknowledge Pertamina UTC for the financial support as well as good collaboration throughout the SLS project at UGM. In addition, we also would like to thank Pusat Studi Energi (PSE) UGM for good arrangement and administration of the project.

References

[1] Froning H R, Fussell D D and Heffern E W 2015 *Enhanced Oil Recovery in Encyclopedia of Chemical Technology* vol 17, ed Kirk Othmer (Canada : John Wiley & Sons Inc)
[2] Rieger M M 1985 Surfactant in Cosmetic in *Surfactant Science Series* (New York: Marcell Dekker Inc)
[3] Rosen M J dan Kunjappu J T 2012 *Surfactants and Interfacial Phenomena* (Canada: John Wiley & Sons Inc)
[4] Inwood J P 2014 *Sulfonation of Kraft Lignin to Water Soluble Value Added Products* (United States : ProQuest LLC)
[5] Pollak A 2015 Lignin in *Encyclopedia of Chemical Technology* vol 17, (Canada : John Wiley & Sons Inc)
[6] Lim L 1994 *Isolasi Lignin dari Lindi Hitam Pabrik Kertas Leces* (Indonesia : Gadjah Mada University)
[7] LuongN D, Binh N T T, Duong L D, Kim D O, Kim D S, Lee S H, Kim B J, Lee YS and Nam J D 2012 An eco-friendly and efficient route of lignin extraction from black liquor and a lignin-based copolyester synthesis *Polym. Bull.* **68** 879-90
[8] Teke J 2015 *Pengaruh Penambahan Epoksidasi Asam Risinoleat Minyak Jarak (Castor Oil) dan Kosurfaktan terhadap Kinerja Sodium lignosulfonate (SLS) dalam Menurunkan Interfacial Tension (IFT) pada Proses Enhanced Oil Recovery (EOR)* (Indonesia : Gadjah Mada University)
[9] Matsushita Y and Yasuda S 2005 Preparation and evaluation of lignosulfonates as a dispersant for gypsum paste from acid hydrolysis lignin *Bioresour Technol.* **96** 465-470
[10] Yasuda S and Asano K 2000 Preparation of strongly acidic cation exchange resins from gymnosperms acid hydrolysis lignin *J. Wood Sci.* **46** 477 - 479
[11] Ouyang X, Ke L, Qiu X, Guo Y and Pang Y 2009 Sulfonation of alkali lignin and its potential use in dispersant for cement *J. Dispersion Sci.Technol.* **30** 1-6
[12] Meister J J 2002 Modification of lignin *J. Macromol. Sci. - Poylemer Reviews C* **42** 235-289

Optical properties of Cu and Ru doped BST thin films with additive glycerol and MESA surfactant

N G Pamungkas[1], M Dahrul[2], Irzaman[2] and H Alatas[2]

[1] Biophysics Program, Department of Physics, Bogor Agricultural University, Indonesia
[2] Departement of Physics, Bogor Agricultural University, Indonesia

Email : novita.galih@gmail.com

Abstract. In this paper, barium strontium titanate (BST) thin films were deposited by chemical solution deposition (CSD) with addition glycerol and methyl ester sulfonic acid (MESA) surfactant as additive materials. The effect of variation spinning speed (5500 and 8000 rpm), different dopant (Cu and Ru dopants), and also additive materials on optical properties were studied by Spectrophotometer Vis-NIR. As a result, BST thin films with additive MESA have higher absorbance and similar spectrum pattern than additive glycerol. Generally, BST thin films with MESA surfactant had band gap smaller than glycerol of BST thin films. Absorption and extinction coefficient of BST with glycerol were observed and we reported Ru doped BST at spinning speed 5500 rpm had higher value than other films. On other hand, Cu doped BST at 8000 rpm had absorption and extinction coefficient higher than other films.

1. Introduction

$Ba_{0.5}Sr_{0.5}TiO_3$ (BST) is a ferroelectric perovskite materials has been well studied in thin film and bulk form. BST as thin film has been under investigation for about 15 years starting from 90'[1]. The initial interest in BST thin films was due to its high dielectric constant, low dielectric loss, high dielectric breakdown and composition dependent Curie temperature [2]. These advantages make the thin film BST as a candidate replace SiO_2 as the dielectric mass storage (capacitor), dynamic random access memory (DRAM), application of acoustic waves, and tunable microwave [3-6]. Besides having ferroelectric properties, BST thin films have the sensitivity to light (opto-electronic), sensitive to temperature stimuli (pyroelectric), and sensitive to the influence of pressure (piezoelectric) [7-9]. The pyroelectric characteristics of BST have been applied as a switch thermal infrared and infra-red sensors [7, 8]. Opto-electronic properties of BST has been applied as photodiode [10, 11]. In addition, its sensitivity to light has been applied in the model of automatic drying in agriculture [12].

Content from this work may be used under the terms of the Creative Commons Attribution 3.0 licence. Any further distribution of this work must maintain attribution to the author(s) and the title of the work, journal citation and DOI.
Published under licence by IOP Publishing Ltd

Dopant material is a material that is expected to change the lattice constant, dielectric constant, pyroelectric properties, opto-electronic properties, piezoelectric properties of the ceramic and thin film BST become well. Addition of hard dopant and soft dopant in BST thin films increase optical and electrical properties of BST thin films [5, 11]. Another factor affecting the characteristics of the thin film is deposition process on substrate. Chemical solution deposition (CSD) method is of particular interest because of its good control of stoichiometry, ease of fabrication and low temperature synthesis. There are mainly two methods for the coating of solutions in CSD process: dip coating and spin coating. Spin speed of spin coating influence coating process and thickness film [13].

Here, utilization of material additives in deposition process especially spin coating technique such as polymer or surfactant, increase homogeneous in thin film surface [14,15]. Surfactant is material agent that reduces the surface tension of water or solution by forming a molecular or atomic layer on the surface. The addition of surfactant to the solution can enhance various properties of the films, such as increasing optical absorbance, energy gap, and the thickness of films [15]. Methyl Ester Sulfonic (MES) surfactants produced from palm oil and has been investigated for *Enhanced Oil Recovery* (EOR) applications in the petroleum industry [16]. Materials for methyl esters are palm olein, methanol and potassium hydroxide (KOH). Sulfonation process methyl ester from palm olein produce methyl ester sulfonate surfactant acid (MESA), dark and blackish. Glycerol is the residue of transesterification process palm olein into methyl ester [17].

In this study, BST thin films were deposited by chemical solution deposition on Silicon substrates using spin-coating. Glycerol and MESA surfactant additive were used to BST solution. BST thin films use doping variation, Copper (Cu) and Ruthenium (Ru) with variation spinning speed, 5500 and 8000 rpm. Optical properties, namely absorbance, band gap, absorption coefficient, and extinction coefficient were investigated to study the potential of using glycerol, MESA surfactant, and also dopants for optical properties in thin film.

2. Research and Method

2.1. Material Preparation

BST solution with 1 M concentration have been synthesized by CSD method. Barium acetate [$Ba(CH_3COO)_2$, 99%] and strontium acetate [$Sr(CH_3COO)_2$, 99%] were dissolved in acetic acid [CH_3COOH, $\geq$ 99.7%] and stirring for 60 minutes. Titanium isopropoxide was added in solution and stirring for 30 minutes. Then, Ruthenium oxide and copper acetate were added to the different solution and stirring again for 30 minutes as dopants. Finally, glycerol and MESA was mixed to the solution and continuously stirred for 1 hours. Silicone as substrate was cleaned in acetone solution. Films were prepared by depositing the solution on the surface silicone. Coating thin film use spin coating technique. Thin layer was spin-coated with variation spinning speed 5500 and 800 rpm on substrate for 40 s. Thin layer was dried on hotplate. This process was repeated for the 3 layers. BST Films were annealed for 8 hours in an air atmosphere using furnace Vulcan model 3-130. BST films were annealed 850°C.

2.2. Analysis

Spectroscopy is one of analysis to observe the optical properties of the surface material. Optical properties were observed in this research are absorbance, band gap, absorption coefficient, and extinction coefficient. The Optical properties of BST thin films were characterized using Spectrophotometer Vis-NIR USB 1000 Oceanoptic in range 339 – 1022 nm.

3. Results and discussion

3.1. Absorbance BST thin films

Absorbance BST thin film aims to observe the maximum absorption and wide absorption spectra of thin films of BST. It is based on that electromagnetics wave radiation from the sun has the greatest intensity in the visible region (black box radiation). From the Figure 1, BST thin films with MESA surfactant additive have higher absorbance than BST thin films with glycerol additive. Glycerol additive leads surface of thin films inhomogeneous. Surfaces formed glossy but not flat. This is due to viscous solution. Viscous solution causes centrifugal process in the spin-coating was not optimal. MESA surfactant additions in the deposition process to result thin films with higher absorbance than glycerol. The addition of additive ethylene has been done in some previous studies. The addition of the additive in the film coating process causes the film surface to form more homogeneous. Homogeneous layer formation with MESA additive result increasing film thickness.

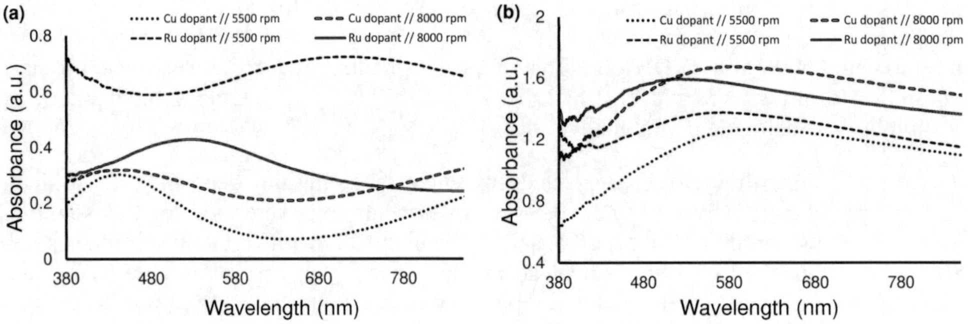

Figure 1. Absorbance of BST thin films with (a) glycerol additive and (b) MESA additive

Figure 1. (a) showed the absorbance curve BST thin films with dopants variation (Cu and Ru), variation speed of spin coater (5500 and 8000 rpm), and with glycerol additive. BST thin film with dopant Ru has a higher absorbance than dopant Cu. Absorbance peaks of BST thin films with Ru dopants shifted towards smaller wavelengths on a speed of 5500 rpm, from 500 nm to 700 nm. However, BST thin film with Cu dopants no shifting (400 nm). Figure 2. (b) showed variation of BST thin films with MESA surfactant additive. Generally, the absorbance peaks for all types of films have the same curve shape and there is no shifting. Speeds of 8000 rpm have higher absorbance than 5500 rpm. MESA surfactant leads films surface homogeneous and deposition thin film (3 layers) well formed.

3.2. Band gap of BST thin films

Determination of the gap energy BST thin films using optical absorbance characteristics, by plotting (αhv) n with hv. The resulting gap energy using indirect transition value that is n = 2 [18,19]. This can be explained that the transition of electrons between the valleys of the conduction band and valence band peak occurs indirectly but absorb or emit energy phonons. The smaller the gap energy of a semiconductor material, the more easily an electric current flows. Figure 2 showed gap energy for all films type with glycerol additive. Addition of Ru dopants lowering gap energy. Speeds of 8000 rpm lead gap energy decreasing at Cu doped BST and increasing at Ru doped BST. Figure 3 showed BST thin films with MESA surfactant additive. Addition of Ru dopants lead gap energy increasing. Variation speeds of spin-coating leads the gap energy constant at Cu doped BST and increasing at Ru doped BST. Smaller gap energy, electrical and optical properties increasing.

IOP Conf. Series: Earth and Environmental Science **65** (2017) 012031 doi:10.1088/1755-1315/65/1/012031

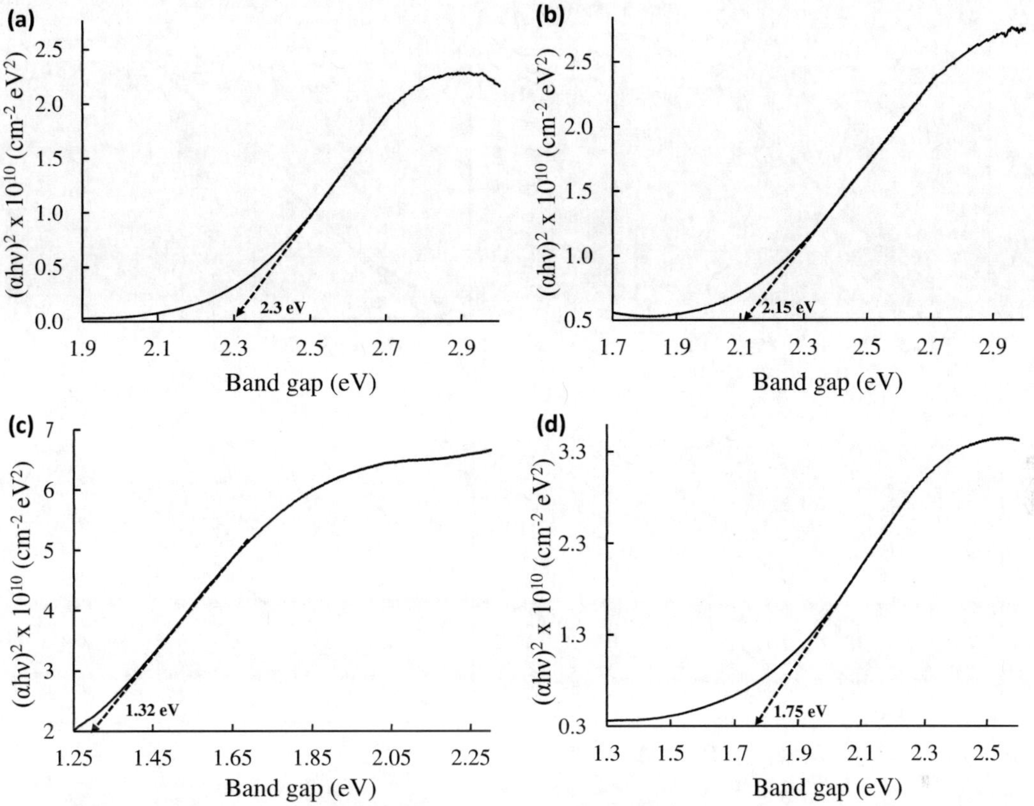

Figure 2. Band gap of BST thin films with glycerol additive and variation a) Cu dopant at 5500 rpm b) Cu dopant at 8000 rpm c) Ru dopant at 5500 rpm d) Ru dopant at 8000 rpm

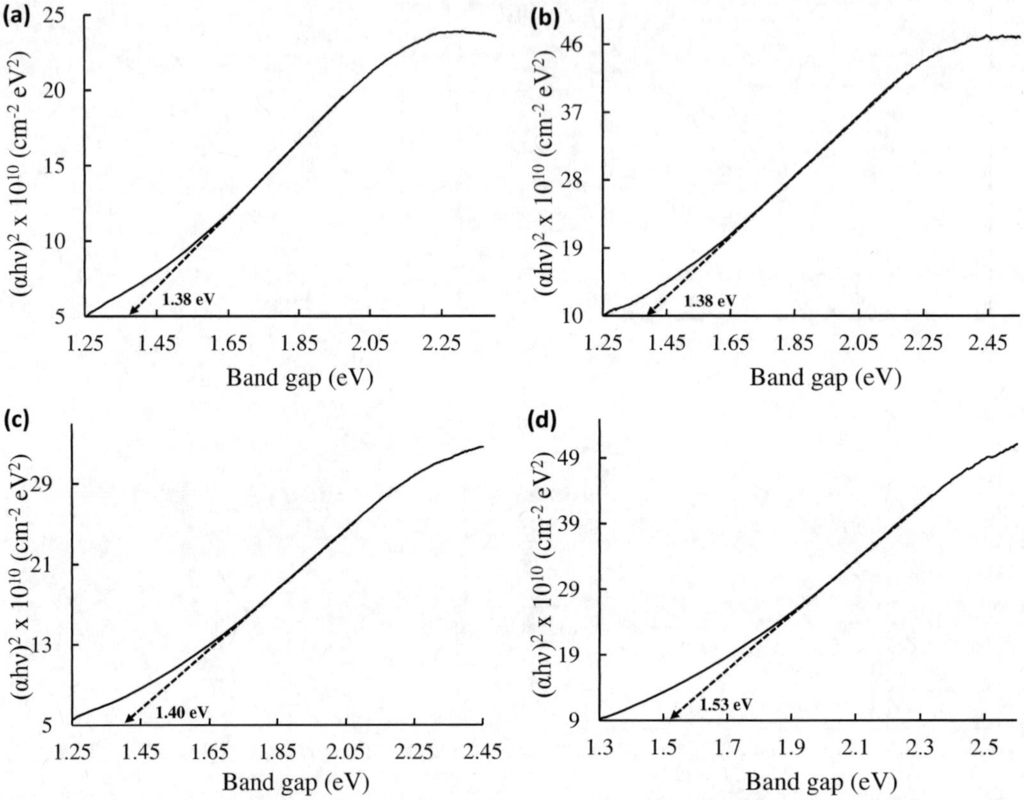

Figure 3. Band gap of BST thin films with MESA additive and variation a) Cu dopant at 5500 rpm b) Cu dopant at 8000 rpm c) Ru dopant at 5500 rpm d) Ru dopant at 8000 rpm

3.3. Absorption coefficient and extinction coefficient of BST thin films

The others parameter in optical properties are absorption coefficient and extinction coefficient. The absorption coefficient is a function of the absorbance and thickness so that the thickness difference factor can be ignored. Extinction coefficient have been used to calculate the damping or reduction factor of light intensity when passing through thin films. Figure 4 and 5 showed absorption coefficient and extinction coefficient. Absorption coefficient of BST thin films with MESA surfactant additive higher than glycerol additive. Ru doped BST thin films higher absorption coefficient than Cu dopants. Figure 5 showed small extinction coefficient for all films.

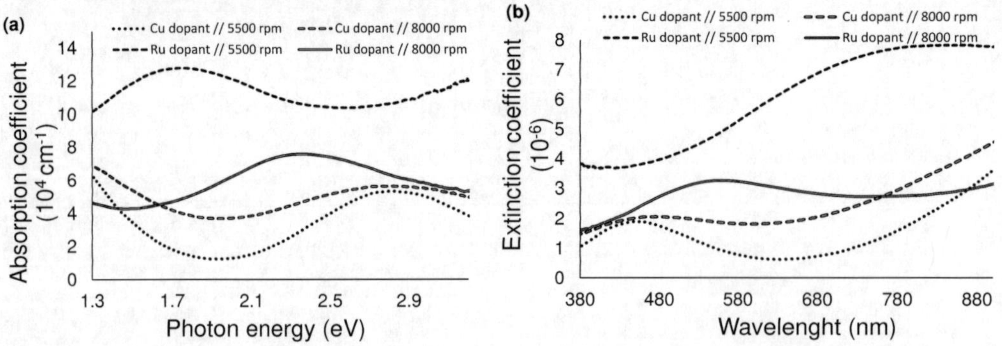

Figure 4. (a) Absorption and (b) extinction coefficient of BST thin film with additives glycerol

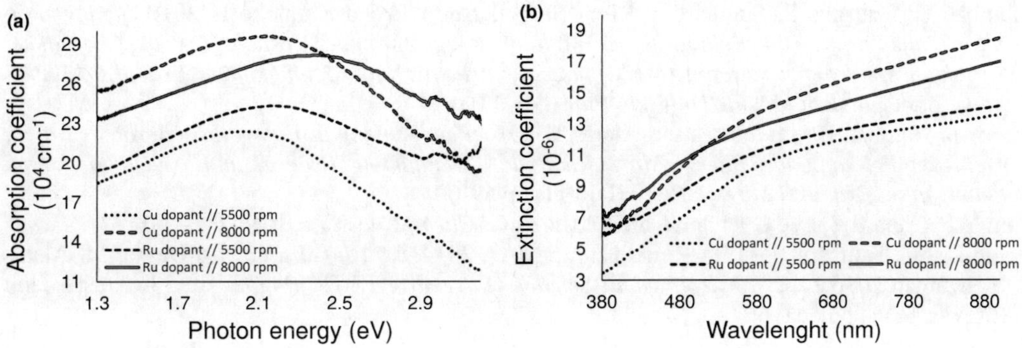

Figure 5. (a) Absorption and (b) extinction coefficient of BST thin film with additives MESA

4. Conclusions

BST thin films were successfully deposition onto silicone substrate by CSD and sol-gel method with glycerol and MESA surfactant additive. BST thin films with MESA surfactant additive have higher absorbance and absorption coefficients than BST thin films with glycerol additive. Ru dopants in BST increase absorbance and absorption coefficients. BST Thin films with MESA surfactant additive, absorption coefficients and extinction coefficients increase with Ru dopants. Absorbance BST thin films with MESA surfactant additive and spinning speed 5500 rpm higher than 8000 rpm. These result indicate that MESA additive has the potential to be used as surfactant at deposition thin film.

References

[1] Baumert B A, Chang L -H, Matsuda A T, Tsai C JT -L, Tracy C J, Gregory R B, Fejes P L, Cave N G, Chen W, Taylor D J, Otsuki T, Fuji E, Hayashi S and Suu K 1998 *J. Appl. Phys* **13** 197-204

[2] Adem U 2003 *Preparation of BaxSr1-xTiO3 Thin Films by Chemical deposition and their Electrical Characterization (Thesis)* (Ankara: The Middle East Technical University) p 34-36

[3] Manavalan S G 2005 *Structural and Electrical Properties of Barium Strontium Titanate Thin Films for Tunable Microwave Application (Thesis)* (South Florida: University of South Florida) chapter 2 pp 20-27

[4] Gurumurthy V 2007 *Barium Strontium Titanate Films for Tunable Microwave and Acoustic Wave Application (Thesis)* (South Florida: University of South Florida) p 19

[5] Irzaman, Syafutra H, Darmasetiawan H, Hardhienata H, Erviansyah R, Huriawati F, Akhiruddin, Hikam M, Arifin P 2011 *Atom Indoneia* **37** 133-138

[6] Bhalachandran R, Ong B H, Wong H Y, Tan K B and Rasat M M 2012 *Int. J. Electro chem. Sci.* **7** 11895-11903

[7] Okatan M B, Cole M W and Alpay S P 2008 *J. App. Phys.* **104** 104-107

[8] Irzaman and Barmawi M 2008 *J. SAINS MIPA* **14** 5-7

[9] Kanareykin A, Nenasheva E, Kozyrev A, Kazakov S, Yakovlev V 2010 *Proc. Int. Particle Accelerator Conf.* (May, 23-28; Kyoto) vol 7 (Genewa: JACoW) pp 3896-3898

[10] Kocanda M, Mohiudin S F and Abdel-Motaleb I 2012 *CSTA* **12** 17-20

[11] Dahrul M, Syafutra H, Arif A, Irzaman, Indro M N, Siswadi 2010 *AIP Conf. Proc. (October 10-13; Bandung)* vol 1325 (New York: AIP Publishing) pp 43-46

[12] Irzaman, Putra I R, Aminullah, Syafutra H, Alatas H 2016 *Proc. Environmental Science (November 17-18; Bogor)* vol 33 (Amsterdam: Elsevier B.V.) pp 607-614

[13] Bornside D E, Macosko C W and Scriven L E 1987 *J. Image Tech.* **13** 122-129

[14] Doshi D A, Gibaud A, Goleeto V, Lu M, Gerung H, Ocko B, Han S M and Brinker CJ 2003 *J. Am. Chem. Soc.* **125** 11646-11655

[15] Halin D S C, Talib I A, Daud A R and Hamid M A A 2014 *Int. J. Pho.* **2014** 1-6

[16] Hambali E, Suarsana P, Sugihardjo, Rivai M, Zulchaidir E and Handoko H 2010 *Peningkatan Nilai Tambah Minyak Kelapa Sawit Melalui Pengembangan Teknologi Proses Produksi Surfaktan Metil Ester Sulfonat (MES) dan Aplikasinya untuk Meningkatkan Produksi Minyak Bumi Menggunakan Metode Huff dan Puff* (DIKTI) p 14

[17] Irzaman, Hambali E, Suryani A and Alatas H 2016 *Pengembangan Fotovoltaik Berbasis Bahan Alam untuk Kebutuhan Energi Masyarakat dan Mendukung Penerapan Biosurfaktan untuk Peningkatan Produksi Minyak Bumi* (DIKTI) pp 59-62

[18] Lim Y F, Chua C S, Lee C J J and Chi D 2014 *Phys. Chem. Chem. Phys.* **14** 241-250

[19] Halin D S C, Talib I A, Daud A R and Hamid MAA 2014 **2014** *Int. J. Pho.* 1-6 Dahrul M, Alatas H, Irzaman 2016 *Proc. Environmental Science (November 17-18; Bogor)* vol 33 (Amsterdam: Elsevier B.V.) pp 661–667

IOP Conf. Series: Earth and Environmental Science **65** (2017) 012032 doi:10.1088/1755-1315/65/1/012032

Optimalization of silicon extraction from husk ashes by excessive magnesium addition on increasing rate of temperature reduction

I Sintha[1], M Dahrul[2] S S Ismawati[2], M Kurniati[3], Irmansyah[3], Irzaman[3]

[1] Biophysics program, Department of Physics, Bogor Agricultural University, Indonesia
[2] Physics Programed, Department of Physics, Bogor Agricultural University, Indonesia
[3] Department of Physics, Bogor Agricultural University, Indonesia

Email : inggriediphone@gmail.com

Abstract. The Silicon can be extracted from rice husk ash by reducing silica with magnesium. In this study the use of Magnesium and Silica was 49:60 with various rate of increase in the reduction temperature (1 °C/minutes, 3 °C/minutes, and 5°C/minutes). It was done in order to obtain silicon with a higher purity than previous studies. The results of EDX showed that the purity silicon with the increase rate of 1 °C/minutes and 5 °C/minutes were 1.78% and 29.39%, respectively. Then, silicon was not formed by increasing rate of 3 °C/minutes. Moreover, XRD results of the samples by using the increase rate of 1 °C/minutes and 5 °C/minutes showed that the pattern correspond to Silicon while the sample with increase rate of 3 °C/min corresponds to Silica. The FTIR spectra indicate the presence of stretching vibration an harmonic for functional groups of Si-O-Si with spring constant for the increase rate of 1 °C/minutes, 3 °C/minutes, and 5 °C/minutes respectively were 977.23 N/m, 1003 N/m, and 982.3 N/m. FTIR result was consistent with EDX and XRD result proving by the existence of Si-O-Si functional group which showed the purity silicon produced was not optimum and presence of OH-functional group indicate that sample contained water

1. Introduction

Rice husk is waste rice cultivation results that are hard and rough, and nutrient levels are low and not worth economically. However, the current waste rice husks can already be used to fuel the red brick, cement base material, even the husk furnace fuel [1]. The results will appear burning other waste is waste rice husk [2]. Processing and utilization is not proportional to the amount of chaff is increasing every year. Therefore, the use of chaff should be maximized. Rice husk is the outer layer of the grain, which is a byproduct of rice milling process is currently done. From the rice milling process is usually obtained husk around 20-30% of the grain weight [3]. Husk waste is widely used as an alternative source of energy, one as rice husk furnace fuel.

Rice husk ash can be a source of synthesis of amorphous Silicon Dioxide bio ceramics at low cost because of its availability in abundance. Rice husk in general burned in the temperature range of 500 °C to 600 °C to obtain amorf [4]. Silicon dioxide Silicon dioxide produced can be extracted into silicon to reduce Magnesium. Silicon dioxide is a compound with a tetragonal structure that can be

Content from this work may be used under the terms of the Creative Commons Attribution 3.0 licence. Any further distribution of this work must maintain attribution to the author(s) and the title of the work, journal citation and DOI.
Published under licence by IOP Publishing Ltd

found all around us and are widely used as raw material for the electronics industry. Pure silicon dioxide contained in three forms, namely quartz, tridymite, and cristobalite. Reduction magnesium silicon dioxide with a reductor at a temperature of 620 °C to 650 °C years. Silicon powder can be obtained from rice husk with ashing temperature of 1000 °C with a gray color that has such a shiny metal [5].

Rice husk ash research has been able to produce silica. In addition to silica, it turns ash from rice husk can also be extracted into a silicon purity can reach 60.87%. To obtain Silicon, Silica is reduced with magnesium at a temperature of 650 °C [6]. Silicon currently has a high selling power, the use of silicon is now no doubt. In the field of silicon technology is already widely used as a base for the manufacture of semiconductor components that are environmentally friendly. So that the waste generated does not damage the environment. Therefore, a challenge of exploiting the potential of Indonesia's natural resources one of which is rice husk. This research will be carried variation reduction rate of temperature increase between silica and magnesium, it is aimed to determine the rate of temperature rise is most optimum in producing high-purity silicon.

2. Material and Method

2.1. Materials
Bahan yang digunakan dalam penelitian sekam padi dari limbah pertanian, asam klorida (HCl), Magnesium (Mg) dan aquades. This study is divided into several stages, rice husk charcoal manufacture, extraction silica, and silicon extraction.

2.2. Rice husk charcoal manufacture
Making rice husk rice husk begins to weigh. Rice husk using sun dried and then weighed and then inserted into a furnace chaff. Rice husk produced subsequently weighed. Rice husk using sun dried and then weighed and then inserted into a furnace chaff. Rice husk produced subsequently weighed.

2.3. Extraction silica
The process of making silica, rice husk charcoal burned in a furnace at 400 °C for 2 hours. Furthermore, the heating temperature was increased to 900 °C for 1 hours with a heating rate of 0.9 °C/minutes. Rice husk ash was weighed and washed using hydrochloric acid (HCl 3%) to reduce the impurities present in rice husk ash so that only the silicon dioxide. Furthermore, washed using hot distilled water repeatedly until acid-free (tested using litmus paper), then filtered through ash-free filter paper. Results of screening (residue + filter paper) is heated in a furnace to temperature of 900 °C until white remaining silicon dioxide [7].

2.4. Silicon extraction
Silicon dioxide has been obtained is filtered with a micro filter so that the same particle size with Magnesium powders. Furthermore, Silicon dioxide is reduced by magnesium using comparison magnesium and Silicon dioxide at 49:60 which mass of magnesium is 0.82 grams and Silicon dioxide 1 gram. The reduction process in furnace at temperature of 650 °C with rise temperature variation 1 °C/minutes, 3 °C/minutes, and 5 °C/minutes. After the reduction process, a sample of the reduction was washed using HCl 3% to remove impurities. The washing process is done twice, washing first early samples of the reduction put in a beaker, then mixed with HCl 3% 80 ml, then heated above the bath for the process of stirring for 2 hours at temperature of 200 °C and a stirrer speed of 240 rpm. After that, followed by washing the latter. The same sample mixed with HCl 3% of 300 ml, then heated above the bath at a temperature of 200 °C for 1 hour with a stirrer speed of 240 rpm. Then, sample washed with hot distilled water repeatedly until acid-free (tested with litmus paper, pH 7, and then filtered with ash-free filter paper. Results of screening (residue) is heated in a

furnace to a temperature of 110 ° C for 12 hours for drying. The results of this screening suspected as Silicon.

2.5. *Analyzed EDX, XRD, and FTIR*

Silicon analyzed using EDX (Energy Dispersive X-ray) to identify the composition of the elements contained in the sample so as to determine the purity of the silicon. Analysis of XRD (X-Ray Diffraction) to observe the silicon structure. This method to identify crystalline phases in the material by determining the parameters of the lattice structure as well as to obtain particle sizes. Analysis of FTIR (Fourier Transform Infra-Red) to characterize the functional groups and molecular bonding.

3. Results and discussion

EDX analysis results of samples Si of the silica reduction process with initial purity of 87.48% with magnesium were showed in Table 1. Table 1 showed different chemical composition on the rate of temperature rise is different. The ratio between magnesium and silica at 49:60, purity silicon obtained from variations in the rate of temperature rise of 1 °C/minutes, 3 °C/minutes, and 5 °C/minutes different from each other. At the rate of temperature rise of 1 °C/min, purity silicon by 1.78% because there are impurities Rubidium. But purity silica increased from 87.48% to 97.23%. At the rate of temperature rise of 3 ° C/minutes percentage of oxygen element very much, but there are still remains of the magnesium reduction process, and there are impurities rubidium and potassium. This causes the sample yet to form pure silicon, even a silica purity of the sample was reduced to 49.23%. At the rate of temperature rise of 5 °C/minute percentage of the silicon element more than the element oxygen that purity silica increased to 99.99%. However, because there are still remaining Magnesium and there are impurities in the form of Rubidium, purity silicon obtained only by 29.39%. Purity silica and silicon calculated using the percentage of atoms detected on FTIR instrument.

Table 1. EDX Analysis of Silica and silicon

Elements	Comparison Mg and SiO_2			
	(49:60)			(50:60)[a]
	1°C/minutes (%)	3°C/minutes (%)	5°C/minutes (%)	1°C/minutes (%)
Oxygen	63.37	73.31	42.98	26.09
Magnesium	-	8.84	4.72	-
Silicon	32.41	16.41	47.26	73.91
Rubidium	4.22	1.23	47.26	-
Potassium	-	0.22	5.04	-
Purity of silica	97.23	49.23	99.99	99.99
Purity of silicon	1.78	-	29.39	60.78

[a] Muzikarno [10]

Purity silicon produced in the study by Muzikarno [8] at the same rate of temperature rise is 5 °C / minutes compared to silicon in this study with a ratio of 49:60 has a smaller purity is 29.39%. Therefore, it can be concluded that the mass ratio of magnesium and silica used in this study was not a comparison that is optimum to produce high purity silicon. But with the variation of the rate of rise in temperature reduction can be seen that the greater the reduction in the rate of temperature rise, the purity of the silicon produced even greater. Although the rate of temperature increase rate of reduction used are not optimum for the reduction of temperature rise resulting silicon has not had a high purity. The purity of the silicon element rice husk ash depends on the content of the elements contained in rice husk and the geographical situation of the region of origin of rice husk. Similarly, the presence of rubidium, depending on the type of rice used. In this study, after the extraction process had appeared silica impurities in the form of Rubidium, so that the process of extracting the silicon looks still remaining rubidium. This is due to the use of rice husks contains rubidium. Silicon which has the fewest impurities or without impurities is a good silicon, fewer impurities, the purity of the silicon will be higher. EDX results on the three variations of this study indicate rubidium and potassium is a metal residual silica in the sample. These metals are naturally present in the husk.

a b. c

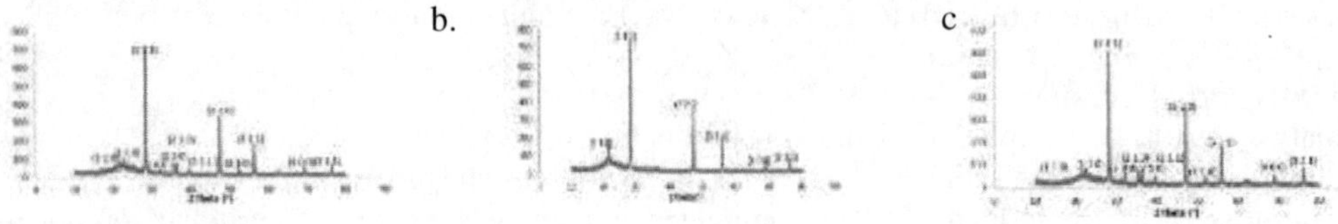

Figure 1. XRD pattern of (a) Silikon 1 °C/minutes (b) Silikon 3 °C/minutes (c) Silikon 5 °C/minutes

Figure 1 showed the x-ray diffraction pattern of silicon with a variation of 1 °C/minutes, 3 °C/minutes, and 5 °C/minutes. The third picture above showed the diffraction pattern with a peak of almost the same. The highlight appears indicating their compounds x-ray exposure when tested XRD, the peak of silica and silicon compounds. After compared with the data of JCPDS ICDD 1997, the diffraction pattern in Figure 1 (a) seen the peak of silicon at an angle of 28.5°, 47.56°, 56.4°, 69.58°, and 76.8°. But there is still a silica compound indicated by peaks at 22o. The same is shown in Figure 1 (c), the diffraction pattern showed peaks of silicon and silica. Silicon peak at 28.4°, 47.3°, 56.1°, 69.1°, 76.3°. Silica peaks shown by angle of 22°, 23°, 32.2°, 35.89°, 36.2°, 40°, 52.2°. Figure 8 (b) shows the X-ray diffraction pattern of the same, namely the silicon peak at an angle of 28.3°, 47.4°, 56.1°, 69.1° and 76.2°, while silica peaks at 21.9o, 23°, 32.5°, 36°, 36.5°, 39.8°, 52.2°. Third diffraction pattern shows that the peak of the silica is still there. This indicates silica or silicon dioxide does not dissolve in HCl and do not react with magnesium [1,2,11,12].

Peak compounds of silicon dioxide (SiO_2) is not dominant, but in the data EDX oxygen atoms exist. It is suspected that the oxygen atoms do not form a compound of silicon dioxide (SiO_2) as shown in Figure 1, but is thought to form a compound of water (H_2O), in which the hydrogen atom is not detected in the data EDX. Moreover, the existence of rubidium and potassium were detected in EDX the data support the possibility of the formation of other compounds that cause the many peaks that form the XRD spectra of figure 8 (b) and (c). To convince them, it is necessary FTIR analysis. Table 2 showed Lattice constant of silicon sample. That table conclude structure of silicon is cubic.

Table 2. Lattice constant of silicon

Silicon sample	Lattice parameters Å			Structure
	a	b	c	
1 °C/minutes	5.4416	5.4416	5.4416	Cubic
3 °C/ minutes	5.4467	5.4467	5.4467	Cubic
5 °C/ minutes	5.4416	5.4416	5.4416	Cubic

Infrared transmittance silicon pattern shown in Figure 2 and the functional groups were detected shown by Table 2. In Figure 2 third IR spectra have a similar shape. The spectra are formed from silica each treatment has a dominant peaks same. At the rate of temperature rise 1°C/minutes absorption band appearing at wave number 494 cm^{-1} indicate the presence of Si-O bending vibration of a siloxane (Si-O-Si) (20). Absorption at 791 cm-1 and an absorption band at 1095 cm^{-1} shows the asymmetric stretching vibration of Si-O-Si-O Si (21). Absorption around 3421 cm^{-1} to 3823 cm^{-1} shows the OH stretching vibration of Si-OH or air (20). Main absorption peak is in an absorption band in 1095 cm^{-1}, which is reinforced by the absorption band at 791 cm^{-1} indicating the asymmetric stretching vibration of Si-O siloxane groups (Si-O-Si) on a wave number of 1277.8 cm^{-1} with a spring constant of 977.23 N/m.

Similarly, the pattern of absorption at the FTIR spectra of silicon at a rate of 3 °C/ minutes and 5 °C/minutes has a peak identical to the main absorption at 1092 cm^{-1}. The second graph shows the cluster Si-O from the group siloxane (Si-O-Si) that is reinforced by the group Si-O in an absorption band 795 cm-1 for the rate of rise 3 °C/minutes, and the absorption band at 791 cm-1 for the rate of temperature rise of 5 °C/minutes. The presence of an absorption band at 791 cm-1 and 795 cm-1 indicates the asymmetric stretching vibration of Si-O and allow inharmonic oscillations that occur between the main absorption peak and an absorption band amplifier12-14. Where's the rate of temperature rise of 3 ° C/minutes had a constant vibration that occurs has a spring constant is 1003 N/m and for the rate of temperature rise of 5 °C/min was 982.3 N/m. Table 5 shows the wave number, frequency and spring constant of the vibration of Si-O an harmonic group of siloxane groups (Si-O-Si).

Table 3. Functional groups of an harmonic vibration

Silicon	Wavenumber (cm^{-1})	Functional groups
1 °C/minutes	463-493	Bending Si-O
	791	Stretching asymmetry Si-O (Si-O-Si)
	976	Si-O-Metal
	1095	Stretching asymmetry Si-O (Si-O-Si)
	1339	Stretching Si-C_6H_5
	3421-3541	Stretching –OH dari Si-OH
3 °C/minutes	471	Bending Si-O (Si-O-Si)
	795	Stretching asymmetry Si-O (S-O-Si)
	957	Si-O-Metal
	1092	Stretching asymmetry Si-O
	3259-3352	Stretching Si-OH
5 °C/minutes	471	Bending Si-O (Si-O-Si)
	791	Stretching asymmetry Si-O (S-O-Si)
	1092	Stretching asymmetry Si-O
	3379	Stretching Si-OH

Tabel 4. Constant spring of Si-O an harmonic groups

Silicon	υ_1 (cm^{-1})	υ_2 (cm^{-1})	Wavenumber ($\overline{\omega}_e$) (cm^{-1})	Frequency(f) (Hz)	Constant spring (N/m)
1 °C/ minutes	1095	791	1277.8	$3.83x10^{13}$	977.23
3 °C/ minutes	1092	795	1293	$3.88x10^{13}$	1003
5 °C/ minutes	1092	791	1281	$3.84x10^{13}$	982.3

IOP Conf. Series: Earth and Environmental Science **65** (2017) 012032 doi:10.1088/1755-1315/65/1/012032

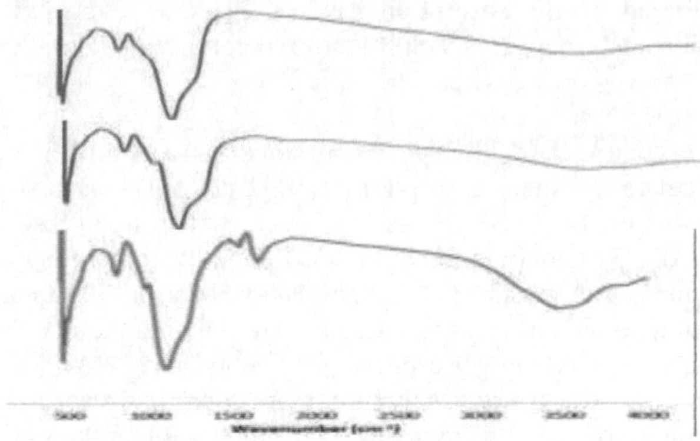

Figure 2. FTIR spectra of (a) Silicon with rate temperature 1 ^{o}C/minutes, Silicon with rate temperature 3 ^{o}C/minutes, 5 ^{o}C/minutes Silicon with rate temperature

Absorption peaks at wave number 3421 cm-1 to 3823 cm-1 showed absorption of Si-OH group and an OH group15. Absorption band group Si-OH and OH on the data FTIR supports the XRD data, where the samples are oxygen atoms that do not form compounds of silicon dioxide (SiO2), but form a compound of water (H2O), in which the hydrogen atoms are not detected in the data EDX but in FTIR visible absorption peaks OH group

4. Conclussions

The results were obtained silicon with a purity of 1.78% at a rate of temperature rise of 1 °C/minutes and 29.39% in the rate of temperature rise of 5 °C/minutes. While the rate of temperature rise of 3 °C/minutes has not been formed silicon. The results of XRD analysis with rate of temperature rise variations showed that XRD diffraction pattern which is not much different. Rate of temperature rise of 1 °C/minutes and 5 °C/minutes, the peak of silicon is more dominant than the peak of silica. Rate of temperature rise of 3 °C/minutes more dominant at the peak of silica. The data is the same as the EDX results showed that silicon is formed at a rate of temperature rise of 1 °C/minutes and 5 °C/minutes, while the rate of rise of 3 1 °C/minutes has not been formed is still in the form of silicon or silica. In addition, all three variations have a cubic structure. FTIR analysis results of three samples tested showed patterns of uptake that is not much different. The FTIR test results prove that the silicon produced in the sample still contained water compounds are not detected by the test EDX and XRD.

References

[1] Beiser A 1992 *Konsep Fisika Modern* 4th Edition (Indonesia : Erlangga)
[2] Lau WS 1999 *Karakterisasi inframerah untuk mikroelektronik* World Scientific
[3] Rohaeti E 1992 *Pemisahan Silikon Dioksida (Sio$_2$) dari Sekam Padi dan Karakterisasinya, Thesis* (Indonesia: Institute Technology of Bandung)
[4] Irzaman, Alatas H, Darmasetiawan H, Yani A and Musiran 2007 Tungku Sekam Padi Sebagai Energi Alternatif dalam Peningkatan Kesejahteraan Masyarakat (Kajian Ekonomi dan Finansial Tungku Sekam Padi: Skala Rumah Tangga) Report of Kegiatan Pengembangan IPTEK (Indonesia)
[5] Harsono H 2002 Pembuatan silikon dioksida amorf dari limbah sekam padi *Jurnal Ilmu Dasar* **2** 98-103

[6] Nayak J P and Bera J 2010 Effect of sintering temperature on mechanical behaviour and bioactivity of sol-gel synthesized bioglass-ceramics using rice husk ash as a silica source *J. Appl. Surface Science* **257** 458-462

[7] Ikram N and Akhter M 1998 X-Ray diffraction analysis of silicon prepared from rice husk white ash *J. Materials Science* **23** 2379-2381

[8] Hikmawati 2010 *Produksi Bahan Semikonduktor Silikon dari Silikon diDioksida Limbah Arang Sekam Padi sebagai Alternatif Sumber Silikon, Thesis* (Indonesia : Bogor Agricultural University)

[9] Larbi K K 2010 *Synthesis of High Purity Silikon From Rice Husks,* Thesis Departement of Materials Sicence and Engineering (Toronto: University of Toronto) 1-116

[10] Muzikarno O, Irzaman and Rohaeti E 2013 *Addition of Magnesium in Excess Produce Silicon of Pure Rice Husk as Semiconductor Materials, Thesis* (Bogor: Bogor Agricultural University)

[11] Vlack, Lawrence H and Van 2004 *Elemen-elemen Ilmu dan Rekayasa Material* 6th *Edition* (Jakarta: Erlangga)

[12] Ahmad L 2012 *Uji Struktur dan Sifat Listrik Silikon Dioksida dan Silikon dari Sekam Padi, Thesis* (Indonesia : Bogor Agricultural University)

Palm oil anionic surfactants based emulsion breaker (Case study of emulsions breaker at Semanggi Field production wells)

Muhpidah[1], E Hambali[2,3], A Suryani[2,3], I A Kartika[2]

[1]Post Graduate Program of Agroindustrial Technology, Bogor Agricultural University, Bogor, Indonesia
[2]Department of Agroindustrial Technology, Faculty of Agricultural Engineering and Technology, Bogor Agricultural University, Bogor, Indonesia
[3] Surfactant and Bioenergy Research Center (SBRC), Bogor, Indonesia

Email : muhpidah@gmail.com

Abstract. The presence of emulsion in oil production process is undesirable. The emulsion will increase the production costs, transportation and costs related to emulsion separation process between water and oil. The development of palm oil-based surfactant as an emulsion breaker needs to be conducted given the availability of abundant raw materials in Indonesia and as an alternative to petroleum-based surfactant. The purpose of this study is to produce palm oil-based emulsion breaker, assessing the effect of additive application to the emulsion breaker and analyze the performance of the emulsion breaker. This research was conducted by formulating palm oil anionic surfactant in water formation with the addition of co-surfactant additive and co-solvent. Palm oil anionic surfactant-based emulsion breaker with 0.5% concentration in water can reduce 50% of emulsions with the interfacial tension (IFT) of 2.33×10^{-2} dyne/cm. The addition of co-solvent (toluene: xylene) is able to remove the emulsion formed with a lower IFT namely 10^{-3} dyne / cm. The resulting emulsion breaker is capable to remove the emulsion between water and oil. The performance test of emulsion breaker show that the emulsion is able to maintain its performance at reservoir temperature with no indicate of plugging and the value generated incremental oil recovery values is 13%.

1. Introduction

Semanggi field is one of active oil fields in Indonesia, located in Cepu, Central Java. Semanggi field reservoir temperature is 50 °C and composed of Sandstone rock formations. The main problem of the petroleum industry especially at Semanggi field is the decrease in production caused by several factors, among other the old age of oil fields, formation damage and emulsions formation between oil and water formation. The emulsion of oil and water will cause a decrease in productivity. The emulsion would lead to increase the costs and production time as well as low oil production. Problems of productivity decline and emulsion formation at Semanggi field has been addressed by oil wells stimulation. Oil wells stimulation is injection of a certain amount of chemical compounds on the production wells to increase the productivity of produced oil. Oil well stimulation consists of three stages, namely reinjection, soaking and reproduction of the same oil wells. The injection process is expected able to increase the oil production. The process of stimulation in this study was conducted by injecting the emulsion breaker into production wells, a mixture of chemical compounds that can break the emulsion formed between oil and water formations.

Content from this work may be used under the terms of the Creative Commons Attribution 3.0 licence. Any further distribution of this work must maintain attribution to the author(s) and the title of the work, journal citation and DOI.
Published under licence by IOP Publishing Ltd

The application of emulsion breaker in the petroleum industry has been developed. The applied emulsion breaker is include TOMAC (trioctyl methyl ammonium chloride), trietanilamin, polyethylene glycol, and fatty acid ethoxylate with effectiveness reaches 49-55% in emulsion separation [1]. Some of the emulsion is petroleum-based surfactant. Such surfactants are not environmental friendly, having limited raw materials at a high price. The development of palm oil-based surfactants can be used as a solution for petroleum-based surfactants as emulsion breaker formula in oil well stimulation process. High oil production in Indonesia led to the abundant availability of raw materials, where the palm oil plantations in 2015 are 11 million ha with a total production of 31.3 million tonnes [2]. Due to the reasons, palm oil-based surfactant is necessary to be developed.

In this study, co-surfactant and co-solvent was used as additives in the emulsion breaker. Addition of additive was conducted to improve the emulsion breaker effectiveness to break the emulsion formed between oil and water formations. The selection of non-ionic co-surfactants is because of its use as an emulsion breaker in petroleum industry and the selection of aromatic hydrocarbons co-solvent is due to the ability in destabilizing rigid layer as barrier between the oil and water formations. The produce of emulsion breaker must be able to show a good performance as indicated by thermal resistance, filtration ratio value and oil recovery in the performance test of emulsion breaker formula.

2. Materials and Methods

2.1 Materials and Tools

Materials used in this research is oil and formation water from Semanggi oil field (SMG-P12), palm oil anionic surfactant, nonionic surfactant (APG and ethoxylate), Whatman 41 filter paper (20-25 μm), stainless steel strainer 20 μm, core Berea 100 md, xylene, toluene and chemicals materials for laboratory analysis. The tools used in this study are the analytical balance, hot plate stirrer, rotary oven, microscopes, pH meters and centrifuges.

2.2 Selection of Surfactant Concentration

Formulation process of emulsion breaker was started by dissolving the amount of palm oil anionic surfactant in the interval of 0.5, 1, and 1.5% into formation water as solvent media. Surfactant solution is stirred with a speed of 100-200 rpm at room temperature for 30 minutes. The best concentration was determined with compatibility test (pH, density, droplet size, absorbance and transmission), IFT measurement and observation of emulsion formation (at interval of 5 and 10 minutes). The concentration of surfactant that completely dissolves, have the lowest IFT values, and able to eliminate the emulsion formed was selected as the best surfactant concentration to dissolve in water formation.

2.3 Selection of Additives

Additives mixing into surfactant formulation are intended to eliminate the emulsions formed in crude oil. This study used two types of additives namely co-surfactant and co-solvent. Determining of the best additive is based on IFT measurement and observation of the emulsion formation after the additives addition.

2.4 Performance tests

Performance tests were conducted to obtain the best emulsion breaker performance from the mixing of anionic surfactant with additives. The resulting performance test will illustrate the effectiveness of the emulsion breaker before implemented in Semanggi Field. Performance test of emulsion breaker formula is includes thermal stability test, filtration test and core flood test.

3. Result

3.1 Surfactant Concentration

The first stage to produce emulsion breaker is the selection of surfactant concentration to get the best surfactant concentration as emulsion breaker. Surfactants used in this study are anionic surfactant. The best surfactant concentration was selected based on the compatibility test, IFT measurement, and observation of the percentage of the emulsion.

3.1.1 Compatibility test

Compatibility test is performed to determine compatibility between the surfactant solution with formation water from the reservoir. The results of compatibility test are presented in Table 1.

Table 1. Compatibility test of anionic surfactant in formation water

Surfactant Concentration (%)	pH	Density (g/cm3)	*Droplet* Size (μm)	Absorbance	Transmission (%T)
0.5	8.6±0.01	1.00±4.00E-05	1.61 ± 0.28	0.85 ± 0.01	14.3 ± 0.35
1	8.6±0.01	1.00±5.00E-05	2.36 ± 0.62	1.35 ± 0.07	4.5 ± 0.71
1.5	8.5±0.01	1.00±4.00E-05	3.55 ± 0.13	2.28 ± 0.05	0.54 ± 0.05

Results of compatibility test showed that the tested surfactant concentration is compatible to water formation. It is characterized by perfect dissolved in formation water and there is no sediment or lumps in the surfactant solution before and after heating at a reservoir temperature of 50°C. The compatibility test is also showed that anionic surfactant concentration of 0.5% is the best concentration for emulsion breaker formula. Use of anionic surfactant 0.5% was generating alkaline solution, low density, and very small droplet size that has good solubility in formation water. Absorbance and transmission measurements is also show that the concentration of anionic surfactant 0.5% generating clear and not cloudy solution as indicated by low absorbance values and high transmission values. A surfactant solution that clear, compatible and does not produce sediment or floating material is a solution that can be used in oil wells stimulation. In a compatibility test, the more compatible surfactant solution produced, the more effective to be used in oil wells and the decline in the IFT [3].

3.1.2 Interfacial Tension (IFT)

Results of IFT testing indicate that increasing of anionic surfactant concentration greater than 0.5% will increase the IFT produced. Increasing of the surfactant concentration shows that surfactant concentrations 0.5% are the concentration of micelle formed and known as CMC (critical micelle concentration). When CMC condition is reached, IFT measurements will stable or increase. This is because the addition of the surfactant more than 0.5% is no longer able to improve the interface performance. Surfactant has been saturated and prefer to form micelle at water phase that influence on IFT [4].

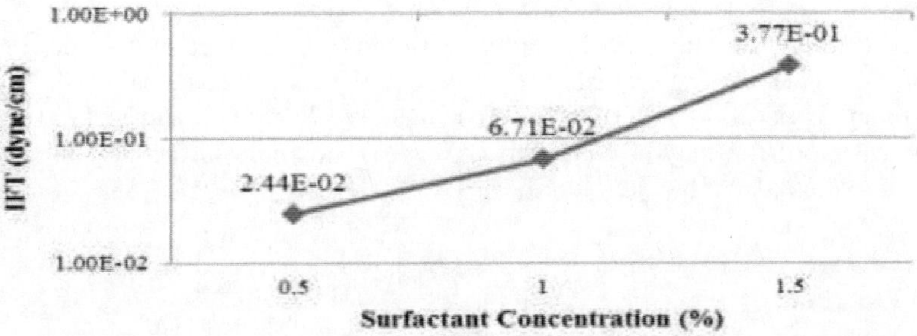

Figure 1. The Effect of anionic surfactant concentration on IFT

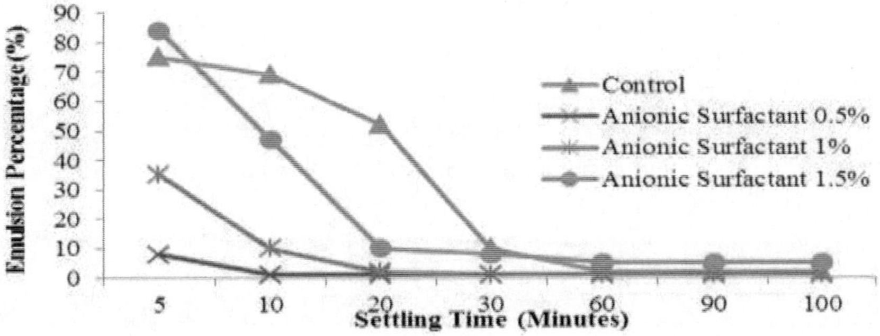

Figure 2. Effect of settling time on the emulsion percentage

Observations of emulsion on the resulting formulation solution was performed at reservoir temperature 50°C. Observation of the emulsion formed (Figure 2) shows that the control (oil and formation water) produce an emulsion of 0.2 mL or 2% after storage at rotary oven. The addition of anionic surfactant by 0.5 and 1% in formation water indicates the emulsion percentage 1% and 5% with the addition of surfactant 1.5%. Anionic and nonionic surfactants can be used as demulsifier [5]. Surfactants that dissolved into formation water when in contact with oil, slowly will destabilize the emulsion between oil and water, which leads to changes in the free energy around the oil surface and surfactant solutions and in turn will reduce rigidity of the film on the surface of oil and water emulsion.

The results of compatibility test, IFT measurement and observation of the emulsion showed that surfactant concentrations 0.5% is the best surfactant concentrations to emulsion breaker. The concentration of 0.5% is compatible to water formation. Measurement of IFT showed the lowest measurement and observation of the emulsion showed that in just 5 minutes was able to produce the lowest percentage of emulsion. Under these conditions, surfactant concentrations 0.5% serves as the best concentration of palm oil anionic surfactants based emulsion breaker.

3.2 Selection of Additive

Additives are chemical compounds that added to emulsion breaker solution. The additives used in this study are co-surfactant and co-solvent. The addition of additives to emulsion breaker solution is intended to eliminate the emulsion. The first stage is begins with co-surfactant addition and then followed by co-solvent when co-surfactant addition does not show better results. Selection of the best additives is based on IFT measurement and emulsion observations.

3.2.1 Co-surfactant

Interfacial tension (IFT) observation shows that the addition of co-surfactant into the emulsion breaker solution does not provide the better results (Figure 3). IFT is increased after the addition of co-surfactant either ethoxylate or APG surfactants namely $7.48x10^{-2}$ and $4.17x10^{-2}$ dyne / cm, respectively. Increasing of IFT may be caused by the type of nonionic surfactant added was not able to improve detergency of emulsion breaker.

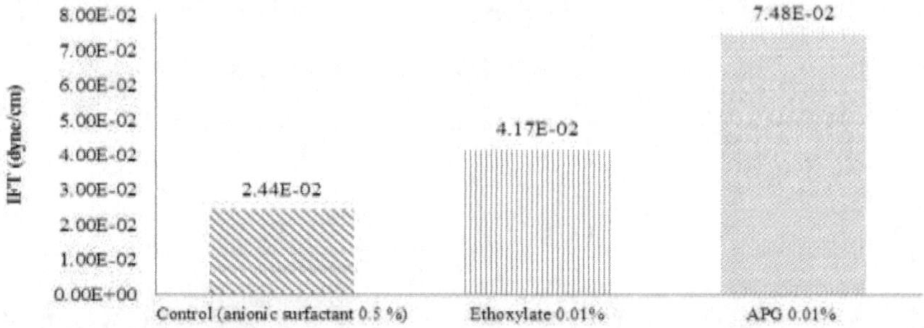

Figure 3. The Effect of surfactant formulation 0.5% and co-surfactant additive on IFT

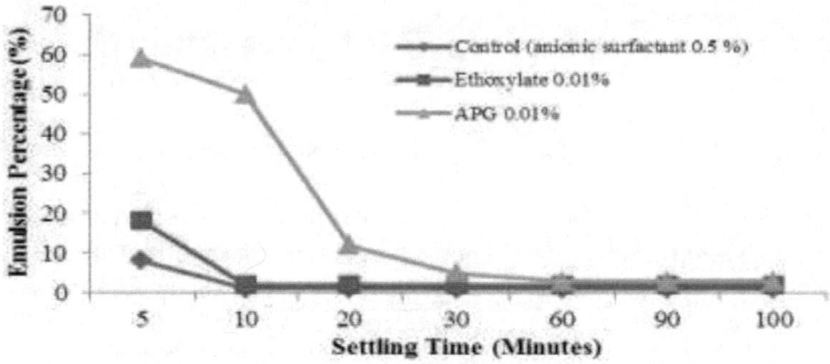

Figure 4. Effect of settling time on the emulsion percentage

Observation of the emulsion formation (Figure 4) shows that the addition of APG and ethoxylate co-surfactant has no significant effect compared to anionic surfactants. IFT measurement also does not give better results with the addition of the co-surfactant. The use of ethoxylate surfactants is produce quite low of interfacial tension values and emulsions by 3%. The use of APG provides emulsion observations 2% but does not produce low IFT. It can be concluded that the addition of co-surfactant does not increase the effectiveness of the resulting emulsion breaker.

3.2.2 Co-Solvent

The use of co-solvent is expected to provide better results than co-surfactants. Co-solvent used is a mixture of toluene and xylene with several different mixing ratios. Fan *et al.* [4] explains that demulsification process with surfactants can be improved the effectiveness with the use of polymer and solvent mixtures. Type of aromatic solvents such as benzene and toluene can be used in the preparation of demulsifier process [6].

IFT measurements (Figure 5) on the emulsion breaker with the addition of co-solvent showed different results compared to the use of co-surfactants. IFT measurements were conducted to the whole of solution mixing of emulsion breaker. The results of IFT measurement show a value of 10^{-3} dyne/cm, which is lower than the use of anionic surfactant alone. The measurement results showed a tendency of

IFT increased along with increased concentrations of co-solvent in various ratios of toluene: xylene. The lowest IFT was produced at the co-solvent concentration of 0.1% with a ratio of toluene: xylene 1: 3. IFT value generated at $1.33x10^{-3}$ dyne / cm.

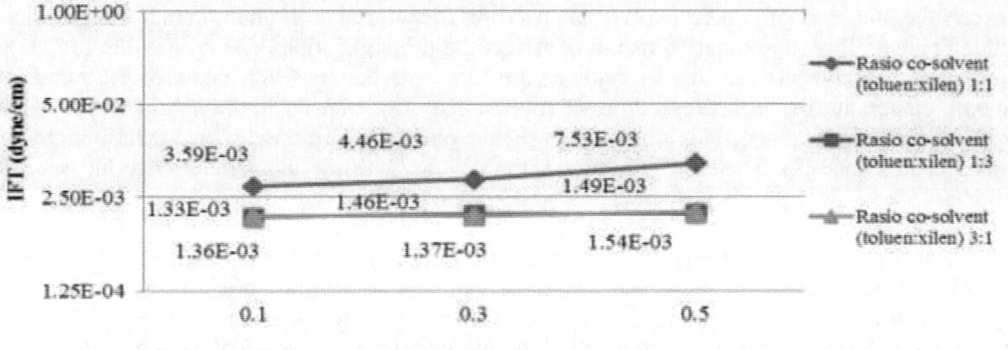

Figure 5. The effect of co-*solvent* concentration at different ratio of toluene: xylene on IFT

The results of IFT measurement showed that the addition of co-solvent gives lower IFT, especially on the use of co-solvent by a ratio of toluene: xylene 1:3 and 3 : 1. But if the three co-solvent ratio compared, the ratio of 1: 1 has high IFT value compared to the two co-solvent ratio. The value generated by the ratio of 1:1 is quite different namely 10^{-2} dyne/cm, but the resulting value is still in the scale of 10^{-3} dyne/cm, which is still within the limits of applicable surfactant solution for injection process, namely $\leq 10^{-2}$ dyne/cm [7]. In addition, the use of co-solvent capable to lowering the IFT compared to surfactant alone. The decline of IFT occur due to the co-solvent is able to completely dissolved in the emulsion breaker that consisting of anionic surfactant and formation water.

Emulsion breaker formula with the addition of co-solvent is able to lower the IFT. This is because the co-solvent disrupts asphalthene aggregation that form films on the surface of emulsion droplet through interactions with asphalthene aggregate polar groups. When asphalthene aggregation disrupted, the emulsion stability will be disrupted because the film gradually thins and breaks. The film break causes the emulsion breaker adsorbed on the surface of oil and water that change the IFT. Adsorption of emulsion breaker on the surface of the water and oil will affect the free energy in the interface area that decreases IFT [8]. Interface free energy or interfacial tension is the minimum amount of work required to expand the area of the interface.

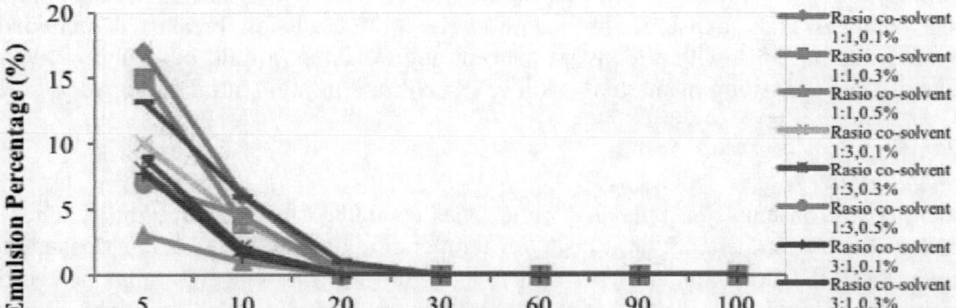

Figure 6. The effet of settling time on emulsion percentage at *co-solvent* addition with different *solvent* ratio

Observations (Figure 6) show that the emulsion percentage 0% can be produced within 10 minutes, especially in solvent concentrations of 0.1 and 0.3% of various solvent ratios. The use of solvent proved to be effective in removing the emulsion formed between oil and water. The longer settling time, the better separation of oil and water. Although the first minute observations will produce a fairly high percentage, but gradually there is perfect separation of water and oil and no emulsion produced until 100 minutes. This shows that temperature affects the destabilization of emulsion, the longer an emulsion system in contact with the reservoir temperature, the better effectiveness of the emulsion separation. Temperatures cause increased brown motion and mass transfer in the interface area of the two droplets of dispersed phase. This situation causes the momentum of the dispersed phase increases and causes coalescence [9]. Therefore, the longer the contact with the temperatures produces better separation.

The effectiveness of co-solvent addition to break the emulsion is due to the use of aromatic hydrocarbon solvents. As it was explained earlier, asphalthene as natural emulsifier is the cause of emulsions formation. According to earlier studies, it is known that asphalthene increasing to dissolved in line with the increase in aromatic compounds. The use of aromatic co-solvent can generate a good emulsion separation. Aromatic solvent mixture will be able to separate the emulsion formed by disturbing π-π bond and polar bond between asphalthene monomer [10]. The presence of surfactant in emulsion breaker will solved asphalthene monomer so that it will remain dissolved in the oil and does not form aggregates which produce the film. This is the cause of the use of emulsion breaker with the addition of co-solvent is more effective than just anionic surfactants and the addition of co-solvent.

Observations on the emulsion percentage showed that at various ratios and concentrations of co-solvent is able to eliminating the emulsion. While at IFT measurement results it is concluded that there are two co-solvent ratio that showed the lowest IFT value which are 1:3 and 3:1 at consentration of 0.1%. The best concentration based on Duncan test is 0.5% therefore, the concentration that will be choose as the best concentration is 0.1%. The concentration of 0.1% resulted in the separation of emulsions as well as a concentration of 0.5% and the fewer of co-solvent amount will reduce production cost. For the co-solvent ration that will be use as the best ratio is 1:3. Ratio 1:3 showed the lowest IFT value and ratio of toluene that had been using is very low, compared with the other ratio. Using of solvent with a low flash point began to reduce in the field. This is because the solvent with a low flash point is flammable, thus would endanger and disrupt oil production process. Increasing the temperature of the solvent with a low flash point will cause changes into vapor and eventually reach the point of the flame.

3.3 Performance Test

Performance test is conducted to determine the most effective of emulsion breaker acording ot IFT variance analysis and emulsion observations. The best emulsion breaker is obtained from surfactant concentration 0.5% with co-solvent concentration 0.1% at a ratio of toluene : xylene 1:3. Performance tests are consisting of thermal stability test, filtration ratio (filtration test) and core flood test.

3.3.1 Thermal Stability Test

Thermal stability test on emulsion breaker formula was conducted to determine the changes on the formula produced at hot temperature or at reservoir temperature based on IFT value. Measurements to the fifth day produce IFT in the range of 10^{-3} dyne/cm (Figure 7). Although IFT value increased at the seventh day but the values is still $\leq 10^{-2}$ dyne/cm. Measurement of thermal stability is closely related to the soaking time that will be applied in reinjection process of emulsion breakers into oil wells. Based on the results of thermal stability test, it can be concluded that surfactant formula is still able to work well in reservoir temperature until the seventh day.

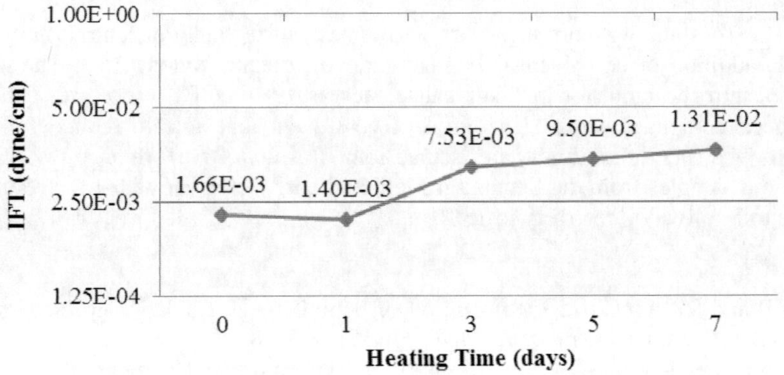

Figure 7. The Effect of Heating Time on *emulsion breaker* IFT

3.3.2 Filtration Ratio

Filtration ratio measurement is intended to determine the suitability of emulsion breaker when applied. Filtration ratio aims to determine the indication of plugging from the emulsion breaker seen from the filtration ratio value that less than 1.2. The emulsion breaker solution was passed through a filter that has a pore size of 1.2 μm that representing pores of reservoir rocks. The filtration ratio value is determined by the calculation of the volume of emulsion breaker solution based on a certain time.

Filtration test showed that the filtration ratio of the emulsion breaker is 1.04 (≤ 1.2). The result showed that the emulsion breaker capable of passing through a filter. The solution that can pass through the filter and has filtration ratio value less than 1.2 indicate that the there is no possibility of the injected solution could be the cause of plugging. Plugging in the oil field is undesirable because it affect the permeability and porosity of the native core.

3.3.3 Core Flood Test

Core flood test was conducted to determine the extent of oil recovery that can be produced through emulsion breaker injection when applied in oil field. The calculation of oil recovery value on emulsion breaker injection was done after soaking for 12 hours. Injection of emulsion breaker is able to give a recovery of 13% (Figure 8). This proves that surfactant injection was able to increase the oil recovery that produce 43.6% of total oil production in the reservoir.

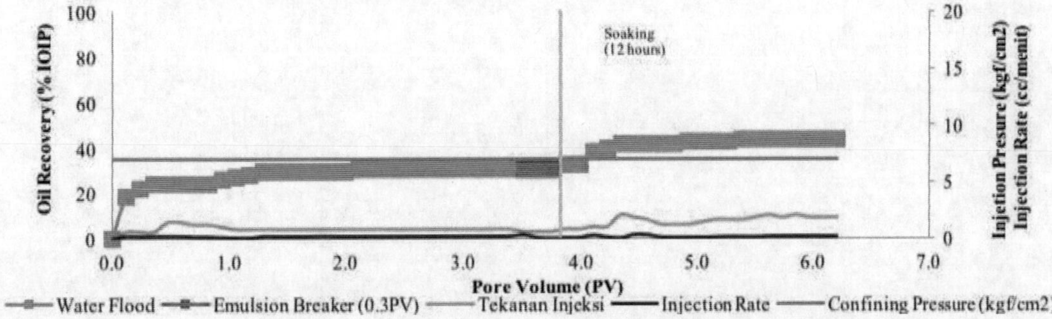

Figure 8. Incremental oil recovery with emulsion breaker

4. Conclusions

Emulsion breaker formula was produced from the best surfactant concentration namely anionic surfactant 0.5%, addition of co-solvent 0.1% and ratio of toluene: xylene 1: 3. The addition of co-solvent additives into the emulsion breaker cause increase in the IFT value and the percentage of emulsion produced compared with the addition of co-solvent that able to reduce IFT value to 10^{-3} dyne/cm and produces perfect oil and water separation. Results of performance test (core flood test) using oil well fluid samples from the Semanggi Field and core Berea show that the resulting emulsion breaker capable to improve oil recovery to 13%.

Reference

[1] Hajivand P and Vazir A 2015 Optimization of demulsifier formulation for separation of water from crude oil emulsions. *Braz. J. Chem. Eng*. **32** 107-18

[2] Directorate General of Estate Crops Ministry of Agriculture 2016 *Tree crop estate statistics of indonesia 2015-2016 palm oil* (Jakarta: Directorate General of Estate Crops)

[3] Hambali E, Suryani A, and Rivai M 2012 *Teknologi Surfaktan dan Aplikasinya.* (Indonesia: IPB Press)

[4] Fan Y, Simon S and Sjöblom J. 2009. Chemical destabilization of crude oil emulsions: effect of noniionic surfactant as emulsion inhibitor. *Energy Fuels* **23** 4575–83

[5] Ojinnaka C M, Ajienka J A, Abayeh O J, Osuji L C and Duru RU 2016 Formulation of best-fit hydrophile/lipophile balance-dielectric permittivity demulsifiers for treatment of crude oil emulsions *Egyp.J.Pet*. **25** 565–74

[6] Abdel-Azim AA-A, Abdul-Raheim A A M, Kamel K R and Abdel-Raouf M E 2011 Demulsifier systems applied to breakdown petroleum sludge. *J Petrol Sci Eng*. **78** 364–70

[7] Anderson D R, Bidner M S, Davis H T, Manning C D and Scriven L E 1976 Interfacial tension and phase behavior in surfactant-brine-oil systems *Soc.Pet.Eng*. 189-200.

[8] Al-Sabagh A M, Nasser N M and Khamis E A 2012 Demulsification of crude oil emulsions using ethoxylated aliphatic amine polyesters as novel demulsifiers. *Int.J.Sci.Res*. **3** 1171-83

[9] Hirasaki G J, Miller C A, Raney O G, Poindexter M K, Nguyen D T and Hera J 2010 Separation on produced emulsions from surfactant EOR processes *Energ Fuels.* Presented at the *11th Int. Conf. on Petroleum Phase* (2010)

[10] Sullivan A P and Kilpatrick P K 2012 The effects of inorganic solid particles on water and crude oil emulsion stability *Ind Eng Chem Res.* **14** 3389-3404

IOP Conf. Series: Earth and Environmental Science **65** (2017) 012034 doi:10.1088/1755-1315/65/1/012034

Palm oil based surfactant products for petroleum industry

P Permadi[1,2], R Fitria [2] and E Hambali[2,3]

1Petroleum Engineering Program, Bandung Institute of Technology, Bandung, west Java, Indonesia
2 Surfactant and Bioenergy Research Center (SBRC), Bogor Agricultural University, Bogor, West Java, Indonesia
3 Department of Agroindustrial Technology, Faculty of Agricultural Technology, Bogor Agricultural University, Bogor, West Java, Indonesia

Email: pudji@tm.itb.ac.id

Abstract. In petroleum production process, many problems causing reduced production are found. These include limited oil recovery, wax deposit, asphaltene deposit, sludge deposit, and emulsion problem. Petroleum-based surfactant has been used to overcome these problems. Therefore, innovation to solve these problems using surfactant containing natural materials deserves to be developed. Palm oil-based surfactant is one of the potential alternatives for this. Various types of derivative products of palm oil-based surfactant have been developed by SBRC IPB to be used in handling problems including surfactant flooding, well stimulation, asphaltene dissolver, well cleaning, and wax removal found in oil and gas industry.

1. Introduction

Many problems causing reduced productivity and environmental pollutions were found in petroleum production process. These problems include limited oil recovery, wax deposit, asphaltene deposit, sludge deposit, emulsion formation, and water and land pollution as a result of spilled oil residue. Efforts need to be done to overcome these problems as the demand for oil is increasing and there is no other source of energy can replace it yet. Pollution handling is an important thing to do as environmental pollution endangers living organisms in land and sea.

Chemical injection is used to overcome several problems in oil industry. The chemical used is usually petroleum-based surfactant. It is known that surfactant plays a role in reducing interfacial tension between fluid and fluid, fluid and rocks, and fluid and hydrocarbon to increase the permeability of formation in an effort to increase production rate. In addition, surfactant can break surface tension of oil emusion bonded to the rocks (emulsion block), reduce the occurence of water blocking, and alter the wettability of rocks from oil-wet to water-wet. The common surfactant used is petroleum-based surfactant. Therefore, innovation to solve these problems using surfactant containing agricultural materials deserves to be developed.

Content from this work may be used under the terms of the Creative Commons Attribution 3.0 licence. Any further distribution of this work must maintain attribution to the author(s) and the title of the work, journal citation and DOI.
Published under licence by IOP Publishing Ltd

One of the innovations to be developed is oil palm-based surfactant. Total crude palm oil (CPO) production in Indonesia in 2015 was 31.3 million tons. In 2020, palm oil production is estimated to reach 40 million tons. With high CPO production in Indonesia, it will be important to further develop oil palm derivative products to increase the added value of palm oil and at the same time overcome problems in oil industry. This way, oil palm plays a role in the improvement of national oil productivity.

2. The Role of Surfactant for Oil and Gas Industry

Surfactant or surface active agent is a molecule which has hydrophilic and hydrophobic groups in one structure. Surfactant reduces interfacial tension between fluid and fluid, fluid and rocks, and fluid and hydrocarbon to increase the permeability of formation in an effort to increase production rate. In addition, surfactant can break surface tension of oil emusion bonded to the rocks (emulsion block), reduce the occurence of water blocking, and alter the wettability of rocks from oil-wet to water-wet. The common surfactant used is petroleum-based surfactant. The role of surfactant in the improvement of production rate is depicted in figure 1.

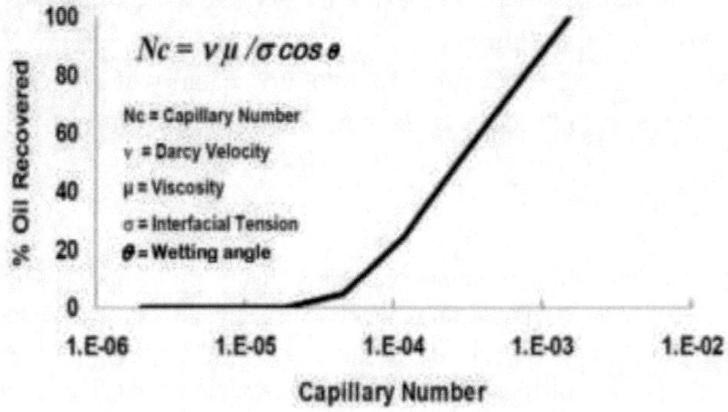

Figure 1. Capillary number versus oil recovery [1]

There are two dominant forces working in the pores of oil reservoir rocks, namely capillary ($\sigma \times \cos \theta$) and viscous ($v \times \mu$) forces. The ratio between these two forces is called Capillary Number as depicted in Figure 1 which shows the oil that can be pushed out (% oil recovered). Therefore, % oil recovered indicates the percentage of residual oil which can be taken out from the pores after water injection. In general, the Capillary Number found in primary production and Water Injection process lies in the range of $10^{-7} – 10^{-5}$ (dimensionless). This means that there is no waterflood Sor (residual oil saturation) which can be pushed by normal water. Therefore, surfactant is used to reduce capillary forces and this means increasing capillary number effectively so that the process become operationally feasible and more economical.

Increasing Capillary Number can also be done by altering wettability (namely contact angle, θ). If contact angle θ can be altered to 90º (and this is Neutral wettability, not water-wet), then $\mathrm{Cos}(\theta) = \mathrm{Cos}(90^o) = 0$ and capillary forces becomes null (0) so that Capillary Number becomes infinite. As a result, residual oil after waterflood can be pushed perfectly up to 100%. It is expected that, in addtion to its function of reducing the IFT of oil-water, surfactant can alter wettability to a netral direction.

If surfactant solution can form microemulsion with particle size smaller than the size of the smallest pores of porous rocks, then the surfactant solution will most probably be able to to push residual oil out as much as possible. Microemulsion can be formed when the IFT is very small ($\sigma < 10^{-2}$ dyne/cm) and surfactant solution has a tendency to dissolve in oil. In a miscible displacement mechanism (IFT = 0, where the pushing fluid is soluble in the pushed fluid), surfactant is able to push 100% Sor (residual oil saturation) out from inside the pores.

Palm oil-based surfactant is expected to be able to substitute petroleum-based surfactant. Palm oil based-surfactant developed by SBRC IPB is biodegradable as it is made from plam oil which is abundant in Indonesia so that its sustainability is ensured. Palm oil-based surfactant is more resistant to Ca^{2+} and Mg^{2+} ions and salinity. It has an IFT value $\leq 10^{-03}$ dyne/cm and adsorption value < 400 μg/g core. It is stable at reservoir temperature for 3 months, produced at Phase III / II(-), and it has a Filtration Ratio ≤ 1.2.

3. Palm Oil-based Surfactant for Various Purposes

Based on the its ionic charge of hydrophilic groups after it is dissociated in liquid medium, surfactant is classified into four types, namely (1) anionic; its hydrophilic group contains negative charge, (2) cationic; its hydrophilic group contains positive charge, (3) non ionic; its hydrophilic group contains almost no charge, and (4) amphoteric; molecules in its hydrophilic group contains positive or negative charge depending on the medium pH [2]. All types of surfactants can be synthesized from palm oil because of its physicochemical properties and fatty acid composition as shown in table 1 and 2 fit the requirements.

Table 1. Physicochemical properties of CPO [3]

Physical and chemical properties	Value
Triglyceride	95%
Free fatty acids	2-5%
Color (5 ¼ " Lovibond Cell)	Red orange
Water and filth content	0.15-3.0 %
Peroxide number	1 – 5.0 (meq/kg)
Anisidine value	2-6 (meq/kg)
β carotene content	500-700 ppm
Phosphorus number	10-20 ppm
Iron (Fe) content	4-10 ppm
Tocopherol level	600-1000 ppm
Diglyceride	2-6%
Acid number	6.9 mg KOH/g minyak
Saponification number	224 – 249 mg KOH/g minyak
Iodium number (wijs)	44-54 mg iod/g minyak
Melting point	21-24 °C
Refraction index (40 °C)	36.0-37.5

Some palm oil-based surfactants developed by SBRC IPB and had potential for applications in oil industry in Indonesia include glycerol ester, diethanolamide, alkyl polyglycoside, and methyl ester sulphonate. Various derivative products of palm oil-based surfactant, whether in their formulation process, surfactant ia used as a single component or there is a another additive substance added in, are potential to be used as a substitute for petroleum-based surfactant. This surfactant can be used in many applications including chemical flooding, stimulation, asphaltene dissolver, wax dissolver, emulsion breaker, sludge breaker, well cleaning, oil spill dispersant, and enhanced bioremediation.

Table 2. Fatty acid composition of palm oil

Fatty Acids	PKO [4]	CPO	RBDPO [5]	PFAD [3]	Olein [6]	Stearin [6]
Lauric (C12:0)	40 – 52	< 1.2	0.12 – 0.28	0.1 – 0.3	0.1 – 0.5	0.1 – 0.6
Myristic (C14:0)	14 – 18	0.5 – 5.9	0.87-1.19	0.9 – 1.5	0.9 – 1.4	1.1 – 1.9
Palmitic (C16:0)	7 – 9	32 – 59	42.46 – 48.54	42.9 – 51.0	37.9 – 41.7	47.2 – 73.8
Palmitoleic(C16:1)	0.1 – 1	< 0.6	0.13 – 0.16	-	0.1 – 0.4	0.05 – 0.2
Stearic (18:0)	1 – 3	1.5 – 8	4.10 – 4.75	4.1 – 4.9	4.0 – 4.8	4.4 – 5.6
Oleic (18:1)	11 – 19	27 – 52	35.23 – 41.67	32.8-39.8	40.7 – 43.9	15.6 – 37.0
Linoleic (C18:2)	0.5 – 2	5.0 – 14	7.74 –11.75	8.6-11.3	10.4 – 13.4	3.2 – 9.8
Linolenic (C18:3)	-	< 1.5	0.06 – 0.24	-	0.1 – 0.6	0.1 – 0.6
Arachidic (C20:0)	-	-	0.27 – 0.44	-	0.2 – 0.5	0.1 – 0.6

3.1 Surfactant for chemical flooding

Chemical flooding is a method used to obtain more oil after the reduction of primary production process (natural) by using natural energy originating from the reservoir itself (natural reservoir drive) [7]. Chemical flooding is obtaining oil by injecting materials from outside the reservoir [8]. Surfactant to be used for chemical flooding should have the following properties: pH 7-8, IFT value ≤ 10^{-3} dyne/cm, adsorption value < 400 μg/g core. It should also be stable for > 3 months at thermal stability test (± 10^{-3} dyne/cm), produced at Phase III / II(-), have a filtration ratio ≤ 1.2, and have incremental oil recovery in a coreflooding test of about 15.0-20.0 % IOIP.

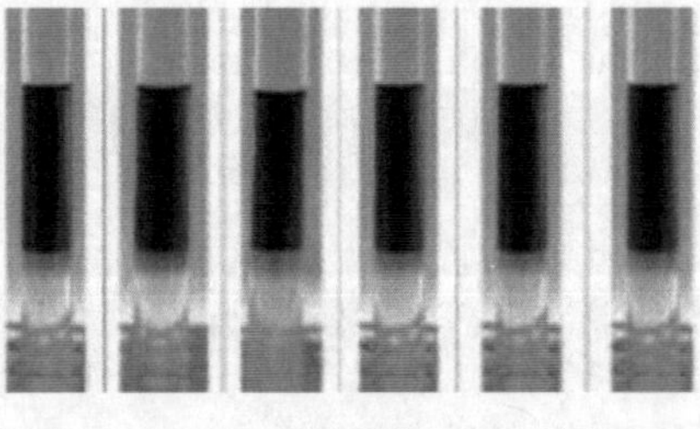

Figure 2. Formation of microemulsion phase at 60°C

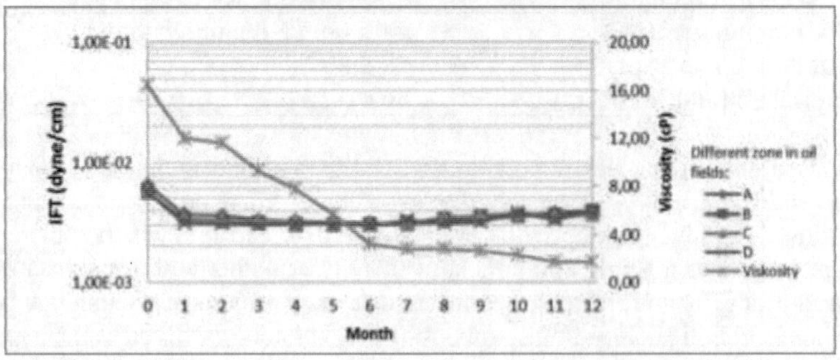

Figure 3. SBRC 100A 0.3% - polymer 0.1% is stable in thermal stability test at reservoir temperature of 60°C until 12 months.

3.2. Surfactant for Stimulation

Stimulation is a process of oil well repairement in order to increase formation permeability in an effort to increase production rate. This process causes dissolution of particles blocking porous space of rocks. Acidizing and surfactant injection are the stimulation methods commonly used in oil industry [9]. In surfactant stimulation process, reservoir rocks are soaked in surfactant to give time to surfactant to penetrate into reservoir rocks. Surfactant penetration is expected to alter rock wettability to water wet, reduce interfacial tension, reduce capillary force, and reduce water cut occurence. This way, permeability and well productivity can be improved. In surfactant stimulation process, surfactant can make contact with high reservoir temperature for a long period of time [10]. SPRC IBP has developed surfactant for stimulating agent with the following properties: IFT value $\leq 10^{-3}$ dyne/cm, stable at reservoir temperature, stable at stability thermal test (IFT value $\leq 10^{-3}$ dyne/cm), and filtration ratio value <1.2.

3.3. Surfactant for Asphaltene Dissolver

The use of solvent in asphaltene deposit handling can only remove about 50% of asphaltene deposit. The addition of surfactant in deposit handling makes surfactant dispersed in asphaltene molecules by breaking down the bonds between asphaltene molecules. This, in turn, will inhibit the reformation of asphaltene deposit. Temperature, pressure, oil composition changes and injection of chemical substances in oil production process result in the disruption of balance and stability of asphaltene miscelles in oil. This disruption is initiated by the release of resin from asphaltene surface. Then there is an interaction among asphaltene molecules and a solid phase is formed through a precipitation process. Further, the solid phase moves up to the surface causing the formation of asphaltene deposition [11].

The asphaltene dissolver produced by SBRC IPB has asphaltene solubility level of 99.38% and it can alter the contact angle of rocks from 48.60° too 80.89°. It was shown from wetting characteristic test that this asphaltene dissolver has an ability to increase wettability on a metal surface. This is characterized by the small amount of deposit remain which is left on the metal surface and asphaltene removal which reachs 99.66%.

3.4. Surfactant for Well Cleaning

Well cleaning is surfactant developed as an agent to clean holes of perforation well, pipe lines, and oil production equipment from blocking oil remains so that oil production process can be optimum. Results of palm the performance test of oil-based surfactant showed as a well cleaning agent show that there are two phases formed and there is no macroemulsion phase is formed in the phase behavior test. This sufactant is also shown to have IFT value of 10^{-2} dyne/cm and it is stable at reservoir temperature in a thermal stability test.

3.5. Surfactant for Wax Dissolver

Wax is formed as a result of lowered temperature under oil pouring point. Wax deposit is formed around tubing and production pipe causing a decrease in oil production by 10-35%. Surfactant plays an important role in increasing oil production in highly depleted wells by reducing interfacial tension, altering wettability of rocks, and cleaning oil wells from organic sediment. The use of surfactant in a solvent is a way of handling wax chemically. Dissolving wax is more effective by using a solvent and surfactant than by using a solvent only. Wax dissolver developed from palm oil-based surfactant by SBRC IPB is able to reduce the viscosity of crude oil with high wax content from 149.05 cP to 14.96 cP at 40^0 C.

4. Conclusion

Various palm oil-based surfactants have been specifically developed at SBRC IPB for oil and gas industry needs. Surfactants can be specifically produced for either improved oil recovery (IOR) or enhanced oil recovery (EOR), well stimulation, asphaltene deposit removal, well cleaning agent and wax removal agent.

Acknowledgement
Funding, facility, and equipment assistance from Surfactant and Bioenergy Research Center Bogor Agricultural University was acknowledged.

References

[1] Chatzis and Morrow 1994 Vicolastic surfactant for EOR (Society of Petroleum Engineers)
[2] Perkins W S1988 *Surfactants-A Primer*. www.p2pays.org/ref/03/02960.pdf
[3] Hui Y H 1996 Bailey's industrial oil abd fat product 5th Ed **1-5** (New York: John Wiley & Sons, Inc.)
[4] Basiron Y 1996 Bailey's industrial oil and fat product 5th Ed **2**, ed. Hui YH (New York: John Wiley & Sons, Inc.)
[5] Hasibuan H A 2012 *J. Standarisasi* **14**
[6] Swern D 1979 Bailey's Industrial Oil and Fat Product 14th Ed **1** (New York: John Wiley & Sons, Inc.)
[7] Hyne, Norman J 1991 Dictionary of petroleum exploration drilling & production (Tulsa, Okla: PennWell Pub.Co.)
[8] Larry W L 1989 *Enhanced Oil Recovery* (Englewood Cliffs: Prentice-. Hall, Inc.)
[9] Gomaa E E 1997 *Enhanced Oil Recovery: Modern Management Approach* (IATMI-IWPL/MIGAS Conference)
[10] Hu P C, Tuvell M E 1998 *JAOAC* **65** (6) 1007
[11] Zewen L and Angsong G 2000 Asphaltene in oil reservoir recovery *Chinese Sci. Bull.* **45** (8) pp 682-688.

Performance of Surfactant Methyl Ester Sulphonate solution for Oil Well Stimulation in reservoir sandstone TJ Field

F R Eris [1], E Hambali [2,4], A Suryani [2,4] and P Permadi [3,4]

[1] Agroekoteknologi Department, University of Sultan Ageng Tirtayasa, Serang – Banten, Indonesia
[2] Department of Agroindustrial Technology, Fakulty of Agroindustrial Technology, Bogor Agricultural University, Indonesia
[3] Department of Petroleum Engineering, Institute Technology of Bandung
[4] Surfactant and Bioenergy Research Center, Bogor Agricultural University, Indonesia

Email: fitria.eris@untirta.ac.id

Abstract. Asphaltene, paraffin, wax and sludge deposition, emulsion and water blocking are kinds ofprocess that results in a reduction of the fluid flow from the reservoir into formation which causes a decrease of oil wells productivity. Oil well Stimulation can be used as an alternative to solve oil well problems. Oil well stimulation technique requires applying of surfactant. Sodium Methyl Ester Sulphonate (SMES) of palm oil is an anionic surfactant derived from renewable natural resource that environmental friendly is one of potential surfactant types that can be used in oil well stimulation. This study was aimed at formulation SMES as well stimulation agent that can identify phase transitions to phase behavior in a brine-surfactant-oil system and altered the wettability of rock sandstone and limestone. Performance of SMES solution tested by thermal stability test, phase behavioral examination and rocks wettability test. The results showed that SMES solution (SMES 5% + xylene 5% in the diesel with addition of 1% NaCl at TJformation water and SMES 5% + xylene 5% in methyl ester with the addition of NaCl 1% in the TJ formation water) are surfactant that can maintain thermal stability, can mostly altered the wettability toward water-wet in sandstone reservoir, TJ Field.

1. Introduction

In general, oil reservoir has different characteristics depending on its composition, temperature and pressure where the oil accumulation takes place [1], oil reservoir is different in its geological rock condition, water content in the reservoir and type of oil contained.

TJ Fields is an oil field with the main rocks is sandstones and conglomerates, with average depth is in between 800-1200 m. Number of wells operated in TJ Fields in 2006 is 79 production wells (consisting of 50 bobbing pumps and 29 ESP pump wells), and 37 injection wells [2]. TJ field showed an increasing of water production coming from the produced water and fresh water that was used as water injection, the increasing of water causing many problems in related with inorganic scale deposition in reservoirs, hole, wells, pipelines, and pumps.

TJ Fields is an oil field with the main rocks is sandstones and conglomerates, with average depth is in between 800-1200 m. Number of wells operated in TJ Fields in 2006 is 79 production wells

Content from this work may be used under the terms of the Creative Commons Attribution 3.0 licence. Any further distribution of this work must maintain attribution to the author(s) and the title of the work, journal citation and DOI.
Published under licence by IOP Publishing Ltd

(consisting of 50 bobbing pumps and 29 ESP pump wells), and 37 injection wells [2]. TJ field showed an increasing of water production coming from the produced water and fresh water that was used as water injection, the increasing of water causing many problems in related with inorganic scale deposition in reservoirs, hole, wells, pipelines, and pumps. However the scales are not found in the surface facilities, (production flow line, separators, or storage tanks). Other than that, the unique characteristics of TJ field's oils that containing high paraffin is also a problem. Hydrocarbons accumulation is dominated with oil, while gas only found in a small quantity in the form of associated gas and free gas. TJ oil is paraffinic 40.3. OAPI (0.8235 SG) and a wax content is 30% (pour point is approximately 95 OF) (Wibowo et al., 2008) The analysis show oil viscosity of 2.77 cp, API Density (15 ° C) at 0.8340 and API Gravity (15 ° C) of 37.97.

Surfactant solution to be used in the stimulation of oil wells to a reservoir is different to other reservoir. Some test can be used in order to select the surfactants i.e. : oil solubility test, electrolyte influence, the surfactant solution density test, the viscosity test of surfactant solution, the time of solubility test, identification of surfactant co-solvent optimal formulation, and core flood [1]. To obtain the stimulation result, several laboratory researches must be conducted. I.e.: the interface tension measurements, Phase behavior examination and wettability properties of rocks by surfactants as well stimulation agent

The properties of the phase behavior and wettability properties of reservoir rocks are things that need to be understood more deeply due to the properties of rock that will affect the mechanism of fluid flowing within the rock pores. The mechanism microscopically is influenced by the wettability properties of rocks, structure and geometry of rock pores. Utilization of SMES surfactant that is injected into the oil wells is expected to prevent and or to overcome the damage occurred to the well by changing the wettability properties of rocks and also decreases the interfacial tension of oil and water.

This research aimed to examine the performance of the SMES 5 % + xylene 5% + methyl ester (TJ formation water + NaCl 1%) and SMES 5% + xylene 5% + methyl ester (TJ formation water + NaCl 1%) to the heat resistance, phase behavior and wettability properties of rocks at sandstone reservoir.

2. Materials and methods

2.1. *Thermal Stability Test method*

As 20 ml of surfactant solution is poured into a closed tube. The surfactant solution is made in two samples. The test is conducted by heating the SMES surfactant solution in its reservoir temperature for seven days in an oven at 60 ° C according to the reservoir sandstone temperature of TJ Field. The surfactant solution is expected to maintain the value of interfacial tension until the seventh day, or at least remain at a value of 10-2 dyne / cm. A daily observation is conducted to observe the surfactant solution and the measurement of the value of interfacial tension. The observation is conducted to observe the tendency of changes in the value of interfacial tension that occurs during the heating on the temperature of the reservoir takes place.

2.2. *Phase Behavioral Examination*

The stages in conducting the Phase behavioral examination is 20 ml of formation water sample and surfactant solutions which have been prepared is poured into a closed tube, with the ratio of the SMES surfactant solution sample : formation water at 80:20. Then the samples are stored in the oven with the temperature equivalent to the reservoir temperature (60 ° C) for 60 minutes. After that, the rotary mixing is turned on at a speed of 6 rpm at a reservoir temperature. The observation is conducted in the minute of 10, 60, 120, 240, 720, 1440, and 2880.

Types of emulsions that can be formed in the process of the phase behavior examination are: (1) Lower emulsion phase , where the emulsion is formed is in the aqueous phase, there is oil excess phase (excess oil), in a state of two phases, and translucent colored (clear opaque); (2) micro emulsion or the middle phase emulsion, which is the emulsion consisting of three phases (water- micro emulsion - oil) and translucent colored; (3) The upper phase emulsion, where the emulsion formed is in the oil phase and is water excess phase occurred (excess water), the emulsion is consists of two phases; (4) Macro emulsion, the emulsion is viscous and milky white colored (milky).

2.3. *Rocks wettability test*

Test on wettability properties of rocks by using SMES solution is conducted in 4 steps. The first step is to soak the washed formation rocks in a core extraction in the formation water in the temperature equivalent to reservoir temperature for 1 week. This treatment is to put the rocks in the condition of reservoir. Then the measurement is conducted to determine the wettability by dripping oil on them. The measurement is conducted to determine the angle degree of the rocks in the early contact. The second step is the rocks are soaked in TJ field oil for 1 day. And the degree angle is measured again. The test then followed by soaking the rock in a solution of the SMES surfactant for six hours at reservoir temperature and the measurement of the contact angle is conducted as the previous step. The soaking of rocks in the surfactant solution for six hours is due to the process of minimum stimulation in the oil wells which took place is only six hours. After the rocks soaked in SMES surfactant, then the rocks angle contact are re measured. This test is a measurement of contact angle of the third step. The fourth step is the soaking of the rocks (rocks similar to the previous steps) with TJ field formation water at reservoir temperature for one day and its contact angle are re measured. It also aims to condition the rocks back to reservoir conditions after the application of SMES surfactant solution as a well stimulation agent. Oil-wet occurred if a rock has a fluid contact angle (oil) to the rock which is smaller than 90° ($\theta < 90$ °), while the water-wet is occurred if a rock has a fluid contact angle (oil) to the rocks with an angle greater than 90 ° ($\theta > 90$ °).

3. Results and Discussion

Surfactant solution with SMES concentration, salinity and the best concentration of aromatic solvent can be observed in its performance in solution resistance in hot temperatures, the formation of emulsions and the ability to change the wettability properties into a water-wet. The test performed is an examination on the heat resistance / thermal stability, phase behavior test/ phase behavior, and wettability properties of the rocks. The surfactant solution resulted as well stimulation agent is the methyl ester + SMES 5% + 5% of xylene (TJ Formation Water + NaCl 1%) or diesel + SMES 5% + 5% + xylene 5% (TJ Formation Water + NaCl 1%) for field reservoir sandstone of TJ field. Surfactantsolution with SMES concentration, salinity and the best concentration of aromatic solvent can be observed in its performance in solution resistance in hot temperatures, the formation of emulsions and the ability to change the wettability properties into a water-wet. The test performed is an examination on the heat resistance / thermal stability, phase behavior test/ phase behavior, and wettability properties of the rocks. The surfactant solution resulted as well stimulation agent is the methyl ester + SMES 5% + 5% of xylene (TJ Formation Water + NaCl 1%) or diesel + SMES 5% + 5% of xylene (TJ Formation Water + NaCl 1%) for field reservoir sandstone of TJ field.

3.1. *Thermal stability*

The results of the analysis on interfacial tension of the surfactant solution after the thermal stability test can be seen in Figure 1.

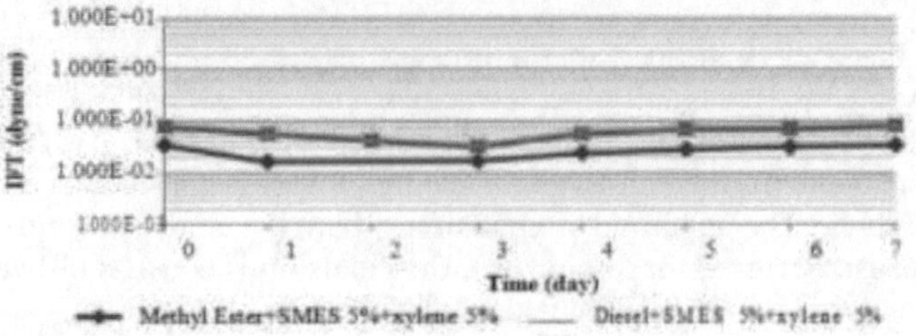

Figure 1. Thermal stability of the surfactant solution oilbased of SMES in sandstone reservoir of TJfield in the temperature of 60°C

Surfactant solution for well oil stimulation by methyl ester carrier medium or diesel on the TJ field, until the third day still maintain the interface tension value, and after the third day there was an increase in the value of interfacial tension. Both of the surfactant solution stability tested due to the heating in each reservoir temperature was still able to maintain its interface tension value until the day of 7 as indicated by the value of the interfacial tension which is still in the range of 10-2 dyne/cm [3] in his research the increasing of temperature which is close to the reservoir temperature is effective in lowering the value of interfacial tension.

3.2. Phase Behavior

The results of the analysis of the phase behavior of surfactant solution made are presented in Figure 2 and Table 1

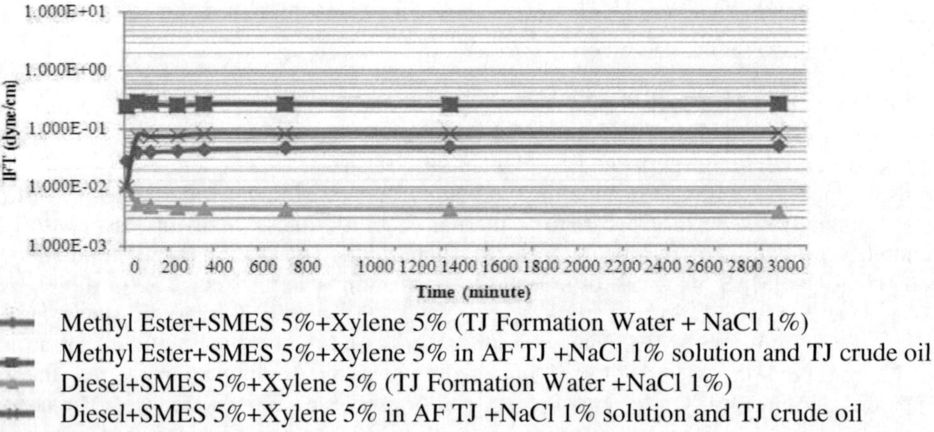

Figure 2. Phase behavior of TJ field sandstone reservoir surfactant solutions on temperature of 60°C

The behavior phase Test is intended to observe the performance of the surfactant solution with formation water fluid from each reservoir. In the process of stimulation of oil wells the surfactant solution is injected into the well, and then soaked for some time (maximum 48 hours) and after that the surfactant solution and the fluid is pumped back through the same well. Observation conducted is visual observation by observing the changes that occurred in mixture of the SMES surfactant solution with formation water fluid from each reservoir.

Table 1. The results of the phase behavior test of surfactant solutions on TJ field reservoir sandstone

Formula	Picture
Methyl ester+ SMES 5%+ xylena 5% (TJFormation Water +NaCl1%)	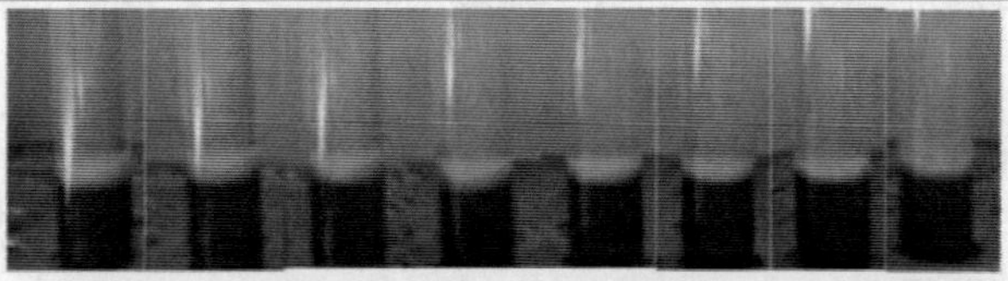 10 minute 1 hour 2 hour 4 hour 6 hour 12 hour 24 hour 48 hour
Diesel+ SMES 5%+ xilena 5% (TJFormation Water +NaCl1%)	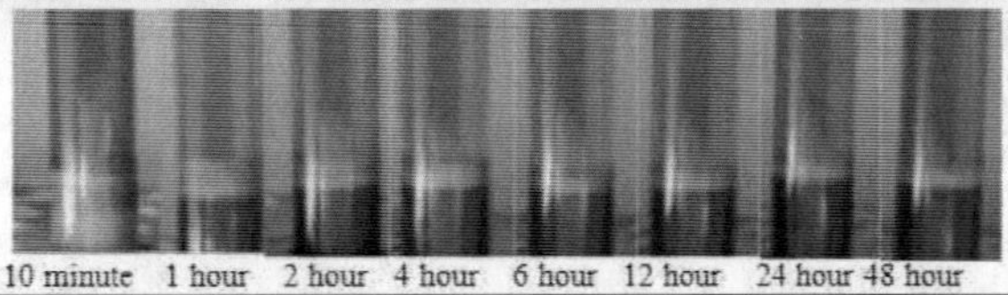 10 minute 1 hour 2 hour 4 hour 6 hour 12 hour 24 hour 48 hour

Visual appearance on phase behavior test in TJ field reservoir sandstone on Table 1 shows that in both of the surfactant solution tested, the macro emulsion of surfactant solution formed with water formation. By observing through a microscope with 1000 times magnification , it can be seen in the middle phase (macro emulsion), the average droplet size that formed is 4:09 - 5:53 μm for Methyl ester + SMES 5% + xylene 5% (TJ Formation Water + NaCl 1%) and 11.70 - 14.61 μm. For Diesel + SMES 5% + xylene 5% (TJ Formation Water + NaCl 1%), When observed from the droplet size and visual appearance under the microscope, it can be seen that macro emulsion formed is an emulsion of oil in water (o/w). Depravity of formation due to water-in-oil emulsion causing a higher depravity than the oil in water. this is caused by the water emulsion in the oil causing a high solution viscosity. Solution with this high of viscosity will impede the flow of fluid within the pores [4]. The forming of macro emulsion showed that SMES surfactant become incompatible with the oil fluid.

3.3. Wettability properties of rocks

Rocks that are water-wet tend to be more easily wettable by water than oil .so is conversely. Rocks with oil-wet properties tend to be more easily wettable by oil than by water. The alteration of wettability properties in general is assumed due to the absorption of polar compounds onto the surface of the rock [5].

Wettability is measured by the contact angle between the liquid and the solid in this case oil with reservoir rocks. This contact angle analysis is conducted with the adjustment of the actual conditions in the reservoir. The initial reservoir condition is the soaking of the rocks with formation water and oil. Histogram of the contact angle analysis result of the surfactant solution made is presented in Figure 3

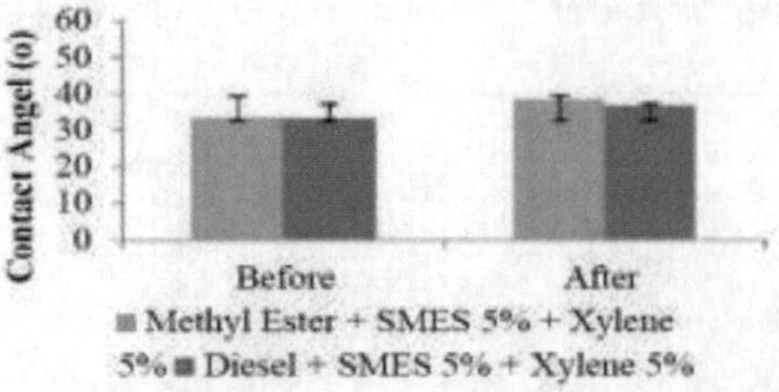

Figure 3. The results of analysis on rock wettability properties of oil–rocks on TJ field reservoir sandstone

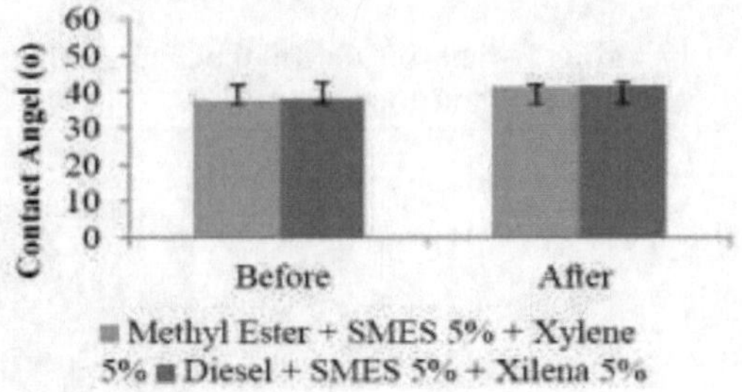

Figure 4. The result of analysis on rock wettability properties of oil–rocks in quartz

The alteration of the surfactant solution's contact angle in the reservoir sandstone of TJ field shows that the surfactant solution has altered the nature of the oil-wet rocks become inclined to water-wet.

This is shown on Figure 25, where at the beginning of treatment, the contact angle formed is 33.33^o and increase to 38.24° for the surfactant solution with methyl ester medium carrier and 33.14° increase to 36.41° for formula with the diesel medium carrier. The same conditions occurred in the analysis results on oil wettability properties of the oil – rocks on the quartz mineral rocks (Figure 4), where there is an increase in the contact angle after the quartz mineral treated.

The existence of paraffin and asphaltene in the TJ field's rocks formation has become the cause of oil-wet the condition of oil wet may cause the decrease in permeability of reservoir. Therefore the properties of rocks wettability must be transformed into water wet to increase the permeability of oil.

Surfactants can cause the ion pair in the formation reacts to the alteration in the wettability properties of rocks due to electrostatic repulsion between the anionic cluster and negative content of oil components adsorbed in the surface. The surfactants may alter the rock wettability by forming a monolayer of surfactant on oil-wet rock layers. The surfactant adsorption occurred through hydrophobic interactions with the layer of hydrocarbons adsorbed in the surface of the rock, and then the hydrophilic cluster of the surfactant will move towards the solution, it would result in the formation covered by water zone and forming a weak capillary pressure during the process of imbibition. This process will occur sequentially in the surfactant-oil-rock interface zone [6]. Low salinity water is able to altered wettability of rock until the oil flow easily. Water injection with low salinity can reduce the adhesion force between the oil and the clay particles on the surface of the rock [7].

The content of the reservoir sandstone is dominated by quartz mineral (± 85%) and the rest are clay minerals and carbonates. negative charge of the quartz surfaces, and when the sandstone surface in contact with the surfactant, then the hydrophobic positive cluster of the surfactant will be adsorbed to the surface with negative charge, then the hydrophilic cluster will move away from the surface so it make the surface had the hydrophilic properties [8].

4. Conclusion

This research resulted that the solution formula from TJ field reservoir sandstone is solution of SMESoilbased with SMES 5% in + xylene + 5 % in the diesel with addition of 1% NaCl at TJ formation water and SMES 5% +xylene 5% in methyl ester with the addition of NaCl 1 % in the TJ formation water. Results ofthe solution formula from TJ field reservoir sandstone show that (i) formula oil based on TJ field reservoir sandstone forming macro emulsion of surfactant solution with formation water; (ii) oil based formula on TJ field reservoir sandstone is capable to alter the properties of the rocks from oil wet into water wet; (iii) the thermal stability test of oil based formula is able to maintain the value of interfacial tension until the third day of observation.

Acknowledgements
The authors would like to thank to SBRC (Surfactant and Bioenergy Research Center) for all the support during the research.

References

[1] Lake LW 1989 *Enhanced Oil Recovery* (Prentice Hall : New Jersey)
[2] Komarudin S B 2006 Aplikasi analisa after-closure untuk menentukan permeabilitas dan tekanan formasi di lapangan tanjung *Prosiding Simposium Nasional & Kongres IX Ikatan Ahli Teknik Perminyakan Indonesia (IATMI) 2006* 15-17 November 2006 (Indonesia- Jakarta)
[3] Golabi E, Azad F S, Ayatollahi S S, Hosseini S N and Dastanian M 2012 Experimental study of anionic and cationic surfactants effects on reduce of ift and wettability alteration in carbonate rock *IJSER* **3** 1050-1053
[4] Bennion D B, Thomas F B, Bennion D W and Bietz RF 1995 Mechanisms of formation damage and permeability impairment associated with the drilling, completion and production of low api gravity oil reservoirs *SPE International Heavy Oil Sym.* June 19-21 Calgary - Canada
[5] Tie H and Morrow N R 2005 Low flood rate residual saturations in carbonate rocks IPTC10470 *The 2005 International Petroleum Technology Conf.* November 21-23 Doha-Qatar
[6] Zhang J 2004 Experimental and computational study of nucletae pool boiling heat transfer in aqueos surfactant and polymer solution (China: University of Science and Technology Beijing)
[7] Berg S, Cense AW, Jansen E and Bakker K 2010 Direct experimental evidence of wettability modification by low salinity *Petrophysics J.* **51** 9
[8] Rosen M J 2004 *Surfactants and interfacial phenomena* (New Jersey :John Wiley & Sons)

Performance evaluation of integrated solid-liquid wastes treatment technology in palm oil industry

J R Amelia[1], S Suprihatin[1], N S Indrasti[1], U Hasanudin[2] and K Fujie[3]

[1]Department of Agro-Industrial Technology, Bogor Agricultural University, Bogor, Indonesia
[2]Department of Agro-Industrial Technology, Faculty of Agriculture, University of Lampung, Bandar Lampung, Indonesia
[3]School of Environmental and Information Sciences, Yokohama National University, Yokohama, Kanagawa Prefecture 240-8501,Japan

Email : fiamel2702@gmail.com

Abstract. The oil palm industry significantly contributes to environmental degradation if without waste management properly. The newest alternative waste management that might be developed is by utilizing the effluent of POME anaerobic digestion with EFB through integrated anaerobic decomposition process. The aim of this research was to examine and evaluate the integrated solid-liquid waste treatment technology in the view point of greenhouse gasses emission, compost, and biogas production. POME was treated in anaerobic digester with loading rate about 1.65 gCOD/L/day. Treated POME with dosis of 15 and 20 L/day was sprayed to the anaerobic digester that was filled of 25 kg of EFB. The results of research showed that after 60 days, the C/N ratio of EFB decreased to 12.67 and 10.96 for dosis of treated POME 15 and 20 L/day, respectively. In case of 60 day decomposition, the integrated waste treatment technology could produce 51.01 and 34.34 m^3/Ton FFB which was equivalent with 636,44 and 466,58 $kgCO_2e$/ton FFB for dosis of treated POME 15 and 20 L/day, respectively. The results of research also showed that integrated solid-liquid wastes treatment technology could reduce GHG emission about 421.20 and 251.34 $kgCO_2e$/ton FFB for dosis of treated POME 15 and 20 L/day, respectively.

1. Introduction

Indonesia is currently the largest producer and exporter of palm oil in the world. The oil palm plantation and palm oil processing sectors have become a key part of Indonesia's economy, but oil palm mill also significantly contributes to environmental degradation if without waste management properly. Manufacturing processes generate large quantities of empty fruit bunch (EFB) and palm oil mill effluent (POME) on the output side. Recycling EFB and POME by conversion into usable by products has been done. Nowadays, EFB is used as mulch in agricutural lands and POME will be treated with the pond system. The ponding system is the most conventional method for treating POME. More than 85% of palm oil mills use ponding systems. Ponding systems are easy operating systems but they have some disadvantages such as methane production which results harmful effect on the environment [1]. The anaerobic ponds emit a huge amount of the strong green house gasses and the treated POME of the ponds (effluent) contains nutrients responsible for pollution of surface and ground water. The major concern about the ponding systems is the uncontrollable release of large

Content from this work may be used under the terms of the Creative Commons Attribution 3.0 licence. Any further distribution of this work must maintain attribution to the author(s) and the title of the work, journal citation and DOI.
Published under licence by IOP Publishing Ltd

amounts of CH_4 and CO_2 into the atmosphere from open ponds, which may worsen the effect of global warming [2]. The large amount of greenhouse gases emission can threat the palm oil industry sustainability.

Total CH_4 emission per open digesting tank was 518.9 kg/day [3] and total methane emission per anaerobic pond was 1043.1 kg/day [4]. Those data calculated from the two equations derived from relationships between methane emission and total carbon removal. the results showed that the production of 1 ton of FFB produced 119 kgCO_2eq using a yield of 20.7 ton oil palm fresh fruit bunches (FFB)/ha [5]. The production of 1 ton of CPO in a mill without and with biogas capture emitted 971 and 506 kg CO_2eq, respectively. For the production of 1 ton of refined palm oil in a refinery which sourced the CPO from a mill without biogas capture and with biogas capture, the GHG emitted was 1,113 kg and 626 kg CO_2eq, respectively.

Palm oil industry in Indonesia with 19324 million of CPO production, was predicted causing about 15.61 x 10^6 ton of greenhouse gasses emission equals to CO_{2n} [6]. The anaerobic ponds emit a huge amount of the strong green house gas methane with the biogas and the treated POME of the ponds contains nutrients responsible for pollution of surface and ground water. Each ton of produced crude palm oil is responsible for the emission of 46 m^3 (32.9 kg) of methane, corresponding 384 m^3 (756 kg) CO_2 equivalent [7]. Emission loose total of 1 ton CPO was 1472.24 kgCO_2eq. The sequence of emission resource per 1 ton CPO production was 1026.4 kgCO_2eq from Palm Oil Mill Treated POME (POME); 179.35 kg CO_2 eq from Empty Fruit Bunch (EFB); 144.18 kg CO_2eq from NPK fertilizer; 44.73 kg CO_2eq fiber 32.25 kg CO_2eq from gas waste; 15.92 kgCO_2eq; from solar and 13.37 kgCO_2 eq from ash waste [8]. The research showed that POME, EFB, and the utilization of chemical fertilizer were the biggest greenhouse gases emission contributor in palm oil plantation industry. The large amount of greenhouse gases emission can threat the palm oil industry sustainability. The newest alternative waste utilization that might be developed is by utilizing the outcome of wastewater treatment from anaerobic digestion with empty fruit bunch (EFB) through anaerobic integrated composting process. The technology hopefully can reduce or resolve carbon emission and greenhouse gasses loose from palm oil plantation industry. The aim of this research was to examine and evaluate the integrated solid-liquid waste treatment technology in the view point of greenhouse gasses emission, compost, and biogas production.

2. Material and Methode

2.1 Equipment and Material

The equipment were 5000 L anaerobic bioreactor with methane capturer system and H_2S *purifier, anaerobic composting digester*, gasflow meter, gas sampler bag, pH meter HM-20P, analytic balance, desicator, furnace, DRB200, HACH spektrofotometer DR/4000U, Gas Chromatography, Elementar Analyzer, etc. The materials were Palm Oil Mill Treated POME (POME), sludge, Empty Fruit Bunches (EFB) from palm oil industry in Lampung, etc.

2.2 Method

The research was started by doing POME characterization. Fresh POME was treated by anaerobic treatment process (loading rate was 1.65 gCOD/L/day). Shredded EFB then was added to anaerobic composting digester. The treated POME from anaerobic treatment process was sprayed to EFB everyday. There were two digesters used during the anaerobic composting. The treated POME with dosis of 15 and 20 L/day was sprayed to the anaerobic digester that was filled of 25 kg of EFB. EFB anaerobic composting was run in two months. Detailed scheme on how the research conducted shown in figure 1.

IOP Conf. Series: Earth and Environmental Science **65** (2017) 012036 doi:10.1088/1755-1315/65/1/012036

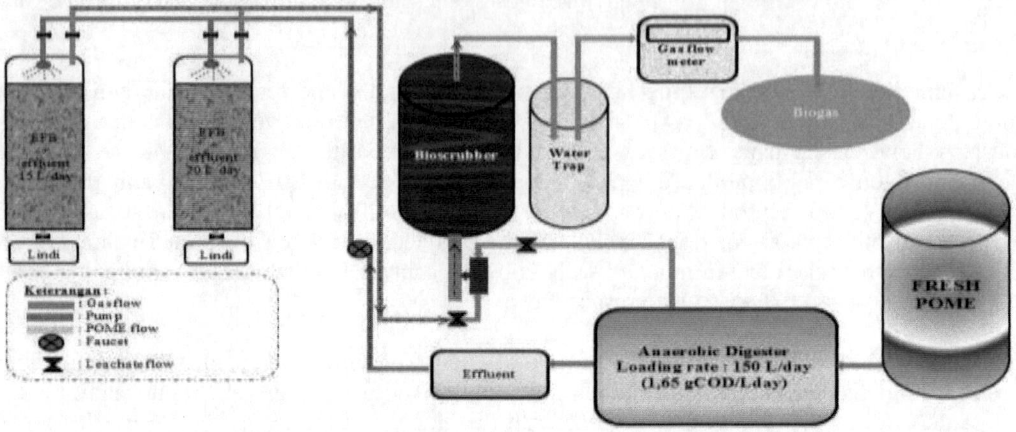

Figure 1. Scheme of research (proposed condition)

2.3 The Emission Reduction Potential

2.3.1 Existing condition

In Indonesia, almost all palm oil mills use open ponding systems to treat POME. Even though the ponding system is economical, it's land and time intensive, and it releases a large amount of methane gas (one of *GHGs)* into the atmosphere primarily from the organic decomposition that occurs in the anaerobic pond. The emission potential of GHG from open ponding systems was calculate based on COD removal.

2.3.2 Proposed condition

The emission potential of GHG from integrated waste treatment system was estimated for each treatment by using carbon mass balance method. Carbon mass balance that was used is showed on figure 2.

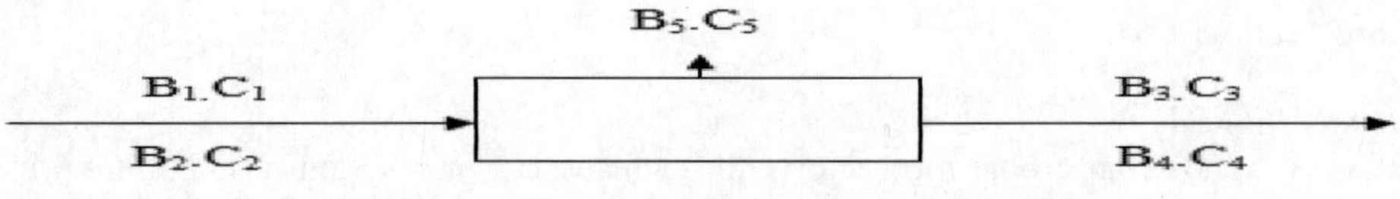

$$B_1.C_1 + (B_2.C_2)n_a. = B_3.C_3 + (B_4\ .C_4)n_b + B_5.C_5 \tag{1}$$

Figure 2. Carbon mass balance

Information:

B_1 = Amount of EFB that is used in composting (kg) C_1 = Concentration of EFB carbon
B_2 = Amount of treated POME (kg) C_2 = Concentration of treated POME carbon
B_3 = Amount of compost produced (kg) C_3 = Concentration of compost carbon
B_4 = Amount of leachate produced (kg) C_4 = Concentration of leachate carbon
B_5 = Amountof evaporated compound C_5 = Concentration of evaporated biogas
n_a = Amount of treated POME sprayer times during one composting cycle
n_b = Amount of leachate retrieval

$$\text{Emission potential} = B_5.C_5 \ x \ \frac{\%CH_4}{\%CH_4 + \%CO_2} \ x \ \frac{16}{12} \ x \ 25 \tag{2}$$

ERP = Proposed condition ($kgCO_2e$/Ton FFB) – Existing condition ($kgCO_2e$/Ton FFB) (3)
Note : ERP = Emission Reduction Potential

3. Result and Discussion

3.1 Characteristics of Palm Oil Waste

Main materials which were used in this research were empty fruit bunches (EFB) and the treated POME of anaerobic digester process. EFB used in the research had 58.52% of moisture content, 41.48% of total solid (TS) , 42.59% of total carbon, 0.93% of total nitrogen, and 45.79 of C/N ratio. The characteristics of treated POME are shown on table 1.

Tabel 1. Treated POME characteristics

Treated POME	Value
pH	7.36-7.65
Chemical Oxygen Demand (mg/L)	22815
Total Solid (mg/L)	19577
Total Suspended Solid (mg/L)	14231.60
Volatile Solid (mg/L)	13780
Volatile Suspended Solid (mg/L)	9153.78
Total Carbon (%)	33.63
Total Nitrogen (%)	4.41
C/N Ratio	8.07

3.2 Biogas Productivity

The anaerobic integrated digestion system with anaerobic composting system can produce biogas as the new energy resource compared to the other palm oil waste treatments. Each anaerobic digestion and anaerobic composting process can produce methane gas (CH_4) as energy resource that is caused by decomposition process of organic compounds. Biogas measurement of anaerobic composting process was done by calculating accumulation of carbon total. Biogas productivity from this research is shown in figure 3. Figure 3 also show that biogas productivity accumulation of each treatment is fluctuated, however the constantly increase occurred until last day treatment.

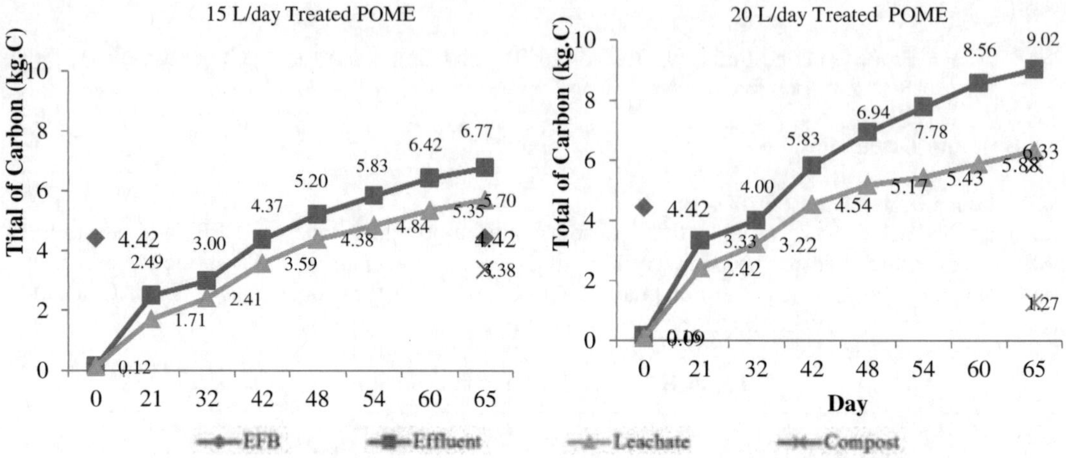

Figure 3. Biogas potential production of various treated POME treatment

Figure 4 shows total carbon in each treatment from the research. Biogas that was produced generally consists of methane gas (CH_4), carbon dioxide (CO_2), and the other gasses. CH_4 and CO_2 content in each biogas treatment were different. Biogas characteristics produced in each treatment show on table 2. The result of research showed that biogas which was produced in 15 and 20L/day treated POME treatment about 38.89 m^3/ton FFB and 22.22 m^3/ton FFB with methane contain about 19.12 m^3/ton FFB and 11.85 m^3/ton FFB. The result of research also shows that biogas which was produced in 15 and 20 L/day of treated POME treatments contained methane gas more than 45% (table 2). If the containing of methane gas in biogas is more than 45%, the biogas will be easy to be burnt and able to use as fuel because the biogas has enough high of heat. It showed that biogas that was produced from 15 and 20 L/day treated POME treatment has good burning level because it contained methane gas more than 45%. The containing of methane gas in this research was higher compared to another same kind research [9]. The concentration of methane gas (CH_4) from EFB anaerobic composting (16 kg) with treated POME was only 40,1% [9]. It shows that the amount of material added in composting process will influence contain of methane gas in produced biogas.

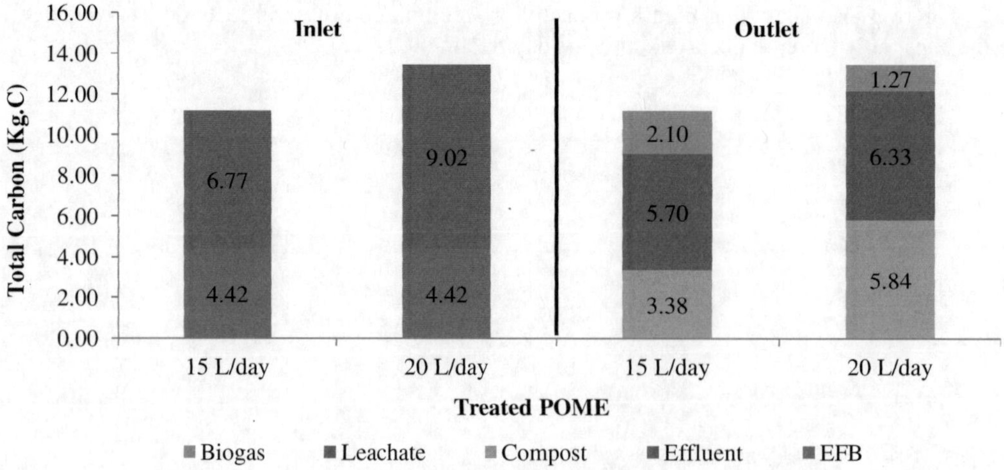

Figure 4. Total carbon from anaerobic composting

Table 2. Characteristic of biogas

Parameter	Unit	Teated POME	
		15 L/day	20 L/day
CH_4	%	49.17	51.08
CO_2	%	39.51	38.89

3.3 The Evaluation of Green House Gas Emission from Waste Integrated Treatment

Waste water treatment in palm oil plantation industry generally is done by using open lagoon. That conventional process will produce carbon dioxide (CO_2) and methane gas (CH_4) that if it dispersed to atmosphere would cause global warming. Green House Gas (GHG) emission that is produced from POME conventional treatment is 215.24 $kgCO_2e$/Ton FFB with 0,7 m^3 ton CPO produced. POME potentially produces GHG emission about 133,95 $kgCO_2eq$/EFB [10]. Those data show that POME conventional treatment is not effective because it can potentially cause global warming.

Treated POME utilization system from anaerobic digestion combined with EFB throughout anaerobic integrated composting process hopefully can be a solution to reduce GHG emission. The CO_2 emission potential from POME anaerobic treatment has been calculated by carbon mass balance of each treatment process. Potential of CO_2e emission from POME anaerobic treatment has been calculated from carbon total measurement, POME as carbon inlet, treated POME as carbon outlet, and biogas carbon total as evaporated carbon, meanwhile anaerobic composting CO_2e emission potential was from carbon total measurement, newest EFB and treated POME as carbon inlet and the carbon outlet was from compost total carbon, leachate and total carbon evaporated. The result of research shows that total emission from POME anaerobic process by using anaerobic digestion lagoon system was 77.26 $kgCO_2e$/ton FFB. The result of research shows that total emission from POME anaerobic process by using anaerobic digestion lagoon system was 77.26 $kgCO_2e$/ton FFB. Anaerobic composting process from each treatment could produce total emission differently, as it is presented on figure 5.

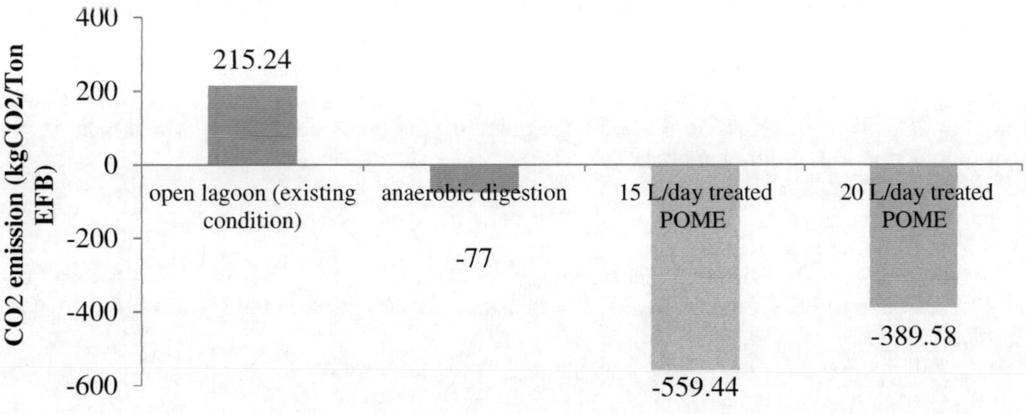

Figure 5. GHG emission potential from each anaerobic composting treatment

The result of research showed that palm oil mill integrated waste treatment techonolgy could reduce GHG emission about 421.20 $kgCO_2e$/ton FFB (15 L/day treated POME) and 251.34 $kgCO_2e$/ton FFB (20 L/day treated POME). POME anaerobic treatment with methane capture techonolgy could reduce GHG emission only 109.41-175.35 $kgCO_2e$/ton FFB [11]. Those data showed that palm oil mill integrated waste treatment could reduce GHG emission higher than other waste treatment in palm oil industry (existing conditions).

3.4 Compost

Composting is an aerobic or anaerobic biological process that uses naturally occurring microorganisms to convert biodegradable organic matter into a humus-like product, can destroy pathogens, converts nitrogen from unstable ammonia to stable organic forms, reduces the volume of waste, and satisfies the needs of fertilizer for agricultural use seasonally [12]. Composting is a biologically based process which is practiced to stabilize the organic matter for soil amendment (producing compost) and to protect the environment from the detrimental effects of these waste products [13]. Composting of the empty fruit bunch (EFB) could simultaneously treat the solid waste from palm oil mill. Composting EFB is a possible way to transform the bulky bunches into a valuable, manageable product for use in the plantation or as market product [14]. C/N ratio is one of the important factors affecting composting process and compost quality. The course of decomposition of organic matter is affected by the presence of carbon and nitrogen. C/N ratio compost from this research is shown in figure 6.

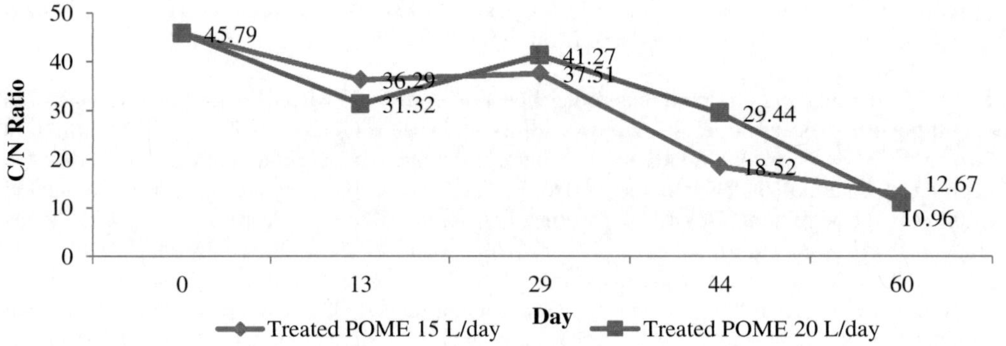

Figure 6. C/N ratio compost from this research

The results of research showed that after 44 days the C/N ratio of EFB decreased rapidly from an initial value (45.79) to 18.52 and 29.44 for dosis of treated POME 15 and 20 L/day, respectively. After 60 days, the C/N ratio of EFB continue decreased to 12.67 and 10.96 for dosis of treated POME 15 and 20 L/day, respectively. The changes in the C/N ratios reflect the organic matter decomposition. The optimum ratio in soil organic matter is about 10 carbons to 1 nitrogen, or a C:N ratio of 10:1, and a C:N ratio of 20, where C and N are the available quantities, is the upper limit at which there is no danger of robbing the soil of nitrogen [15].

4 Conclusion

The conclusion from this research are:

1. Palm oil mill integrated waste treatment techology could reduce greenhouse gasses emission about 421.20 kgCO_2e/ton FFB (15 L/day treated POME) and 251.34 kgCO_2e/ton FFB (20 L/day treated POME).
2. Palm oil mill integrated waste treatment technology could produce biogas about 51.01 m^3/ton FFB (15 L/day treated POME) and 34.34 m^3/ton FFB (20 L/day treated POME).
3. The best treatment in palm oil mill integrated waste technology is 15 L/day treated POME. It could reduce greenhouse gasses emission higher than other treatment.

Acknowledgements

The authors would like to thank The Directorate General of Higher Education of the Ministry of Education and Culture, Republic of Indonesia, and The Japan Society for The Promosition Of Science (JSPS) for financial support. Thanks also go to all leaders of the Directors and Chairmen of PTPN VII Bekri, AgroIndustrial Technology Department of Fateta IPB, Faculty of Agriculture of UNILA, and

Analytical Laboratories of Department Agro-Industrial Technology UNILA, for the provision of data, information and discussion.

References

[1] Rupani P F, Singh R P, Ibrahim M H and Esa N 2010 Review of current palm Oil mill effluent (POME) treatment methods: vermicomposting as a sustainable practic *World Applied Sciences Journal* **11**(1) 70-81

[2] Wu T Y, Mohammad A W, Jahim J M and Anuar N 2010 Pollution control technologies for the treatment of palm oil mill effluent (POME) through end-of-pipe processes J Environ *Manage* **91** 1467–1490

[3] Yacob S, Hassan M A, Shirai Y, Wakisaka M and Subash S 2005 Baseline study of methane emission from open digesting tanks of palm oil mill treated POME treatment *Chemosphere* **59** 1575–1581

[4] Yacob S, Hassan MA, Shirai Y, Wakisaka M, and Subash S 2006 Baseline study of methane emission from anaerobic ponds of palm oil mill treated POME treatment *Science of The Total Environment* **366**(1) 87–196

[5] Choo Y M, Muhamad H, Hashim Z, Subramaniam V, Puah C W and Tan Y 2011 Determination of GHG contributions by subsystems in the oil palm supply chain using the LCA approach *The International Journal of Life Cycle* **16**(7) 669-681

[6] Tong S L 2011 Recent developments on palm oil mill residues biogas recovery and utilization International Conference and Exhibition of Palm Oil Jakarta

[7] Schuchardt F, Wulfert K, Darnoko D and Herawan T 2007 Effect of new palm oil mill processes on the EFB and POME utilisation Proceedings of Chemistry and Technology Conference PIPOC 2007, Kuala Lumpur Malaysia

[8] Hakim H M 2013 Life Cycle Assessment (LCA) produksi crude palm oil (CPO) kebun dan pabrik kelapa sawit pelaihari PT. Perkebunan Nusantara XIII Universitas Gajah Mada Yogyakarta

[9] Apria N E 2014 Produksi biogas melalui proses dry fermentation menggunakan limbah tandan kosong kelapa sawit Universitas Lampung Lampung

[10] Hasanudin U 2012 Potensi penyediaan energi dari limbah industri sawit, tepung tapioka, dan peternakan, FGD peluang dan tantangan pemanfaatan biogas untuk pemenuhan kebutuhan energi sendiri di industri. Kementrian Energi dan Sumber Daya Mineral, Republik Indonesia, 8 Oktober 2012, Jakarta

[11] Hasanudin U, Sugiharto R, Haryanto A, Setiadi T and Fujie K 2015 Palm oil mill treated POME treatment and utilization to ensure the sustainability of palm oil industries *Water Science & Technology* **72**(7)

[12] Zhu N 2007 Effect of low initial C/N ratio on aerobic composting of swine manure with rice straw *Bioresource Technology* **98** 9–13

[13] Mohammad N, Alam M D, Kabbashi N A and and Ahsan A 2012 Effective composting of oil palm industrial waste by filamentous fungi: A review *Resources, Conservation and Recycling* **58**, 69–78.

[14] Schuchardt F, Darnoko D and Guritno P 2002 Composting of empty oil palm fruit bunch (EFB) with simultaneous evapoaration of oil mill waste water (POME) International Oil Palm Conference, Nusa Dua, Bali, Indonesia.

[15] Washington State University 2016 Compost fundamentals : compost needs : materials & methods to ensure quality compost (http://whatcom.wsu.edu/ag/compost/fundamentals/needs_carbon_nitrogen.htm).

International Conference on Biomass: Technology, Application, and Sustainable Development IOP Publishing
IOP Conf. Series: Earth and Environmental Science **65** (2017) 012037 doi:10.1088/1755-1315/65/1/012037

Potential lactic acid production from crude glycerol as the precursor of polylactic acid analog : literature review

D Y Hastati [1,2], E Hambali[3], K Syamsu[3] and E Warsiki[3]

[1]Post Graduate Program of Agroindustrial Technology, Bogor Agricultural University,Bogor 16602, Indonesia
[2]Program Study of Food Quality Assurance, Diploma Program, Bogor Agricultural University, Bogor 16151, Indonesia
[3]Department of Agroindustrial Technology, Faculty of Agricultural Technology, Bogor Agricultural University, Bogor 16602 Indonesia

Email : dwiyuninugraha@gmail.com

Abstract. Biodiesel has gained a significant amount of attention over the past decade as an environmentally friendly fuel. However, the biodiesel production process generates glycerol-containing waste streams which have become a disposal issue for biodiesel plants and generated an abundant of crude glycerol, a low-value byproduct of biodiesel manufacturing. Conversion crude glycerol to valuable chemical such as lactic acid, a precursor of polylactic acid (PLA), has a great potential to substitute traditional feedstocks of PLA, i.e., carbohydrate or sugar sources. Some of the process perspectives and the potential of glycerol to produce lactic acid by chemical transformation or microbial conversion are discussed in this paper, as well as the possibility of extending lactic acid to polylactic acid (PLA).

1. Introduction

Indonesia produces biodiesel 2 million kiloliters per year in 2016 and will raise to be 5 million kiloliters per year. Furthermore, under a mandatory of biodiesel Indonesia is expected to increase from 20 % (B20) in 2016 to 30 % (B30) by 2020 [1]. The increase of biodiesel production has thus resulted an inevitable abundance of glycerol as a by=products. This high amount of crude glycerol must be used for efficiency of biodiesel industry. Almost all industry use only refined glycerol as raw material, consequently unrefined glycerol has become potential environmental pollutant [2]. Purification of glycerol is an expensive process and the low process it has fetched recently have rendered it economically unfeasible. In order to gain sustainability of biodiesel industry, it is important to find feasible solution and efficient approaches to convert waste glycerol into valuable products. An alternative solution to handle the crude glycerol waste is by converting it into more valuable product, such as lactic acid (2-hydroxypropanoic acid) [3] [4] [5]. Lactic acid can be polymerized to produce Polylactic acid (PLA) [6].

PLA is the most intensively developed bioplastic coming from renewable resources and is used as a raw material for biodegradable plastic. Currently, PLA is established from starch. The problems are the limited resources and its effect to food sovereignty. PLA should be made from non-food raw materials and the alternative is glycerol from by-product of biodiesel production. PLA has already been industrially

Content from this work may be used under the terms of the Creative Commons Attribution 3.0 licence. Any further distribution of this work must maintain attribution to the author(s) and the title of the work, journal citation and DOI.
Published under licence by IOP Publishing Ltd

produced in an amount of 180 kton in 2012 and its manufacture is projected to reach 1 Mton per year by 2020. This means, the demand for lactic acid is increasing along with the increasing demand for PLA.

Synthesis PLA from glycerol, known as PLA analog can be divided into three main stages: ***firstly***, the synthesis of glycerol to lactic acid through a chemical reaction [7][8][9] or through microbial fermentation [10] [11]; ***secondly***, transformation of monomer lactic acid to lactide; ***thirdly***, the lactide is polymerized into PLA [6]. Generally, all stages of the process involves a catalyst during the process to increase the yield and accelerate the reaction rate.

The challenge in producing analog PLA is that it must have the same physical and chemical properties, or at least come near to the physical and chemical properties of conventional PLA. Adopted from [12], the successful of lactic acid into PLA transformation is influenced by: 1) the purity of lactic acid, 2) the acid endurance of microorganism used, 3) the quality of PLA produced must fulfill industrial PLA criteria. This paper reveals the challenges in developing poly lactic acid analog from glycerol as biodiesel by product.

2. Methodology

This research consists of exploratory research aimed at identifying the challenge of poly lactic acid synthesis from glycerol as biodiesel by product. This was followed by a review of journals, magazines, textbooks, etc. available on internet to identify the current conditions. A bibliographic review of national and international journals was carried out to identify available method in producing poly lactic acid and lactic acid from glycerol.

3. Discussion

Lactic acid is conventionally produced through carbohydrate fermentation and Lactonitrile chemical synthesis[13]. These two methods have been used in commercial production. The weakness of these method is the availability of feedstocks. Even though lactic acid is produced from renewable source, carbohydrate, the availability of carbohydrate as the feedstock competes with its use as food source. Whereas lactonictrile chemical synthesis uses unrenewable petrochemical ingredient and produces lactic acid in racemic mixture. The downside of this racemic lactic acid is its low purity value which only enables it to be used as co-monomer of other polymers [9].

Naturally, lactic acid is composed of two optical isomer: D(–)-lactic acid and L(+)-lactic acid. D(–)-lactic acid is hazardous to human, which makes L(+)-lactic acid is more preferable in food industry and pharmacy.

Value-added transformation processes of glycerol to lactic acid can be accomplished by chemically [7], [8] [3] [14], microbiologically [10] [11]. The method is quite widely reported to synthesize lactic acid from glycerol is in an alkaline hydrothermal and hydrothermal oxidation of glycerol. The second drawback of this method is to use a strong acid such as H_2SO_4 to convert the sodium lactate into lactic acid which may cause corrosion on the equipment used. Acidity is one factor affecting the success of the conversion of glycerol into lactic acid. In addition, this method produces lactic acid which is not pure, but in the form of racemic mixture comprising: L (+) - lactic acid, D (-) - lactic acid, and meso-lactic acid. Dimerization of racemic lactic acid will produce a racemic lactide comprising a mixture of: L-Lactide, D-lactide, and meso-lactide. Results of racemic lactide polymerization will produce PLA amourf so limited in its use. On the other side, the advantages of using microbial fermentation method compared to chemical method in producing lactic acid are its ability to produce pure lactic acid, uses low energy input (pressure / temperature), and no needs intensive pretreatment requirements [15].

3.1. The Purity of Lactic Acid

The purity of lactic acid will affect the quality of PLA products, especially on the physical properties. To get a high degree cristallinity and high melting point, the PLA must be made from the pure isomer L (+) - lactic acid or D (-) - lactic acid instead of the racemic of both isomers [16]. PLA with high crystallinity value and high melting point can be used to produce fiber, oriented films, and liquid crystals [17].

In the production of lactic acid from glycerol, the purity of isomer, i.e., L-lactic acid or D-lactic acid, depends on the strain of microbes used. The appropriate microorganism can produce desirable isomer. [18] Reported that lactic acid could be produced with glycerol fermentation by using some *E. coli* strains as an alternative to NAD+ regeneration in the absence of external electron acceptors. Moreover, high chiral purity of D-lactate can be produced by fermenting glycerol using a recombinant strain that overexpresses enzymes that respond to glycolytic intermediates and inactivates fumarate reductase, phosphate acetyltransferase, alcohol/acetaldehyde dehydrogenase, and D-lactate dehydrogenase [10], and reported that 32 g/L of D-lactate (99.9% chiral purity) could be produced from 40 g/L of glycerol by using E. coli D-specific LDH. Furthermore, [19] reported that L-specific LDH from *Streptococcus bovis* could replace the native *E. coli* D-specific LDH from the previous study, e.g., 50 g/L of L-lactic acid (99.9% chiral purity) was produced from 56 g/L of waste glycerol.

Other researchers found that using *Rhizopus* strain as alternative lactic acid producer microbe has many advantages because it produces only L(+)-lactic acid isomer, does not require special nutrition, and also more easier in separating biomass from fermentation media. These could reduce the cost of downstream process [20]. The use of *R. oryzae* NRRL 395 on 75 g/l concentration of crude glycerol produce 3.72 g/g L(+)-lactic acid and the concentration of lactic acid is assumed to be 48 g/l. Interestingly, in this research the increase of lactate productivity is done by adding Lucerne green juice into fermentation media without addition of inorganic nutrient on crude glycerol [21].

3.2. Acid Endurance of Microorganism

Generally, the desirable characteristics of microbe strain in glycerol fermentation is the ability to rapidly converse substrate and to produce high yield. For lactic acid production, specific microbes that can stand at low pH and high temperature are needed. Conventional lactic acid production is conducted via glucose fermentation using homofermentative LAB (Lactic Acid Bacteria) which requires weak acidic to neutral pH. Low pH can give inhibitory effect on cell metabolism which will result in lactic acid production [22]. Generally, LAB cannot grow in under pH 4 even though lactic acid pK value is 3.78. To keep neutral pH to maintain cell growth, lime is frequently added into fermentor which will produce over 90% of lactic acid, however it will result in insoluble calcium sulfate (gypsum) [22]. The use of high concentration sulfuric acid causes corrosion problem. The amount of gypsum is usually higher than the amount of lactic acid produced. Roughly more than 1 ton of gypsum were produced. This will result in waste management which will impose high cost to other commercial product. The cost of purifying lactic acid is relatively high cost component from production cost total.

The advantage of using fungi is that it can grow in low pH medium while able to significantly reduce the use of neutralizing agent ($Ca(OH)_2$) during fermentation stage and can avoid gypsum formation ($CaSO_4$). But there is still a need to emphasize metabolic engineering research in yeast and fungi which can naturally produce pure lactic acid with high productivity.

3.3. Meeting PLA Industrial Criteria

The purity of lactic acid is influenced by fermentation medium composition which depends on the nutrition requirement of microbe strain used. Traditionally, actic acid can be produced by using bacteria and fungi. Lactic Acid Bacteria (LAB) has been widely used for having high growth rate and good product yield. However, LAB has complex nutrition requirement since it has limited ability in synthesizing vitamin B and amino acid. Thus, there is a need for adequate nutrition supplement on the media used such as yeast extract or additional supplementary complex, for example dried blood needed to catalyze active cell mass, in order to produce lactic acid [23]. This relatively high cost downstream process increases the whole cost of lactic acid production if lactic acid bacteria were to be used.

The use of relatively cheap main ingredients in lactic acid industry will reduce production cost and maintaining economic viability of microbial process. Using starch and glucose as main ingredients increases of lactic acid cost, because lactic acid fermentation process takes up 40-70% of production cost total [24].

Several parameters must be considered in choosing main ingredients in lactic acid industry. There are low main ingredient price, minimum contaminant, high fermentation rate, high lactic acid production, few or no by-product formation, and always readily available. Ever since the increase of glucose price for its usage in food industry, production of lactic acid begins to favour other alternative renewable resources and have low price substrate, for example the use of glycerol and lignocellulosic [25] [26] [27].

4. Conclusion

With the increase of biodiesel industry from palm oil, there has been important development by using different bioengineering and chemical techniques to convert glycerol as by-product of biodiesel into lactic acid. There were several excellent reviews published in the past several years with a focus on microbial species or chemical catalytic processes. However, there is still a challenge to find the best method in producing lactic acid with the high purity and high yield (productivity). Nowadays, the price of lactic acid is still high and it has small market capacities. This is likely due to its cost production. The ability to make this compound at a lower cost could help to bring it into a larger market.

References

[1] Ministry of Industry - March 2016 Indonesia produsen utama biodiesel. Article [Online]. Available: (www.kemenperin.go.id/artikel/1903/Indonesia-Produsen-Utama-Biodiesel)

[2] Leoncti A B, Leoncti V A, De Olievera S V W B 2012 Glycerol as a by-product of biodiesel production in brazil: alternatives of the use of unrifened glycerol *Renew. Energy*. **45** 138-45

[3] Ramirez-Lopez C A, Ochoa-Gómez J R, Fernández-Santos M, Olga Gómez-Jiménez-Aberasturi O G J, Alonso-Vicario A and Torrecilla-Soria J 2010 Synthesis of lactic acid by alkaline hydrothermal conversion of glycerol at high glycerol concentration *Ind. Eng. Chem. Res*. **49** 6270–78.

[4] Dibenedetto A, Angelini A, Aresta M, Ethiraj J, Fragale C and Nocito F 2011 Converting wastes into added value products: fromglycerol to glycerol carbonate, glycidol and epichlorohydrin using environmentally friendly synthetic routes *Tetrahedron* **67** 1308–13

[5] Balaraju M, Jagadeeswaraiah K, Prasad, P S S and Lingaiah N 2012 Catalytic hydrogenolysis of biodiesel derived glycerol to 1,2-propanediol over Cu–MgO catalysts *Catal. Sci. Technol.* **2** 1967–76

[6] Dusselier M, Wouwe P V, Dewaele A, Jacobs P A and Sels B F 2015 Shape-selective zeolite catalysis for bioplastics production *Science* **349** 78-80

[7] Kishida H, Jin F, Zhou Z, Moriya T and Enomoto H 2005 Conversion of glycerin into lactic acid by alkaline hydrothermal reaction *Chem. Letters* **34** 1560-61

[8] Shen Z, Jin F, Zhang Y, Wu B, Kishita A, Tohji K and Kishida H 2009 Effect of alkaline catalysts on hydrothermal conversion of glycerin into lactic acid *Ind. Eng. Chem. Res.* **48** 8920–25

[9] Ye P and Ren S 2014 Value-Added Chemicals from Glycerol *Am. Chem. Society.* ACS Symposium Series; American Chemical Society: Washington, DC.

[10] Mazumdar S, Clomburg J M and Gonzales R 2010 *Escherichia coli* strains engineered for homofermentative production of d-lactic acid from glycerol *App. And Env. Microbiology* **76** 4327–36.

[11] Chen X Z, Algasan G, Tian K M, Singh S, Niu D D, Wang Z X and Shen W 2014 Efficient bioconversion of crude glycerol from biodiesel to optically pure D-lactate by metabolically engineered Escherichia coli *Green Chem.* **16** 342–50

[12] Upadhyaya B, DeVeaux L C and Christopher L P 2014 Metabolic engineering as a tool for enhanced lactic acid production *Trends in Biotech* **32** 637-44

[13] Wee Y J, Kim J N and Ryu H W 2006 Biotechnological Production of Lactic Acid and Its Recent Applications *Food Technol. Biotechnol* **44** 163–72

[14] Kishida H, Hasegawa T, Moriya T, Ohara H and Nomura N 2014 Operation of bench plants producing racemic lactic acid and lactide isomers from glycerol, a by-product of biodiesel fuel production process *Energy Procedia* **56** 187 – 94.

[15] Yang F X, Hanna M A and Sun R C 2012 Value-added uses for crude glycerol—A byproduct of biodiesel production *Biotech. Biofuels* **5** 13

[16] Hofvendahl K and Hans-Hagerdal B 2000 Factors affecting the fermentative lactic acid production from renewable resources *Enz.Microb. Technol.* **26** 87-107

[17] Soedegard A and Stolt M 2002 Properties of lactic acid based polymer and their correlation with composition *Prog. in Polymer Sci.* **27** 1123-63.

[18] Hong A A, Cheng K K, Peng F, Zhou S, Sun Y, Liu C M and Liu D H 2009 Strain isolation and optimization of process parameters for bioconversion of glycerol to lactic acid *J. Chem. Technol. Biotechnol* **84** 1576–81

[19] Mazumdar S, Blankschien M D, Clomburg J M and Gonzalez R 2013 Efficient synthesis of L-lactic acid from glycerol by metabolically engineered *Escherichia coli Microb. Cell Fact* **12** 1-11

[20] Yin P M, Nishina N, Kosakai Y, Yahiro K, Park Y and Okabe M 1997 Enhanced production of l(+)-lactic acid from corn starch in a culture of *Rhizopus oryzae* using an air-lift bioreactor *J. Ferment. Bioeng.* **84** 249–53

[21] Vodnar D C, Duft F V, Pop O L and Socaciu C 2013 L (+)-lactic acid production by pellet-form Rhizopus oryzae NRRL 395 on biodiesel crude glycerol *Microb. Cell Factories* **12**

[22] Datta R and Henry M 2006 Lactic acid: recent advances in products, processes and technologies - a review *J. Chem. Technol. Biotechnol.* **81** 1119–29

[23] Koller M, Salerno, A, Dias M, Reiterer A and Braunegg G 2010 Modern biotechnological polymer synthesis: a review *Food Technol. Biotechnol.* **48** 255–69.

[24] Yazdani S S and Gonzalez R 2007 Anaerobic fermentation of glycerol: A path to economic viability for the biofuels industry *Curr. Opin. Biotech* **18** 213–19

[25] Ye L, Zhou X, Hudari M S, Li Z and Wu J C 2013 Highly efficient production of L-lactic acid from xylose by newly isolated Bacillus coagulans C106 *Bioresour. Technol.* **132** 38–44

[26] Zhang Y, Xiangrong Chen X , Luo J, Qi B and Wan Y 2014 An efficient process for lactic acid production from wheat straw by a newly isolated Bacillus coagulans strain IPE22 *Bioresour. Tech.* **158** 396-9

[27] Morales M, Dapsens P Y, Gioninazzo I, Witte J, Modelli C, Papadokonstantakis S, Hungerbuhler K and Ramirez JP 2014 Environmental and economic assessment of lactic acid production from glycerol using cascade bio and chemocatalysis *Energy & Env. Science.* **8** 558-67

Potential production of palm oil-based foaming agent as fire extinguisher of peatlands in Indonesia: Literature review

P Subekti[1,2], E Hambali[3,4], A Suryani[3,4], P Suryadarma[3]

[1]Program Study of Mechanical Engineering, Pasir Pengaraian University, Regency of Rokan Hulu, Riau, Indonesia
[2]Post Graduate Student, Program Study Agricultural Industrial Technology, Bogor Agricultural University, Bogor, Indonesia
[3]Department of Agricultural Industrial Technology, Faculty of Agricultural Industrial Technology, Bogor Agricultural University, Bogor, Indonesia
[4]Surfactant and Bioenergy Research Center, Bogor Agricultural University, Bogor, Indonesia

E-mail: purwos73@gmail.com

Abstract. This study aims to analyze the potential aplication of of palm oil-based foaming agent as peat fires fighter in Indonesia. From literature review, it has been known that the foaming agent able to form foam to extinguish fire, wrap and refrigerate the burning peat. It is necessary to develop the production and application of foaming agent in Indonesia because peat fires occur almost every year that caused smoke haze. Potential raw material for the production of environmental friendly foaming agent as foam extinguishing for peat fires in Indonesia aong other is palm oil due to abundant availability, sustainable, and foam product easily degraded in the environment of the burnt areas. Production of foaming agent as fire-fighting in Indonesia is one alternative to reduce the time to control the fire and smog disaster impact. Application of palm oil as a raw material for fire-fighting is contribute to increase the value added and the development of palm oil downstream industry.

1. Introduction

Alternative ways to accelerate peat fires fighter is using foam. In peat fires fighting, foam serves to wrap, put out the fire and cooling down the burnt area so that the fire did not initiate again. Some foaming agent has been used as fire fighter such as : Halon but but since 2000 has been banned because it produces phosgene gas which is harmful to health [1,2]; Halotron give good results but emit gases that reduce the ozone layer and in 2015, its use to be stopped [3]. As an alternative, Film-Forming Fluoro Protein (FFFP) is also able to produce a good foam to extinguish the fire and more environmental friendly [4]. These materials are made overseas, so it is necessary to use local materials and abundant that are environmentally friendly in Indonesia.

Foaming agent is a foam-forming material when mixed with a certain concentration of water then stir to form foam solution. Foam is a dispersion system of gas/air in the bubbles fluid that can float on the liquid surface and flows over the surface of solids [5,6]. Pure liquid will not be foamy except surfactants that mixed in the liquid. The presence of surfactant will reduce interfacial tension of gas/liquid thus simplifying the gas dispersion in liquid to form bubbles [6,7]. Foam can be used to extinguish the fire class B such as gasoline, diesel, benzene and butane [8]. In other hand, foam can also be used to extinguish the fire class A such as material on land and forest fires [9].

Content from this work may be used under the terms of the Creative Commons Attribution 3.0 licence. Any further distribution of this work must maintain attribution to the author(s) and the title of the work, journal citation and DOI.
Published under licence by IOP Publishing Ltd

In some countries such as the USA, Australia, Russia and Japan, the application of foaming agent from natural material to extinguish forest fire continues to develop as a substitute for synthetic materials derived from petroleum that tends to be difficult to be degraded in the environment [9,10]. Firefighters using foam is more quickly than water because foam wrap the burning material so that air does not enter into it. In addition, the use of foam is also make more efficient in the use of water [4,9,10,11,12,13,14].

In some parts of Indonesia, forest and peat fires occurs almost every year during the dry season that resulting smoke disaster. Peat fires contribute around 85-90% of smoke disaster due to more smoke than the fire of other soil types [15,6]. Smoke generated from peat fires cause health problems, especially respiratory infections [15], the increase in carbon emissions, disruption of economic activity, environmental, political relations with neighboring countries and ecosystems [17,18,19,20]. In other hand, some activities also will be disrupted such as educational activities, transportation and health of both humans and animals.

From the description above it is necessary to study on the use of foaming agent as a material made of foam for the fire in peat lands in Indonesia as well as other countries use it, so expect smoke haze caused more quickly resolved.

2. Research method

This article is a review of previous research finding related to development and production of environmental friendly and sustainable fire-fighting foaming agent, using analytical description of the relevant literature and secondary data from previous researchers. Associated literatures are among others Mizuki *et al.* (2010) about the foaming agent environmental friendly material from fatty acids, Oguike (2013) on foam made from red palm oil, Kawahara, *et al.* (2016) on the development of soap - based Vfire fighter agent with the basic ingredients of fatty acids, Joshua and Tickell (1999) about the potential of vegetable oil, GAPKI (2017) about the production of CPO in Indonesia and some other relevant researchers. The discusion is focused on the potential production of fire-fighting palm oil based foaming agent in Indonesia.

3. Results dan discusion

3.1. Production process of foaming agent

The production process of foaming agent firefighter with palm oil as raw material generally use palmitic, stearic and oleic acid as product from palm oil refining through etilester hydrolysis process [24]. The reaction temperature in the manufacturing process of foaming agent is between 80 - 95 °C, and at operating pressure of 1 atm [25]. Potassium chloride is also added in the reaction to reduce the viscosity of the reaction product to facilitate transportation of reaction through the pump. The neutralization reaction takes place in a circulation reactor consisting of Turbodisper and mixer. Turbodisper serves to homogenize the reactant mixture while the mixer serves to provide sufficient residence time for the reaction to react completely. Turbodisperser stirring rotation speed is between 40-50 rps and in mixer is between 5-20 rps [26]. Conversion of fatty acids reaction obtained in this way can reach more than 99.9% [25]. After the reaction occurs then the foaming agent can be directly added with additives to improve foam quality. Substances that are added to the foaming agent is include silica and acetic acid, then stir until evenly distributed [27].

To obtain firefighter foam, the foaming agent must be mixed with water in foam generator regulator. Foam will come out from regulator exhaust after the mixture of foaming agent, water and air occur. For the purpose of application on peatland, the foam should be stable, able to cover and cooling the burnt area so the fire is die and did not return. Foam stability is refers to the ability to maintain the main parameters in a constant state for a certain time. These parameters include the bubble size, fluid content, and the total volume of the foam. Foam lifetime is the simplest measure to show the stability of the foam [28]. For general application of foam as firefighter, foam is often collapse or evaporates due to pressure from the surface of the burnt areas. In addition, the main cause of foam collapse is thinning film and coalescent. Thinning occurs because the foam tends to rise to the top but also pulled down due to fluid flow (drainage) as a result of gravity force. Because drawn from two directions, the foam film thinned so much easier to rupture .In addition, varies foam size causes the gradient of gas pressure. As a result, the gas diffusion can occur, where small bubbles will merge into a larger foam (coalescent). The larger size of foam means the greater the surface tension, so that the foam more easily to rupture [6,29].

Increasing the bulk viscosity of the fluid can prevent the foam collapse, for example with the addition of glycerol or polymers. Increasing the viscosity will reduce the drainage speed. When the drainage speed decreases, the thinning can be minimized [6]. In addition, the foam will create a steric hindrance that inhibits foam for mutually joining [29]. In addition, foam stability may also be supported by increase in the viscosity and elasticity of the surface by mixing several kinds of surfactants, in order to get a dense surfactant film and not easily to collapse[6].

3.2. Developent of Fire Fighter foaming agent

In some countries, land and forest fire fighting are already using more advanced technology by utilizing environmental friendly and sustainable materials. Researchers who have large contribution to develop thus material among others is Timpson LGM (1938) which developed fire fighter foam from coconut oil and palm kernel oil. The resulting foam is stable and resistant to moisture. Mizuki *et al.* (2007) has develop fire fighter foam for large areas using soap-based environmental friendly liquid foam with the raw material such as sodium oleic (10% w / w), potassium laurate (6.87% w / w), potassium palmitate (0.1% w / w), glutamatic acid diacetic acid (GLDA) (40% w / w) and diluents (43.03% w / w). The foaming agent concentration is 1% and life time of the foam is shorter than the synthesis foam. Besides as subtitute of chemicals substance that tend to damage the environment, the fire fighter foam is environmental friendly because it is easily decomposed by water and land. Mizuki *et al.* (2010) is also developing environmental friendly foaming agent using long chain fatty acids (LCFA), a biodegradable chelating agent and the additive that produces biodegradable foaming agent. Iwamoto *et al.* (2013) has developed foaming agent fire fighter from the seeds of guar gum. The foams from these materials use less water concentrations when compared to non-vegetable synthetic materials and also able to reach a wider area.

Meanwhile, Oguike (2013) has developed foaming agent fire fighter from ethyl ester of red palm oil (78.84 grams), sodium hydroxide (68.89 gr), toluene (125 ml) and ethanol (50 ml) by the hydrolysis process. By ratio of foaming agent with water at 1: 8 is able to produce foam lifetime around 345 hours. The resulting foam is able to distribute uniformly cover the surface of burning liquid. Joseph *et al.* (2014) has developed fire fighter foam agent from potato starch, in which the test material coated with potato starch extract formulated with water is able to survive longer to burn out than uncoated test material. In 2015, Yue and Dong has developed foaming agent firefighter from rice protein, peanut and cottonseed. Foam produced from the formulation can be stored longer, low viscosity and high foam expansion power.

Likewise, Vinogradov *et al.* (2015) has developed the silica based foaming agent which able to extinguish fire at temperature of 800 °C. The test show that the foam improve the efficiency of nearly 50 times higher than ordinary water, and 15 times better than the synthesis material. Silica can also be produced by using palm oil mill boiler ash. The ash is a combustion residue of fibers and shells of palm oil fruit, with a silica content of 86.7% in weight [34]. Kawahara *et al.* (2016) developed a soap-based foaming agent from chelating agent raw materials (66.84% w / w), potassium laurate (4% w / w), potassium oleate (6% w / w), sodium oleate (6 % w / w), methyl gliycine diacetic acid (MGDA) (9.7% w / w), ethylene dinitrilotetra acetic acid (EDTA) (9.7% w / w), nitrilotriacetic acid (NTA) 6.3% w / w) and others additive. The resulting foaming agent is capable to produce a stable foam that survive at temperatures -10 °C, which has been applied for forest fire and the resulting foam is very biodegradable and environmental friendly.

Nevertheless, peat fire is differing than canopy and litter fire so that applicable foaming agent needs to pay attention on the environmental impact. Peat environmental analysis is needed to determine the extent of the effect of palm oil based foaming agent for the ecosystem after application [35,36,37,38].

The use of foam for forest fires fighter have increased the effect of fire fighter, surfactant in foam agent is significantly lowers the surface tension. In other hand, the use of surfactant allows more efficient water use, easy to stick on the burning material and allow water to penetrate into the combustible material. In some countries, including Indonesia, concerns about environmental damage from the use of synthetic foam cause the switch to using foam derived from natural ingredients and has extensive forest fire fighting [9]. The use of soap based foam is has much lower toxicity for aquatic organisms in rivers, lakes and the ocean compared with the synthetic foam [13,39,40,41]. Applications of peat lands firefighters in Indonesia needs to be examined to obtain foaming agents which capable form foam for firefighter, wrap, and cooling peat burning area. Fire propagation on peat land surface is around 58.1-160 m^2/min with flame heights is around 1.3-2.1 m, while fire propagation in the inner layer of peat is around 0.1-0.15 m/day [19]. With the fast fire propagation in the surface, it is required fire fighting foam that covers the surface of the burning area and can survive for some time until the fire was no longer burning. Besides the foam is resistant to hot temperatures that occur on peat fires [42].

In general, the foam used to fight the fire is easily evaporate because surface temperatures of fire is higher than foam temperature [19,43] so it is required a lot of foam volume. The evaporation of the foam is caused by bubbles rupture, film layer thinning and coalescent that cause foam collapse. Thinning occurs because the foam is pulled down by fluid flow (drainage) due to gravity but once the foam tends to rise to the top. Foam film thinning is pulled from two directions, making it easier to rupture. Greater foam dimension cause greater surface tension, thus facilitating rupture [6,28]. By increasing viscosity of the fluid, for example with the addition of glycerol or polymers, it can prevent the foam rupture. Increasing the preparation viscosity will make drainage speed decreases, and in turn thinning can be minimized [6]. Increased surface viscosity and elasticity through mixing several kinds of surfactant to produce dense and not easily collapse surfactants film can also increase the stability of the foam [6]. In addition, it is expected that produced foaming agent is capable to increase the amount of foam and performance as well as wider range of fire fighting. Substance that are used to lower the water surface tension is surfactant that serves to increase the formation of a stable foam [44,45,46], and because it is not easy to evaporate, the foam liquid is able to penetrate the surface pores of burning [7,9].

3.3. Potential production of fire fighter foaming agent

Foaming agent is a foam material for fire fighter from vegetable fat oils and other synthetic materials. Vegetable fat oils can be used as raw material for fire fighter foaming agent [47,48,49]. Almost every country has potential plant-based material that can be used as raw material for the production of fatty oil. Each type of vegetable raw materials which processed into fat oil produce different yield per hectare as presented in Figure 1.

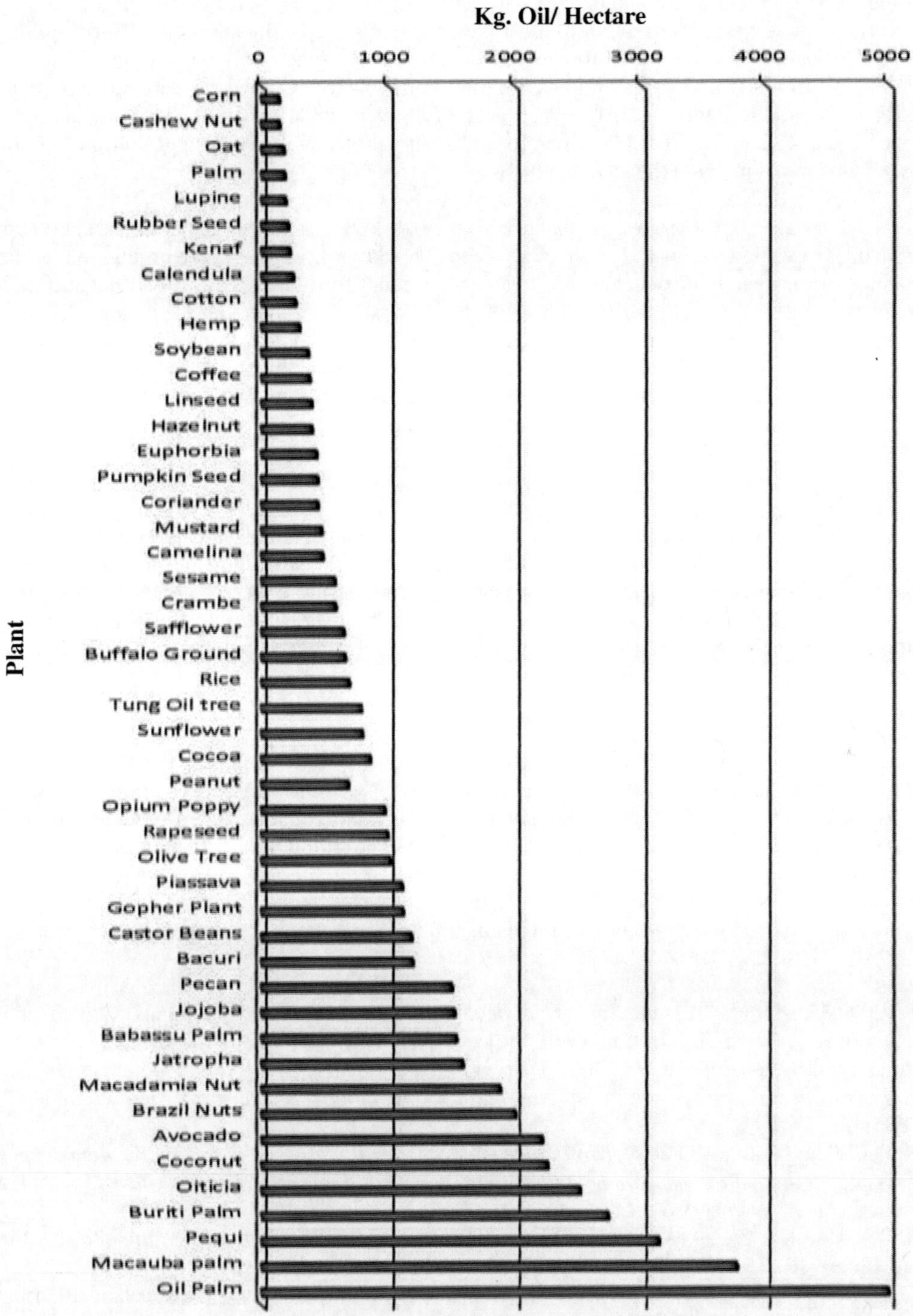

Figure 1. Oil production crops [22]

In Indonesia, there are different types of fat oil-producing plant materials that can be used as raw material for forest fighter foaming agent. Utilization of bio material as raw material for fire fighter foam is potential to be developed in Indonesia, but consideration must be paid on the potential disrupt of food supply. Selection of palm oil as a raw material for fire fighter foam is the right choice and the opportunity is quite large compared with other vegetable materials. In addition, the material is environmental friendly because of easily degraded and the existence will be sustainable due to abundant supply. In 2016, Indonesia produced Crude Palm Oil (CPO) and its derivative products amounted to 31.5 million tons, mostly for export (79.68%), while the remaining for domestic consumption amounted to 20.32% [23]. Therefore, the supply of raw materials will not interfere the domestic food even will increase the added value.

Figure 1 shows that the yearly production of fatty oil is 5000 kg/Ha from raw materials of palm fruit, the biggest oil production compared with other vegetable crops [22]. Potential production of fire fighter foaming agent need to be developed in Indonesia because forest and peat fires still continues to occur each year. Forest and peat land fires are expected to continue in line with the ongoing opening of new plantations [50]. Firefighter foaming agent will continue to be needed in Indonesia because of fires will recur from year to year, especially in the dry season, along with an increase in the number of land for plantation and drought in some forests and peat lands [51]. Raw materials of foaming agent from fatty acid of palm oil is abundant, environmental friendly and rapidly degraded to soil and water [12,13,39,40].

4. Conclusion

Peat land firefighter with foaming agent as a medium to form foam is able to stop fire significantly. In other hand, the use of foam allows water use more efficient, easy to stick on the burning material and allow water to penetrate into the deep layers of peat land. Indonesia's concerns about environmental damage from the use of synthetic foam for the fire, causing the necessity to use foam derived from natural ingredients and with larger effects. The use of foam from fatty acid of palm oil is lower environmental toxicity on peat land than synthetic foam.

In Indonesia, the most potential material to be developed as fire fighter foam agent is fatty acid derived from palm oil and able to be produced larger than other oil sources. Palm oil is also a potential commodity that does not affect the use of food products because large part of the Indonesian palm oil is for export and only smaller percentage for domestic consumption. Production of fire-fighting foaming agent from palm oil is potential to be developed in Indonesia. In addition to speed up the distribution of fire fighting and minimizing smoke disaster risk, it s also increasing the value added of palm oil products. Peat fire fighting using foam-based soaps from palm oil allows fire fighting more quickly than water because foam blanketing burning that prevent air to enter into it.

References

[1] WMO 2007 *Scientific assessment of ozone depletion: 2006* World Meteorological Organization Global (Ozone Research and Monitoring Project—Report No. 50) pp 8.15-8.32

[2] Peraturan Mentri Perindustrian RI, 2007 *Tentang: Larangan memproduksi bahan perusak lapisan ozon* serta *memproduksi barang yang menggunakan bahan perusak ozon* Permenrin RI 33 (Jakarta)

[3] Wuebbles DJ and Patten KO 2009 *Three-dimensional modeling of HCFC-123 in the atmosphere: Assessing its potential environmental impacts and rationale for continued use* Journal of Environmental Science & Technology **43**(9) 3208-3213 (doi:10.1021/es802308m)

[4] Krol B, Prochaska K. and Chrzanowski L 2012 *Biodegradability of firefighting foams* Fire technology **48** 173-181 (doi: 10.1007/s10694-010-0204-7)

[5] Rosen MJ 2004 *Surfactants dnd interfacial phenomena, third edition* (Surfactant Research Institute Brooklyn College The City University of New York) p 277

[6] Tadros TF, 2005 *Applied surfactants: Principles and applications* (WILEY-VCH Verlag GmbH and Co. KgaA Weinheim) pp 155-260

[7] Schmitt TM 2001 *Analysis of surfactants* 2rd ed Revised and Expanded (Marcel Dekker Inc. New York) p 61

[8] Bourgeois A, Bergendahl J and Rangwala A 2015 *Biodegradability of fluorinated fire-fighting foams in water* Chemosphere **131** 104–109 (doi.org/10.1016/j.chemosphere.2015.02.042)

[9] Kawahara T, Hatae S, Kanyama T, Ishizaka I and Uezu K. 2016 *Development of eco-frindly soap-based firefighting from for forest fire* Evironmental Materials and Protocols Section: Short Communication **54**(1) 75-78 (doi. 10.2525/ecb.54.75)

[10] Mizuki H, Ueza K, Kawano T, Kadono T, Kobayashi M, Hatae S, Oba Y, Iwamoto S, Mitumune S, Nagatomo Y, Owari M, Umeki H and Yamaga K. 2007 *Novel environmental friendly soap-based fire-fighting agent* J.Environ. Eng. Manage. **17**(6) 403-408

[11] Rewet D, Smith R and Karvainis G 1996 *A Comparison of water additives for mopping-up after forest fires* Int. J. Wilrllarrd Fire **6**(1) 37-43

[12] Oguike R S. 2013 *Study of fire fighting foam agent from palm oil for extinguishing of petrol fires* Material Science Research Laboratory Department of Chemistry Abubakar Tafawa Balewa University Bauchi Nigeria **1**(1) 1-7 (doi:10.14340/spp.2013.12A0002)

[13] Kawano T, Otsuka K, Kadono T, Inokuchi R, Ishizaki Y, Dewancker B and Uezu K. 2014 *Eco-toxicological evaluation of fire-fighting foams in small-sized aquatic and semiaquatic biotopes* Advanced Materials Research 875-877 699-707 (doi:10.4028/www.scientific.net/AMR.875-877.699)

[14] Noriyasu A, Otsuka K, Ishizaki Y, Tanaike Y, Matsuyama K, Uezu K and Kawano T 2014 *A novel wild-land fire-fighting foam for minimizing the phytotoxicity of wood burning-derived smoke tested in living plant cells* Advanced Materials Research. Trans Tech Publications, Switzerland **875-877** 725-733 (doi:10.4028/www.scientific.net/AMR.875-877.725)

[15] Hinwood AL and Rodriguez CM 2005 *Potential health impacts associated with peat smoke: A review* Journal of the Royal Society of Western Australia **88** 133-138

[16] Permadi DA and Oanh NTK 2013 *Assessment of biomass open burning emissions in Indonesia and potential climate forcing impact* Atmospheric Environment **78** 250-258 (doi.org/10.1016/j.atmosenv.2012.10.016)

[17] Andreae MO and Metlet P 2001 *Emission of trace gases and aerosols from biomass burning Global Biogeochemical Cycles* **15**(4) 955-966 (doi: 10.1029/2000GB001382)

[18] Rahmayanti M 2007 *Kontribusi kebakaran lahan gambut terhadap pemanasan global* Kaunia **3**(2) 101-117

[19] Rachmawati N 2008 *Karakteristik bahan bakar dan perilaku api pada kebakaran hutan dan lahan rawa gambut* Jurnal Hutan Tropis Borneo **22** 55-64

[20] Toriyama T, Takahashi T, Nishimura S, Sato T, Monda Y, Awaya Y, Limin SH, Susanto AR, Darma F, Krisyoyo, Saito H and Kiyono Y 2014 *Estimation of fuel mass and its loss during a forest fire in peat swamp forests of Central Kalimantan Indonesia* Forest Ecology and Management **314** 1–8 (doi.org/10.1016/j.foreco.2013.11.034)

[21] Mizuki H, Uezu K, Toyomura M, Yasui H, Kawano T, Akiba I, and Mizota C 2010 *Microbial degradation of a shoap-based fire-fighting agent in activated sludge* J. Environ. Eng. Manage **20**(2) 109-113

[22] Joshua and Tickell K 1999 *From the fryer to the fuel tank: The complete guide to using vegetables oils as an alternative fuel* 2rd ed Tickell Energu Consulting (Tallahassee) pp 47-144

[23] GAPKI 2017 Reflection palm oil industri outlook for 2016 and 2017 Indonesian palm oil association (reflection 2016)

[24] Moody CA and Jennifer AF 2000 *Critical review: Perfluorinated surfactants and the environmental implications of their use in fire-fighting foams* Environmental Science and Technology **34**(15) 3864-3870

[25] Othmer HG 1976 *The qualitative dynamics of a class of biochemical control circuits* Journal of mathematical biology **3**(1) 53-78

[26] Spitz dan Luis. 1996. Soap and Detergents-A Theorical and Practical Review. AOCS Press. Champaign.

[27] Vinogradov A V., Kuprin D, Abduragimov L, Kuprin G, Serebriyakov E and Vinogradov VV 2015 *Silica foams for fire prevention and firefighting* ACS Applied Materials & Interfaces is published by the American Chemical Society (doi: 10.1021/acsami.5b08653)

[28] Schramm LL 2005 *Emulsion, Foams and Suspensions* (Wiley-VCH Verlag GmbH & Co.KGaA, Weinheim) p 7

[29] Exerowa D, Kruglyakov P M and Mobius R 1998 *Foam and foam film: Theory, Experiment, Application* (Elsevier)

[30] Timpson LGM. Patent: Pyrene Minimax Corp 1938 *Method of and agent for producing fire extinguishing foam* Patent United States US2135365A

[31] Iwamoto S, Ohba Y, Miki E, Mochigase H, Suzuki Y, Kurita K and Okuzaki H. Patent: Furukawa Techno Material Co., Ltd, Furukawa Electric Co. Ltd. National University Corporation University of Yamanashi 2013 *Fire-extinguishing agent*. Patent: Jepang WO2013141367 A

[32] Joseph P, Bakirtzis D, Richard Q. 2014 . Preliminary studies on the efficiency of an environmentally-friendly fire-fighting agent based on starch. School of the Built Environment and the Built Environment Research Institute, University of Ulster, Newtownabbey, Antrim, Northern Ireland, UK.

[33] Yue H, Dong PL, Patent: Fire Anhui Tianyuan Technology Co., Ltd. 2015 *Vegetable Protein Foam Extinguishing Agent*. Patent China CN104984508A Civil Aviation Organization (ICAO).

[34] Zahrina I, Yelmida and Akbar F 2012 *Sintesis ZSM-5 dari fly ash sawit sebagai sumber silika dengan variasi nisbah molar Si/Al dan temperatur sintesis* Jurnal Rekayasa Kimia dan Lingkungan **9**(2) 94-99

[35] Utami SNH, Maas A, Radjagukguk B and Purwanto BH 2009 *Sifat fisik, kimia dan FTIR spektrofotometri gambut hidrofobik Kalimantan Tengah* Jurnal Tanah Trop **14**(2) 159-166

[36] Tahrun M, Wawan and Amri AI 2015 *The change of physical peat soil characteristics caused by fire in Teluk Binjai Village District of Teluk Meranti Regency of Pelalawan* JOM FAPERTA **1**(2) (Departement of Agroteknologi Faculty of Agriculture University of Riau)

[37] Vaasma T, Karu H, Kiisk M, Pensa M, Alliksaar KIT and Tkaczyk AH 2016 *Pb-210 and fly ash particles in ombrotrophic peat bogs as indicators of industrial emissions* Journal of Environmental Radioactivity 1-9 (doi.org/10.1016/j.jenvrad.2016.07.027)

[38] Setiawan Y, Pawitan H, Prasetyo LB, Parlindungan M, Lubis MI and Nurdiana A 2016 *Characterizing spatial distribution and environments of Sumatran peat swamp area using 250 multi-temporal MODIS data* (Procedia Environmental Sciences 33) (The 2nd International Symposium on LAPAN-IPB Satellite for Food Security and Environmental Monitoring LISAT-FSEM 2015) pp 117–127

[39] Lin C, Kadono T, Yoshizuka K, Uezu K and Kawano T 2006 *Assessing the eco-toxicity of novel soapbased fire-fighting foam using medaka fish (Oryzias latipes, Red-orange variety) adopted to river and sea water conditions* ITE Letters on Batteries, New Technologies and Medicien **7**(5) 499-503

[40] Goto K, Lin C, Kadono T, Hirono M, Uezu K dan Kawano T *2007 Eco-toxicity of a soap component (sodium oleate) and a synthetic detergent cocktail using green paramecia assayed in East Asia* J. Environ. Eng. Manag **17**(6) 377-383

[41] Kawano T, Lin C, Kadono T dan Uezu K. 2007. Ecological risk assessment of fire-fighting chemicals using medaka fish Oryzias latipes) in different water conditions. ITE-IBA Lett. Batter. New Technol. Medic. 8: 306-311.

[42] Huang X and Rein G 2014 *Smouldering combustion of peat in wildfires: Inverse modellingof the drying and the thermal and oxidative decomposition kinetics* Combustion and Flame **161** 1633-1644 (doi.org/10.1016/j.combustflame. 2013.12.013)

[43] Guitart NP, Rein G, Hadden RM, Belche CM and Yearsley JM 2016 *Effects of spatial heterogeneity in moisture content on the horizontal spread of peat fires* Science of the Total Environment (doi.org/10.1016/j.scitotenv.2016.02.145)

[44] Varade D, Carriere D, Arriaga LR, Fameau AL, Rio E, Langevin D and Drenckhan W 2011 *On the origin of the stability of foams made from catanionic surfactant mixtures* Soft Matter **7** 6557–6570 (doi: 10.1039/c1sm05374d)

[45] Arriaga LR, Drenckhan w, Salonen S, Rodrigues JA, PalomaresRI, Rio E and Langevin D 2012 *On the long-term stability of foams stabilised by mixtures of nano-particles and oppositely charged short chain surfactants* Soft Matter **8** 11085–11097 (doi: 10.1039/c2sm26461g)

[46] Maestro A, Rio E, Drenckhan W, Langevin D and Salonen A 2014 *Foams stabilised by mixtures of nanoparticles and oppositely charged surfactants: relationship between bubble shrinkage and foam coarsening* Soft Matter **10** 6975–6983 (doi: 10.1039/c4sm00047a)

[47] Shrestha LK, Shrestha RG, Solans C and Aramaki K 2007 *Effect of water on foaming properties of diglycerol fatty acid ester-oil systems* Langmuir **23**(13) 6918-6926

[48] Fameau AL, Ventureira J, Novales B and Douliez JP 2012 *Foaming and emulsifying properties of fatty acids neutralized by tetrabutylammonium hydroxide* Colloids and Surfaces A: Physicochemical and Engineering Aspects (dx.doi.org/10.1016/j.colsurfa.2012.03.059)

[49] Figueredo RCR and Sabadini E 2003 *Firefighting foam stability: the effect of the drag reducer poly (ethylene) oxide* Colloids and Surfaces A: Physicochem Engineering Aspects **215** 77-86

[50] Anderson IP and Bowen MD 2000 *Fire zones and the threat to the wetlands of Sumatra Indonesia* (Forest fire prevention and control project Departemen Kehutanan kantor Wilayah Propinsi Sumatera Selatan. European Union Ministry of Forestry) p 41

[51] Adinugroho WC, Suryadiputra INN, Saharjo BH and Sibro L 2011 Manual for the Control of Fire in Peatlands and Peatland Forest (Wetlands International Indonesia Program, Indonesia) pp 2-20

IOP Conf. Series: Earth and Environmental Science **65** (2017) 012039 doi:10.1088/1755-1315/65/1/012039

Preparation and evaluation adsorption capacity of cellulose xanthate of sugarcane bagasse for removal heavy metal ion from aqueous solutions

D A Iryani[1,4], N M Risthy [1], D A Resagian[1], S D Yuwono[2,4], U Hasanudin[3,4]

[1] Department of Chemical Engineering, Engineering Faculty, University of Lampung, Lampung, Indonesia.
[2] Department of Chemistry, Mathemathic and Science Faculty, University of Lampung, Lampung, Indonesia
[3] Department of Agroindustrial Technology, Agriculture Faculty, University of Lampung, Lampung, Indonesia
[4] Research and Development Center for Tropical Biomass, University of Lampung, Lampung Indonesia

Email: dewi.agustina@eng.unila.ac.id

Abstract. The discharge of heavy metals from industrial effluents into aquatic system in surrounding area of Lampung bay become a serious problem today. The data shows that the concentrations of heavy metals in this area are above allowable limits for the discharge of toxic heavy metals in the aquatic systems. The most common of heavy metal pollutant is divalent metal ions. Cellulose xanthate is one of the selective adsorbent to solve this problem, since xanthate contains two negative sulfur atoms that is capable to catch divalent metal ions. Preparation of cellulose xanthate was conducted by reacting carbon disulfide (CS2) and cellulose from sugarcane bagasse. The morphological characteristics of cellulose xanthate were visualized via Scanning Electron Microscope (SEM) and the presence of sulfur groups on sugarcane bagasse xanthate were identified by FTIR spectroscopic study. The degree of substitution (DS), degree of polymerization (DP), and adsorption capacities of cellulose xanthate for Cu^{2+} and Pb^{2+} metal were studied. The results of study reveals that the maximum adsorption capacities of Cu^{2+} and Pb^{2+} metal on cellulose xanthate are 54.226 mg Cu^{2+}/g, and 51.776 mg Pb^{2+}/g, respectively. This study reveals that cellulose xanthate could be a solution to reduce environmental pollution caused by industrial wastewater.

1. Introduction

The discharge of heavy metals from Industrial wastewater effluents into aquatic system in surrounding area of Lampung Bay and become a serious problem today. The data shows that the concentrations of heavy metals in this area are above allowable limits for the discharge of toxic heavy metals in the aquatic systems. Even at low concentrations, heavy metal ions are highly toxic and not biodegradable[1]. The most common of heavy metal pollutant is divalent metal ions such as copper, lead, zinc and chromium. Many studies have shown that these heavy metals may damage human health seriously[2]. Heavy metal accumulation in the human body can cause a health problems such as brain

Content from this work may be used under the terms of the Creative Commons Attribution 3.0 licence. Any further distribution of this work must maintain attribution to the author(s) and the title of the work, journal citation and DOI.
Published under licence by IOP Publishing Ltd

damage, metabolic disorders, and death. Toxicity in small doses can cause neurotoxic (neurotoxin) and abnormal behavior[3].

There are many methods for removal of heavy metals ion from aqueous solutions. The most simple and effective technique for removal of heavy metals ion from aqueous solutions is adsorption process, and ion-exchange are very often used in adsorption processes[4]. Nowadays, there has been considerable interest in the use of agricultural by-products as an adsorbents to solve this problem due to have some advantages include relative cheapness, renewability, high adsorption capacity, and abundance in nature especially in Lampung Province. Lampung is one of the provinces with abundant sugarcane bagasse resources[5]. Sugarcane bagasse is the by-product of sugar industry, which is constituted of cellulose, lignin and hemicellulose. Since sugarcane bagasse contains a large amount of hydroxyl groups in the structure[6], its absorption capability can be improved by chemical modification[7].

The purpose of this study was to modify sugarcane bagasse with carbon disulfide (CS_2) to enhance its adsorption properties for the removal of Cu^{2+} and Pb^{2+} from aqueous solutions. Sugarcane bagasse can uses as an adsorbent of heavy metal waste by converting it to cellulose xanthate. Cellulose xanthate is a reaction product of cellulose with carbon disulfide (CS_2) that forms a salt with the chemical formula $ROCS_2^-M^+$ (R = alkyl; $M^+ = Na^+$)[8]. Basically, xanthate contains two negative sulfur atom that is capable to catch divalent metal ions such as Cu^{2+}, Pb^{2+}, and others[9]. Effect of concentrations of CS_2 in cellulose xanthate synthesis was studied. The morphological characteristics of cellulose xanthate were visualized via Scanning Electron Microscope (SEM) and the functional groups present in the adsorbent were characterized by a Fourier Transform Infrared (FTIR) spectrophotometer. The degree of substitution (DS), degree of polymerization (DP), and adsorption capacities of cellulose xanthate (CX) for Cu^{2+} and Pb^{2+} metal were also studied.

2. Material and Methods

2.1 Chemicals

All chemicals Purchased from Merck were NaOH, carbon disulfide (CS_2), HNO_3, H_2SO_4, $BaCl_2$, distilled water and heavy metal solutions $PbSO_4$ and $CuSO_4$.

2.2 Sugarcane Bagasse Purification

Sugarcane bagasse (SB) provided from PT. Gula Putih Mataram (GPM) – Central Lampung was ground using a cutting mill to form powder with a maximum particle size of 1.0-1.5 mm, washed and dried in an oven at 105 °C for 3 h before purification treatment. Purification treatment method was adopted from Cerqueira *et al.*[10]. About 150 g of sugarcane bagasse were treated by soaking in 750 mL of NaOH 0.25 M for 18 hours at room temperature and continued in 750 mL of HNO_3 20% (v/v) for 3 hours at room temperature. After the treatments the purified sugarcane bagasse (PSB) was filtered and washed with distilled water, then dried at 105°C for 3 hours.

2.3 Synthesis Cellulose Xanthate

About 15 g of PSB was soaked in 100 mL of 18% NaOH solution and stirred for 3 hours at room temperature[11]. About 160, 180, 200% of CS_2 (w/w) by the amount of cellulose used was added to the mixture. It means about 24 g (19.2 mL) of CS_2 was used on the CS_2 concentration of 160%, 27 g (21.6 mL) of CS_2 on the CS_2 concentration of 180%, and 30 g (24 mL) of CS_2 on the CS2 concentration of 200%. After CS_2 was added to the mixture, the reaction was continued for 100 minutes at 35°C[8]. The product of cellulose xanthate (CX) was washed with distilled water to remove excess alkali, and then the sample was dried at 105°C for 3 hours and stored at low temperature (5-8°C)[12].

2.4. *Biosorption Experiments And Analysis*

All of the adsorption tests carried out by the batch technique. About 0.1 g of adsorbent was put into a 100 mL conical flask together with a 100 mL single metal ion (Cu^{2+} or Pb^{2+}) solution with pH range 2.0-6.5. The concentration of metal ion solution was 100 mg/L. The mixture was shaken at 120 rpm on a shaker for 2 hours[12]. The residual metal ion concentrations in the supernatant liquor were determined by using an Atomic Absorption Spectrophotometer (AAS), the amount of metal adsorbed (Q) were determined by using the following equations:

$$Q = \frac{V\,(C_0 - C_a)}{m} \tag{1}$$

Where Co (mg/L) and Ca (mg/L) are the initial and final metal ions concentrations, respectively. V (L) is the volume of solution, and m (g) is the weight of cellulose xanthate (adsorbent)[12].

2.5. *Determination of degree of substitution (DS)*

The degree of substitution is the ratio between the number of xanthate groups per cellulose unit in the cellulose xanthate compound. Before calculating the value of DS, the ratio of sulfur to cellulose ((% sulfur)/(% cellulose)) has to be determined first. Amount of sulfur in the cellulose xanthate was determined by using gravimetric methods[13], whereas amount of cellulose in the cellulose xanthate was determined by using gravimetric methods[14]. The ratio of sulfur to the cellulose of 0.395, stating there is one xanthate group per cellulose unit in the cellulose xanthate.

$$\frac{\text{Sulfure}}{\text{Glukosa}} = \frac{2S}{C_6H_{10}O_5} = \frac{64}{162} = 0.395 \tag{2}$$

Thus, the value of DS can be determined by using the following equation[15]:

$$DS = \frac{\text{ratio of sulfure to glucose}}{0.395} \tag{3}$$

2.6. *Determination of degree of polymerization (DP)*

The degree of polymerization can be calculated by comparing the molecular weight of cellulose xanthate (CX) to the molecular weight of structural unit[16]:

$$DP = \frac{\text{Molecular weight of CX}}{\text{Molecular weight of structural unit}} \tag{4}$$

The molecular weight of cellulose xanthate can be determined with the viscosity methods by using Viscometer Ostwald. This methods has been conducted by Agnemo[17]. Molecular weight of cellulose xanthate was calculated by using Mark Houwink equations:

$$[\eta] = K.M^a$$

$[\eta]$ = viscosity of cellulose xanthate
M = Molecular weight of cellulose xanthate.
K and a are the constants of Mark-Houwink, K = 12 x 10-3 mL/g and a = 0,52.

3. Results and Discussion

3.1 Characterization of the Adsorbent

Cellulose xanthate was synthesized by reacting cellulose with an amount of carbon disulfide (CS_2) in NaOH solution. In this study, cellulose xanthate was synthesized by reacting purified sugarcane bagasse (PSB) with 160, 180, and 200% of CS_2 (w/w, by the amount of PSB used) in NaOH solution (18%). figure 1 shows the hydroxyl groups on the cellulose backbone combined with CS_2 in the xanthation process.

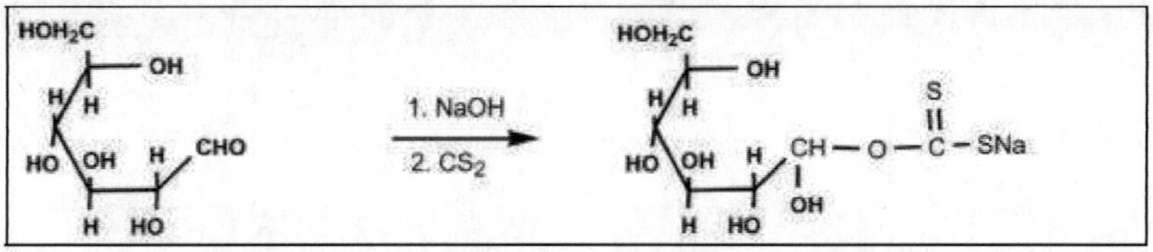

Figure 1. Scheme of cellulose xanthate synthesis[18]

The FTIR spectra of sugarcane bagasse (SB), purified sugarcane bagasse (PSB), alkalized cellulose (AC) and cellulose xanthate (CX) were carried out as a qualitative analysis to determine the main functional groups present in the adsorbent. FTIR spectra of SB, PSB, AC, and CX are shown in figure 2. In the spectrum of SB and PSB the peaks around 3,450 cm^{-1} correspond to the hydroxyl groups stretching vibrations. The peaks observed around 1,070 cm^{-1} are due to the C-O stretching vibrations that is attributed to the characteristic of carboxylic acids and alcohols. The peaks around 1,650 cm^{-1} can be assigned to bending vibration of the C=C group. The peaks at 2,900 cm^{-1} is stretching vibrations of C-H group[12]. There is a new peak at 1,208 cm^{-1} in the spectrum of AC that is attributed to the presents of Na. There are new peaks around 580 cm^{-1}, 1030 cm^{-1}, and 1156 cm^{-1} at FTIR spectrum of CX 160%, CX 180%, and CX 200%, that correspond to C-S, C=S, and S-C-S, respectively, which indicates the existence of sulfur groups in cellulose xanthate (CX). Compared to the spectrum of SB and PSB, there are some different functional groups in the spectrum of CX. The strong O-H band at 3,448 cm^{-1} in the spectrum of SB is shifted to 3,343 cm^{-1}, 3,347 cm^{-1}, and 3,348 cm^{-1} in the spectrum of CX 160%, CX 180%, and CX 200%, respectively, which shows that the hydroxyl groups have combined with CS_2 in the xanthation process.

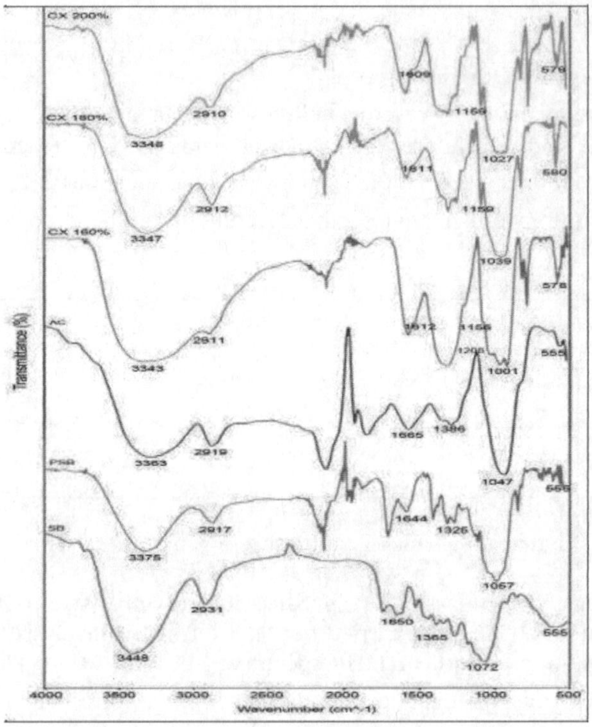

Figure 2. FTIR spectra of SB, PSB, AC, and CX

The SEM of purified sugarcane bagasse (PSB) and cellulose xanthate (CX) were carried out as a qualitative analysis to describe the morphological characteristics of adsorbent. The morphology of PSB and CX are shown in figure 3 and figure 4. Figure 3 shows the disorder pattern and rough surface of the cellulose structure in the PSB. Figure 4 shows that the cellulose structure is finer and has uniform pattern in the SEM of CX. The changing of morphology of PSB due to the fibrous was liberated from the matrix caused alkalization process. Additionally, alkalization process (soaking treatment of NaOH) could make the structure of cellulose expand and increase the porosity and specific surface area. This structure is needed to makes CS_2 react easily with cellulose to form cellulose xanthate in the xanthation process[8], and could improve the adsorption capacity of adsorbent for heavy metal ions in aqueous solution[12].

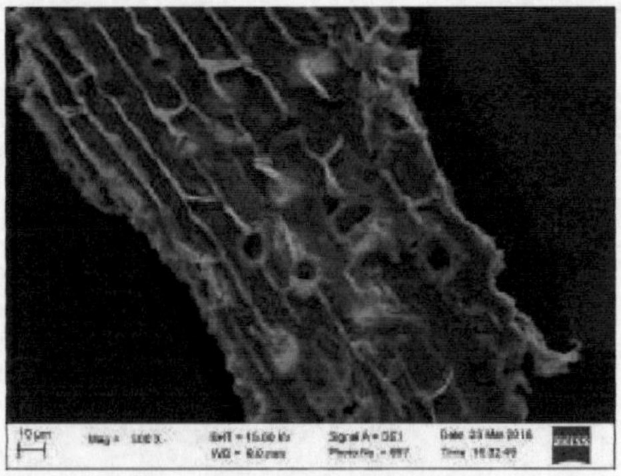

Figure 3. SEM of PSB

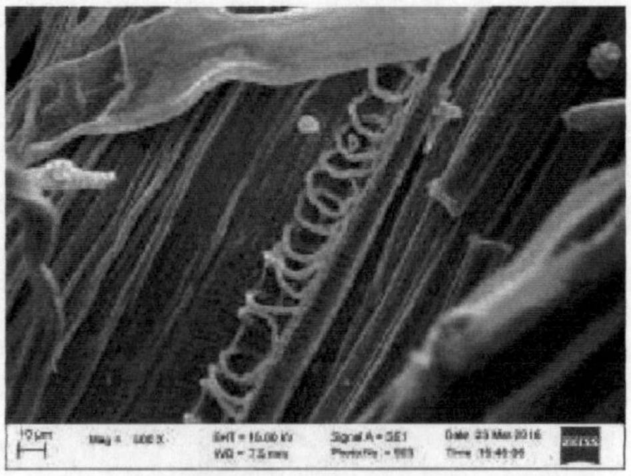

Figure 4. SEM of CX

The degree of substitution (DS) and degree of polymerization (DP) of cellulose xanthate (CX) were also carried out as a quantitative analysis. DS represents the number of xanthate groups per cellulose unit in the CX compound, whereas DP represents the amount of structural unit in the CX.

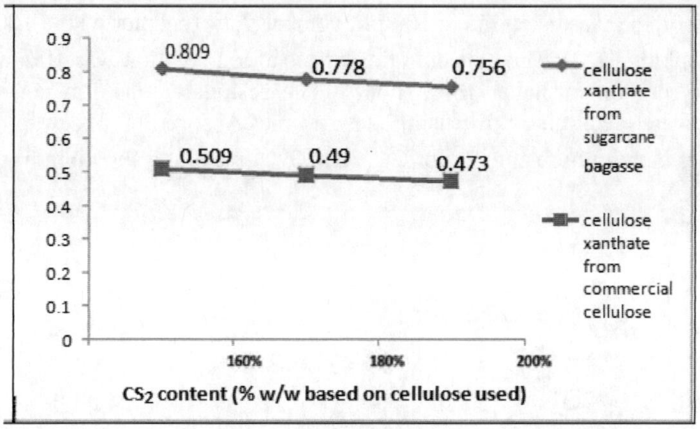

Figure 5. DS of cellulose xanthate

Figure 5 and figure 6 shows the effects of CS_2 content used in CX synthesis to the value of DS and DP. The CX made with 160% of CS_2 showed higher value of DS and DP. The highest value of DS are 0.809 for CX from sugarcane baggase and 0.509 for CX from commercial cellulose, whereas the higher value of DP is 325.757. If CS_2 content was higher than 160%, the value DS and DP was decreased. However, the reason for the decreasing value of DS and DP at higher CS_2 content is not clear. We speculate if CS_2 content was too higher in xanthation process, the trend of side reaction between CS_2 and sodium hydroxide become higher. Another speculation is some CS_2 molecules might be adsorbed physically on the hydrocarbon backbone if using an excess amount of CS_2 in CX synthesis[19].

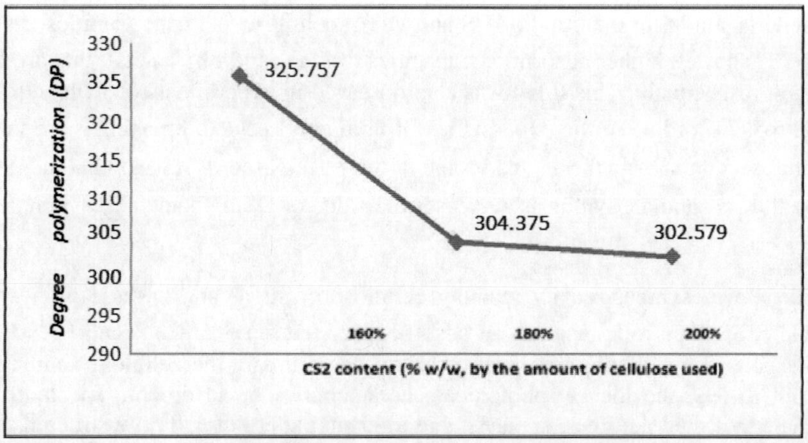

Figure 6. DP of cellulose xanthate

3.2 Adsorption Capacity
There are some important parameters affecting the adsorption process such as pH, temperature, etc. pH range of aqueous solution are from 2.0 to 6.5 on adsorption of Cu^{2+} and Pb^{2+} at room temperature for 120 min. 0.1 g of adsorbent was taken into a 100 mL conical flask together with a 100 mL single metal ion (Cu^{2+} or Pb^{2+}) solution [12]. The concentration of metal ion solution was 100 mg/L. The mixture was shaken at 120 rpm on a shaker for 2 hours. The residual metal ion concentrations in the supernatant liquor were determined by using an Atomic Absorption Spectrophotometer (AAS). Adsorption capacities of cellulose xanthate for Cu^{2+} and Pb^{2+} metal are shown in figure 7.

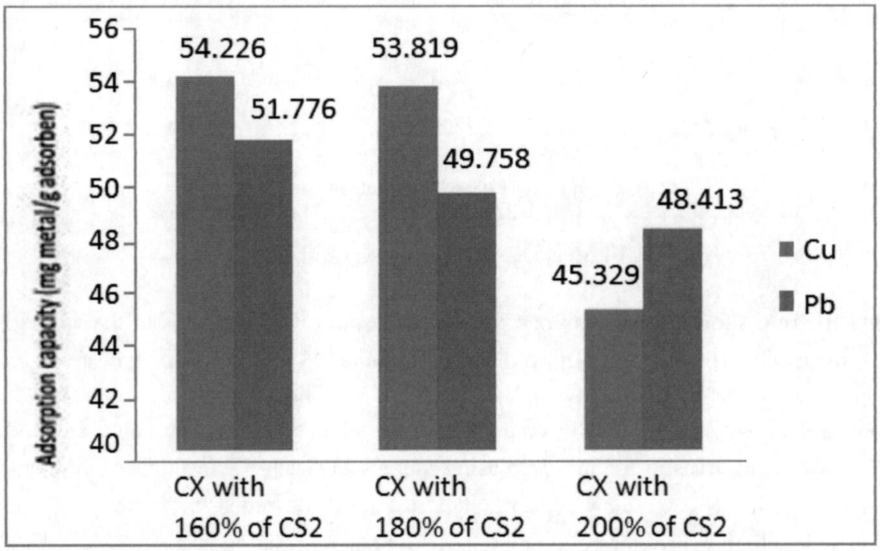

Figure 7. Adsorption capacity of cellulose xanthate

Adsorption capacity of cellulose xanthate decreases when CS_2 content in xanthation process was higher. This trend is similar to the trend of DS and DP of cellulose xanthate. Cellulose xanthate made with 160% of CS_2 showed higher adsorption capacities of Cu^{2+} and Pb^{2+} metal, that are 54.226 mg/g and 51.776 mg/g, respectively. This trend was also observed in experiments of Kim and Lee[19] with Pb^{2+} under pH 6.5. According to the experiments of Kim and Lee[19], by using more than 14 mL of CS_2 in the cellulose xanthate synthesis, lead removal was decreased. It reveals that an excess amount of CS_2 in the cellulose xanthate synthesis decreases the value of DS, DP, and adsorption capacity.

4. Conclusions
Cellulose xanthate was synthesized by reacting cellulose of sugarcane bagasse with an amount of carbon disulfide (CS_2) then to be used as an adsorbent for the removal Cu^{2+} and Pb^{2+} from aqueous solution. FTIR spectra showed the presence of the sulfur groups in the cellulose xanthate. The SEM were carried out to describe the morphological characteristics of adsorbent. The morphological of cellulose xanthate indicated the finer, expands, and uniform pattern of cellulose structure, which could increase the porosity, specific surface area, and improve the adsorption capacity of adsorbent for heavy metal ions in aqueous solution. The degree of substitution (DS), degree of polymerization, and adsorption capacity of cellulose xanthate decreased when CS_2 content was higher than 160% in cellulose xanthate synthesis. Cellulose xanthate made with 160% (19.2 mL) of CS_2 showed the highest

value of DS, DP, and adsorption capacities of Cu^{2+} and Pb^{2+} metal. The highest DS and DP of cellulose xanthate from sugarcane bagasse are 0.809 and 325.757, whereas the highest adsorption capacities of cellulose xanthate for Cu^{2+} and Pb^{2+} metal are 54.226 mg/g and 51.776 mg/g, respectively. This study reveals that sugarcane bagasse modified with carbon disulfide (CS_2) is a promising biosorbent for the removal of heavy metals and could be a solution to reduce environmental pollution caused by industrial wastewater.

References

[1] Argun M and Dursun S 2008 A new approach to modification of natural adsorbent for heavy metal adsorption *Bioresour. Technol.* **99** 2516-2527

[2] Liang S, Guo X, Feng N, Tian Q 2010 Enhanced Cu (III) adsorption by orange pell modified with sodium hydroxide *Journal of Elsevier. Trans. Nonferrous Met. Soc. China* **20** s187-s191.

[3] Darmono 1995 *Logam dalam Sistem Biologi Mahluk Hidup* (Jakarta: UI Press)

[4] Wang S and Peng Y 2010 Natural zeolites as effective adsorbents in water and wastewater treatment *Chem. Eng. J.* **156** 11

[5] Silviani 2014 *Pemanfaatan Lahan Kosong Untuk Perkebunan Tebu dan Pabrik Gula.* (Semarang, Indonesia : Semarang University)

[6] Iryani D, Kumagai S, Nonaka M, Nagashima Y, Sasaki K and Hirajima T 2014 The hot compressed water treatment of solid waste material from the sugar industry for valuable chemical production *International Journal of Green Energy.* **11** 577-588

[7] Vaughan T, Seo C, Marshall W 2001 Removal of As (V) from aqueos solutions by waste crab shell *Bioresour. Technol.* **78** 133

[8] Heuser E 1943 *The Chemistry of Cellulose* (London: John Wiley & Song INC)

[9] Riwayati I, Hartati I, Purwanto H, Suwardiyono 2014 Adsorsi logam berat Timbal dan Kadmiun pada limbah batik menggunakan pulpa kopi terxanthasi *Prosiding Seminar Nasional Aplikasi Sains & Teknologi* 15th November 2014 Yogykarta Indonesia.

[10] Cerqueira D, Filho G and Meireles C 2007 Optimization of sugarcane bagasse acetylation *Carbohydrate Polymers* **69** 579-582

[11] Gilbert R 1994 *Cellulose Polymers, Blends, and Composites* (Germany: Hanser Publications)

[12] Tian A, Jiang X, Yu H, Pan D and Liu Q 2015 Equilibrium, kinetic and mechanism studies on the biosorption of Cu^{2+} and Ni_2^+ by sulfur-modified bamboo powder *Korean Journal Chem. Eng.* **32** 342-349

[13] SNI (Standar Nasional Indonesia) 2004 Prosedur analisa sulfat dengan metode gravimetri 06-6989.20-2004

[14] Sluiter A, Hames B, Ruiz R, Scarlata C, Sluiter J, Templeton D and Crocker D 2005 Determination of structural carbohydrates and lignin in biomass. *The US National Renewable Energy Laboratory Technical Report* (US: NREL)

[15] Vincent D 1953 Xanthate methyl esters of simple alcohols and of cellulose *Thesis for The Degree of Doctor of Philosophy* (Canada: McGill University)

[16] Habibah R, Nasution D, Muis Y 2013 P b k d d j d p α-selulosa yang berasal dari alang-alang ((Imperata cylindrica) dengan metode viskositas *Jurnal Saintia Kimia* **1**

[17] Agnemo R 2009 Methods to analyze cellulose pulps for viscose production *Paper presented at 4 Workshop on Cellulose, Regenerated Cellulose and Cellulose Derivates* (Sweden: Karstald University)

[18] Homagai P, Ghimire K and Inoue K 2010 Preparation and characterization of charred xanthated sugarcane bagasse for the separation of heavy metals from aqueous solutions *Separation Science and Technology* **46** 330-339

[19] Kim T and Lee K 1999 Application of insoluble cellulose xanthate for the removal of heavy metals from aqueos solution *Korean J. Chem. Eng.* **16** 298-302.

Process design of *in situ* esterification-transesterifica tion for biodiesel production from residual oil of spent bleaching earth (SBE)

A Suryani[1,2], Z Mubarok[1], Suprihatin[1], M Romli[1] and E N Yunira[1]

[1]Department of Agro industrial Technology, Faculty of Agricultural Engineering and Technology, Bogor Agricultural University, Bogor, Indonesia
[2]Surfactant and Bioenergy Research Center (SBRC), Bogor, Indonesia

Email : anisuryani.sbrcipb@gmail.com

Abstract. Indonesia is the largest producer of Crude Palm Oil (CPO) in the world. CPO refining process produces spent bleaching earth (SBE), which still contains 20-30% oil. This residual oil is very potential to be developed as a biodiesel feedstock. The purpose of this research was to develop an *in situ* biodiesel production process of residual oil of SBE, which covered stirring speed of esterification and transesterification and also transesterification time to produce biodiesel with the best characteristics. The production was conducted in a 100 L reactor. The stirring speeds applied were 650 rpm and 730 rpm, and the transesterification time varied at 60, 90 and 120 minutes. The combination of 730 rpm stirring speed for 90 minutes transesterification resulted in the best biodiesel characteristics with the yield of 85%, the specific energy of 6,738 kJ/kg and the heater efficiency of 48%. The physico-chemical properties of biodiesel was in conformity with the SNI of Biodiesel.

1. Introduction

Bleaching earth is an adsorbent that used to adsorb pigments color containing in CPO to produce bleached cooking oil. Manufacture of refining oil to produce cooking oil is the biggest consumer of bentonite. Every year, Indonesia need almost 200.000 ton bentonite for cooking oil refining industry. In refined CPO, Bentonite basically is a single use material. Bentonite is a natural resource that cannot be renewable and spent bleaching earth (used bentonite) still contains 20-30% vegetable oil [1]. The high content of vegetable oils in spent bleaching earth is potential to be utilized as biodiesel. Spent bleachieng earth produced in refinery process. Chemical compounds of SBE are phosporic acid [2], high quantities of water-insoluble substances, such as fatty acids, macroelements (sodium, potassium, calcium, magnesium, phosphorus, sulphur and chlorine) and microelements (iron, manganese, zinc, copper, io-dine, cobalt and selenium), plant pigments and heavy metals (cadmium, mercury, arsenic, lead, chromium and nickel) [3,4].

Biodiesel is an alternative liquid fuels from vegetable oil which is formulated especially for diesel engines. Biodiesel is a methyl ester or ethyl ester from fatty acids which is produced by transesterification reaction between vegetable oil and methanol or biodiesel ethanol. Biodiesel production is urgently needed since the demand for diesel fuel is increased meanwhile the availability of diesel derived from petroleum are limited.

Content from this work may be used under the terms of the Creative Commons Attribution 3.0 licence. Any further distribution of this work must maintain attribution to the author(s) and the title of the work, journal citation and DOI.
Published under licence by IOP Publishing Ltd

The current developed method in the process of biodiesel production is directly taken from materials containing oils and fats. This process is called *in situ* esterification or transesterification. Despite of being more efisien, This process will also shorten time for raw material converting process into biodiesel which is done simultaeously with oil extraction process. Biodiesel usages as a diesel substitution fuel must be followed by an efficiently and economically process technology development.

The purpose of this study was to obtain the best conditions of the biodiesel production process in the form of the stirring speed and time of *in situ* transesterification in the 100 L reactor. The best production process was chosen by considering performance of production process design like yield, specific energy and heater efficiency in 100 L reactor. The purpose of this research is to determine the physical and chemical properties of biodiesel production from optimization production process and in accordance with the SNI biodiesel.

2. Materials and Methods

2.1 *Materials and tools*

The main material used in this study was spent bleaching earth from PT. Asianagro Agung Jaya Jakarta. The chemicals materials used in the study were methanol, H_2SO_4, NaOH, HCl, 95% neutral alcohol, saturated KI solution, phenolphthalein indicator (PP), an indicator of starch 1%, HNO_3, distilled water, and several other chemicals.

The main equipment used in the production process was the reactor with 100 L capacity and vacuum filter. In addition, to the purification process, the equipment used was the deposition gourd. The equipment used for the analysis was the measuring cup, cup, porcelain cup, erlenmeyer, filter paper, spatula, funnel, pipette, the pipette volumetric, magnetic stirrer, rotary evaporator, viscometer Brookfield, burette, centrifuges and vacuum pumps, furnaces, pH meter , soxhlet apparatus, hot plate etc.

2.2 *Methods*

2.2.1 *Raw Materials Characterization*

Characterization of raw materials carried on the preliminary research stages include the determination of water content (SNI 01-3555-1998), fat (SNI 01-2891-1992) and free fatty acid (SNI 01-3555-1998).

2.2.2 *Biodiesel Production Process*

This stage aims to get the best treatment in the production process. Based on the previous studies of residual oil contained in the SBE tend to have a high FFA content (> 2%), so in this study biodiesel production design consists of two phases, *in situ* esterification and transesterification. *In situ* esterification was carried out by reacting 7.5 kg of spent bleaching earth with 45 L methanol and H_2SO_4 catalyst. The amount of catalyst added was 1.5% (v / w) of solids. *In situ* esterification process was carried out at a temperature of 65°C for 3 hours. After the reaction time was reached followed by a reaction of *in situ* transesterification process varied at 60, 90, and 120 minutes by adding NaOH catalyst. The amount of NaOH catalyst added was 1.5% (w / w) of the spent bleaching earth. Before it was added into the reactor, NaOH was dissolved in the methanol. The esterification and transesterification processes were carried out *in situ* process with variation of the stirring speed of 650 and 730 rpm. Transesterification reaction was stopped after reaching the specified time, with stopping the process of heating and stirring. After the reaction was stopped, the process of separation between the bleaching earth and methanol containing oils can be done. Separation was done by using a vacuum filter. Solvents were also separated from the biodiesel by the evaporation. Then Biodiesel was deposited for 24 hours and then purified after that it was heated. Furthermore, biodiesel yield was calculated by the following equation:

$$\text{Biodiesel yield} = \frac{m_1}{m_2}\, x\ 100\ \% \tag{1}$$

Note:

m1 = mass of pure biodiesel after washing and drying (g)
m2 = mass of oil in the SBE (g)

2.2.3 Energy Analysis
Energy requirements were calculated on the heating process, *in situ* esterification, *in situ* transesterification, filtration, evaporation, and biodiesel heating. The amount of energy required was the result of multiplying the power of the motor stirrer, heater, and vacuum pumps. The energy requirements are calculated with the following equation.

$$E = \frac{Pxt}{m} \tag{2}$$

Where: E = required energy (kJ)
P = installed appliance power (kW)
t = Time use of heater (s)

The magnitude of the heating efficiency produced by the heater can be calculated by comparing the useful heat Q with power input or power installed on the heater or may be calculated by the equation.

$$\eta = \frac{Q}{Pxt}\ x\ 100\% \tag{3}$$

Where: η = Efficiency of heater (%)
Q = useful heat (kJ)
P = installed heater power (Watt)
t = Time use of heater (s)

Beneficial heat is heat that is used to raise the temperature of the material until it reaches 65 ° C. The beneficial heat is calculated by similarities:

$$Q = m\ cp\ \Delta T \tag{4}$$

Where: Q = beneficial heat (kJ)
cp = Specific heat of material (kJ / kg.K)
ΔT = temperature difference (K)

2.2.4 Characterization of Biodiesel
Biodiesel which was produced from the best production processes was then analyzed to determine the characteristics of the biodiesel. The analysis include density (AOCS Cc 10c-95), the kinematic viscosity (ASTM D 445), a flash point (ASTM D 93), cetane number (ASTM D 613), acid value (SNI 01-3555-1998), saponification value (AOCS Cd 3-25), total glycerol content (AOCS Ca 14-56), alkyl ester value, as well as water and sediment content (ASTM D-2709).

2.2.5 Design of Experiments and Data Analysis

Testing method used in this study was completely randomized factorial design models. In this models, the factors involved were cross each other as well as environmental conditions encountered either homogeneous or uniform. The interaction model for completely randomized factorial design is described as follows:

$$Yijk = \mu + Ai + Bj + (AB)\,ij + \varepsilon ijk \quad (5)$$

Note:

Yijk : measured response
μ : the general average
Ai : the influence of stirring speed of the-i
Bj : the influence of the time factor transesterification j
ABij : the effect of the interaction between factors stirring speed level of –i with transesterification time of stage j
εijk : errors

3. Result and Discussion

3.1 The Characteristics of Raw Material

The raw materials used in this research is spent bleaching earth (SBE) known as used bentonite or secondhand bleaching earth. SBE is taken from P.T. Asianagro Agungjaya, Jakarta. SBE characteristics that will be analyzed include moisture content, free fatty acid value and fat content. The SBE characteristics analysis results can be seen in table 1.

Table 1. Characteristics of SBE

Quality characteristics	Value (%)
Moisture content	5,20
Free fatty acid value	3,29
Fat content	14,67

Moisture content of the material affects the yield of biodiesel. The presence of water in esterification and transesterification reactions can cause oil hydrolysis into free fatty acids. In addition, methyl esters which was formed will caused hydrolysis reaction with free fatty acids, thereby it reduce biodiesel yield.

The high number of free fatty acids will affect the fluency during the transesterification and may decreasing biodiesel yield. Maximum yield of biodiesel can be obtained from raw materials which have the FFA content less than 0.5% [5]. High number of free fatty acids can formed soap because free fatty acid reacts with a base catalyst in transesterification process so it will be potential to reduce biodiesel yield. The production process of biodiesel with high free fatty acid levels are conducted into two steps, it is esterification and transesterification [6].

The next analysis of SBE is the fat content. This test was conducted to determine the oil content in materials that can be extracted and converted into biodiesel. The fat number in SBE obtained is 14.67% (bb).

3.2 Stage Production of Biodiesel

Biodiesel production process was conducted using *in situ* esterification and transesterification. *In situ* Esterification was carried out by reacting material contained fatty acids using alcohol solvent with the aid of acid catalysts which was done simultaneously in a reactor.

In situ transesterification method is a simple step to produce biodiesel by eliminating extraction process and oil refinement that could saves production costs [7]. Triglycerides used *in situ* transesterification process are triglycerides derived from raw materials sources and not from oil extraction and refinement (Qian et al., 2008). It is better than conventional techniques for *in situ* process since it does not need to perform oil extraction and refinement separately. Extraction process, esterification and transesterification of oil residue in SBE run simultaneously within a single reactor. Thus process is more effective and save energy.

3.3 *Mass Balance*

Material heating process and *in situ* esterification get input materials such as methanol 45000 mL, SBE 7500 grams, and $H_2SO_4$112,5 mL for all treatments and replicates. Later during *in situ* transesterification process occurs catalyst addition as much as 256 grams of NaOH dissolved in 3060 mL of methanol. The results of three series processes produce a mixture of methanol, biodiesel, glycerol, impurities and residual solids SBE which is still difficult to quantify amount of mixtures. The amount of the mixture can be calculated after the separation phase of the liquid and solids in the filtration process. This filtration process outputs liquid phase as much as 41700-45000 mL and solid waste 6400-6750 grams.

The next process is evaporation to separate methanol from liquid phase because methanol has the lowest boiling point. From methanol input of 41700-45000 mL produced 40000-42800 mL and mixture of 1365-1550 mL biodiesel, glycerol, and impurities or 1325.03 to 1487.25 grams. After that, the mixture was precipitated to separate biodiesel and glycerol. This process can be separated from crude biodiesel amount 833.63 to 983.25 grams and glycerol 491.40 to 567 grams. Crude biodiesel was washed using water to remove impurities, which dissolves in the water soluble such as catalysts and residual methanol. From washing process, it produced pure biodiesel as much as 786.60 to 940.50 grams and waste in the form of impurities as much as 21.38 to 68.40 grams. *Losses* that occur in biodiesel production process was the methanol amount to 5060 - 8060 mL. The losing of methanol occurs in esterification, transesterification, and evaporation process. Methanol is evaporated into the environment through a gap of the tool.

3.4 *Biodiesel Yield*

This process produced 72.65 to 84.5% yield. Variance analysis indicates that the treatment of transesterification time significantly influence ($P < 0.05$) to biodiesel yield. Meanwhile, the mixing speed is not significant ($P > 0.05$) to biodiesel yield production. The interaction between the stirring speed and transesterification time also did not significantly affect the generated yield. Furthermore, the Duncan test which was conducted to know the effects of time indicated that the yield of the transesterification time of 60 minutes and 120 minutes were not significantly different. Meanwhile, the transesterification time of 90 minutes was significantly different from the transesterification time of 60 minutes and 120 minutes.

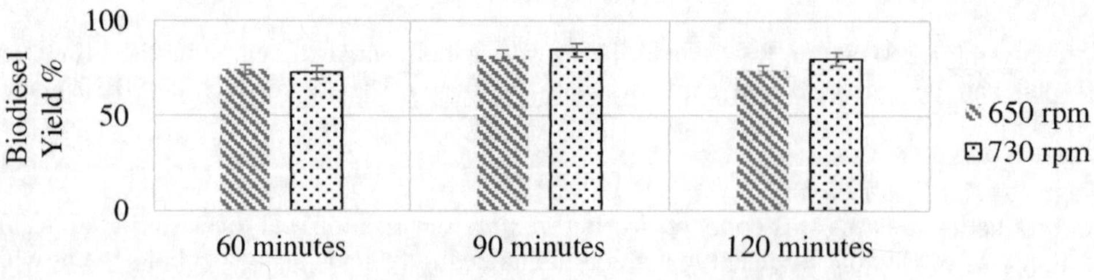

Figure 1. Biodiesel Yield in Various Treatment

The mixing speed affects yield production, but after the equilibrium, the mixing speed will not be significant. The stirring speed of biodiesel production process is closely related to the homogeneity of mixture reaction to reach the complete reaction. The higher stirring speed, the higher collisions between particles of the reactants so that the reaction will proceed more perfect. It can be seen from the results of the study (figure 1) that increasing stirring speed will increase the biodiesel yield. However, according to the research of Noureddini and Zhu [9], it shows that there are significant differences to produced yield on the stirring speed of 150-300 rpm but on the stirring speed between 300 rpm and 600 rpm shows the slight or inconsiderable difference. Time is one factor that determines the amount of yield production. The longer the reaction time will influence the chance methyl esters formed that automatically increase significant result of high biodiesel yield. However, if it reaches a point of equilibrium reaction, the reaction time have no effect on the yield of methyl ester. The conversion of triglycerides into methyl esters increased rapidly during the first 180 minutes to reach 98% yield and more than 180 minutes of time did not affect biodiesel yield [10]. The results showed that the longer time of *in situ* transesterification reaction, the higher result of yield. Optimum yield was reached at reaction time of 90 minutes and dropped back to the reaction time of 120 minutes (figure 1). These result is consistent with the study of Qowim's [11] which states that the optimal time of *in situ* transesterification biodiesel process of SBE is at 104.73 minutes after *in situ* esterification for 3 hours and yield will fall back after that time.

The highest yield obtained from the production process using 100 L reactor was at 84.5%. The Yield was obtained from transesterification treatment time for 90 minutes with a stirring speed of 730 rpm.

3.5 Analysis of Energy Needs

Energy measurement was the direct energy in the form of fuel energy or electrical energy. Observation and measurement energy needs were in material heating process, *in situ* esterification, *in situ* transesterification, filtration, methanol evaporation, and biodiesel heating. The energy measurement on biodiesel production process was the specific energy that required to process for each kilogram of material.

The specific energy required in the biodiesel production process was in the range of 6596.55 to 6832.99 kJ per kilogram of material. The variance analysis results showed the significant transesterification time ($P < 0.05$) to the total specific energy needs. Meanwhile the stirring speed was not significant ($P > 0.05$) to the total specific energy needs. The further Duncan test on the treatment period showed that the specific energy needs of transesterification total treatment time of 60 minutes and 90 minutes were significantly different from the transesterification time of 120 minutes.

Table 2. Specific energy need for a production process of biodiesel

Process	Average needs of The specific energy (kJ / kg of material)
Material Heating	391,25 ± 15,85
***In situ* esterification**	933,81 ± 0,00
***In situ* Transesterification**	439,72 ±131,10
Filtration	196,34 ± 4,46
Evaporation	3521,42 ± 91,16
Biodiesel Heating	1268,03 ± 78,63

The production process with 650 rpm stirring speed and transesterification time of 60 minutes took the smallest specific needs energy. Otherwise, the production process with the stirring speed of 650 rpm and transesterification time of 120 minutes took the highest total specific energy needs. Time is one of determinant factor in the energy use. Total energy needs equals to transesterification time. The longer the time of transesterification process, the more electrical energy was required to drive the agitator and heating materials. The energy required by each kilogram of material in the production process of all treatments was still too small compared to the energy that can be produced by biodiesel as much as 40 100 kJ / kg biodiesel [12]

The highest specific energy needs was in the evaporation process because this process need heat and mechanical energy in a long time. Meanwhile, the smallest specific energy was required by the filtration process because this process just applied a vacuum pump with the power 373 W. The energy needs of each process can be seen in table 2.

3.6 The specific energy needs in heating raw materials process, in situ esterification and transesterification

The process of raw material heating was used to raise material temperature like methanol, SBE, and a catalyst until reached temperature of 65°C. In this process, used energy was the heat energy that coming from heater and mechanical energy from motor to move the agitator. The stirring was done in order to get heat more evenly so that the heat transfer from the heating medium to the material became faster and the material reached the optimal temperature too. After the temperature reached 65°C, the esterification process were ready to carry out by maintaining the temperature and stirring for three hours. After the esterification process was completed, it was followed by the transesterification process at temperature of 65°C with stirring. The total specific energy in the three processes of all the treatments were in the range of 1595.29 to 1922.16 kJ / kg of material. The results of variance analysis showed the transesterification time and stirring speed significantly ($P < 0.05$) against the specific energy needs of these three processes. Meanwhile the interaction between the transesterification time and stirring speed did not significantly affect the specific energy needs of these three processes. Furthermore, the Duncan test on the treatment period showed that the specific energy needs of the three treatments were significantly different for all transesterification.

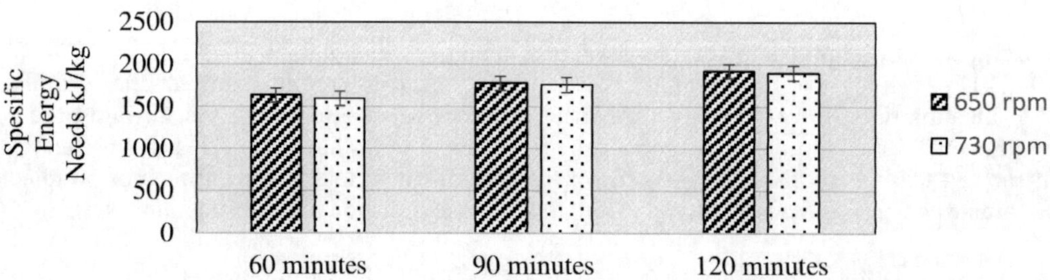

Figure 2. The specific energy needs on the various treatments

The time of transesterification gave significant effect on specific energy in the heating process, esterification and *in situ* transesterification because time was a power multiplier factor to get the value of the energy needs. The greater time, the greater energy needs would be. Meanwhile the stirring speed significantly affected the specific energy needs to these three processes, especially in the heating process materials for the higher rate of stirring, the better heat transfer occurs, and the heat was more evenly spread to the material so that the time required to reach a temperature of 65 ° C was faster and energy took getting smaller. It can be seen in figure 2.

3.7 Energy Balance
The energy balance was used to determine the total energy input used in a process. In calculating the total energy, the energy needs was not divided by the mass of fuel. In the process of heating materials, esterification, transesterification, and filtration of the total energy produced by the heater, the motor, and a vacuum filter was 77732-93466 kJ. Based on the energy input, of 68587-83113 kJ, the energy was lost to the environment only in the form of heat. The next process was the process of evaporation. In this process was amounted to 121 370 kJ of energy required in the form of heat energy from the heater and the mechanical energy of the motor stirrer. The energy was used in the form of heat to vaporize methanol in the amount of 38213-41453 kJ. The rest was wasted as heat into the environment. The final process was drying which required heat energy of 1,080 kJ used separately to evaporate the residual water in biodiesel.

3.8 Heater Efficiency
The heater functions as heat providers to accelerate the reaction between methanol with triglycerides in SBE. The heater efficiency showed how well this equipment to withstand the heat produced by the heater. The heater efficiency was calculated by comparing the power used to heat the material from room temperature until it reached a temperature of 65 ° C with power that was attached to the heater.

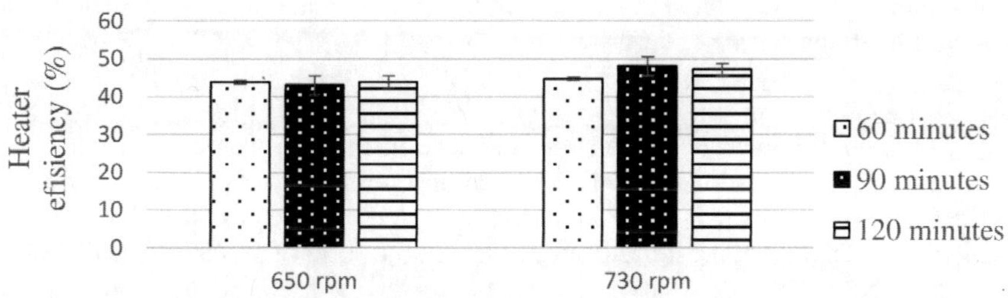

Figure 3. Heater Efficiency on Various Treatment

The heater efficiency of the equipment used in this study ranged from 42.99% - 48.05%. This condition showed that there were about 51.95% - 57.01% of heat was not used to heat the material. The results of the variance analysis showed the stirring speed and interaction between treatment time with the stirring speed affected significantly (P <0.05) on the efficiency of the heater. While *in situ* transesterifikai time is not significant (P> 0.05) on the efficiency of the heater. Duncan test showed the combination treatment transesterification time of 90 minutes and the stirring speed 730 rpm and transesterification time of 120 minutes and the stirring speed 730 rpm were significantly different from other treatment combinations.

The production process with a stirring speed of 730 rpm had a thermal efficiency that was greater than the process speed of stirring 650 rpm (figure 3) because the higher speed, the heat transfer was better and spread evenly on the material so that the time required to heat the material could be minimized. On the other side, the efficiency of the best heater use was on the treatment of stirring speed of 730 rpm with a time of 90 minutes transesterification. Meanwhile, the efficiency of the smallest heater use was on the treatment with stirring speed 650 rpm with a time of 90 minutes transesterification.

3.9 Biodiesel Characterization
Characterization of biodiesel was conducted to the best treatment derived from the yield parameters, energy needs, and the heater efficiency. The best treatment of the biodiesel production process was the production process with a stirring speed of 730 rpm with time of transesterification 90 minutes.

3.9.1 Density
Density showed the ratio of sample weight with the weight and volume of water at the same temperature. The density was one of the most important characteristics in biodiesel. It was associated with power and calorific value produced by the diesel engines per unit volume of biodiesel. The density of best treatment result biodiesel was 0.855 g / ml.

3.9.2 Viscosity
Viscosity was the resistance possessed fluid which was flowed within the capillary against the force of gravity. It was usually expressed in the time to flow at a certain distance. This characteristic was very important because it affected the performance of injectors in the diesel engines. The viscosity value obtained from biodiesel on best treatment was 5,863 cSt.

3.9.3 Cetane Number
Cetane number indicated how fast the diesel engine fuel injected into the combustion chamber could ignite spontaneously after mixed with air. The cetane number of biodiesel on best treatment was amounted to 54.5. This number met the requirements of ISO Biodiesel which was at least 48.

3.9.4 Flash Point
Flash point was the lowest temperature point during the formation of the flame ignition test (*flame test*). These characteristics were related to security in the storage and handling of fuel. The flash point of biodiesel on best treatment was 172 ° C.

3.9.5 Water Level and Sediment
Water level and sediment showed how well the purification process was carried out after biodiesel produced. The water level and sediment of biodiesel on best treatment was less than 0.005 ml that was considered undetectable or very small. This condition indicated that the circuit of biodiesel purification process succeeded in reducing the presence of impurities in the form of water and sediment.

3.9.6 Saponification
Saponification showed the molecular weight of triglycerides tested to estimate the largest component in the material. The saponification obtained was amounted to 205.5 mgKOH / g.

3.9.7 Acid Number
Acid numbers indicated the biodiesel acid content of free fatty acids derived from the degradation of ester or free fatty acids that were not contaminated with methyl ester in the esterification process. The resulted acid number reduced from the prior acid number before the SBE oil was converted into biodiesel. The esterification and transesterification process successfully reduced the level of free fatty acids. The acid number of the biodiesel reached 0,711 mg NaOH / g.

3.9.8 Total Glycerol Content
The content of total glycerol was calculated from the sum total of bound glycerol and free glycerol contained in the fuel. The existence of residual glycerol and glycerides which were not converted might harm engine especially for the -OH group that was chemically aggressive toward non-ferrous metals and the mixture of chromium led deposits in the combustion chamber [13]. Total glycerol content was 0.16% (w / w).

3.9.9 Alkyl esters value
Level of alkyl esters represented the number of organic acids which were fused as esters and had a relationship with the acid number, and saponification glycerol levels [14]. The biodiesel alkyl esters level indicated the level of purity. The level of alkyl esters of biodiesel on best treatment was amounted to 99.296% (w / w).

The results of biodiesel physical and chemical characterization of the nature on the best treatment was in the transesterification time for 90 minutes with a stirring speed of 730 rpm in accordance with SNI. The results showed that biodiesel produced was suitable for the use as fuel for diesel engines. The physical and chemical characterization and the comparison with SNI can be seen in table 3.

The best treatment for biodiesel production was transesterification time of 90 minutes and stirring speed of 730 rpm.

Table 3. Comparison of characteristics of biodiesel results on best treatment combination with Indonesian Standard of Biodiesel

No.	Parameter	Biodiesel A2B2a	Standar SNI	Unit
1	**Kinematic viscosity**	5,863 ± 0,028	2,3 - 6,0	cSt
2	**Density**	0,855 ± 0,001	0,850 - 0, 890	g / ml
3	**Cetane Number**	54,500 ± 0,283	min 48	
4	**Flash Point (closed cup)**	172,000 ±11,314	min 100	C
5	**Water dan Sediment**	Trace	max 0,05	% vol
6	**Acid Value**	0,711 ± 0,038	max 0,8	mg NaOH/gr
7	**Total Glycerol**	0,160 ± 0,014	max 0,24	% mass
8	**Saponification**	205,500 ± 4,950		mg KOH/gr
9	**Alkyl Esters Value**	99,296 ± 0,007	min 96,5	% mass

4. Conclusion

The raw materials of SBE had an acid value more than 2% so that the biodiesel production process was performed through *in situ* esterification process and was followed by *in situ* transesterification. The best treatment biodiesel production process in 100 L reactor with the yield parameters, specific energy needs, and the efficiency of the production process with the heater was 730 rpm stirring speed and transesterificasion time 90 minutes with the yield of Biodiesel 84.5%, specific energy needs6,738.35 kJ/kg and heater efficiency 48.05%. The biodiesel produced from the process had the value of the density of 0.855 g/ml, 5,863 cSt viscosity, flash point 172°C, cetane number 54.5, acid value 0,711 mgNaOH/g, saponification value 205.5 mgKOH/g, total glycerol 0, 16% mass, alkyl ester value of 99.296% mass, and trace content and sediment. The physico-chemical properties of biodiesel was in conformity with the SNI of Biodiesel. This research need to have future research about reusing evaporated methanol because of big proportion of used methanol and economic analyses: feasibility of biodiesel production process from residual oil of SBE.

Reference

[1] Kheang L S, Cheng S F, Choo Y M and Ma A N 2006 A Study Of Residual Oils Recovery From Spent Bleaching Earth: Their Characteristics And Applications *J App Sci* **3** 2063-2067
[2] V. Gibon, Greyt W D, Kellens M 2007 Palm oil refining. Eur *J. Lipid Sci Technol* **109** 315–335.
[3] Chandra M, Sathiavelu S 2009 Waste management in the oil industry in the Mediterranean region by composting. *Clean Techn Environ Policy* **11** 293–298.
[4] Piotrowska Cyplik A, Chrzanowski Ł, Cyplik P, Dach J, Olejnik A, Staninska J, Czarny J, Lewicko A, Marecik R Powierska-Czarny J (2013) Composting of oiled bleaching earth: fatty acids degradation, phytotoxicity and mutagenicity changes *Int Biodeterior Biodegrad* **78** 49–57.
[5] Knothe G, Gerpen J V and Krahl J 2005 *The Biodiesel Handbook* (Illinois : AOCS Press).
[6] Zhang Y, Dube M A, Mc Lean D D and Kates M 2003 Biodiesel production from waste cooking oil: 1. Process design and technological assessment *Bioresour Technol* **89** 1-16.

[7] Haas MJ, Karen MS, William NM, Thomas AF 2004 *In Situ* Alkaline Transesterification: An Effective Method For The Production Of Fatty Acid Esters From Vegetable Oils *J Am Oil Chem Soc* **81** 83-89.

[8] Qian J, Wang F, Liu S, Yun Z 2008 *In situ* Alkaline Transesterification Of Cottonseed Oil For Production Of Biodiesel And Nontoxic Cottonseed Meal *J Bioresour Technol* **99** 9009-9012.

[9] Noureddini H, Zhu 1997 Kinetics of Transesterification of Soybean Oil *JAOCS* **74** 1457 – 1463

[10] Jain S, Sharma M P 2010 Kinetics Of Acid Base Catalized Transesterification Of Jatropha Curcas Oil *J Bioresource Technology* **101** 7701-7706.

[11] Qowim A S F 2012 Biodiesel production process engineering of *in situ* Esterification-Transesterification from residual oil of spent bleaching earth *Thesis* (Bogor : Postgraduated Pogram, Institut Pertanian Bogor)

[12] Ismail 2008 Performance test and energy analysis of reactor type static-mixer for catalycally biodiesel production *Bachelor Degree Thesis* (Bogor : Institut Pertanian Bogor)

[13] Soerawidjaja T H, Tahar A, Siagian U W, Prakoso T, Reksowardojo I K and Permana K S 2005 Study of Usage Policy of Biodiesel in Indonesia In : Hariyadi P, Andarwulan N, Nuraida L, Sukmawati Y, ed. Policy Studies and Research Articles Collection (Bogor: Ministry of Research and Technology RI, MAKSI, SEAFAST, IPB) p 3-107.

[14] Ketaren S 1986 *Intoduction : Food Technology of Oil and Fats* (Jakarta : UI Press)

IOP Conf. Series: Earth and Environmental Science **65** (2017) 012041 doi:10.1088/1755-1315/65/1/012041

Processing of oil palm empty fruit bunch as filler material of polymer recycles

D R Saepulloh[1], S Nikmatin[1] and H Hardhienata[1]

[1]Departmennt of Physics, Faculty of Mathematics and Science, Bogor Agricultural University, Bogor, Indonesia

Email : dika.rahmat.s@gmail.com

Abstract. Oil palm empty fruit bunches (OPEFB) is waste from crude palm oil (CPO) processing plants. This research aims to process OPEFB to be a reinforcement polymer recycle with the mechanical milling method and identify each establishment molecular with the orbital hybridization theory. OPEFB fibers were synthesized using a mechanical milling until the size shortfiber and microfiber. Then do the biocomposite granular synthesis with single screw extruder. TAPPI chemical test shows levels of α-cellulose fibers amounted 41.68%. Based on density, the most optimum composition contained in the filler amounted 15% with the size is the microfiber. The test results of morphology with SEM showed deployment of filler OPEFB fiber is fairly equitable distributed. Regarding the molecular interaction between matrix with OPEFB fiber, described by the theory of orbital hybridization. But the explanation establishment of the bond for more complex molecules likes this from the side of the molecular orbital theory is necessary complete information of the hybrid levels.

1. Introduction

Palm oil is the source of the state's non-oil and gas foreign exchange, the working field procurement, the source of state's income, and the national economy driver. Currently, Indonesia has became the biggest palm oil's producer in the world, where Indonesia contributes 48% of the total volume of palm oil production in the world. Palm oil production in Indonesia in 2014 reached 31,5 million ton, where the quantity has shown a stable increase for the last 20 years which is 11% every year. In 2020, Indonesia's Crude Palm Oil (CPO) is estimated will reach 40 million ton [1].

The high productivity of palm oil is accompanied with the overflowing production of waste as the side product. Every ton of oil palm fresh fruit brunches (OPFFB) that has been processed, produced about 50% of the total waste, while the oil palm empty fruit brunch (OPEFB) produced about 23% of the total waste. If not processed well, it will be hazardeous to the environment [2].

OPEFB is solid waste that is produced by the CPO factory or industry. At the factory, OPEFB was only burned and now is prohibited due to the recent concern of the environment defilement or is dumped therefore caused complaines and problems because it can decrease soil's ability to absorb water. One of the solution to solve the problem is by reusing or reproducing OPEFB's waste into other usable materials [3].

Content from this work may be used under the terms of the Creative Commons Attribution 3.0 licence. Any further distribution of this work must maintain attribution to the author(s) and the title of the work, journal citation and DOI.
Published under licence by IOP Publishing Ltd

OPEFB is widely known to contains lot of fiber and cellulose that are useful as recycle polimer reinforcer in biocomposite application [4]. OPEFB's fiber is bind-able with *Acrylonitril Butadiene Styrene*/ABS ($(C_8H_8 \cdot C_4H_6 \cdot C_3H_3N)_x$) polymer through chemical binding in a composite structure. The bind between OPEFB's fiber dan ABS polymer can be explained by the molecular orbital theory [5].

The molecular orbital theory sees that the bonds between the molecules occured due to the hybridisation between the orbitals of each molecules. When the OPEFB's fiber molecules placed near with the ABS molecules, the wave function of each molecules wil be supepositioned liniery, forming hybrid binding that has the hibrid energy levels characteristics, which principally can be found by solving the multiatomic Schrodinger equation. Even so, the solving of the equation for complex molecule such as lignocellulose or ABS polymer required mathematical approachment and computated modelling which are quite complicated and is outside this research's field. In this research, in addition to process the OPEFB's waste to be a filler, there will also qualitative analysis on how the hybridisation bonds between the cellulose and ABS polymer were formed [5].

Molecular Orbital is the result of a linear superposition of the orbital wave functions and the marges of atomic orbitals on molecule. Approach the wave function can be formed as a linear combination of

$$\psi = \sum_{i}^{n} c_i \phi_i \tag{1}$$

Where the coefficient optimal uses variation principle. So that the expectation value of the energy is minimum. The function is called molecular orbital. From basis functions n, the n that differs in their orthogonal molecular orbital can be constructed accordinng to the rules of atomic orbitals combination, amount of molecular marges equal to the atomic orbital marges. When two atoms are marges, each atom give an atomic orbitals then produced two molecular orbitals, one of which is a combination of the sum the atomic orbital is a superposition constructive associated with energy levels of bonding and the rest is a superposition of destructive associated with energy levels antibonding. The strength of the bond depends on the degree of superposition (overlap) of orbital.

The strength of the bond between the two molecular orbitals are influenced by molecular symmetry and type of bonding orbitals hybridize. Two types of bonding that occurs is a σ bond and π. σ bond is a bond where the lines symmetrical electron density is concentrated along the line connecting the two atoms. For example, Figure 1 shows the formation of σ bonds between two molecules [5].

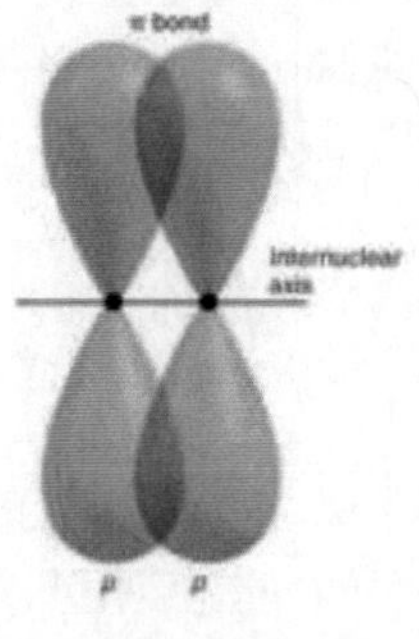

Figure 1. π Bonding

While the π bond is a bond where there superposition area above and below the internuclear axis (the line of nodes located along the internuclear axis). Examples of π bond can be seen in Figure 2.

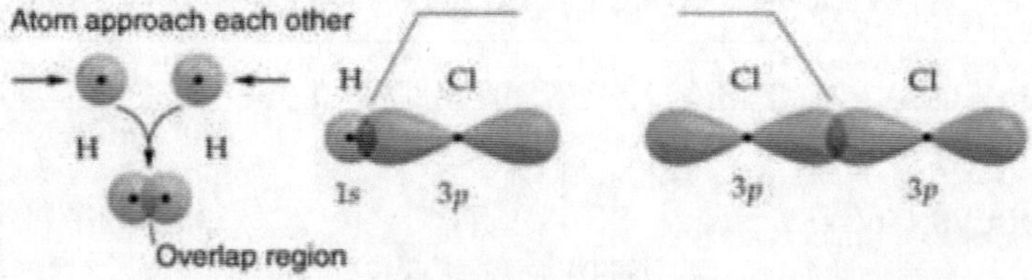

Figure 2. σ Bonding

The amount of the bond that forms is determined by the number of valence electrons in bonding and antibonding levels and can be calculated using the equation.

$$N_{bonding} = \frac{1}{2}(Ne^{-}{}_{orbital\ bonding} - Ne^{-}{}_{orbital\ antibonding}) \quad (2)$$

2. Methods

2.1 Sample Preparation

OPEFB as the main materials of research were obtained from PT Perkebunan Nusantara VIII Cikasungka- Bogor, while the matrix ABS recycle obtained from PT MUB Jaya Cibinong. In addition, there are additives used for the manufacture of granules, antioxidants, Scavenger Acid and Maleic Acid.

The initial stage of this research is the preparation of palm bunches to be cleaned of impurities and enumeration manually to get OPEFB in chip size. Then warmed by drying and oven at 100 °C for 8m hours to reduce the water content OPEFB. Furthermore, the initial testing includes testing the chemical composition, density test, and FTIR (Fourier Transform Infrared Spectroscopy).

2.2 Filler Production

Chopped OPEFB was gradually milled using miller at the speed of 25.000 rpm. Milling for 5x1 minutes. The results of the first milling sieved to 20 mesh sieve to separate the fibers are too large in size with a tiny fibers. Fiber large size (retained 20 mesh) be milled back until the size is able to pass through a sieve. While for the fibers that pass 20 mesh sieving will be continued with a 60-mesh sieve. Just like before, for a fiber that can pass through 60 mesh sieve will be continued with 100 mesh sieve. While the fiber cannot pass through the sieve are reprocessed until smooth. Fibers used in this research is a fiber that can pass through 60 mesh sieve size which is 250 μm fiber can be referred to as the short fiber and fiber that can pass through a 100 mesh sieve size which is 150 μm fiber can be referred to as micro fiber.

2.3 Manufacture of composite granules

The next phase is to make granular biocomposite by combining components of biocomposites are fillers, polymer recycle (matrix) and additive materials on an industrial scale machine single screw extruder. The composition of each buffer material shown in Table 1.

Tabel 1. Biocomposites Composition

Identity	Composition (gram)					
	OPEFB Fibers		ABS Recycle	Antioxidants	Additives Scavenger Acid	Additives Coupling Agent
1a	Short fiber	10%	87%	1%	0.3%	2%
1b		15%	82%	1%	0.3%	2%
1c		20%	77%	1%	0.3%	2%
2a	Micro fiber	10%	87%	1%	0.3%	2%
2b		15%	82%	1%	0.3%	2%
2c		20%	77%	1%	0.3%	2%

3. Result and Discussion

3.1 Chemical analysis composition of OPEFB fiber with TAPPI method

Tabel 2. Chemical composition with TAPPI method

OPEFB	Chemical composition (%)			
	α-cellulose	Hemicellulose	Lignin	Holocellulose
Without treatment	35.63	32.10	16.38	67.73
Treatment 100 °C, 8 Hours	41.68	20.35	16.78	62.03

The chemical composition of the OPEFB can be identified by standard methods of the Technical Association of the Pulp and Paper Industry (TAPPI). TAPPI T222 om-88 (1988) is a method for determining the composition of lignin, TAPPI T203 OS-61 (1961) method is used for determination of α-cellulose and TAPPI T9m-54 (1988) for the determination of holocellulose.

From the data presented, the composition of the cellulose is 41.68% in the treatment of 100 ° C for 8 hours, while it OPEFB without heat treatment had a composition of 35.63% cellulose. Based on this, it can be seen that an increase in cellulose after heat-treated so that the crystallization and decomposition of cellulose. On the other hand, the amount of hemicellulose decreased from 32.10% to 20.35% and holocellulose decreased from 67.73% to 62.03%. Meanwhile lignin, are relatively fixed in the range of 16%, because lignin will be decompose at a temperature 280-500 °C [6]. A decrease in the presence of hemicellulose and holocellulose caused by heat and it is adequate to make hemicellulose and holocellulosa degraded. The composition cellulose of the OPEFB, has sufficient minimum levels to be used as a polymer filler. Because in previous research of composite-based cellulose, cellulose content in the range of 40% [7]. In addition, OPEFB has a density of 770 kg / m^3, high tensile strength and elastic making the OPEFB as filler composites can improve the mechanical properties of a material polymer [8].

3.2 Density analysis of OPEFB fiber

Density testing can be done based on the principle of Archimedes. That is explain the objects that are partially or wholly in the fluid will receive upward lift force equal to the weight of the displaced fluid. The fiber density is measured chip sized fibers with a length of 2-3 cm, short fiber and micro fiber. Each fiber density will be measured and compared. From the graph above it can be seen that the fiber-sized chip has the least density of 770 kg/m^3, for short fiber sized have density of 938.363 kg/m^3 while the microfiber sized have a density of 1350.073 kg/m^3. The difference in the density of fibers because the size is different. When the particle is smaller, the surface area will be greater so that the volume constant number of particles that occupy more and more space. Therefore, the smaller the particle size will make greater density. With this will certainly affect the density of biocomposite after the fiber is inserted into the matrix.

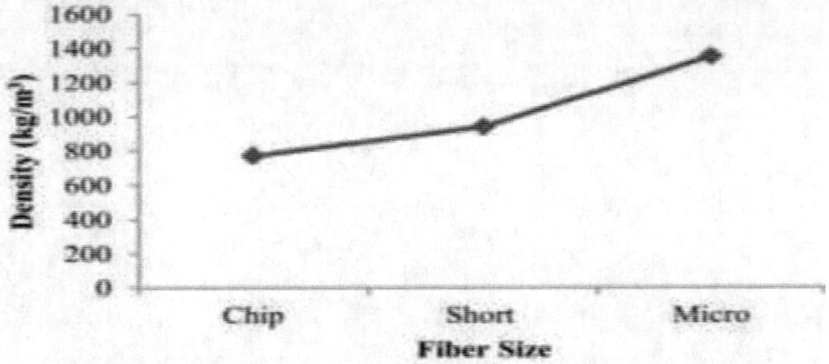

Figure 3. Graph relationship the particle size with density

3.3 Result of Granular Biocomposit Syntesis with Exstrusion Method

OPEFB fibers, recycled polymer and additives were weighed according to Table 4. Then the materials inserted into the mixing machine to do the mixing until the materials mixed with average. Mixing is done in order to bind the material that forms the optimum between the filler, the matrix and additives as well as the homogeneity of all three constituent can be achieved.

The use of additives in this research are useful to maintain the condition of the plastic / polymer used. The addition of the antioxidant additive serves to prevent the oxidation reaction by oxygen that can cause polymer degradation [9]. Acid scavanger while the addition of serves to neutralize the acid formed during the extrusion process so that the polymer chains are not degdradation [10]. While the coupling agent serves to increase the degree of dispersion, where the coupling agent serves as the interface that affect the interfacial adhesion and interfacial tension. Interfacial adhesion will be greater and the interfacial tension will drop with the addition of coupling agent [11].

The extrusion is a polymer material process establishment by pressing the polymer material through a mold cavity arranged magnitude. On the machine there is a screw extruder that rotates so that the material can lead to the extrusion machine parts ranging from the feed zone, melting zone to the melt-pumping zone which has the function of each. Machines used in this study is single screw extruder who only had one screw is placed in the barrel will move toward the exit [12]. Polymer which will be shaped like an elongated "noodles", then pulled through the water for cooling and hardening towards the mower. Inside this machine composite cut into small pellet shaped as shown in.

Figure 4. Granular with different compositions (Table 1)

Figure 4 is a granular result of mixing and extrusion. In the visible shape and size is not much different because it has the same basic ingredients. Only differ in the composition. To know the differences of each granular necessary to test the more that will be discussed in further discussion.

3.4 Density anlysis of Granular

Tests a granular density did not differ by testing the density of the fiber by using archimedes principle. The sample being measured is granular which is a blend of ABS with OPEFB fibers with different fiber composition. From the picture above can be seen that with the granular filler has a density microfiber TKKS size or greater density in each composition. This is because the density of the filler matrix polymer recycle OPEFB microfiber which is greater than the OPEFB shortfiber as described previously. However, the addition of fiber composition actually makes the density of the shrinking. It is caused by a number of empty spaces in granular compositions along with increase fiber. Empty spaces were caused air trapped in the biocomposite exstrusion process in which air is trapped as a result of fiber agglomerated [13].

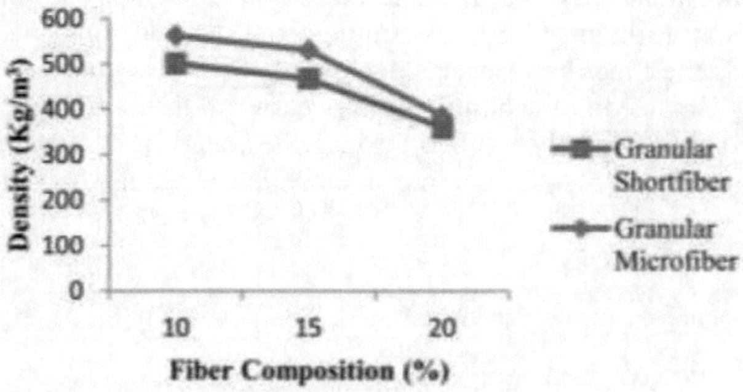

Figure 5. Test result of granular composite density

IOP Conf. Series: Earth and Environmental Science **65** (2017) 012041 doi:10.1088/1755-1315/65/1/012041

3.5 Molecular anlysis from FTIR data

Results of FTIR spectra (Figure 6) can be known about the compound structure of the fiber OPEFB (filler), ABS polymer recycle (matrix) and biocomposites mixture of both. FTIR spectra can be seen in biocomposites dominated by the peaks of the matrix that is the wave number 2237 cm-1 (N≡C), 30, 1600-1900 cm-1 which is the carbon bonds (C = O) of the polymer ABS and in the range of 670-1000 cm-1 identified a bond and CH = CH_2 Benzene, which is part of the ABS polymer matrix as well. While the peak that appears at wave number 2360 cm -1 and 1627 cm-1 are both derived from the peak OPEFB fiber. Their dominance of matrices ABS recycle is because the composition of its own biocomposite mostly ABS polymers recycle. While fiber OPEFB only ranged from 10-20% of the

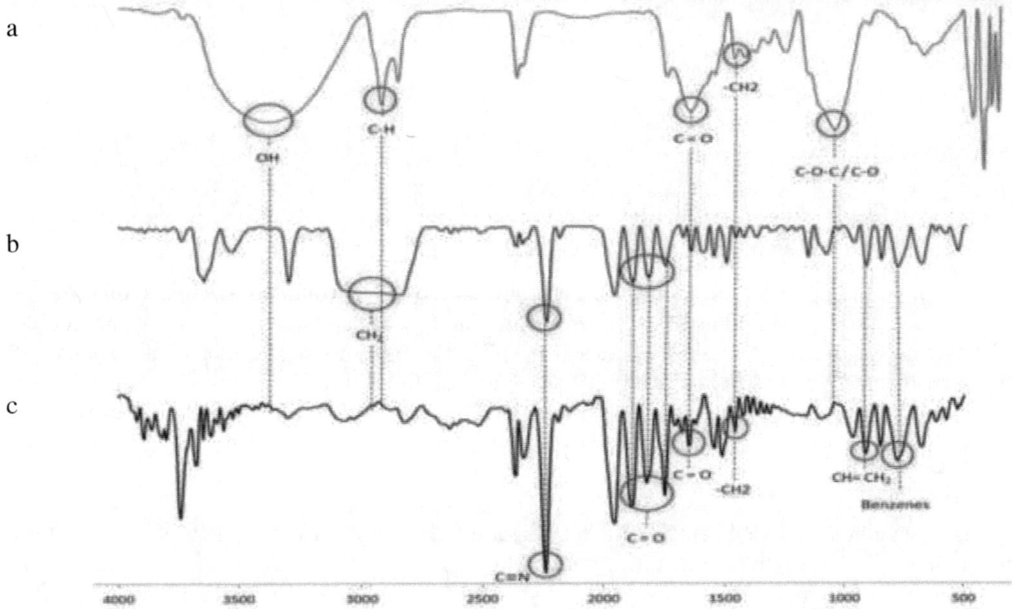

Figure 6. Test Result of FTIR (a) OPEFB fiber (b) Polymer ABS (c) Granular of Biokomposit

total biocomposites. In addition, it can be seen that an increase in the intensity of absorption at some bonding especially on the C = O bond. Where the bond could have come from the fiber and polymer ABS. It can be assumed that the formation of the bond between the filler interface OPEFB fibers with polymer matrices ABS recycle.

Related bond formation cellulose molecules and polymers ABS can be explained by the theory of orbital hybridization. In the figure 6 can be seen there are several kinds of many bonds, ie single bonds, double and triple bonds. This is influenced by the charging of valence electrons in the level of hybridization. If hybridization involves only the s and p orbitals the bond is called hybridization sp. The amount of the bond is determined by the difference between the numbers of valence electrons in the antibonding and bonding level can be calculated using equation (2). The problems that arise for complex molecules such as ABS and lignocellulose is the complexity in determining the levels of hybrid should be done using ab initio quantum calculations that are beyond the discussion of this research. However, in general the type of bond that is formed as bonding π or σ determined from atomic orbitals that contribute to bonding (bonding orbital).

To understand the formation of bonds in complex molecules necessary to understand the process of the formation of a hybrid bond in the molecule is much simpler. For example, carbon monoxide CO molecule. Based on the calculation of molecular orbital theory, the energy levels of CO by Figure 7.

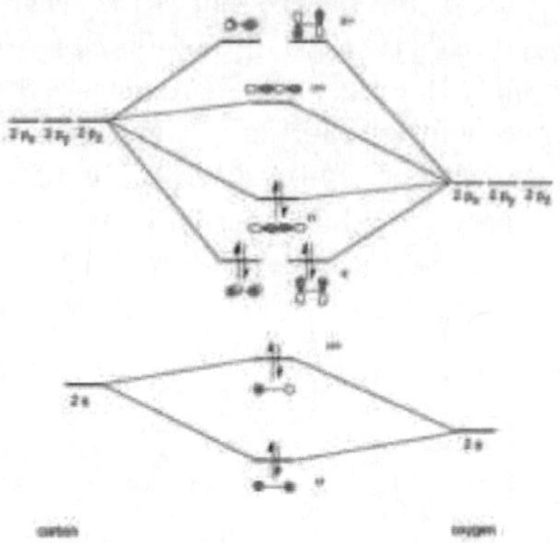

Figure 7. Enrgy's level diagram of CO molecul

Bonding in CO involving one σ bond and two π bond (Figure 16). C-O bond has a distance of 1.1 Å so consistent with a triple bond. In comparison, a single binding CO has a distance of 1.43 Å and double bind CO within Average 1.23 Å. Based on data from FTIR absorption spectra of CO at wavelength 2143. CO bond formed on the kind of p orbitals serve as bonding orbitals. As for knowing how the process of forming a bond between CO with other molecules such as metal required an understanding of the levels HOMO (Highest Occupied Molecular Orbital) and LUMO (Low Unoccupied Molecular Orbital).

From Figure 8 it can be concluded prior to the hybridization, there are 4 valence electrons in the atom C, two in the 3p level and two at level s with a total molecular orbital amounted 4. As for O there are 6 valence electrons, 4 in 3p level and two at s lvel with total number of orbital level 4. After hybridization level, the combined energy is 4 + 4 = 8 by charging the lowest energy level following the first in which only two valence electrons can fill the energy level s in accordance with the rules Pauli. HOMO levels of the CO molecule is filled by two electrons in this case the sigma orbital manifold. From the description of the previous CO can be concluded that the explanation bond formation for more complex molecules of the molecular orbital theory of the necessary information.

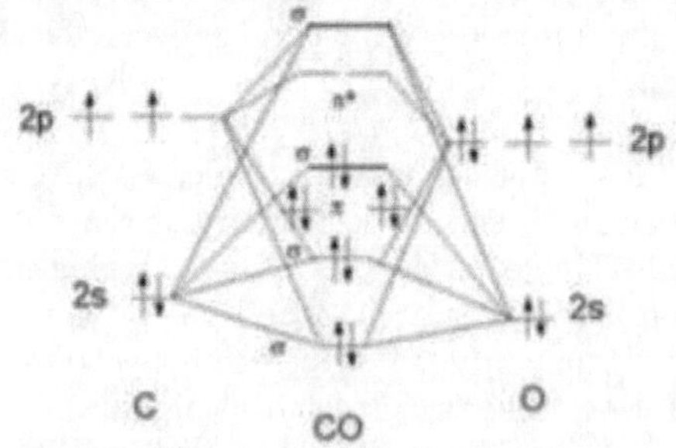

Figure 8. Energy's level diagram of CO molecul

about the levels of the complete hybrid. The number of hybrid orbitals level ABS- OPEFB fiber formed is the summation of the quantities of each molecular orbital ABS and fiber TKKS before bind. The type and amount of the bond that is formed depends on the electron valence electrons fill the energy level of the HOMO and LUMO levels of availability in ABS or OPEFB fiber. Thus further research is needed to determine the levels of hybridization and the number of valence electrons fill each orbital.

4. Conclussion

Results of the analysis composition showed that the cellulose of OPEFB has ties 41.68% which has a high purity level that can make the better quality materials.

Based on the results of density, particle size capable of affecting the density. The smaller of particle size, the density will be higher as evidenced by the value of the greatest density in the fiber size microfiber. It can also be seen in the granular density testing. Granular premises sized microfiber filler density is higher than the filler-sized granular shortfiber. But with the addition of fiber composition, density will experience deterioration. So the most optimum composition contained in the filler as much as 15% the size of the microfiber because density is quite high and able to absorb quite a lot of waste.

While the results of FTIR can confirm the presence of functional groups contained in granular biocomposites. And it is known that the polymer matrix is seen dominating role as compared to filler. Basically, the molecular interaction between the matrix with fiber filler OPEFB explained by the theory of orbital hybridization. But the explanation bond formation for more complex molecules like this from the side of the molecular orbital theory of the necessary information on the levels of the complete hybrid.

References

[1] Badan Pengelola Dana Perkebunan Kelapa Sawit [BPDPKS] 2015 *Gambaran Umum Riset Sawit Indonesia* Jakarta-Indonesia

[2] Sutarta E S, Winarna P L, Tobing dan Sufianto 2000 Aplikasi limbah cair pabrik kelapa sawit pada perkebunan kelapa sawit Paper at the Teknis Kelapa Sawit II, PPKS Medan-Indonesia, June 13-14th 2000

[3] Yunita D S 2015 Pembuatan Komposit dari Serat Tandan Kelapa Sawit (Elaeis Guineesis) Menggunakan Penguat Serat Recycled Polypropylene (RPP) dengan Variasi Massa Report Teknik Kimia. Palembang (ID): Politeknik Negeri Sriwijaya.

[4] Nayiroh, Nurun 2013 "Bahan Ajar Teknologi Material komposit". http://lecturer.uin-malang.ac.id. Diakses pada 15 Mei 2016. Malang (ID): Universitas Islam Negeri Malang

[5] De Luca N 2000 Molecular Orbital. Inorganic chemsitry. Slide presentation in spring camp, Spring 2000. Boston (UK): University of Massachusetts.

[6] Park W C 2008 A Study of Pyrolysis of Char-ring Materials and Its Application to Fire Safety and Biomass Utilization. [Disertasi]. Michigan: Mechanical Engineering-University of Michigan.

[7] Nikmatin S, Lisdar I S, Mersi K 2014 Pengembangan Teknologi Proses Produksi Bionanokomposit Filler Biomassa Rotan. Jurnal Ilmu Pertanian Indonesia **19** 163-168.

[8] Ali, Mahrus 2014 Sifat Mekanik Bahan Komposit Selulosa Bakterial dengan Penguat Serbuk Kayu Sengon. [Skripsi]. Jember (ID): Universitas Jember.

[9] Hanser 2003 Modern additive Handbook. Munich (DEU) : 231-242

[10] William D C J 2004 *Material Science and Engineering an Introduction* (New York: Jhon Wille & Sons)

[11] Arif D 2008 Analisis Pengaruh Ukuran Bahan Pengisi terhadap Karakteristik Komposit Polipropilena Serbuk Kayu. [Skripsi]. Depok (ID) : Universitas Indonesia

[12] Pratama RI 2007 Kajian mengenai prinsip-prinsip dasar teknologi ekstrusi untuk bahan makanan dan beberapa aplikasinya pada hasil perikanan. Bandung (ID): Universitas Padjajaran.

[13] Ardani HK 2013 Pengembangan Serat Kenaf (Hibicus Cannabinus L.) sebagai Filler Komposit Bermatriks Polimer (ABS) pada Aplikasi Helm.[skripsi]. Bogor (ID) : Institut Pertanian Bogor

IOP Conf. Series: Earth and Environmental Science **65** (2017) 012042 doi:10.1088/1755-1315/65/1/012042

Production of furfural from palm oil empty fruit bunches: kinetic model comparation

J R H Panjaitan[1], S Monica[1] and M Gozan[1]

[1]Chemical Engineering Department, Faculty of Engineering, University of Indonesia, Depok, Indonesia

Email : jabosarronggur@gmail.com

Abstract. Furfural is a chemical compound that can be applied to pharmaceuticals, cosmetics, resins and cleaning compound which can be produced by acid hydrolysis of biomass. Indonesia's demand for furfural in 2010 reached 790 tons that still imported mostly 72% from China. In this study, reaction kinetic models of furfural production from oil palm empty fruit bunches with submitting acid catalyst at the beginning of the experiment will be determine. Kinetic data will be obtained from hydrolysis of empty oil palm bunches using sulfuric acid catalyst 3% at temperature 170°C, 180°C and 190°C for 20 minutes. From this study, the kinetic model to describe the production of furfural is the kinetic model where generally hydrolysis reaction with an acid catalyst in hemicellulose and furfural will produce the same decomposition product which is formic acid with different reaction pathways. The activation energy obtained for the formation of furfural, the formation of decomposition products from furfural and the formation of decomposition products from hemicellulose is 8.240 kJ/mol, 19.912 kJ/mol and -39.267 kJ / mol.

1. Introduction

Furfural is a chemical compound that is applied to pharmaceuticals, cosmetics, resins, and cleaning compound. In addition, furfural is widely used as an extracting agent, solvent, vulcanizing agents, flavoring agents, commercial components of pesticides, antiseptics, disinfectants, and so on. Furfuryl alcohol compound is a chemical produced from hydrogenation process of furfural and consuming 62% furfural world market [1].

Industries that produce furfural in Indonesia are still few in number. The consumption of furfural in Indonesia is steadily increasing. In 2010, the needs of furfural in Indonesia reached 790 tons. During this time, the needs of furfural in Indonesia met through import, especially from China, which is the largest furfural manufacturer in the world [2].

Furfural can be produced from biomass through hydrolysis. Palm oil as one of the largest agricultural commodities in Indonesia can be used in furfural production. Around 22-23% of the total fresh fruit bunches become solid waste in the form of Palm oil empty fruit bunches (POEFB) . POEFB is a lignocellulose biomass (cellulose, hemicellulose, and lignin) that can be hydrolyze into a variety of chemical compounds.

Furfural is generally produced with hydrolysis of hemicellulose into pentose which is then dehydrated into furfural. Acids, such as H_2SO_4 and HCl, can catalyze hydrolysis reaction of hemicellulose. Studies

Content from this work may be used under the terms of the Creative Commons Attribution 3.0 licence. Any further distribution of this work must maintain attribution to the author(s) and the title of the work, journal citation and DOI.
Published under licence by IOP Publishing Ltd

have been conducted on sulfuric, hydrochloric, phosphoric, and nitric acids that may improve hydrolysis of hemicellulose and cellulose [3]. However, sulfuric acid is commonly used to hydrolyze polysaccharides into monosaccharaides [4].

Researches on kinetic reactions of furfural production have been done by Dussan et al., Hua et al., and Liu et al. [5,6,7] at different lignocellulosic materials, but there is no research that focus on kinetic model comparation of furfural production. This research studies furfural kinetic models based on POEFB with submitting sulfuric acid catalyst at the beginning of the experiments.

2. Method

2.1. Experiment

The raw material used was POEFB with the composition of the cellulose - alpha, hemicellulose, and lignin 40.3%, 31.2% and 23.2%, respectively. POEFB particle size was reduced to 30 mesh, washed with water, dried, soaked with aqueous ammonia solution 5% for 24 hours, washed with water until neutral, dried and then hydrolyzed to produce furfural. Hydrolysis process for the formation of furfural was done by pressurized reactor with 1 liter volume. Mass ratio of POEFB and 3% sulfuric acid solution was 1:20. The time to reach the desired temperature was about ± 30 minutes. The reaction times start when the desired reaction temperature was reached. The experimental data were taken every 5 minutes at 170°C, 180°C and 190 °C for 20 minutes with 3% sulfuric acid catalyst was add at 30°C. Furfural concentration analysis was performed using HPLC with Aminex HPX - 87H Ion Exclusion Column. The eluent was H_2SO_4 0.008 N with flow rate 1 ml/min. The detector was Refractive Index Detector with the temperature of detector and column were 35°C.

2.2. Kinetics Model

Reaction kinetic models to be tested in this study consist of three models. In the first model hemicellulose is converted into furfural and then furfural is converted into decomposition products. The second model is modification of the first model in which hemicellulose is also produced decomposition products. While the third model assumes the hemicellulose decomposition products are divided into two products.

Xylose which is a degradation compound from hemicellulose before furfural is not included in this reaction models. This is due to research methods that add acid catalyst in the beginning of the experiment. With this method, there isn't xylose concentration detected at the desire reaction temperatures. Furfural kinetic models are shown in table 1. C_{HC}, C_F, and C_{DP} are hemicellulose concentration, furfural concentration, and decomposition products concentration.

Tabel 1. Furfural Kinetic Reaction Models

	Furfural Kinetic Reaction Models	Kinetic Reaction Equations
First Model	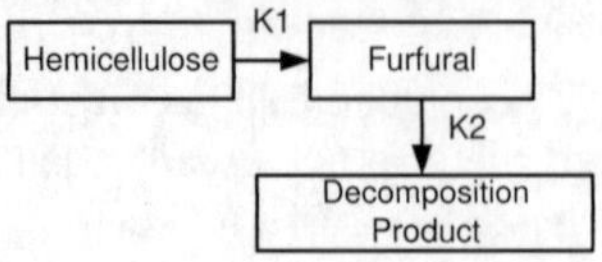	$\frac{dC_{HC}}{dt} = -K_1C_{HC}$ (1) $\frac{dC_F}{dt} = K_1C_{HC} - K_2C_F$ (2) $\frac{dC_{DP}}{dt} = K_2C_F$ (3)
Second Model	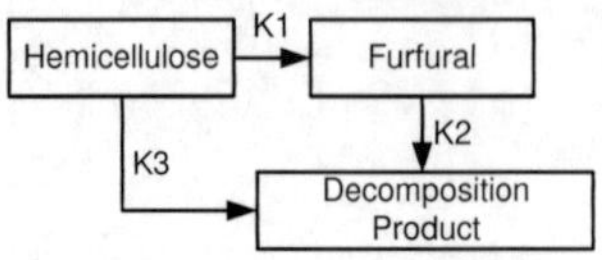	$\frac{dC_{HC}}{dt} = -K_1C_{HC} - K_3C_{HC}$ (4) $\frac{dC_F}{dt} = K_1C_{HC} - K_2C_F$ (5) $\frac{dC_{DP}}{dt} = K_2C_F + K_3C_{HC}$ (6)
Third Model	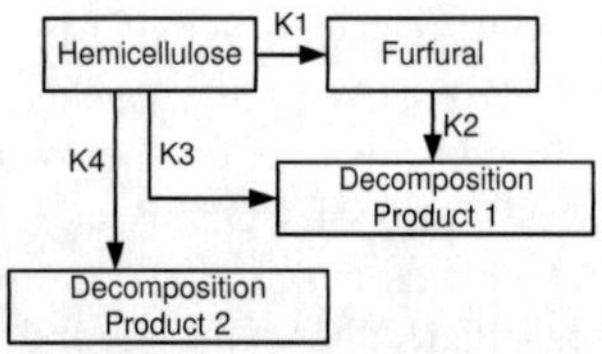	$\frac{dC_{HC}}{dt} = -K_1C_{HC} - K_3C_{HC} - K_4C_{HC}$ (7) $\frac{dC_F}{dt} = K_1C_{HC} - K_2C_F$ (8) $\frac{dC_{DP1}}{dt} = K_2C_F + K_3C_{HC}$ (9) $\frac{dC_{DP2}}{dt} = K_4C_{HC}$ (10)

2.3 Calculation of Kinetic Data

Reaction rate constants (k_1, k_2, k_3, and k_4) obtained by optimized experimental data with kinetic models using optimization toolbox fminsearch on MATLAB software. In the calculation process, the concentration of hemicellulose in POEFB was assumed 0.0354 mol/L at the desired reaction temperature. Reaction rate constants at temperature of 170°C, 180°C and 190°C will be plotted with the Arrhenius equation to get the activation energy.

3. Results and Discussion

3.1. Determination of Kinetic Model

Data was taken every 5 minutes for 20 minutes at 170°C, 180°C and 190°C. Reaction kinetic models selected by optimization value sum of square error of experimental data on kinetic models using MATLAB software with optimization toolbox fminsearch. Error occured in the first and third model simulation. First model isn't optimum because this model doesn't consider the hemicellulose degradation reactions that can produce variety of products beside furfural. The third model predicted that will be other kinds of decomposition products from hemicellulose. However, this model is not very significant to determine kinetic model because basically hemicellulose consists of pentose and hexose sugars [8]. Generally, pentose and hexose sugars in acid hydrolysis process will produce the same decomposition products.

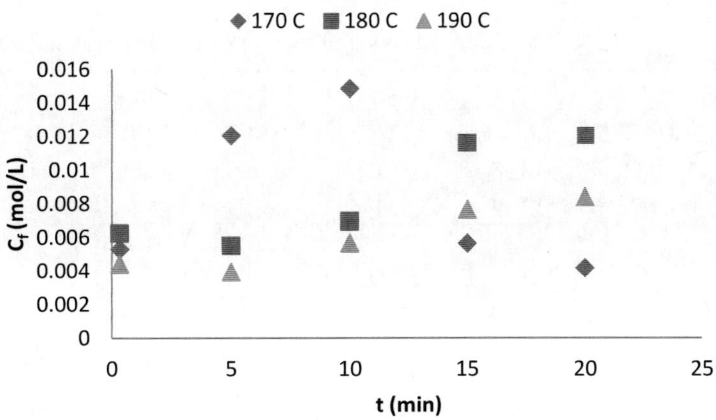

Figure 1. Furfural Concentration Data

Based on the optimization results, second model more suitable to describe the kinetic of furfural production. Second models can describe the general prediction that occur in the conversion process of hemicellulose into furfural, furfural into decomposition products and hemicellulose into decomposition products. Hemicellulose consist of pentose sugars and hexose sugar that can produce a variety of decomposition products from hydrolysis reaction with acid catalyst. Hydrolysis process with acid catalyst will convert pentose sugars in hemicellulose into furfural compound [8]. Danon et al. [9] showed that furfural will be decomposed into two major products which is formic acid and resinous tars. On the other side, hexose sugars will be converted to HMF (5-hydroxymethylfurfural) which then will be converted into levulinic acid and formic acid [10,11]. From the various studies mentioned above, it shows that hemicellulose can be converted to formic acid from different pathways. The detail description on second kinetic model can be seen in figure 2.

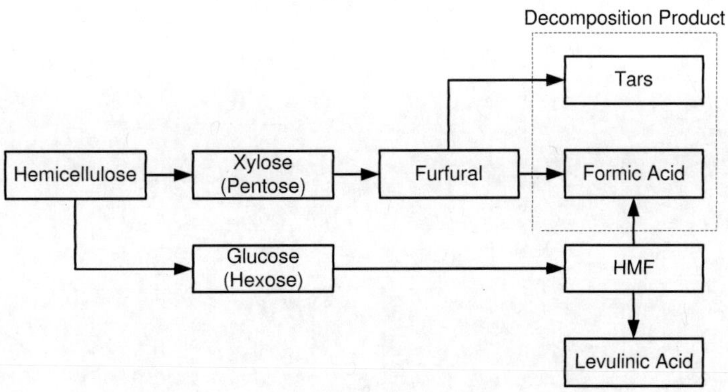

Figure 2. Second Model Detail Scheme

3.2. Energy Activation Calculation

Second model reaction rate constants was calculated from optimization furfural data with the model in each temperature using fminsearch MATLAB optimization toolbox. Activation energy (E) was calculated using Arrhenius equation with plot In K versus 1/T [12].

$$Ink_A = In\,A - \frac{E}{R}\left(\frac{1}{T}\right) \tag{11}$$

Tabel 2. Second Model Reaction Constants

Temperature (°K)	1/T	k_1	In k_1	k_2	In k_2	k_3	In k_3
443	0.002257	0.12	-2.12026	0.067	-2.70306	0.1	-2.30259
453	0.002208	0.138	-1.9805	0.056	-2.8824	0.087	-2.44185
463	0.00216	0.132	-2.02495	0.085	-2.4651	0.063	-2.76462

Tabel 3. Second Model Activation Energy

Reaction Rate Constants	Activation Energy (kj/mol)
k_1	8.240
k_2	19.912
k_3	-39.267

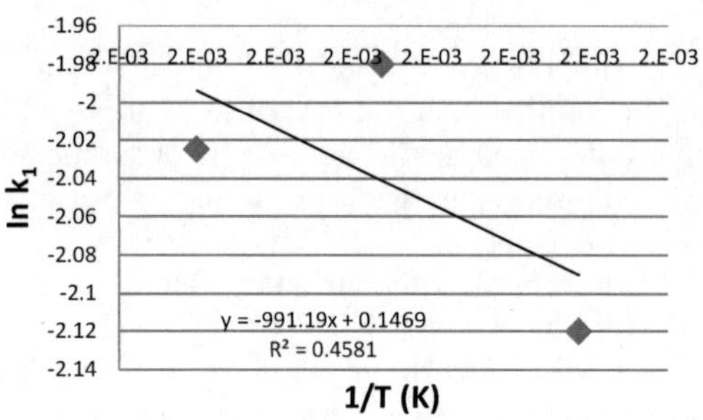

(a)

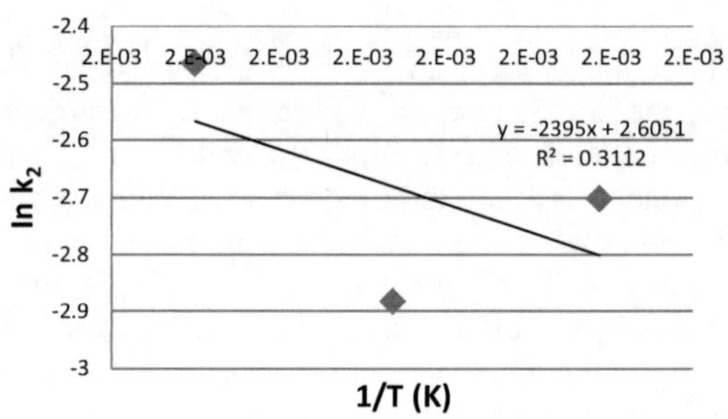

(b)

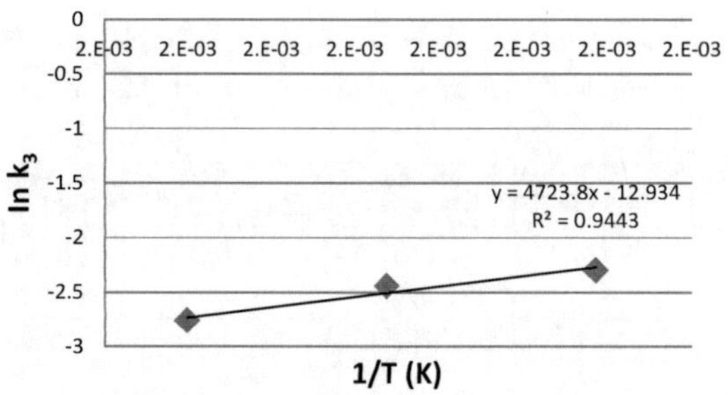

(c)

Figure 3. Arrhenius Plot In k versus 1/T (a) k_1 (b) k_2 (c) k_3

Based on table 3, the activation energy of k_2 (furfural into decomposition products activation energy) was 19.912 kJ/mol which was less than the value of activation energy from previous studies such as 60.3 kJ/mol [13], 132 kJ/ mol [14] and 106 kJ/mol [15]. The low activation energy happened because the effect of submitting sulfuric acid catalyst in the beginning of the experiments. Acid catalyst made the formation of furfural and decomposition products started early which was resulted in low furfural concentration at the desired temperature. The low furfural concentration effects smaller activation energy value to convert furfural become decomposition products. The same phenomenon also applied to the activation energy of k_1 (hemicellulose into furfural activation energy) which was 8.240 kJ / mol. The activation energy of k_3 (hemicellulose into decomposition products activation energy) is in negative value. This indicates that the reaction rate decreases with increase the temperature. Therefore, further researches need to investigate chemical reactions of hemicellulose especially hexose sugars into decomposition products.

4. Conclusion

Kinetic model that describes the production of furfural with submitting sulfuric acid catalyst in the beginning of the experiments was the kinetic model where in this model hemicellulose and furfural will tend to produce the same decomposition products namely formic acid. Hexose sugars in hemicellulose were converted into formic acid at the end of the reaction pathway. The activation energy obtained by this study for the formation of furfural from the hemicellulose, the formation of decomposition products from furfural and the formation of decomposition products from hemicellulose are 8.240 kJ/mol, 19.912 kJ/mol and -39.267 kJ/mol. Formic acid as final decomposition product of hexose and pentose sugars must have detail examination in future.

Acknowledgements

The authors acknowledge financial support from Indonesia Estate Crop Fund for Palm Oil (BPDPKS Research grant number Peng-01/DPKS.4/2015).

References

[1] Peleteiro S, Rivas S, Alonso J L, Santos V and Parajo J C 2016 Furfural production using ionic liquids: A review *Bioresource Technology* **202** 181-191

[2] Wijarnarko A, Wintono J, and Wiguna M 2006 Tinjauan komprehensif perancangan awal pabrik furfural berbasis ampas tebu di Indonesia *Journal of The Indonesian Oil and Gas Community* 1-7

[3] Singh R, Shukla A, Tiwari S, and Srivastava M 2014 A review on delignification of lignocellulosic biomass for enhancement of ethanol production potential *Renewable and Sustainable Energy Review* **32** 713-728

[4] Mood S H, Golfeshan A H, Tabatabaei M, Jouzani G S, Najafi G H, Gholami M and Ardjmand M 2013 Lignocellulosic biomass to bioethanol, a comprehensive review with a focus on pretreatment *Renewable and Sustainable Energy Review* **27** 77-93

[5] Dussan K, Girisuta B, Haverty D, Leachy J J and Hayes M H B 2013 Kinetics of Levulinic Acid and Furfural Production From Miscanthus x giganteus *Bioresource Technology* **149** 216-224

[6] Hua D R, Wu Y L, Liu Y F, Chen Y, Yang M D, Lu X N and Li J 2015 Preparation of furfural and reaction kinetics of xylose dehydration to furfural in high-temperature water *Pet. Sci.* **13** 167-172

[7] Liu H, Hu H, Baktash M M, Jahan M S, Ahsan L and Ni Y 2014 Kinetics of Furfural Production From Pre-Hydrolysis Liquor (PHL) of a Kraft-Based Hardwood Dissolving Pulp Production Process. *Biomass and Bioenergy* **66** 320-327

[8] Morone A, Apte M, and Pandey R.A, 2015. Levulinic Acid Production From Renewable Waste Resource: Bottlenecks, Potential Remedies, Advancements And Applications *Renewable and Sustainable Energy Review* **51** 548 - 565

[9] Danon B, Aa L.V.D, and Jong W.D, 2013. Furfural Degradation in a Diluted Acidic and Saline Solution in The Presence of Glucose. *Carbohydrate Research* **375** 145-152

[10] Yan K, Jarvis C, Gu J, and Yan Y, 2015. Production and Catalytic Transformation of Levulinic Acid: A Platform for Speciality Chemicals and Fuels. *Renewable and Sustainable Energy Review* **51** 986-997

[11] Rackemann D W and Doherty W O S 2012 A Review On The Production of Levulinic Acid And Furanics From Sugars *Proc Aust Sugar Cane Technol* **34** 1-9

[12] Fogler H S 2006 *Elements of Chemical Reaction Engineering Fourth Edition* (New Jersey : Prentice Hall International Inc.)

[13] Lavarack B P, Griffin G J and Rodman D 2002 The acid hydrolysis of sugarcane bagasse hemicellulose to produce xylose, arabinose, glucose and other products *Biomass and Bioenergy* **23** 367-380

[14] Nabarlatz D, Farriol X and Montane D 2004 Kinetic modeling of the autohydrolysis of lignocellulosic biomass for the production of hemicellulose-derived oligosaccharides *Industrial & Engineering Chemistry Research* **43** 4124–4131

[15] Morinelly J E, Jensen J R, Browne M, Co T B and Shonnard D R 2009 Kinetic Characterization Of Xylose Monomer And Oligomer Concentrations During Dilute Acid Pretreatment Of Lignocellulosic Biomass From Forests And Switchgrass *Industrial & Engineering Chemistry Research* **48** 9877–9884

IOP Conf. Series: Earth and Environmental Science **65** (2017) 012043 doi:10.1088/1755-1315/65/1/012043

Study on structure and morphology (*Boehmeria nivea*) in the irregular and regular parts of the fiber after biodegumming

A P Wulandari[1], D Septarini[1] and A Zainuddin[2]

[1]Biology Department, Faculty of Mathematic and Sciences , University of Padjadjaran, Bandung, Indonesia
[2]Chemistry Department, Faculty of Mathematic and Sciences , University of Padjadjaran

E-mail : wulandariasripeni@gmail.com

Abstract. Ramie is a natural fiber that is very potential to be developed in Indonesia. Decorticated-fiber which has been known as china grass produce different structures irregular part but shows a long straight section in the middle. This study aims to determine differences in chemical components, morphology and microstructure of two different parties after biodegumming process. China grass has been processed to remove gum using pectinolytic fungus. The microstructure of the treated was further tested by Fourier Transform InfraRed (FTIR), X-Ray Diffraction (XRD), and Scanning Electron Microscope (SEM). The FTIR study indicated that during the biodegumming process, chemical bonding of non-cellulose components most removed by the activity of pectinase from the fungus. XRD analysis reflects an increase in the crystallinity of the fiber after biodegumming. Scanning electron microscopy (SEM) was used to confirm a reduction in the size of the fiber after biodegumming either in the irregular and regular part of the fiber after biodegumming.

1. Introduction

Gum removal process (degumming) on natural fiber in plants has been studied by various methods and it is important to increase the content of cellulose if the application will be used in the textile and other industrial. Degummed fiber produced would facilitate for further processing. By knowing the morphology and microstructure of natural fibers of the fiber is expected to obtain the basic knowledge essential for the further development of the natural fiber polymer, considering the utilization can vary widely as well as the need for raw materials with a high content of high purity of cellulose. A novel biodegumming process with pectinases for ramie fiber was developed in combination with oxygen plasma, in order to improve the efficiency of enzymes in impurity degradation with a temperature of 75 °C, a pH value of 8.0 in subjected solution for 60.0 min [1].

Application of biodegumming had reported by researchers to use *Pectobacterium* sp. [2], a mutant strain T66 from *Bacillus subtilis* [3], with a mixed culture of four bacteria 72 h followed by boiling in 7 g/L NaOH solution for 4 h [4], a large-scale degumming using *Bacillus pumilus* DKS1, a newly isolated strain with high pectate lyase activity, was reported [5]. The galacturonic acid from the ramie previously treated with chemical reagents could be removed by *Grewia optiva* and *Bacillus* sp. [6]. Some enzymes for the degumming of ramie were obtained from Actinomycete [7], Amycolata sp. [8]. A rapid process of ramie biodegumming by *Pectobacterium* sp. CXJZU-120 has been established. Factory practical tests for the rapid process had reported [9].

Content from this work may be used under the terms of the Creative Commons Attribution 3.0 licence. Any further distribution of this work must maintain attribution to the author(s) and the title of the work, journal citation and DOI.
Published under licence by IOP Publishing Ltd

Many treatment sets of degumuming and bleaching reduce the extractive content on lignocellulosic fiber to produce microcrystalline, pulp, and cotton. Change in the physical properties of character from the fibers usually investigated by FTIR spectroscopy and X-ray diffraction to estimate the crystallinity and amorphous degree [10] [11]. The crystalline fraction of fiber is given only by cellulose, since the component of hemicellulose and lignin identified as an amorphous. With the purpose of supplying of biomaterial, the factors of physical and chemical properties of the fiber have a significant effect by crystallinity[12].

China grass is coarse fibers produced from the decortication that has different structures with no homogeneous in structure and morphology, especially at the middle and end of the fiber. The structural of fiber with crystallinity and amorphous component could be responsible for making effective a process of degumming. In this study particular objective was to characterize the difference structure and morphology ramie (*Boehmeria nivea*) in the irregular and regular parts of the fiber after biodegumming.

2. Method

2.1. Materials

China grass was supplied from Central of Ramie, Wonosobo, Central Java, Indonesia. The fiber was cut as part regular shape and an irregular shape. Biostarter for bio degumming, a pectinolytic fungus R-apw1 as culture collection from Microbiology Laboratory, Department Biology, University of Padjadjaran, Indonesia.

2.2. Biodegumming

Two part of the china grass as ramie bast fibers was soaked in 3 L medium modified [8] in reactor degumming containing 10% of decorticated dried raw fibers and 10 % of activated biostarter from 3-day-old liquid cultures. The reactors were incubated at room temperature for 42h. Biodegummed fiber was rinsed with water, finally air-dried.

2.3. Analysis Sample

China grass and bio degummed fiber were characterized in terms of changes in the functional group's fiber components by using Fourier Transform Infra Red (FTIR). The FTIR spectra of the cellulosic samples were measured on a FTS 2000 spectrometer Series DIGILAB with a frequency range of 4600-400 cm. X-Ray Diffraction (XRD) was used to observe changes in the diffraction pattern of the cellulose /crystallinity. Data diffraction patterns were collected on a RIGAKU RINT 2500 apparatus. To draw main graphics, the instrument was equipped with a transmission type goniometer. Analysis changes in appearance (morphology) of fiber microscopic observation by Scanning Electron Microscope (SEM).

3. Results

3.1. Fibre Morphology

Morphology of ramie fiber can be observed at the macro structure of china grass before and after biodegumming. The microfibrils shape only can be observed by SEM to observe changes in the size and effectiveness of the treatments biodegumming.

3.1.1. Macro fiber

Decorticated fiber (china grass) generally has a length of ± 175 - 215 cm, two distinct parts in the middle with regular long straight and at both ends are tangled because of mistreatment mechanically on the decortication (separation of bark from the trunk. Two type china grass as a regular and irregular part was degummed by pectinase produced by isolate R-apw1 showed differences character microfibril ramie as shown in Figure 1.

IOP Conf. Series: Earth and Environmental Science **65** (2017) 012043 doi:10.1088/1755-1315/65/1/012043

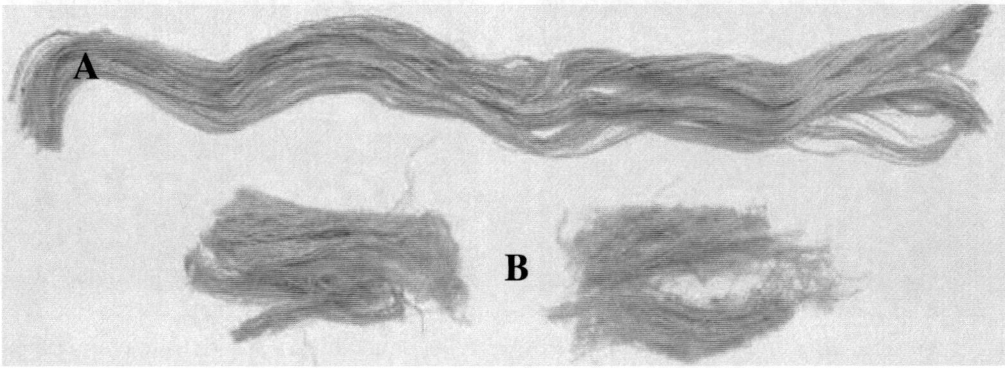

Figure 1. *China grass* (A: middle part- regular and B: the end fiber- irregular)

Gum spreading around the anatomic structure of the bark was composed of pectin as a building block of fiber. Used pectinolytic fungus for biodegumming effective for microfibril degradation cause shrinkage diameter fiber to be microfibrils. As well as physical and visual observations indicate that the results biodegumming fiber become more delicate and brightly colored than before treatment biodegumming. (Figure 2)

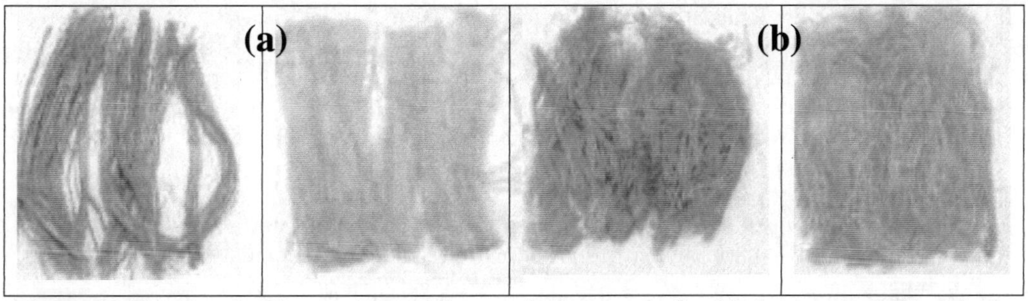

Figure 2. The Ramie Fiber : (a) middle part (before and after biodegumming) (b) end of the fiber (before and after biodegumming)

3.1.2. Micro fiber

Confirmation of loss of pectin and reduced diameter fiber at the regular and irregular fiber, before and after biodegumming shown as results of SEM (Figure 3)

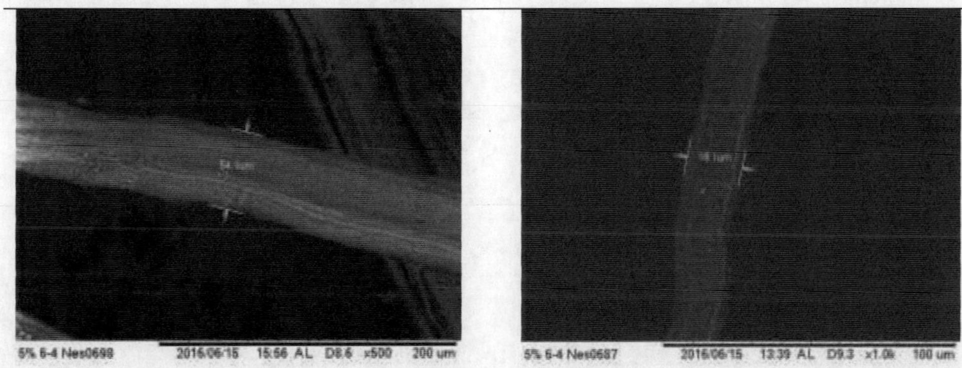

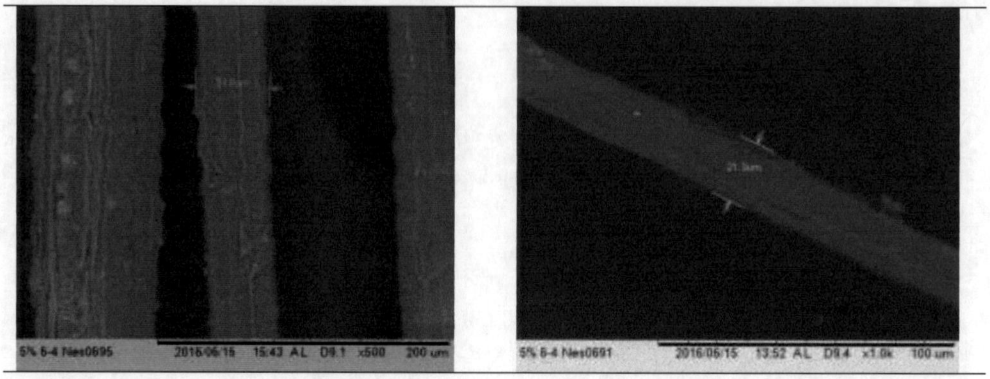

Figure 3. SEM Image Cellulose Microfibrils (A) middle part and (B) the end part (before -(left) and after biodegumming - right))

The size reduction indicated that the fiber fibrillation with a perfect process degradation organic components that change the character of rigid and mutually bonded strongly with each other. the fiber strands become smaller. Activities pectinase produced by pectinolytic fungus capable of hydrolyzing pectin which is a compound of the content of the gum in the fiber.

Changes in the size of the diameter of the microfiber being after biodegumming prived that the process can decipher biodegumming coarse flax fiber decortication result of rigid and mutually bonded strongly with each other, the fiber strands become smaller. It is caused by uses microfungus as biostarter for biodegumming. This fungus was selected with ability pectinolytic to hydrolyze the pectin. Biodegumming of the china grass for both of part of the fiber middle and the ends part leads to remove component non-cellulose and reduced the diameter size china grass before treated by biodegumming had a diameter >50 μM, while after the treatment, the micro fibril of cellulose was average ± 20 μM.

3.1.3. FTIR Spectra

FTIR is an effectiveness technique for characterizations of surface or interface lignocellulosic fibers. Figure 4 and Figure 5 were shown FTIR spectra of regular and irregular fiber before and after biodegumming.

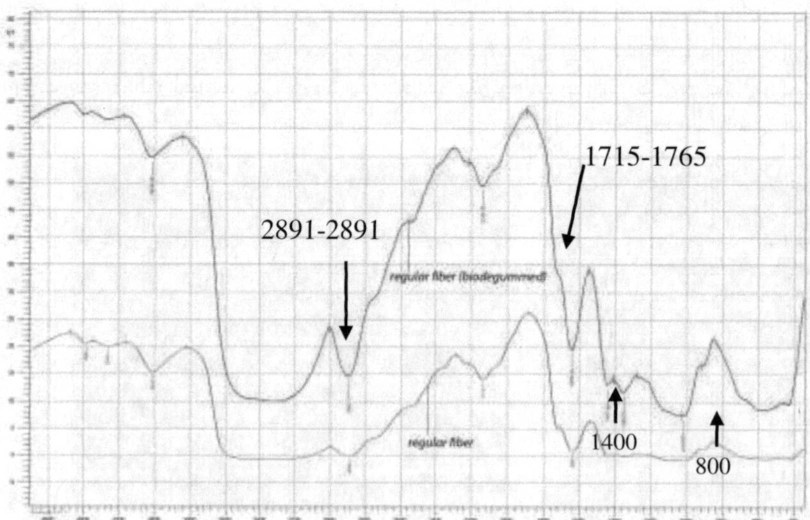

Figure 4. FTIR Spectra of regular fiber before and after biodegumming

IOP Conf. Series: Earth and Environmental Science **65** (2017) 012043 doi:10.1088/1755-1315/65/1/012043

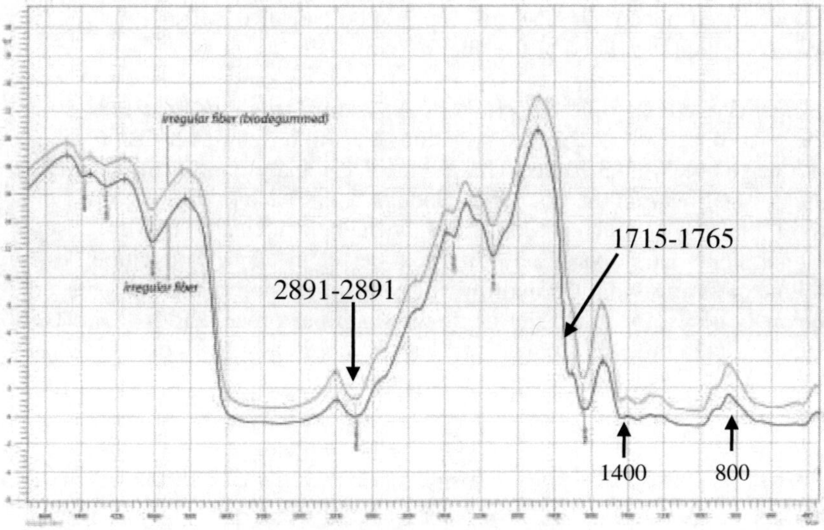

Figure 5. FTIR Spectra of irregular fiber before and after biodegumming

The FTIR was conducted to determine the presence of a functional group of cellulose and no cellulosic, also its changes after biodegumming. Cellulose molecule consists of the elements C, H, and O which has the chemical formula $(C_6H_{10}O_5)n$ hydrogen bonding. Based on data from FT-IR spectra of the fiber (middle and end part) either on the fiber before and after in-biodegumming (Figure 4 and 5), there is the C-H bond in the area of wave numbers between 2891 to 2894 cm-1 were identified as a constituent cellulose. FTIR spectra show that the biodegumming did not damage the chemical bonds biodegumming constituent of cellulose.

C = O at 1765-1715 cm-1 known as chemical bonding of non-cellulose components. FT-IR spectrum the fibers either in the middle or end one showed on the infrared absorption wave number region between 1765-1715 cm-1 in the FT-IR spectrum is an area of vibrational strain C = O of the acetyl group and the ester group of compounds pectin, uronat , hemicellulose or carboxylate ester bond ferulic acid (ferulic) and p-coumaric of lignin or hemicellulose. Disorder structure was examined by the IR absorption after biodegumming this region showed increasing in the intensity. It was found that the absorption spectra both in fiber regular and irregular part of the fiber had changed with the loss of pectin components.
Biodegumming treatment produced some disposition absorption bands in the 1400-800/cm-1 region, especially in the 1000-1200 cm-1 region was found to be very sensitive to change amount of ratio crystalline and amorphous structure of cellulose.

3.1.4. X-ray Diffraction Analysis
Figure 6 showed the spectra-X-ray Diffraction (XRD) showed a pattern of the character of cellulosic fiber as a natural fiber. Both spectra of fiber regular and irregular part before and after bio degumming exhibited a primary peak at the diffraction angle $2\theta = 23.12°$ and two peaks at $2\theta = 15.70°$ and $16.68°$, however, those small peaks tended to merge.

XRD overview of fiber central part of the regular and irregular in the end showed the same diffraction pattern, however, high intensity are different, with the presence of amorphous components in part irregular. These figures also featuring high-peak difference crystallinity between china grass and biodegummed fiber. Figure 6 (A) in the regular shape in the middle position of the fiber, the maximum relative intensity of the crystallinity of the fiber at position 2θ - 23° was 11,142.68 cuts, while after biodegumming, increased only 12,106.66 cts. The middle part of the fiber is basically a homogeny polymer cellulose compared to the end of the fiber.

In the irregular, increased fiber crystallinity at the position of 2θ -22.68 ° height peak at 4480.80 cts, while after biodegumming was observed changed position spectra be come 2θ-23.76 cause increasing crystallinity with height peak 6543.73 cts. The results of the analysis of X-ray diffraction pattern showed an increasing crystallinity on the fiber after biodegumming, due to the loss of the gum which is one component of the non-natural cellulose fibers, which incidentally is the cause of formation properties of amorphous (non-crystalline) on the fiber. Part amorphous more easily hydrolyzed than the crystalline portion, so that the hydrolysis treatment causing the fibers to become more crystalline. Therefore it is necessary to do the next stage of fiber processing to completely eliminate non-cellulose components in order to produce pure natural fibers with high level of isolated cellulose.

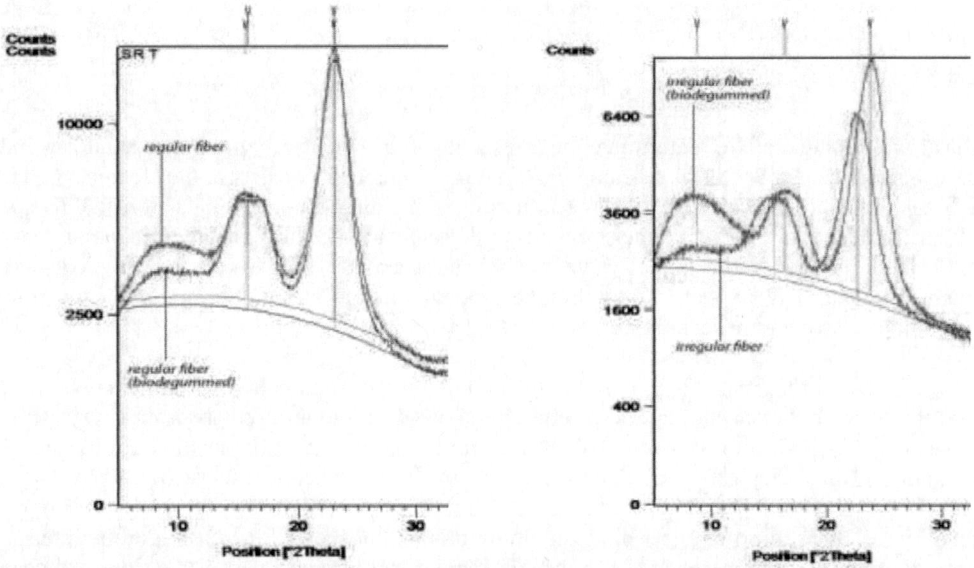

Figure 6. XRD Spectra: (A) Regular part and (B) Irregular part of the fiber before and after biodegumming

Biodegumming of the china grass for both of part of the fiber middle and the ends part leads to remove component non-cellulose proved by characteristic morphology and absorption spectra by FTIR dan XRD.

Acknowledgement
The present work was supported by a Grant from the Ministry of Research Technology and Higher Education, with skim funding Penelitian Unggulan Nasional (PUSNAS), 2016.

References

[1] Ming S, Ling W, Jia-Jie L 2015 Journal of Cleaner Production **101** 395-403
[2] Liu Z C *et al. Textile Research Journal* **82**1553–1559
[3] Sun Q X, Wang M S and Qin C Y 1985 Patent ZL85103481(China)
[4] Paul NB and Bhattacharyya S K 1980 *The Indian Textile J* **10** 311–316
[5] Basu S, Saha M N, Chattopadhyay D, Chakarbati K 2009 *J. Ind. Microbiol Biot.* **36** 239–245
[6] Kashayp D R, Vohra P K and Soni S K 2001 *Biol. Lett.* **23** 1297–1301

[7] Bru¨hlmann F, Kim K S, Zimmerman W, and Fiechter A 1994 *Appl. Environ. Microbiol* **60** 2107–2112

[8] Bru¨hlmann F, Leupin M, Erismannb K H, Fiechter A 2000 *J. Biotech* **76** 43–50

[9] Liu Z C 2007 *et al. China J. Textile Res.* **40** 91–94

[10] Park S, John O B, Michael E H, Philip A P, and David K J 2010 *Biotechnology for Biofuels* **3** 10

[11] Poletto, M, Ornaghi Júnior, Heithor L,and Zattera, Ademir J 2014 *Materials* **7** 6105-6119

[12] Lionetto F, Roberta D S, Donato Cannoletta, Giuseppe V and Alfonso M 2012 *Materials* **5** 1910-1922

IOP Conf. Series: Earth and Environmental Science **65** (2017) 012044 doi:10.1088/1755-1315/65/1/012044

Surface free energy analysis of oil palm empty fruit bunches fiber reinforced biocomposites

G S Suryadi[1], S Nikmatin[1], Sudaryanto [2], Irmansyah[1] and S G Sukaryo[2]

[1] Department of Physics, Bogor Agricultural University, Bogor, Indonesia

[2] Centre for Science and Advanced Materials Technology, National Nuclear Energy Agency of Indonesia, South Tangerang, Indonesia

Email : gemasukmawati@gmail.com

Abstract. Study of the size effect of natural fiber from oil palm empty fruit bunches (OPEFB) as filler, onto the contact angle and surface free energy of fiber reinforced biocomposites has been done. The OPEFB fibers were prepared by mechanical milling and sieving to obtain various sizes of fiber (long-fiber, medium-fiber, short-fiber, and microparticle). The biocomposites has been produced by extrusion using single-screw extruder with EFB fiber as filler, recycled Acrylonitrile Butadiene Styrene (ABS) polymer as matrix, and primary antioxidant, acid scavanger, and coupling agent as additives. The obtained biocomposites in form of granular, were made into test piece by injection molding method. Contact angles of water, methanol, and hexane on the surface of biocomposites at room temperature were measured using Phoenix 300 Contact Angle Analyzer. The surface free energy (SFE) and their components were calculated using three previous known methods (Girifalco-Good-Fowkes-Young (GGFY), Owens-Wendt, and van Oss-Chaudhury-Good (vOCG)). The results showed that total SFE of Recycled ABS as control was about 24.38 mJ/m^2, and SFE of biocomposites was lower than control, decreased with decreasing of EFB fiber size as biocomposites filler. The statistical analysis proved that there are no statistically significant differences in the value of the SFE calculated with the three different methods.

1. Introduction

Natural fibers have been used as reinforcement in polymer composites during the last decades [1-5]. Oil palm empty fruit bunches (OPEFB) as one of natural fiber biomass offer some advantages when considered for biocomposites applications. They are available in large amounts, low cost, low density, renewable and biodegradable [6]. A thermoplastic polymer, Acrylonitrile–butadiene–styrene (ABS), was used as a matrix polymer to compound with the OPEFB natural fillers for the purpose of manufacturing the bio-composite. ABS is an important engineering copolymer widely used in industry due to superior mechanical properties, chemical resistance, ease of processing and recyclability [7].

Biocomposites as fiber reinforced polymeric material can be used as an alternative material in some industries such as furniture, automotive, sports or household equipment. Studies on the properties of surface materials, including surface free energy, are the subject of intensive scientific research for over

Content from this work may be used under the terms of the Creative Commons Attribution 3.0 licence. Any further distribution of this work must maintain attribution to the author(s) and the title of the work, journal citation and DOI.
Published under licence by IOP Publishing Ltd

forty years. These quantities are being assumed as important criteria for evaluation of adhesion properties of solid polymers [8]. Then the evaluation of surface free energy parameters was become a very useful tool for theoretical studies of the surface behavior of biocomposites as well as for practical developments in many technological applications, such as painting, coating, or other surface treatments of materials [9]. Thus the effect of filler on the surface free energy of biocomposites important to be studied.

Surface free energy (SFE) is one of the thermodynamic quantities describing the state of atom equilibrium in the surface layer of materials [10-11]. This quantity is characteristic for each substance. It reflects the specific state of unbalance in intermolecular interaction which is present at the phase boundary of two mediums [11]. It is measured in [mJ/m^2]. The surface free energy of biocomposites were measured by the contact angle measurements in order to study effect of filler fiber size on the wetting phenomena. One of the most popular methods for measuring the contact angle is the sessile drop method, which involves depositing a liquid drop on a smooth solid surface and measuring the angle between the solid surface and the tangent to the drop profile at the drop edge [12].
The basic methods of calculating SFE of a solid (γ_s) from measurements of the contact angle (θ) is due to Young equation [13]. It is given as:

$$\gamma_l \cos\theta = \gamma_s - \gamma_{sl} \quad (1)$$

where γ_l denotes SFE of liquid in contact with the solid and γ_{sl} means the interfacial free energy between the solid and the liquid. If the values of γ_l and θ are known, it is impossible to determine SFE directly from the equation (1) because of the two unknowns, γ_s and γ_{sl}. In order to solve this equation for γ_s, one has to add a second equation that correlates γ_{sl} with γ_s and γ_l. Until recently, many types of correlations have been employed.

Girifalco and Good [14] showed that interfacial free energy between the solid and liquid can be calculated from the SFE of the solid (γ_s) and the liquid (γ_l). These equations can be solved for:

$$\gamma_{sl} = \gamma_s + \gamma_l - 2(\gamma_s \gamma_l)^{1/2} \quad (2)$$

Fowkes [15] observed that, for solid-liquid interaction of polar materials with nonpolar ones, in van der Waals systems, only dispersion forces could effectively operate across the interface. He suggested separating the surface energy of the polar material into dispersion (γ^d) and polar (γ^p) contributions.

$$\gamma = \gamma^d + \gamma^p \quad (3)$$

Therefore, he suggests that eq. (2) be written as:

$$\gamma_{sl} = \gamma_s^d + \gamma_l - 2(\gamma_s^d \gamma_l^d)^{1/2} \quad (4)$$

where γ_l represents the combined contributions [equation (3)] of the dispersion and polar components of the aqueous phase; γ_l^d is the dispersion component, and γ_s^d is the surface energy of the solid having only a dispersion component. Combining eq. (4) with Young's equation (1), obtains an expression known as the Girifalco-Good-Fowkes-Young (GGFY) equation [16].

$$\cos\theta = -1 + \frac{2(\gamma_s^d \gamma_l^d)^{1/2}}{\gamma_l} \quad (5)$$

Owens and Wendt [17] argued that the polar interaction could be computed using the same geometric mean mixing rule as for the dispersion force interaction. If the contact angle of at least two liquids, usually a polar and a non-polar liquid, with known γ_l^d and γ_l^p parameters are measured on a solid surface, the γ_s^d and γ_s^p parameters of that solid can be calculated [16]. Thus, Owens and Wendt suggested the following form [18]:

$$\gamma_{sl} = \gamma_s + \gamma_l - 2(\gamma_s^d \gamma_l^d)^{1/2} - 2(\gamma_s^p \gamma_l^p)^{1/2} \quad (6)$$

With substitution eq. (6) into Young's equation (1), then can be written:

$$(1 + \cos\theta)\gamma_l = 2(\gamma_s^d \gamma_l^d)^{1/2} + 2(\gamma_s^p \gamma_l^p)^{1/2} \tag{7}$$

By measuring θ of two different liquids against a solid, simultaneous equations are obtained which can be solved for γ_l and γ_l^d. Thus the components of surface free energy due to various forces can be approximated. The sum of these components by analogy with eq. (3) should yield a reasonable approximation of the total solid surface energy γ_s [17], where γ_s^d and γ_s^p are the dispersion and polar force components of γ_s, respectively [19].

The latest version of SFE calculation is the Lifshitz-van der Waals/acid-base approach, initiated by van Oss, Chaudhuri, and Good [20]. SFE is a sum of two components, while the first component γ^{LW} is connected with long-range interactions (dispersive, polar and inductive, referred to as Lifshitz-van der Waals electrodynamic interactions) and the second component γ^{AB} describes the acid-base interactions [11]. This vOCG method split the polar component (γ^{AB}) to the acid component (electron-acceptor: γ^+) and the base component (electron-donor: γ^-), in such a way that $\gamma^{AB} = 2(\gamma^+\gamma^-)^{1/2}$. This method requires contact angle from three liquids, then the following correlation expressed as:

$$\gamma_{sl} = \left(\sqrt{\gamma_s^{LW}} - \sqrt{\gamma_l^{LW}}\right)^2 + 2\left(\sqrt{\gamma_s^+\gamma_s^-} + \sqrt{\gamma_l^+\gamma_l^-} - \sqrt{\gamma_s^+\gamma_l^-} - \sqrt{\gamma_s^-\gamma_l^+}\right) \tag{8}$$

which, upon combination with Young's equation (1) becomes:

$$(1 + \cos\theta)\gamma_l = 2\left(\sqrt{\gamma_s^{LW}\gamma_l^{LW}} + \sqrt{\gamma_s^+\gamma_l^-} + \sqrt{\gamma_s^-\gamma_l^+}\right) \tag{9}$$

For any flat solid, using three liquids, γ_s^{LW}, $\gamma_s^+$and γ_s^- can be found by contact angle determination, provided γ_l^{LW}, $\gamma_l^+$and γ_l^- are known for all three liquids, using eq. (9). At least two of the three liquids must be polar. It usually is most expedient to determine γ_s^{LW} first, with the help of a high-energy nonpolar liquid [20].

In this paper the effect of fiber size as filler on the surface free energy of recycled ABS/OPEFB fiber biocomposites at room temperature was investigated and compared with recycled ABS as control. Surface free energy analysis using three methods GGFY, Owen-Wendt, and vOCG, through the sesile drop contact angle measurements, three liquids (water, methanol, hexane) were used in these determinations.

2. Experimental

2.1. Materials

The oil palm empty fruit bunches (OPEFB) was obtained from PTPN VIII Cikasungka, Bogor, Indonesia. The OPEFB were obtained from fresh fruit bunches, after steam treatment and stripping the fruit. The OPEFB were washed with water, drained under the sun over 2 days and then chopped into chips form. Chip samples then dried in a drying oven (Type YNC-OV, YENACO, China) at 100 °C for 8 hours. Fibers were prepared by mechanical milling (Model MDY-1000, FOMAC, China) and sieving to obtain long-fiber (ϕ>20 mesh), medium-fiber (20>ϕ>100 mesh), short-fiber (100>ϕ>200 mesh), and microparticle, fibers that have pass 200 mesh with additional treatment, hammer milling (Model HMV-4W-5.5, Vertical Hammer Mill, Indonesia) for 2×10 minutes. The average OPEFB fiber dimension was calculated from images captured by an light microscope (Model BX51, Olympus, Japan) with the software Olympus DP2-BSW integrated with DP25 Olympus Microscope Camera. The average particle size of microparticle was 468.34 nm, which was measured using Vasco particle size analyzer. The recycled ABS polymer (melt flow index 12.1 g/10 min) was purchased from PT MUB Jaya, Bogor, Indonesia.

Table 1. OPEFB fiber dimension average

	Long-fiber	Medium-fiber	Short-fiber
Length (µm)	3867.56	1535.97	230.12
Diameter (µm)	291.82	147.79	58.53

2.2. *Biocomposites preparation*

The biocomposites has been produced by extrusion using single-screw extruder (Model HXSJ-125/125, Kai Xin, China) with 15 wt% EFB fiber as filler, 81.7 wt% recycled ABS polymer as matrix, and 1 wt% primary antioxidant (China), 0.3 wt% acid scavanger (Germany), and 2 wt% coupling agent (Germany) as additives. In the barrel, samples blended with gradient temperature 195-215-220-220-220-225-225-225 °C. The obtained biocomposites in form of granular, were made into test piece by injection molding machine (Model HC-250, Hwa Chin, China) with gradient temperature 170-185-200 °C. The biocomposites test piece about 16.4 cm (long), 13 cm (wide) and 4 cm (thick).

2.3. *Contact angle measurements of biocomposites*

The sessile-drop method was chosen for contact angle measurements and the static contact angle (CA) were determined by Phoenix 300 Contact Angle Analyzer (Surface Electro Optics, Korea). Test liquids used in this study are described in Table 2 and 3. Water (equademineralized) were used as the polar liquids, methanol as semi-polar liquid, and hexane as nonpolar liquid. Three probe liquids were placed on the surface of the biocomposites test piece. Drops of 5 uL volume were employed and a total of three droplets were analyzed for each liquid. The contact angle determinations were performed through the capture of the droplet images using Surfaceware 8 software and a camera based contact angle analysis system. Image capturing continued from the beginning of the test, at 5 minutes, and 10 minutes. The angle was measured at both sides of the droplets. All tests were carried out at room condition (temperature of 25 °C).

2.4. *Surface free energy determination*

The surface free energy (SFE) and their components were calculated for all samples using Surfaceware 8 software, with three methods, Girifalco-Good-Fowkes-Young (GGFY), Owens-Wendt, and van Oss-Chaudhury-Good (vOCG), which uses equations (2)-(9). For GGFY, the liquid used was water (aquademineralized), with known total liquid SFE value 72.8 mJ/m^2. The probe liquid parameters used for the determination of the biocomposites surface energy are listed in tables 2 and 3, for Owen-Wendt and vOCG, respectively. Analysis of variance (One way ANOVA) was used to determine the statistically significant differences between these methods. The ANOVA was also used to prove the differences between the SFE values of different fiber size filler on biocomposites.

Table 2. SFE and their components (in mJ/m^2) for the probe liquids used for contact angle determination according to the Owen-Wendt method

Liquid	γ_l	γ_l^p	γ_l^d
Water	72.8	21.8	51
Hexane	18.4	18.4	0

Table 3. SFE and their components (in mJ/m^2) for the probe liquids used for contact angle determination according to the vOCG method

Liquid	γ_l	γ_l^{LW}	γ_l^{AB}	γ_l^+	γ_l^-
Water	72.8	21.8	51	25.5	25.5
Methanol	22.5	18.2	4.3	0.06	77
Hexane	18.4	18.4	0	0	0

3. Results and discussion

3.1. Contact angle values

Table 4. Contact angles of the test liquids on the biocomposites

Samples	Contact Angle (°)		
	Water	Methanol	Hexane
RABS	87.34	28.01	15.69
RABS/LF	87.95	26.86	13.11
RABS/MF	88.24	32.47	11.03
RABS/SF	102.58	40.05	11.09
RABS/μ	104.75	41.76	14.48

Contact angles of the test liquids (water, methanol and hexane) on the biocomposites filled OPEFB long-fiber (RABS/LF), medium-fiber (RABS/MF), short-fiber (RABS/SF), microparticle (RABS/μ), and recycled ABS (RABS) as control are given in Table 4. The value of each contact angles was the average of three droplets. As it can be seen in Table 4, decrease of fiber size increased the contact angle. The evolution of contact angle versus time is shown in figure 1. Depletion water contact angle was slower than methanol. Furthermore, hexane contact angle depleted fastly, even at 5 minutes the droplet has been wetting perfectly, zero contact angle. Its showed the wettability of the samples decreased with decreasing of EFB fiber size as biocomposites filler. Either recycled ABS or biocomposites was nonpolar and hydrophobic materials, due to the droplets of nonpolar liquid (hexane) on its surface was wet quickly, reacted to the composite surface.

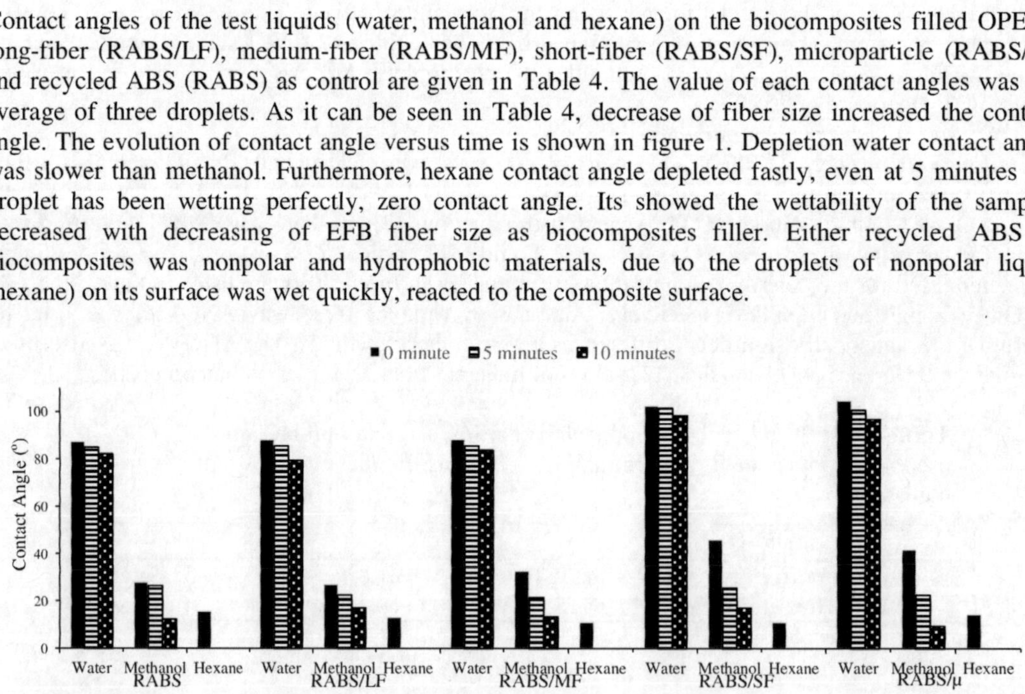

Figure 1. Contact angle depletion in biocomposites versus time (using three probe liquids: water, methanol and hexane)

IOP Conf. Series: Earth and Environmental Science **65** (2017) 012044 doi:10.1088/1755-1315/65/1/012044

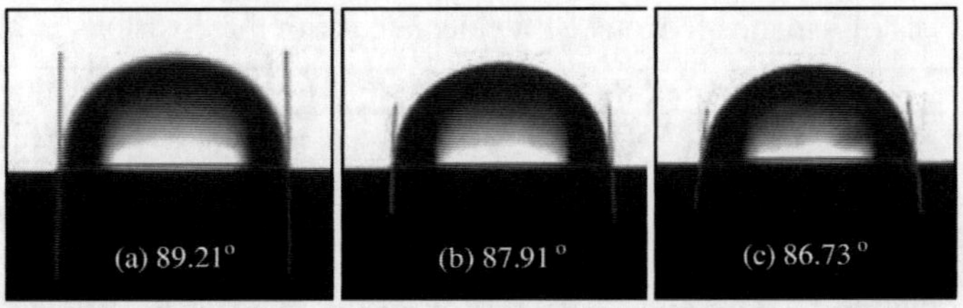

Figure 2. Representative figures of contact angle measurements of RABS/SF (a) 0 minutes; (b) 5 minutes; (c) 10 minutes with water as liquid

3.2. Surface free energy

According to the contact angle values the surface free energy (SFE) of biocomposites and their components were calculates using three previous known methods, Girifalco-Good-Fowkes-Young (GGFY), Owen-Wendt (OW), and van Oss-Chaudhury-Good (vOCG). The values of the SFE and its components are presented in Table 5.

The values of total SFE (γ_S) of biocomposites was lower than control (RABS), decreased with decreasing of EFB fiber size as biocomposites, except for RABS/LF total SFE calculated by vOCG method which higher than RABS. The dispersive components (γ_S^d) obtained by the OW are similar with vOCG (γ_S^{LW}) method and higher than its polar components, incresed with decreasing fiber size. In the case for ABS/NF, the dispersive component of the surface free energy was higher than its polar components even 16 times. It can be observed as well from vOCG method calculation values that component of SFE connected woth long range interaction (polar, dispersive, polar and inductive) is higher than the component describing acid-base inteactions γ_S^{AB}. Lewis base (γ_S^-) surface free energy component is higher than Lewis acid (γ_S^+) for all biocomposites, where Lewis base values of biocomposites lower than RABS and decreased with decreasing fiber size filler. Owing to the insignificant γ_S^{AB} value it may be presumed that these surfaces will show properties of nonpolar materials.

Table 5. SFE and its components (in mJ/m^2) of recycled ABS and biocomposites calculated with various methods

Samples	GGFY	Owen-Wendt			vOCG				
	γ_S	γ_S	γ_S^d	γ_S^p	γ_S	γ_S^{LW}	γ_S^{AB}	γ_S^+	γ_S^-
RABS	24.61	24.39	17.72	6.67	19.57	17.72	1.85	0.07	11.42
RABS/LF	24.10	24.23	17.92	6.31	19.73	17.92	1.81	0.08	10.74
RABS/MF	23.87	24.19	18.06	6.12	19.42	18.06	1.35	0.04	10.86
RABS/SF	13.75	19.52	18.06	1.46	18.54	18.06	0.48	0.02	2.42
RABS/μ	12.48	18.90	17.82	1.08	18.21	17.82	0.39	0.02	1.75

A comparison of the total SFE calculated with three various method for RABS and biocomposites and their grouping information by using the fisher LSD method and 95% confidence, is presented in Table 6 and 7. The analysis proved that there are no statistically significant differences in the values of the SFE calculated with either the GGFY, Owens-Wendt or the vOCG method when the RABS filled with OPEFB fiber with are degreased fiber size.

Table 6. Analysis of variance for total SFE calculated with three various method

Source	DF	Adj. SS	Adj. MS	F-value	*P*-value
Factor	2	27.56	13.78	0.91	0.427
Error	12	181.03	15.09		
Total	14	208.59			

Table 7. Grouping information using the fisher LSD method and 95% confidence

Factor	N	Mean	Grouping
Owen-Wendt	5	22.24	A
GGFY	5	19.76	A
vOCG	5	19.09	A

4. Conclussions

Contact angle values of the test liquids on the RABS, biocomposites filled OPEFB long-fiber (RABS/LF), medium-fiber (RABS/MF), short-fiber (RABS/SF), microparticle (RABS/µ), and recycled ABS (RABS) were increased with fiber size increasing. The values of total SFE (γ_S) of biocomposites was lower than control (RABS), decreased with decreasing of EFB fiber size as biocomposites. Owing to the insignificant γ_S^{AB} value of biocomposites, it may be presumed that these surfaces showed properties of nonpolar materials. The smaller the fiber size of biocomposites filler, the higher water contact angles, which mean the more hydrophobic the surface. It is highly advantageous since in general applications, the coating/painting materials for ABS composite were oil based (nonpolar). Therefore, the more hydrophobic material will increase the wettability of their coating. The total surface free energy values of RABS and biocomposites obtained from the different methods are similar and no statistically significant different (P-value > 0.05).

References

[1] Singha A S. and Thakur V K 2009 *Polym. Plast. Technol. Eng.* **48** 736
[2] Hassan A, Salema A A, Ani F N and Bakar A A 2010 *Polym. Compos.* **31** 2079
[3] Nikmatin S, Syafiuddin A and Kueh A B H 2015 *J. Teknologi* **77** 181
[4] Solikhin A, Hadi Y S, Massijaya M Y and Nikmatin S 2016 *BioResources* **11** 2224
[5] Nikmatin S, Syafiuddin A, Nugroho N, Utama W and Wismogroho A S 2017 *MATEC Web. Conf.* **95** 03001
[6] Nikmatin S, Syafiuddin A and Irwanto D A Y 2017b *BioResources* **12** 1090
[7] Yang S, Castilleja J R, Barrera E V and Lozano K 2004 *Polym. Degrad. Stab.* **83** 383
[8] Żenkiewicz M 2007 *J. Achiev. Mater. Manuf. Eng.* **24** 137
[9] Oporto G S, Gardner D J, Bernhardt G and Neivandt D J 2007 *J. Adhes. Sci. Technol.* **21** 1097
[10] Żenkiewicz M 2005 *Polimery* **50** 365
[11] Rudawska A 2012 Adhesive Properties *Scanning Electron Microscopy* ed V Kazmiruk (Rijeka: InTech) chapter 7 pp 101-126
[12] Njobuenwu D O, Oboho E O and Gumus R H 2007 *Leonardo El. J. Pract. Technol.* **10** 29
[13] Young T 1805 *Phil. Trans. R. Soc. Lond.* **95** 65
[14] Girifalco L A and Good R J 1957 *J. Phys. Chem.* **61** 904
[15] Fowkes F M 1964 *Ind. Eng. Chem.* **56** 40
[16] Grundke K 2005 Surface-energetic properties of polymers in controlled architecture *Molecular interfacial phenomena of polymers and biopolymers* ed P Chen (Cambridge: Woodhead Publishing) chapter 10 pp 323-374
[17] Owens D K and Wendt R C 1969 *J. Appl. Polym. Sci.* **13** 1741

[18] Cwikel D, Zhao Q, Liu C, Su X and Marmur A 2010 *Langmunir* **26** 15289
[19] Sudaryanto, Nishino T, Asaoka S and Nakamae K 2001 *Int. J. Adhes. Adhes.* **21** 71
[20] Van Oss C J, Chaudhury M K and Good R J 1988 *Chem. Rev.* **88** 927

IOP Conf. Series: Earth and Environmental Science **65** (2017) 012045 doi:10.1088/1755-1315/65/1/012045

Sustainable supply chain design for waste cooking oil-based biodiesel in bogor using dynamic system approach

A D Syahdan[1], Y Arkeman [1,2], H Wijaya[1,2]

[1] Department of Agroindustrial Technology, Bogor Agricultural University, Bogor, Indonesia

[2] Surfactant and Bioenergy Research Center, Bogor Agricultural University, Bogor, Indonesia

Email: yandra_ipb@yahoo.com

Abstract. Biodiesel is one of the alternative fuels that are environmentally friendly. Besides palm oil, biodiesel can also be produced from waste cooking oil. Since 2007, the government of Bogor has been utilizing waste cooking oil into biodiesel for use as Transpakuan bus' fuel. However, in practice, the amount of waste cooking oil supplied is never sufficient the needs of fuel of 30 units Transpakuan bus. The main objective of this research is to analyze the availability of waste cooking oil that will be converted into biodiesel within the next ten years as well as providing policy advice to support the program. The method used is a dynamic system that is followed by simulation of multiple scenarios that have been defined. The system is divided into three subsystems, namely supply subsystem, demand subsystem, and production subsystem. The results showed that the current system is not able to guarantee the sustainability of the supply chain of waste cooking oil as a raw material of biodiesel. From some of the scenarios tested can be concluded that biodiesel needs would increase in line with the trend of the use of environmentally friendly fuels. It takes a new system and a new policy relating to the biodiesel supply chain. Policy suggestions that can be proposed from this research is to increase supplier participation, objectify the program of converting angkot into Transpakuan bus, and support the development of biodiesel industry.

1. Introduction

One of the most potential renewable energy to be exploited in Indonesia is biodiesel. Biodiesel is a biofuel or fuel made from vegetable oils, either it is fresh or a leftover frying results through esterification and transesterification stages. The government has targeted the use of biodiesel as a fuel mixture for industry, transportation, power generaton, and commercial respectively by 30% in 2025. While in Blueprint Pengelolaan Energi Nasional 2006-2025, the cumulative capacity of biodiesel is expected to reach 222.000 barrels per day [1]. According *to Kebijakan Energi Nasional (KEN)*, biodiesel supply in 2014-2050 will grow in an average of 9,7% per year [2]. The use of palm oil as the raw material for bodiesel is more focused on the food sector. So, the raw materials of biodiesel is diverted to waste cooking oil as an alternative. Real form of the use of biodiesel as an alternative fuel has been carried out by the government of Bogor city. It has launched a program of utilizaton of waste cooking oil into biodiesel used as fuel of Transpakuan, one of the public transportation in Bogor.

This program was cretaed as a real contribution from Bogor who are members of International Climate Environmental Initiative (ICLEI) as an effort to prevent global warming. The number of public

Content from this work may be used under the terms of the Creative Commons Attribution 3.0 licence. Any further distribution of this work must maintain attribution to the author(s) and the title of the work, journal citation and DOI.
Published under licence by IOP Publishing Ltd

transportation vehicle and transport systems that are uncontrolled in Bogor making an alarming increase of air pollution. It happened because of the increased activity and mobility of people in and around Bogor. According to Badan Pengelolaan Lingkungan Hidup (BPLH) Bogor in 2012, 16.600 liters of waste cooking oil can be converted to 13.280 liters of biodiesel per year [3]. Waste cooking oil obtained from citizens, schools, restaurants, hotels, hospitals, and food processing industry [4]. Waste cooking oil is the cooking oil that have been used for two or three times. It has a bad impact to human health if it is used for more than three times. So, in addition to reducing vehicle pollution , the use of waste cooking oil as a raw material for biodiesel can also improve food security and establish a healthy lifestyle for the people of Bogor [5].

The process of converting waste cooking oil into biodiesel is done by PT. Bumi Energi Equatorial (PT. BEE). Previous research states that the bussiness run by PT.BEE is not financially feasible assessed from it's NPV, B/C ratio, and Internal Rate of Return (IRR). It is caused by the gap existed between production capacity of the plant and the amount of supply of the waste cooking oil. PT. BEE has 30 tons of production capacity while the raw material's supply is just 2 tons per month. For the success of this program, it required an integrated system that can ensure the sustainability of biodiesel production from raw materials to the the hands of consumers. The system involves many stakeholders with different functions and tasks but have the same goal. Therefore, a good supply chain management is needed so that every stakeholder is able to carry out their duties properly and make a profit from these activities.

2. Metodology

The study begins with a literature study on supply chain and dynamic system. Dynamic system was first introduced by Jay W. Forrester in 1961. The dynamic system is used to describe a complex system in which the variables contained therein are interconnected to one another. The relationship between these variables will generate feedback information that changes over time [6]. A system can be referred to complex and dynamic if it is always changing every time, all the variables in the system are interrelated and influence each other significantly, as well as works based on the feedback between the variables in it [7].

Based on the relationship between the variables, the purpose of the system and the factors that can accelerate or inhibit the process of achieving the goals of the system can be defined. Therefore, the dynamic system is suitable for the formulation of future policy in a complex system that is changing continuously. There are two stages used in the dynamic systems approach, namely Causal Loop Diagram (CLD) or a causal diagram and the Stock and Flow Diagram (SFD) or diagram arrow box. Furthermore, the identification of problems is done by analyzing the existing supply chain condition and interviews with relevant stakeholders. Based on the information obtained, we can determine the scope of issues and stakeholders that will be involved in the model. Restrictions on system is very important because it will affect the model that will be used for system simulation. In addition, the analysis was also conducted to look at the role of each stakeholder along the supply chain. After the system is identified, the variables involved in the system are determined and analyzed how they affect each other.

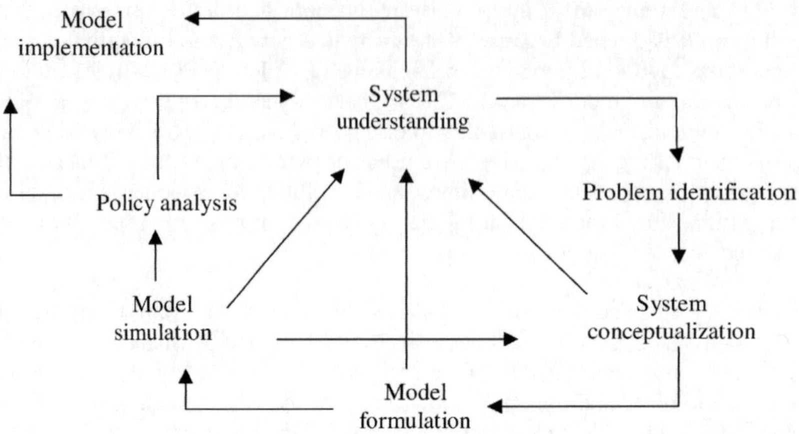

Figure 1. System dynamic approach [8]

The relationship between variables is described by the causal loop diagram. In the causal diagram, the relationship between variables is described by the symbol (+) and (-). The symbol (+) illustrates the proportional relationship between variables, otherwise the symbol (-) illustrates the inverse relationship between variables. Once the causal diagram is formed, the relationship between variables is described qualitatively using a mathematical formulation of the stock and flow diagram. Stock describes the status or condition of a variable at a certain time while the flow indicates the status or the state of variables at certain intervals. Stock is the accumulation of inflow and outfow a variable.

To connect the stock to flow, the type of the variable used is called the auxiliary. By using these three types of variable, relationships between variables in dynamic systems can be better explained [6].

Stock and flow diagrams created with Anylogic 7.3.1 Personal Learning Edition. By using the software, the stage of verification of the model can be done. Verification is required as evidence that the model, which is based secondary data, is technically correct and there is no error at all so it is valid to use. Furthermore, the simulation can be done using Anylogic 7.3.1 Personal Learning Edition based on the scenarios that have been determined for a period of 10 years, namely 2016 as the start time and in 2026 as a stop time.

Model validation is the process of comparison between the output generated by the simulation model with the values contained in the real system. This is done to see if the model is able to represent the real system appropriately [9]. One of the validation methods by comparing the model with real systems (quantitative behavior pattern comparison) is the method of Mean Average Percentage Error (MAPE) [10]. Percentage difference or error between the real system and the model becomes the reference of validation.

3. Result and Discussion

3.1. System Description

Dynamic system modeling is used to describe the relationship between variables that will affect the availability of waste cooking oil as a raw material of biodiesel. Models of dynamic systems that are used are divided into three subsystems, namely supply subsystem, demand subsystems and production subsystem. Supply subsystem consists of supply potential of citizens, the supply of the restaurant, the supply of hotels, the supply of hotels and inns, as well as the supply of school. Demand subsystem consists of a fleet of Transpakuan and mandatory biodiesel blend with diesel yield. While the production subsystem is a biodiesel plant capacity.

Tabel 1. Stakeholders and their needs

No	Stakeholders	Needs
1	BPLH	Absorp all of the potential of waste cooking oil
2	Suppliers	Reward and appreciation
3	Biodiesel plant (PT. BEE)	The sustainability of the raw material and profitt oriented
4	Consumer (PT. PDJT)	Availability of biodiesel

Relationships between variables in the system and how they affect each other can be explained in the causal loop diagram. Feedback between variables is the basis of dynamic system modeling can be done. The availability of waste cooking oil is influenced by the supply of waste cooking oil from citizens, schools, hotels, and hospitals. The greater the number of the supply the more waste cooking oil is available. Total supply of waste cooking oil increase if the number of residents, schools, restaurants, hotels, and hospitals in the city of Bogor was growing too. On the other hand, the increasing demand for biodiesel will reduce the availability of waste cooking oil. Increased demand for biodiesel is caused by the increase of the fleet ofTranspakuan and increase of mandatory biodiesel blend with diesel yield. The increasing demand for biodiesel should be offset by increased factory capacity as a producer of biodiesel. These three factors are influenced by policies of Bogor and national government on the public transport system and the use of renewable energy up to 10 years.

3.2. System Formulation

Based on the causal loop diagram that has been created, the relationship between variables in the system qualitatively described using stock and flow diagram in Anylogic 7.31 PLE software. The relationship between the variables described in a mathematical equation based on several assumptions derived from BPLH, PDJT, and BPS Bogor as follows:

1. Supply of waste cooking oil in the system is limited only from citizens, schools, restaurants, hotels, and hospitals in the city of Bogor.
2. Waste cooking oil that have been collected are given only to BPLH Bogor and biodiesel produced will be sold to one party, PT. PDJT (Perusahaan Daerah Jasa Tansportasi) Bogor.
3. The consumption of cooking oil / capita / year = 7.96 liters / capita / year.
4. The potential of waste cooking oil at every school in the city of Bogor = 50 liters / month.
5. The potential of waste cooking oil at every hotel in the city of Bogor = 2 g / guest / day, 10% of guests ordering food.
6. The restaurant is divided into three categories based on the number of visitors, top (200 visitors / day); medium (100-200 visitors / day); lower (<100 visitors / day). Potential waste cooking oil 5 g / visitors / day.
7. The hospital is divided into three categories based on number of in-patient visits, top (> 12,000 patients / day); medium (4800-12000 patient / day); lower (<4,800 patients / day). Potential waste cooking oil 5 g / patient / day.
8. The density of waste cooking oil is 0.9 kg / liter.
9. The yield of waste cooking oil to the cooking oil = 0.3

10. TransPakuan B20 biodiesel Consumption is 6 liters / day / unit.
11. Bogor City population growth rate is 1.98% per year.
12. The rate of accretion of the school in the city of Bogor is 3.35% per year.
13. The rate of addition of restaurants in the city of Bogor is 20.7% per year.
14. The rate of addition of hotels in the city of Bogor is 3.94% per year.
15. The rate of addition of the hospital in the city of Bogor is 15.3% per year.
16. The rate of increase in the number Transpakuan is 20% per year.
17. The capacity of the plant is 30 tons per month, or 400,000 liters per year.

Mathematical equation that are the main objectives in this model is:

The availability of waste cooking oil = Supply of waste cooking oil – Demand of biodiesel fuel
Supply subsystem defined by the following mathematical equation:

The supply of waste cooking oil = household waste + school waste + hotel waste + restaurant waste + hospital waste
Household waste = total population * potential of waste cooking oil
Total population = 1047922 + dt * population growth rate
Potential of waste cooking oil = cooking oil consumption per capita * waste cooking oil yield
School waste = number of schools * waste cooking oil potential in school
Number of schools = 463 + dt * the growth rate of the number of school
Waste restaurants = number of restaurants * the potential of waste cooking oil in restaurants
Number of restaurants = 224 + dt * the growth rate of the number of restaurants
Hotel waste = Number of hotel * potential of waste cooking oil in hotel
Total hotel = 53 + dt * the growth rate of the number of hotel
Hospital waste = Number of hospital * waste cooking oil potential in hospitals
Number of hospitals = 17 + dt * rate of increase in the number of hospital

Growth rate of population, schools, restaurants, hotels, and hospitals respectively is 1.98%; 3.35%; 20.7%; 3.94%; and 15.3% per year [3, 11, 12]. Data show that all positive growth rate, which means the number of suppliers is always increasing and the potential of waste cooking oil produced is always growing every year. The unit used is a unit / year. While dt is the change in time or a time interval of simulation.

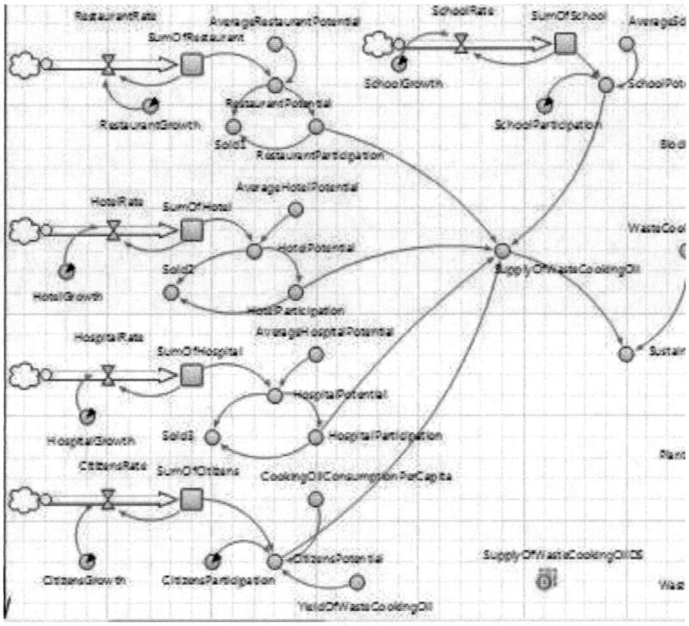

Figure 2. Supply subsystem

Demand subsystem formulated by the following equation:

Demand of waste cooking oil = fuel consumption per unit TransPakuan * mandatory yield * number of fleets * 0.8

The number of fleet = 30 + dt * the growth rate of TransPakuan
Mandatory yield = 0.2 + dt *the growth rate of mandatory yield

The increment of Transpakuan amount and yield of biodiesel blends used will increase the demand of waste cooking oil. It depends on the policy implemented by the central government and local government in line with the developments taking place within the next 10 years.

Production subsystem
Biodiesel production activities done by only one plant, namely PT. Bumi Energi Equatorial (PT. BEE). PT. BEE has a production capacity of 30 tons per month, but the realization of the supply of waste cooking oil that there were only 2 tons per month [13]. Production subsystem defined by the following mathematical equation:

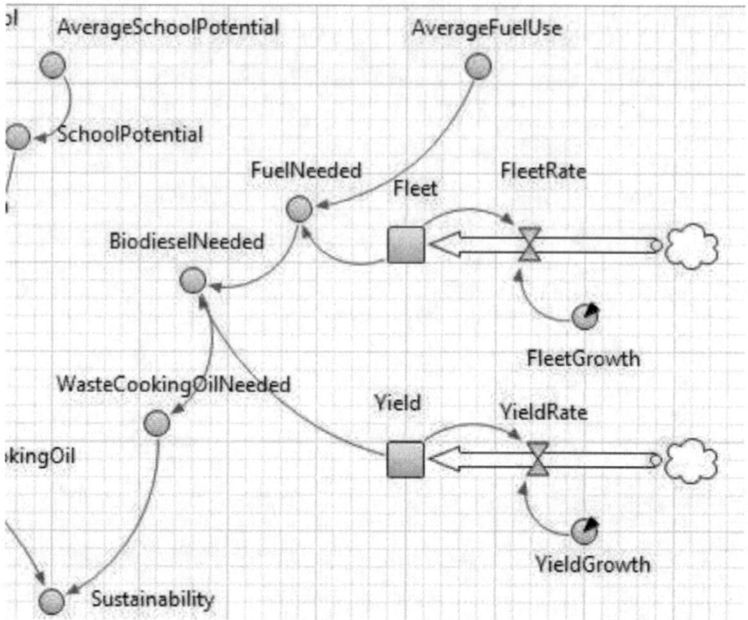

Figure 3. Demand Subsystem

The capacity of the plant = 400,000 + dt * growth rate of plant capacity

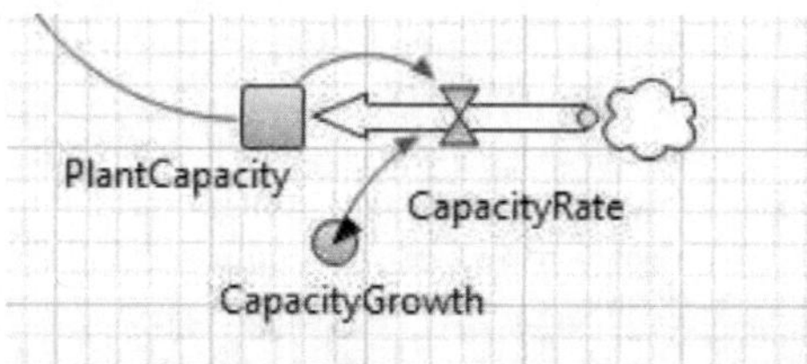

Figure 4. Production subsystem

3.3. Scenarios

Model of availability of waste cooking oil for the next 10 years is analyzed in accordance with the possible conditions and situations that may occur in this period. Selection of scenarios adapted to the stakeholders' targets who influence the system during the simulation period. Each simulation scenarios show different results. These results are used as a reference to answer the research objectives that have been defined. The scenarios used in this study include:

3.3.1. Scenario without alteration / Business as Usual

In this scenario, the simulation is done in accordance with the current state without any change in the system variables. Supplier participation by 0.4%, the rate of Transpakuan, the rate of mandatory yield, and the rate of increase in the production capacity of each is 0. The simulation results show that the supply of waste cooking oil is not able to cover the demand for biodiesel up to 10 years into the future.

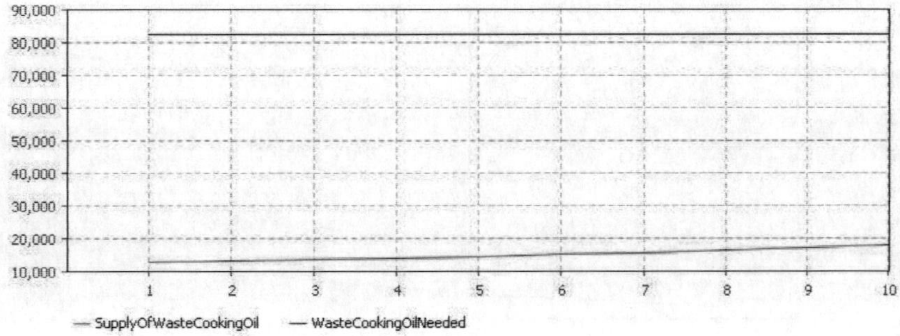

Figure 5. Scenario 1 result

3.3.2. Enhancement of supplier participation and mandatory biodiesel yield scenario
In this scenario, the participation of suppliers increased to 10% and the growth rate of mandatory biodiesel yield increased to 12.5% per year. The simulation results show that during the next 10 years of supply chain waste cooking oil into biodiesel can be assured of sustainability.

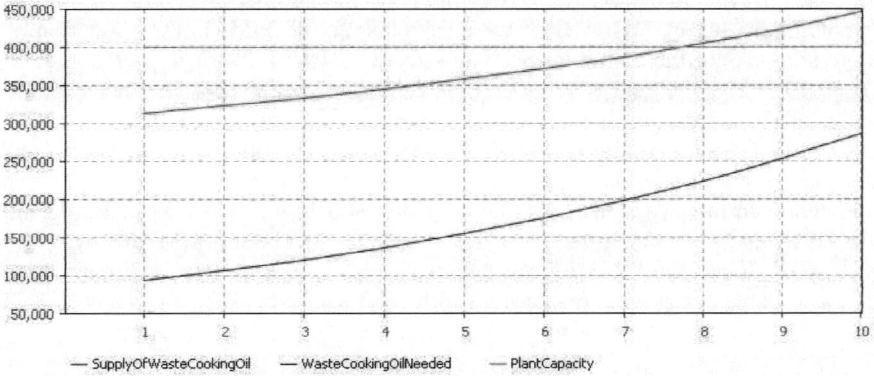

Figure 6. Scenario 2 result

3.3.3. Scenario an increase in supplier participation and the rate of increase in the bus fleet Transpakuan
In this scenario, the situation is assumed that the participation rate is 50% per year, the growth rate of mandatory biodiesel yield remained 12.5% per year, an increase in the growth rate of Transpakuan to 10% per year, and an increase in production capacity by 10% per year. It refers to the target of PT. PDJT Bogor to operate 7 Transpakuan bus corridors in Bogor. The simulation results also indicate that the supply of waste cooking oil can meet the demand for biodiesel from first year to the tenth year. Although there is a change in the growth rate of Transpakuan to 10% per year, but it can be handled by the increase of supplier participation to 50%.

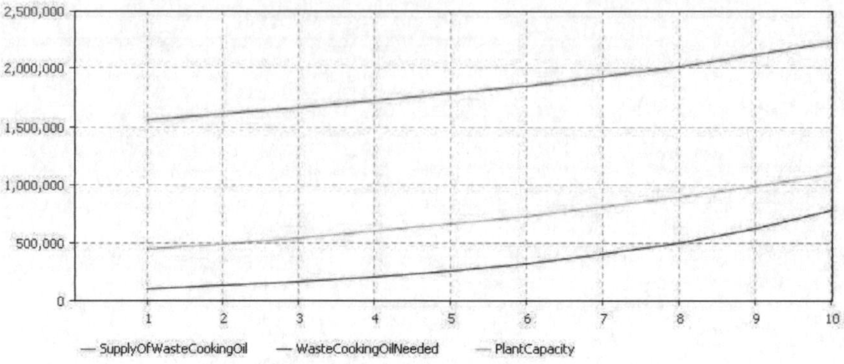

Figure 7. Scenario result

3.4. Model Validation

Model validation conducted by comparing the results of the simulation model of the real system. The samples are population and number of schools in the city of Bogor. The beginning of the simulation adjusted to the start time in simulation. It is 2010 for a variable number of people and 2011 for the number of schools in the city of Bogor. Both were calculated until 2015. In the validation of Bogor citizens, real system shows the value respectively from 2010 to 2015 was 950.334, 987.315, 1.004.831, 1.013.019, 1.030.720, and 1.047.922. Population growth rate per year reaches 1.98%. As for the validation of the number of schools in the city of Bogor, real system respectively from the years 2011-2015 shows the the value of 407 units, 448 units, 452 units, 453 units and 463 units. Growth rates of the number of schools in the city of Bogor per year is 3.35%. The datas of population and number of schools in the city of Bogor obtained from BPS Bogor through Bogor Dalam Angka (BDA) in 2015 [12]. Both variables tested show results of the calculation of MAPE by 0.8% for the population the city of Bogor and 2.4% for the number of schools in the city of Bogor. The results show that the model is able to represent real system with very precise because it is below 5%.

3.5. Policy Recommendations

3.5.1 Increasing the participation of suppliers.

The low participation of the suppliers, both enterprises and society, can occur for several reasons. To increase the participation of suppliers some policies that we suggest are:

a. Improving socialization and publication of the conversion program of waste cooking oil into Biodiesel to the entire community and relevant stakeholders.

The lack of information and understanding of the program performed by the government of Bogor. So that the socialization of the background, benefits, objectives and technical matters of these activities become very important to be conducted, especially for the stakeholders involved throughout the supply chain.

b. Forming a team of waste cooking oil collectors.

Based on the on interviews with BPLH Bogor, currently collecting waste cooking oil from all suppliers is only done once a week using a mobile unit. Waste cooking oil potential is so big that can not be absorbed optimally with collection systems that exist today. A large number of suppliers that spread around the city was the main reason the collection of waste cooking oil can not be fully utilized

IOP Conf. Series: Earth and Environmental Science **65** (2017) 012045 doi:10.1088/1755-1315/65/1/012045

c. Reward and punishment system

If socialization to stakeholders has been done earlier then the sustainability of the program must be be maintained by a system of reward and punishment by the goverment as the organizer of the program. Reward and punishment is a feedback the stakeholders obtained depending on how their performance along the supply chain in carrying out their respective roles. Reward is an appreciation given to certain parties to obtain the optimum working results and cause a feeling of acceptance and engaged in the work environment. Reward touches aspects of compensation and other aspects of the relationship between the relevant parties. Reward can be in the form of materials as well as award or charter [14]. Punishment is a feedback given to those who violate the rules or agreements so as to improve its performance in the future. Punishment can act as a behavior barrier, motivator, and giver of deterrent effect [15].

d. Build a waste cooking oil collection centers.

Good infrastructure is one of the important points in operating a supply chain. Waste cooking oil collection activities closely related to the storage location of the waste cooking oil before being processed into biodiesel. Currently, the waste cooking oil is stored in BPLH office with low capacity and poor inventory management. BPLH office location as well as less strategic if it is seen from the distance with the plant (PT. BEE) and suppliers that spread around the city. In supply chain management there are several strategic steps that can be applied to optimize supply chain performance. One of them is strategic network optimization, including the number, location, and size of warehouses and distribution centers [16].

e. Converts angkot into Transpakuan as the main mode of public transport.

The inefficient transportation in Bogor can be seen from many points prone to congestion. Irregular traffic, the amount of shadow terminal, the provision of transport services which overlap, and yet the realization of intermodal integration into some of the things that must be considered [17]. Angkot is one of the main causes of these problems. So we need an alternative solution that can address the above problems. Transpakuan emerged as one of the solutions offered by the government of Bogor. As consumers of waste cooking oil-based biodiesel, Transpakuan is a major stakeholder that causes biodiesel product demand is always there every year.

f. Support the development of the biodiesel industry.

The realization of 100% absorption of waste cooking oil potential to be processed in to biodiesel requires the commitment of all stakeholders along the supply chain. Especially the government of Bogor and BPLH as the program organizer. Optimizing the supply of waste cooking oil must be balanced with the development of the biodiesel industry as a manufacturer.

4. Conclusion

Supply chain systems that exist today can not guarantee the sustainability of the supply of waste cooking oil as a raw material for biodiesel up to 10 years into the future. Coupled with the plan of using a mixture of biodiesel and diesel fuel with a higher yield and increase the number Transpakuan. Both of these will increase the demand for biodiesel annually. Several scenarios are assumed to occur during the next 10 years include: (1) a scenario without changes or bussiness as usual, that the level of participation of suppliers by 0.4%. The rate of mandatory yield, rate of increase in production capacity, as well as the rate of increase in the number of Transpakuan respectively is 0; (2) The scenario of increasing participation of suppliers to 10% per year and the growth rate of mandatory yield is 12.5% per year; and (3) a scenario of increasing suppliers participation to 50% per year, the growth rate of Transpakuan as much as 10% per year, and the growth rate of production capacity is 10% per year.

There are targets of adding **Transpakuan bus corridors and peraturan menteri ESDM No. 12 Tahun 2015** tentang Energi Baru Terbarukan within the next 10 years to create higher demand for biodiesel in Bogor [18]. To meet the needs of the biodiesel, it requires continuous supply chain of waste cooking oil as a raw material of biodiesel. This can be achieved by increasing the participation of suppliers. In general, the policies proposed to overcome the problem are consists of doing socialization of converting waste cooking oil into biodiesel for all stakeholders involved throughout the supply chain, forming a waste cooking oil collector team to each district in Bogor, implementing a system of reward and punishment for stakeholders in the entire supply chain based on the performance of each in meeting targets, supporting the development of the biodiesel industry, and build an infrastructure that supports the sustainability of the supply of waste cooking oil effectively and efficiently.

Recommendation

The model can be developed wider by adding several sub models, such as the price of diesel, biodiesel prices, the financial analysis of the project, and the environmental aspects that affect along the supply chain. With more data and more complex models, the simulation results will be more accurate when compared to real system.

References

[1] Kementerian Energi dan Sumber Daya Mineral Republik Indonesia 2006 Blueprint Pengelolaan Energi Nasional 2006-2025 Sesuai Peraturan Presiden No 5 Tahun 2006 Jakarta

[2] Dewan Energi Nasional 2014 *Outlook Energi Indonesia 2014*

[3] Badan Pengelolaan Lingkungan Hidup Kota Bogor 2012 *Laporan Akhir Kajian Program Biodiesel Kota Bogor* Bogor: BPLH Kota Bogor

[4] Munir D 2015 Pemanfaatan limbah minyak jelantah menjadi biodiesel di kota bogor (Asosiasi Pemerintah Kota Seluruh Indonesia)

[5] Imam A, Irman H and Yandra A 2015 Financial feasibility study of waste cooking oil

[6] Asgari B and Hoque Md A 2013 Original Research: A system dynamics approach to supply chain performance analysis og the ready-made-garment industry in bangladesh *Ritsumeikan Journal of Asia Pasific Studies* **32**

[7] Sterman J D 2002 System dynamics: system thinking and modelling for a complex world *Working Paper Series: ESD-WP-2003-01-13 Massachussets Institute of Technology Engineering Systems Division*

[8] Suryani E 2006 *Pemodelan dan Simulasi* Yogyakarta (ID): Graha Ilmu

[9] Walpole *et. al* 2007 *Probability & Statistics for Engineers & Scientists Eight Edition* pp 269 – 320. Pearson Prentice-Hall

[10] Daalen V and Thissen W A H 2001 Dynamics systems modelling continuous models. faculteit techniek, bestuur en management (TBM). Delft (NED): Technische Universiteit Delft

[11] Badan Pusat Statistik Kota Bogor 2013 *Kota Bogor Dalam Angka 2013* Bogor: BPS Kota Bogor

[12] Badan Pusat Statistik Kota Bogor 2016 *Kota Bogor Dalam Angka 2015* Bogor: BPS Kota Bogor

[13] Widodo P S P 2011 Analisis kelayakan usaha pengolahan minyak jelantah *(waste cooking oil)* menjadi biodiesel (kasus: pt. bumi energi equatorial (bee) bogor) [Skripsi]. Bogor: Institut Pertanian Bogor

[14] Febrianti S dkk 2014 Pengaruh *reward* dan *punishment* terhadap motivasi kerja serta dampaknya terhadap kinerja (studi pada karyawan pt. panin bank tbk. area mikro jombang) *Jurnal Administrasi Bisnis (JAB)* ***12*** *(1) Juli 2014*

[15] Siahaan R 2013 Pengaruh *Reward* dan *Punishment* terhadap Disiplin Kerja Karyawan pada PT. Perkebunan Nusantara III Rambutan. *Jurnal Ilmiah Bussiness Progress Oktober 2013* ***1*** *(01) 1726.*

[16] Sinaga B B dkk 2011 Analisis Sistem Rantai Pasok PT. Semen Gresik (Persero) Tbk. *Jurnal Optimasi Sistem Industri* ***10*** *(1) April 2011* pp *113-120*

[17] Mayyanti D 2008 *Evaluasi Kinerja Angkutan Umum di Kota Bogor (Studi Kasus Trayek Angkutan Kota Nomor 03, 08, dan 09).*
[18] Kementerian Energi dan Sumber Daya Mineral Republik Indonesia 2015 Peraturan Menteri Energi dan Sumber Daya Mineral Republik Indonesia Nomor 12 tahun 2015 tentang Perubahan Ketiga atas Peraturan Menteri Energi dan Sumber Daya Mineral Republik Indonesia Nomor 32 tahun 2008 Tentang Penyediaan, Pemanfaatan, dan Tata Niaga Bahan Bakar Nabati (Biofuel) Sebagai Bahan Bakar Lain.

Synthesis of glycerol mono-laurate from lauric acid and glycerol for food antibacterial additive

W B Setianto[1], T Y Wibowo[1], H Yohanes[1], F Illaningtyas[1] and D D Anggoro[2]

[1]Center for Agroindustrial Technology, Agency for the Assessment and Application of Technology (Badan Pengkajian dan Penerapan Teknologi), PUSPIPTEK 610-612 Bld, Tangerang Selatan, Indonesia
[2]Department of Chemical Engineering, Diponegoro University, Tembalang, Semarang 50275, Indonesia

Email : wahyu.bahari@bppt.go.id

Abstract. Synthesis of glycerol mono-laurate (GML) has been performed using esterification reaction of glycerol and lauric acid. The reaction was performed at the condition of temperature of 120-140 °C within 7 hour, variation of molar ratio of glycerol - lauric acid, and was using heterogeneous catalyst of zeolist Y. Without catalyst dealumination the maximum acid conversion was 78%, with GML contained in the sample was 38.6%, and it was obtained at the reaction condition of 140 °C, 15wt% catalyst, and 8:1 molar ratio of glycerol - lauric acid. At the same condition, using dealuminated catalyst, the maximum acid conversion was increased up to 98%, with GML contained in the sample was 50.4%. The GML antibacterial activity was examined. It was observed that the GML has antibacterial activity against gram positive bacterial such as *B. cereus* and *S.aureus*.

1. Introduction

Glycerol Mono-laurate (GML) (2,3-dihydroxypropyl dodecanoic) that also called monolaurin is lauryl esters of glycerol. It is important to use in the food and cosmetic industries due to its properties such as emulsifying properties, complexing, and polyfunctional state[1]. GML is considered as a natural surfactant. It has antimicrobial activity against *enveloped-virus* and bacteria including gram negative bacteria and gram positive bacteria (*Streptococcus* spp., *Staphylococcus aureus* dan *Bacillus* spp)[2]. Unlike most of antibiotic that have antibacterial activity against a certain bacteria. GML has some target antibacterial activity through interaction with plasma membrane.

Synthesis of GML can be performed trhough esterification of lauric acid and glycerol using acid catalyst [3], transesterification of palm kernel oil or coconut oil or glycerolysis of methyl laurate using base catalyst[4]. At industrial scale generally the esterification and transesterification process utilize homogeneous catalyst, such as KOH, $NaHCO_3$ or Ca $(OH)_2$. It is reported that the process which used the homogeneous catalyst resulted unpurified product contain 40-60% monoglyceride, 30-40% diglyceride , 5-15% triglyceride, 1-5% free fatty acid and 2-10% glycerol[5]. Further purification usually using molecular distillation process which need high cost production[6].

Content from this work may be used under the terms of the Creative Commons Attribution 3.0 licence. Any further distribution of this work must maintain attribution to the author(s) and the title of the work, journal citation and DOI.
Published under licence by IOP Publishing Ltd

The greater public interest in natural food products has encouraged manufacturers to develop natural preservative. Fatty acid ester compound is not only act as flavoring and emulsifying agents but also has potential of antimicrobials. Various experiments on GML as an antimicrobial effectiveness test has been conducted by several researchers[7,8,2,9]. Moreover, the study of antibacterial activity of GML have been done for *S. aureus, S. hominis, B. cereus, Streptococcus pyogenes, Haemophillus influenzae, P. aeruginosa, Escherichia coli.* GML more effective against gram positive bacteria than gram negative bacteria[10,11,12,13,14]. GML can be more effective when combined with antibiotics such as gentamycin and streptomycin or natural antimicrobial such as subtilosin, lauric acid and poly L-lysine[15,16,11]. It has been reported that a common mechanism of action for fatty acids that involves disruption of the cell membrane, leading to interference with energy production within the bacterial cell[17]. GML might facilitate to disrupt the matrix-embedded bacteria, resulting the release of cell compounds that affecting the cell metabolism[11].

In this work, the synthesis of GML using esterification of lauric acid and glycerol has been performed. More over the antibacterial activity of the GML has been examined.

2. Materials and methods

2.1. Materials

The materials for glycerol mono-laurates synthesis were lauric acid (94% purity) and Glycerol (98% purity) which obtained from Sinar Mas and Bratachem respectively were used as recieved. The catalyst was Zeolite Y CBV 712 (supplied by Zeolyst International in NH_4 form; mole ratio SiO_2 / Al_2O_3 = 12; surface area = 730 m^2 / g.). Microbe test which is used in the examination were *Staphylococcus aureus, Bacillus cereus, Escherichia coli* and *Pseudomonas aerginosa.* Antibacterial examination materials were Nutrient *agar* (NA), *Nutrient broth* (NB), absolute alcohol and distillate water. The standard microbiological apparatus was used for the examination.

2.2. Experimental procedures

The glycerol mono-laurate was synthesized from lauric acid and glycerol. The lauric acid and glycerol was reacted in a batch reactor using zeolist Y as a catalyst. The reaction was performed with mole ratio of glycerol to lauric acid of 2:1, 5:1 and 8:1 at the temperature of 110-140 C for 7 hour. The catalyst was varied from 10-20wt% of the lauric acid. The reaction conversion was observed with acid number analysis. The reaction conversion percentage was calculated using equation 1. Sample of GML was analyzed using GC-MS analysis. The GC-MS analysis of the GML sample showed that the reaction not only resulted GML but also glycerol di-laurate (GDL) and glycerol tri-laurate (GTL). The synthesis was also performed using modified zeolist Y. The catalyst was modified using dealumination method with difference condition of time, temperature and acid normality.

$$Reaction\ conversion = \frac{acid\ number\ of\ lauric\ acid\ (N_i) - acid\ number\ of\ reaction\ product\ (N_t)}{acid\ number\ of\ lauric\ acid\ (N_i)} \times 100\% \quad (1)$$

Preliminary examination of antibacterial activity of the GML was observed. The purpose of the examination was to identify the potential of the GML antimicrobial activity. The microbe of 10^6 CFU/ml was inoculated at the surface of NA solid media for about 0.1 mL. Then, a paper disc that had been immersed in GML solution (in absolute alcohol) was laid on the NA media. The GML solution concentration were 6,25, 12.5 mg/mL, 25 mg/mL, 50 mg/mL, 100 mg/mL.

Minimum Inhibitory Concentration (MIC) determination was performed using nutrient broth (NB) dilution media method. Various concentration of GML in NB was inoculated with 0.1 mL microbe suspension. The tubes were incubated for 24 hours at 36 ^oC. Then, the incubated tubes were streaked on nutrient agar plate. MIC was determined by observation of the lowest concentration of the GML that did not show the colony formation.

3. Results and discussion

3.1. Glycerol Mono-laurate synthesis

In order to examine the effect of glycerol/lauric acid (2:1, 5:1, 8:1) molar ratio on the synthesis of GML, the reactions were performed at the constant temperature of 140 °C and at the same catalyst loading of 15wt%. It can be seen in the figure 1a that the effect of glycerol/lauric molar ratio on the conversion reaction from 5:1 to 8:1 was insignificant. It has been reported their work on synthesis of GMO using SBA-15 mesoporous catalyst. It was reported that the effect of glycerol/lauric acid molar ratio on the lauric acid conversion was insignificant when the glycerol/lauric acid molar ratio was increased from 2:1 to 4:1, especially after 3–7 h of reaction. It is also reported that the reaction product, monoglyceride yield significantly increased from 58% to 68%, while diglyceride yield experienced a decrease from 35% to 23% while triglyceride yield remained stable at very low level (below 2%)[18].

The synthesis GML was carried out at various temperature of 120, 130 and 140 °C using a constant glycerol/lauric acid molar ratio (8:1) and at a constant of catalyst loading of 15wt%. In the figure 1b, it was observed that the increasing temperature condition from 110 to 140 °C enhanced reaction conversion from 62.9 to 84.1%.

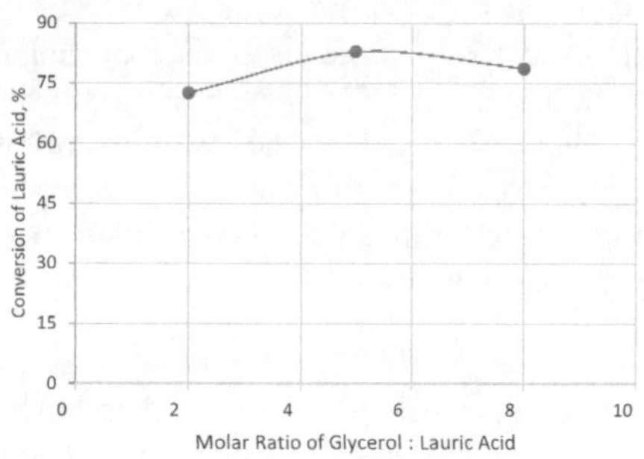

(a)

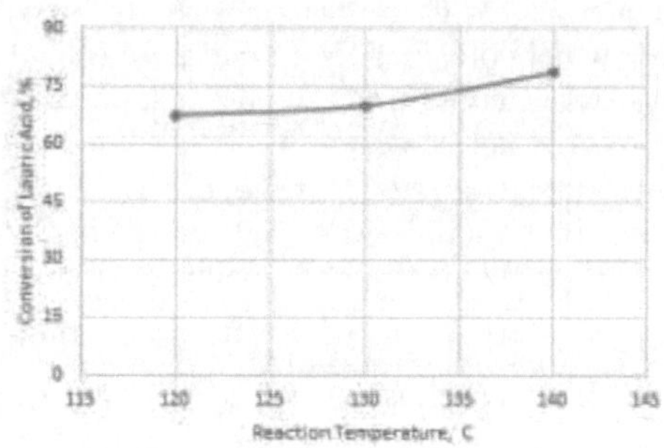

(b)

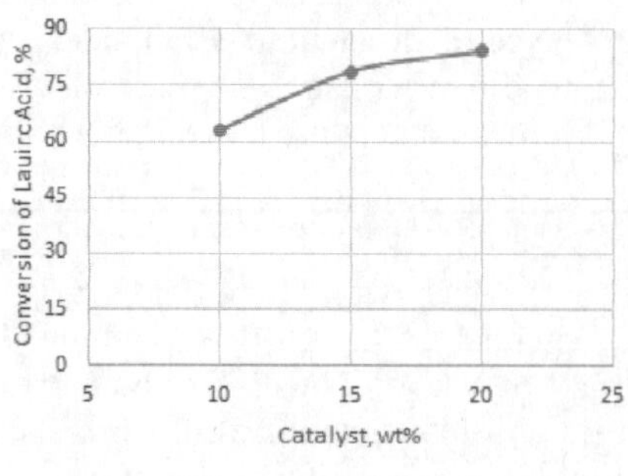

(c)

Figure 1. The effect of reaction conditions on reaction conversion

The effect of temperature to the reaction conversion can be observed at the figure 1b. The lauric acid conversion increased with increasing temperature. This could be due to the increase of kinetic energy of glycerol and lauric acid and it make the number chemisorption on the catalyst active site was increased and enhance the interaction of the molecules of reactant[19].

The effects of loading of catalyst were examined with a constant temperature (140 °C) and glycerol/lauric acid mole ratio (8:1). This observation suggested that the increase in catalysts loading from 10 to 20 wt% increased the number of active site that available for the reaction.

It can be observed from the figure 1c that increasing the amount of catalyst from 10 to 20 wt% increased the reaction conversion from 62.9 to 84.1. However the increasing of the reaction conversion due to the additional catalyst did not give significant improvement of reaction conversion. Moreover, the greater amount of catalyst will make difficulties in the separation process of the GML product and the catalyst.

Due to the low reaction conversion, other experiment of the GML synthesis was conducted using dealuminated catalyst. The catalyst was dealuminated using the method as presented in the International Conference on Chemical Process and Product Engineering 2016 that was organized by Department of Chemical Engineering, Diponegoro University at 14-15th September 2016[21]. In the comparison of GML synthesis using non dealuminated and dealuminated catalyst (table 1), it was observed that conversion of lauric acid that using non dealuminated catalyst was 78%, and according to GCMS resulted GML, GDL and GTL content were 38.6, 19,0 and 9.1% respectively. Meanwhile, the experiment using dealuminated catalyst resulted GML, GDL and GTL content were 59.5, 31.9 and GTL 0.2% respectively. The selectivity of GML on the process using non dealuminated and dealuminated catalysts were 57.9 and 65.0% respectively. The results showed that the dealuminated catalysts could enhance reaction rate. However, reaction time for 7 hour lead the increasing of formation of GDL and GTL.

Table 1. Comparison of the GML synthesis with different treatment of catalyst zeolist Y T = 140 °C, Molar ratio 8:1 (glycerol : lauric Acid), 15 wt% catalysts

Catalyst treatment	Reaction conversion (%)	Selectivity (%)	Glyceride content (%)		
			Glycerol Mono-laurate (GML)	Glycerol Di-laurate (GDL)	Glycerol Tri-laurate (GTL)
Without dealumination	78.0	57.9	38.6	19.0	9.1
With dealumination	97.8	65.0	59.5	31.9	0.2

3.2. Antibacterial activity examination

The antibacterial Activity examination showed that GML has a potential effect to inhibit pathogen bacterial. The biggest inhibition was obtained at *S. aureus* plate with clear zone of 33.44 mm. followed by *B. cereus* plate with clear zone of 32.75 mm. that was found at the concentration of GML was 100 mg/ml (figure 2). The results indicate that GML has more effective on the inhibition of gram positive bacteria. (*S. aureus* and *B. cereus*) than gran negative bacteria (*E.coli* and *P. aeruginosa*). However, the result also showed that E. coli was also inhibited by GML antibacterial activity. Potential antibacterial activity of GML is greater than that of diacyl glycerol[10]. It was also reported that the GML has no antibacterial activity against *S. typhymurium* and *E. coli* which are gram positive bacteria[10].

GML or monolaurin was categorized as indirect antimicrobial or multifunctional food additives, because of this product are added to foods used primarily as emulsifier other than antimicrobial. Monolaurin, the medium-chain fatty acid, is effective at high pH (5.0 – 8.0) different with the sort-chain fatty acids. The application of GML as preservative in food industry will be more effective when combined with another "indirect antimicrobials" because of the fatty acid and esters have a narrow range of effectiveness and generally recognized as safe (GRAS) substances such as *EDTA (*Ethylene diamine tetra acetic acid), citrate, and phenolic antioxidants also have limitations as antimicrobial agents when used alone[22]. GML was more effective inhibit *S. aureus* growth when its combined with pH lowering and *EDTA* addition [14,2]. GML acted synergistically with gentamicin and streptomycin to eliminate detectable viable biofilm bacteria (*S. aureus*)[11].

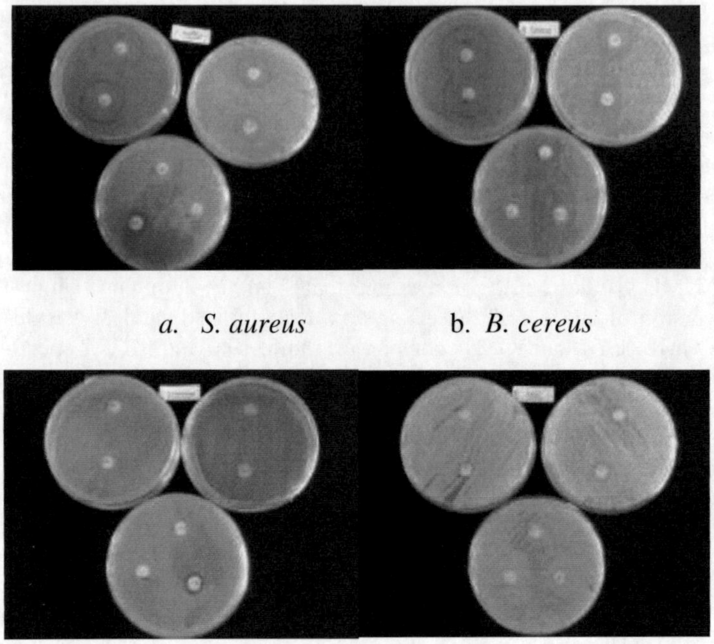

a. S. aureus b. *B. cereus*

c. *P. aeruginosa* d. *E. coli*

Figure 2. Clear zone of GML antibacterial examination

The antibacterial activity examination of GML showed that the minimum inhibitory concentration of GML for *S. aureus* and *B. cereus* growth was 25 mg/ml. *Minimum Inhibitory Concentration* (MIC) value from MDAG-CNO from coconut oil for *S. aureus* and *B. cereus* growth was 17,5 mg/ml[10]. The MIC value for *S. aureus* increase when GML was combined with 1 µg/ml Gentamycin. In the future GML is very applicable for medicinal industry to inhibit biofilm microbial growth[11].

4. Conclusions

Synthesis of glycerol mono-laurate from lauric acid and glycerol could be performed by esterification reaction using zeolist Y catalysts. The catalyst can be used either without dealumination or with dealumination. The conversion reaction of the process using dealuminated zeolist Y catalyst resulted higher than that of the process using non dealuminated one, which were 97.8 and 78.0% respectively. The process using dealuminated zeolist Y at 140 °C, Molar ratio of 8:1 glycerol:lauric acid and catalysts loading of 15wt% for 7 hours reaction time also gave GML selectivity i.e. 59.5%. The GML antibacterial activity was pre-examined. It was observed that the GML has antibacterial activity against gram positive bacterial such as *B. cereus* and *S.aureus*.

Acknowledgements
The research funding from Palm Oil Fund Management Agency (BPDP Sawit) is greatly acknowledged.

References
[1] Sonntag N O V 1982 Glycerolysis of fats status, review and and methyl esters – Critique. *J Am. Oil Chem. Soc.* **59** 795A - 802A
[2] Schlievert P M and Peterson M L 2014 Glycerol monolaurate antibacterial activity in broth and biofilm cultures *PLoS One 7(7):e40350 http://dx.doi.org/10.1371/journal.pone.0040350*
[3] Hamerski F and Corazza M L 2014 LDH-catalyzed esterification of lauric acid with glycerolin solvent-free system *J. Applied Catalysis A: General* **475** 242–248
[4] Abdullah A Z, Wibowo T Y and Zakaria R 2011 Effect of tetramethyl ammonium hydroxide on the activity of LiOH-intercalated montmorillonite catalyst in the transesterification of methyl laurate with glycerol *Chemical Engineering Journal* **167** 328–334
[5] Gupta M 1996 Manufacturing processes for emulsifiers *Bailey's Industrial Oil & Fat Products* ed Y H Hui (New York: John Wiley & Sons, Inc.)
[6] Fregolente L V, Fregolente P B L, Chicuta A M, Batistella C B, Filho R M and Wolf-Maciel M R 2007 Effect of operating conditions on the concentration of monoglycerides using molecular distillation *J. Chemical Engineering Research and Design* **85** 1524–1528
[7] Li Q *et al.* 2009 Glycerol monolaurate prevents mucosal SIV transmission. *J. Nature,* **458** 1034–1038
[8] Strandberg K L, Peterson M L, Lin Y C, Pack M C, Chase, D J and Schlievert P M 2010 Glycerol monolaurate inhibits candida and gardnerella vaginalis in vitro and in vivo but not lactobacillus *J. Antimicrob. Agents Chemother* **54** 597–601
[9] Kab M, Lüthje S, Herweg S and Stepan S 2014 Systematic investigations in the antimicrobial efficacy of glycerine esters with fatty acids of different chain length *J. INFORMS* **25** 390
[10] Nuraida L, Anggraeni D, Mintarto I S and Haryati T 2008 Kajian aktivitas antimikroba MAG dan MDAG dari minyak kelapa dan minyak inti sawit *Prosiding Seminar Tahunan MAKSI* pp 204 – 214
[11] Hess D J, Henry-Stanley M J and Well C L 2014 Antibacterial synergy of glycerol monolaurate and aminoglycosides in Staphylococcus aureus biofilms *J. Antimicrobial Agents and Chemotherapy* **58** 6970-6973
[12] Widiyarti G, Hanafi M, dan Soewarso W P 2009 Kajian awal sintesis monolaurin sebagai antibakteri staphylococcus aureus *Indo. J. Chem.* 2009 **9 (1)** 99 - 106
[13] Schlievert P M, Deringer M H, Kim S J. Projan and Novick R P 1992 Effect of glycerol monolaurate on bacterial growth and toxin production *Antimicrobial Agents And Chemotherapy* **36 (3)** 626-631
[14] Schlievert P M and Peterson M L 2012 Glycerol Monolaurate Antibacterial Activity in Broth and Biofilm Cultures. www.plosone.org. July 2012 | Volume 7 | Issue 7 | e40350
[15] Batovska D I, Todorova I T, Tsvetkova I V and Najrenski H M 2009 Antibacterial activity of the medium chain fatty acid and their 1-monoglyserides: individual effects and synergistic relationships *Polish J. of Microbiology* **58** 43-47
[16] Noll K S, Prichard M N, Khaykin A, Sinko P J and Chikindasa M L 2012 The natural antimicrobial peptide subtilosin acts synergistically with glycerol monolaurate, lauric arginate, and _-poly-L-lysine against bacterial vaginosis-associated pathogens but not human lactobacilli *J. Antimicrobial Agents and Chemotherapy* **56** 1756–1761

[17] Kenny J G, Ward D, Josefsson E, Jonsson I, Hinds J, Rees H H, Lindsay J A, Tarkowski A and Horsburgh M J 2009 The Staphylococcus aureus response to unsaturated long chain free fatty acids: survival mechanisms and virulence implications. *J. PLoS ONE (www.plosone.org)*

[18] Hermida L, Abdullah A Z and Mohamed A R 2011 Synthesis of monoglyceride through glycerol esterification with lauric acid over propyl sulfonic acid post-synthesis functionalized SBA-15 mesoporous catalyst *Chemical Engineering Journal* **174** 668-676.

[19] Connor K A 1990 *Chemical Kinetics: The study of Reaction Rate in Solution* (New York: VCH Publisher Inc.)

[20] Anggoro D A, Setianto W B, Wibowo T Y, Pratama F R, Giovanno A and Buchori L 2016 Characterization and testing of zeolite Y dealuminates catalyst for glycerol conversion to glycerol monolaurate (Semarang, Indonesia: International Conference on Chemical Process and Product Engineering 2016 (ICCPPE 2016))

[21] Jay M J 1996 *Modern Food Microbiology* (New York: Chapman and Hall)

IOP Conf. Series: Earth and Environmental Science **65** (2017) 012047 doi:10.1088/1755-1315/65/1/012047

Synthesis of palm oil fatty acid as foaming agent for firefighting application

M Rivai[1], E Hambali[1,2], A Suryani[1,2], R Fitria[1], S Firmansyah[1] and J Pradesi[2]

[1] Surfactant and Bioenergy Research Center, Bogor Agricultural University, Jl. Raya Pajajaran No. 1, Bogor Indonesia
[2] Departement of Agroindustrial Technology,Bogor Agricultural University, Bogor, Indonesia

Email : mira.rivai2705@gmail.com

Abstract. Many factors including natural factor, human carelessness, new land clearance or agricultural burning/act of vandalism and ground fire are suspected as the causes of forest fire. Foam, which cools the fire down, covers the burning material/fuel, and avoids contact between burning materials with oxygen, is an effective material used to fight large-scale fires. For this purpose, surfactant which can facilitate foam formation and inhibit the spread of smoke is required. This study was aimed at producing prototype product of foaming agent from palm oil and its formulation as a fire fighting material. Before the formulation stage, the foaming agent was resulted from saponification process of oleic, lauric, and palmitic acids by using NaOH and KOH alkaline. Foam stability was used as the main indicator of foaming agent. Results showed that potassium palmitate had the highest foam stability of 82% until the 3rd day. The best potassium palmitate concentration was 7%.

1. Introduction

In 2015, land and forest fires in Indonesia were estimated to damage more than 2 million hectares of land consisting of about 600,000 hectares of peatland and 1.5 million hectares of non peatland. The loss caused by these fires were suffered not only in local but also national, regional, and even international levels. Several factors are indicated as the causes of forest fires including (a) natural factors including high temperature during dry season, lightning strike, and volcanic activity (volcanic mudflow or hot clouds), (b) human carelessness such as careless cigarette butt littering and fire left in camping ground, (c) land-clearing open burning and vandalism, and (d) ground fire which usually occurs beneath the ground in peatland and lits fire especialy during dry season with high air temperature.

Surfactant is the main component of a foaming agent. The use of surfactant in foaming agent improves the effectiveness of water use and the performance of foaming agent in putting out the fire. Surfactant in a foaming agent functions as a surface tension reducer and foaming agent. Foam is used to cool the fire down, cover the burning fuel, and avoid recontact with oxygen that may lit the fire back. In the field of firefighting, many products are available for use in different needs and types of fire. With regard to its use in the field, a foaming agent to be used in firefighting should be environmentally friendly, biodegradable, and nontoxic. Therefore, a foaming agent made of natural

Content from this work may be used under the terms of the Creative Commons Attribution 3.0 licence. Any further distribution of this work must maintain attribution to the author(s) and the title of the work, journal citation and DOI.
Published under licence by IOP Publishing Ltd

feedstock is preferable as it is not only renewable, but also environmentally friendly and not harmful. Palm oil is one of surfactant feedstock which is abundant in Indonesia. Based on the above notion, this study was aimed at producing a prototype product of foaming agent from palm oil fatty acid which could be used as as a feedstock of firefighting agent and assessing the physicochemical properties of the produced foaming agent.

Soap is an alkaline carboxilate salt (RCOONa) with a hydrophobic (non polar) R group and hydrophilic (polar) COONa group. The head part of a soap molecule is a hydrophilic group and the tail part is a hydrophobic group. Soap is widely known as very important as a reducer of surface tension [1] and interfacial tension. Palm oil-based soap is easily degraded and safe for the environment so that it is used in this study as the biggest component in the production of foaming agent. Soap can be synthesized by reacting oil or fatty acid with caustic soda, soda ash, or potassium salt. Soap is resulted through a saponification process of oil/fat and fatty acid. Saponification process is a reaction between triglyceride and alkali resulting in glycerol as a by-product. Meanwhile, fatty acid saponification process which is also called a neutralization process is a reaction between free fatty acid and alkaline without producing glycerol as a by-product [2,3].

Fatty acids produced from the hydrolysis of oil/fat consist of saturated and unsaturated fatty acids. The difference between these fatty acids lies in the existence of double bonds in their hydrocarbon chains. The dominant fatty acids in palm oil are palmitic acid and oleic acid while lauric acid is found to be dominant in coconut oil and palm kernel oil [4]. Oleic acid is long chain unsaturated fatty acid having C18 chain length, one double bond, and a melting point 15.3°C. Lauric acid is a saturated fatty acid having C12 chain length and a melting point 43.2°C. palmitic acid is a saturated fatty acid having C16 chain length and a high melting point 62.9°C. High content of palmitic acid make palm oil more resistant to oxidation (rancidity) than the other types of oil [5]. These three fatty acids have good solubility in water, good detergency property, and good performance in hard water. Saponification (neutralization) results in soap of sodium oleic, potassium lauric, and sodium palmitic types. Studies have been done on the utilization of plant oil derivatives as firefighting agents which are environmentally friendly. Mizuki *et al.* [6] formulated fire fighting agent by using long chain fatty acid (LCFA), biodegradable chelating agent, and additive to produce a firefighting agent which was also biodegradable. The concentration of firefighting agent was 1% and environmentally friendly firefighting agent was produced. However, the resulted foam had shorter lifetime than did synthetic firefighting agent. Kawahara *et al.* [7] formulated a firefighting agent by using a *biodegradable chelating agent* (MGDA) which was different from that used by Mizuki *et al.* [6] who produced a *firefighting agent* which was also biodegradable having an expansion ratio of 9.6 times higher than that of the commercial *firefighting foam for structural fire.*

2. Methods

Materials used in this study included oleic acid, lauric acid, palmitic acid, NaOH, KOH, commercial surfactant, and others. Equipment including saponification reactor, Anton Paar DMA 4500M densitometer for density test, Brookfield DV-III Ultra viscometer for viscosity test, Phoenix 300 contact angle analyzer for wettability test, homogenizer, spinning drop tensiometer TX500C for surface tension dan interfacial tension test, and other glass equipment.

2.1. Fatty acid analysis

Analysis of physicochemical properties of fatty acid was done to determine the properties of fatty acids used in this study. Acid number, iodine number, saponification number, viscosity, and density tests were conducted.

2.2. Synthesis and analysis of fatty acid sodium and potassium

Synthesis of fatty acid sodium and potassium was done by using three fatty acids including oleic acid, lauric acid, and palmitic acid. NaOH and KOH of 30% were used as alkaline reactants. The mole ratio of fatty acid and alkaline was 1:0.5 – 1:1. The process was done at 65°C for 90 minutes. Sodium and potassium analysis of the resulted fatty acid was done on foam stability, foaming ability, surface tension, and interfacial tension.

2.3. Selection of fatty acid sodium/potassium concentration

In this stage, selection was done to get the best sodium/potassium concentration of fatty acids produced in the previous stage. The concentration tested was 3, 5, and 7%. The analysis was done on surface tension, foaming ability, foam stability, viscosity, density, and contact angle tests.

3. Results

3.1. Fatty acid analysis

Three types of palm oil fatty acid feedstock, namely oleic acid, lauric acid, and palmitic acid were used in this study. Results of the analysis showed that these three fatty acids had acid value of 199.2 – 280.6 mg KOH/g, saponification number of 200.2 – 281.6 mg KOH/g, iodine number of 0.1 – 87.7 g/100 g Iodine, viscosity of 4.26 – 7.42 cP, and density of 0.85812 – 0.86650 g/cm^3. These results are listed in Table 1.

Table 1. Results of oleic acid, lauric acid, and palmitic acid analysis

Parameter	Unit	Oleic Acid	Lauric Acid	Palmitic Acid
Acid Number	mg KOH/g	199.2	280.6	218.5
Saponification Number	mg KOH/g	200.2	281.6	219.4
Iodine Number	g/100 g	87.7	0.1	0.2
Viscosity (70°C)	cP	7.25	4.26	7.42
Density (70°C)	g/cm^3	0.85812	0.86650	0.86233

3.2. Synthesis and analysis of fatty acid sodium and potassium

In this stage, saponification process of oleic acid, lauric acid, and palmitic acid by using a reactor of 2L/batch scale. Sodium and potassium of the resulted fatty acids was analyzed for its foam stability, foaming ability, surface tension, and interfacial tension. Results of the analysis are as follows.

3.2.1. Foam Stability

Foam can improve the efficiency of water as firefighting agent. Foam is defined as a dispersion system consisting of gas bubbles covered with a liquid layer. Significantly different density levels found in the bubble and liquid medium make the system quickly separate into two layers with the bubbles go up to the top. When gas bubbles are formed beneath the liquid surface, they will immediately break out as the liquid flow (drainage) occurs as a result of gravity or downward force. Therefore, pure liquid will not be foamy unless a surfactant is added[8]. This added surfactant reduces interfacial tension of gas/liquid making the dispersion of gas in liquid easier to occur [9] which then leads to foam formation. The mechanism of foam formation starts when gas bubbles enter the surfactant solutio. The surfactant wil then be adsorbed in gas/liquid interface to form gas bubbles covered with a film layer. This covered buble is called foam which tends to float upward to the surface because the specific gravity of gas is smaller than that of water. However, surfactant is also found on liquid surface on the border layer between water and air. This surfactant holds the floating foam on the border of liquid surface and avoids it from being released to the air. The accumulating foams on the surface will get closer, make contact one to another, or even resemble to create bigger foams [9].

Foam stability refers to the ability of the foam to maintain the main parameters (buble size, liquid content, total foam volume) in a constant situation for a certain period of time. Foam analysis was done in order to assess the ability of foam to stay in its shape and without disappearing in a period of time. For foaming agent application, foam stability was measured for 3 days. The value of foam stability was obtained from the difference of foam height in minute 0 after the foam was vortexed/homogenized and the foam height in day 3. The foam stability was expressed in cm unit and percentage. Results of foam stability measurement showed that the stability of foam produced from sodium oleic, lauric, palmitic and potassium oleic, lauric, palmitic was about 2-3 days. The highest foam stability was shown by sodium palmitic and potassium palmitic. Sodium palmitic at a molar ratio of 1:0.75 had a foam percentage of 82.2% while potassium palmitic with a molar rasio of 1:1 had a foam percentage of 81.8%. The histogram of the analysis of foam stability of sodium oleic, lauric, palmitic and potassium oleic, lauric, palmitic is depicted in Figure 1.

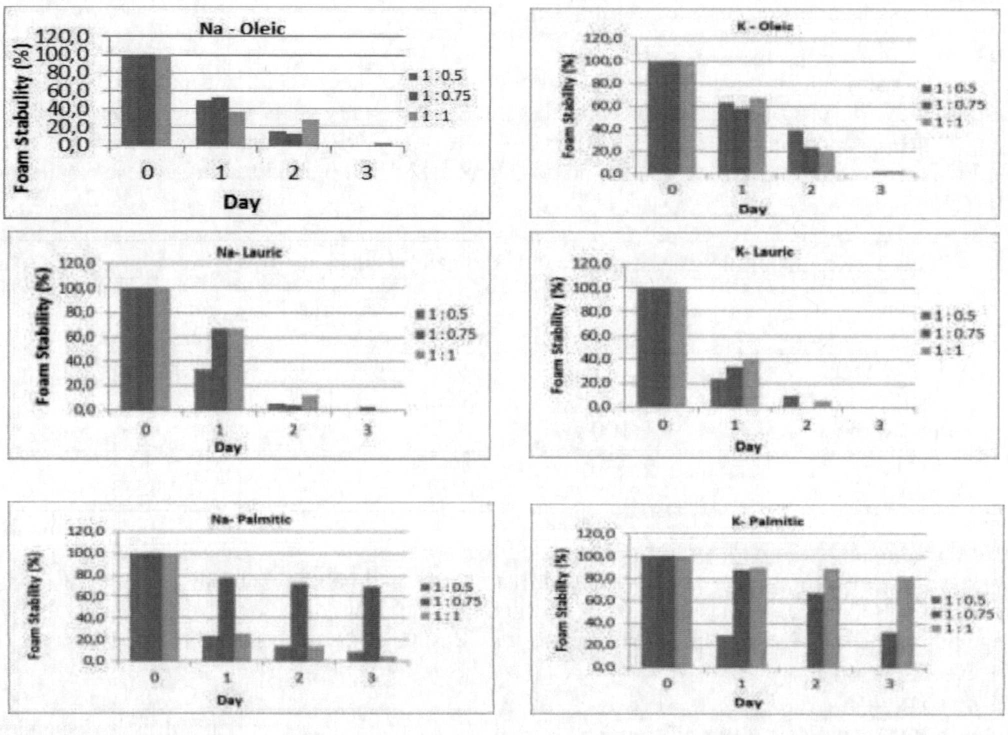

Figure 1. Histogram of the analysis of foam stability of sodium oleic, lauric, palmitic and potassium oleic, lauric, palmitic

Higher molar ratio and longer testing time resulted in reduced foam stability. Yet, potassium palmitic and sodium palmitic had relatively higher foam stability as at testing day 3 their foam stability was about 80% while foam of other samples stayed for 2 days only and quickly collapsed. Foam collapse was caused by thinning in film layer and accumulation of foam to form a bigger foam (coalescent)[8]. Film layer thinning occurs because foam has a tendency to float upward while at the same time it is constantly pulled downward by liquid flow (drainage) occuring as a result of gravity force. Being pulled into two opposite directions, the film of foam gets thinner and easier to break up. Varied foam sizes also create gradient of gas pressure resulting in gas diffusion. This then leads to the accumulation of small foams to form bigger size foam (coalescent) which is easier to break out [10]. This foam break out can be avoided by slowing down the drainage through the improvement of liquid bulk viscosity. Adding materials such as glycerol, polymer, or stabilizer (CMC) increases viscosity and increased viscosity reduces the speed of drainage. Reduced drainage speed minimalizes thinning [8].

3.2.2. Foaming Ability
Foaming ability, which is expressed in % unit, refers to the ability of a liquid to form foam in 0 minute after the liquid is vortexed or homogenized. In this study, results showed that sodium oleic, lauric, and palmitic had foam height of about 19.51 to 170.73% and the two highest foam heights were found in sodium lauric at a molar ratio of 1:0.75 (170.73%) and at a molar ratio of 1:1 (165.85%). Meanwhile, potassium oleic, lauric, palmitic had foam height of about 8.54 to 170.73% and the highest foam heights were found in potassium lauric at a molar ratio of 1:1 (170.73%) and potassium palmitic at a molar ratio of 1:1 (162.2%). The histogram of the analysis of foaming ability of sodium oleic, lauric, palmitic and potassium oleic, lauric, palmitic is depicted in Figure 2. There was a tendency that wider molar ratio resulted in higher foaming ability.

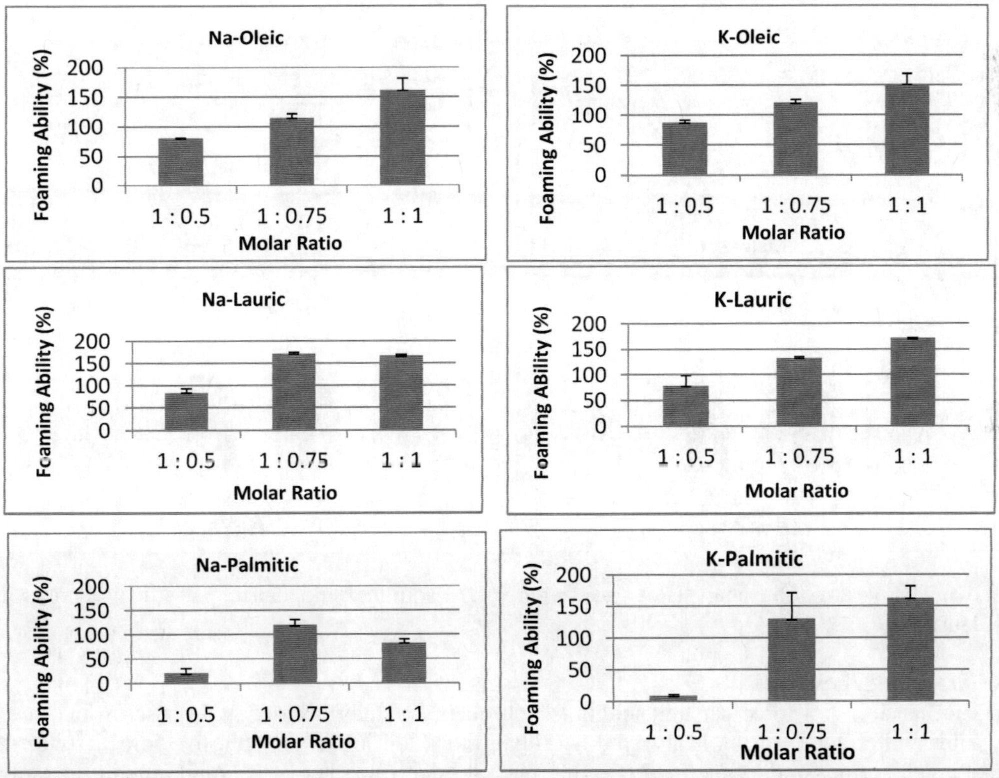

Figure 2. Histogram of the analysis of foaming ability of sodium oleic, lauric, palmitic and potassium oleic, lauric, palmitic

3.2.3. Surface Tension
Surface tension of a liquid is a work done to widen the liquid surface expressed in a unit area. Surface tension occurs because there is a adhessive force between liquid and air. The surface of a liquid tends to get tensed making the surface look like a thin film. This is affected by a cohesive force between molecules of water. Surface tension works only on the surface area and it is evenly spread all over the points. The cohesive force between molecules in a liquid is equal in all directions, yet molecules on liquid surface are more forced downward into the liquid. In this study, it was found that surface tension of sodium oleic, lauric, palmitic was 11.08–30.60 dyne/cm and of potassium oleic, lauric,

palmitic was 11.91–39.30 dyne/cm. The histogram of the analysis of surface tension of sodium oleic, lauric, palmitic and potassium oleic, lauric, palmitic is depicted in Figure 3. There was a tendency that wider molar ratio resulted in lower surface tension. This tendency was in line with that of foam stability but inversely proportional to that of foaming ability.

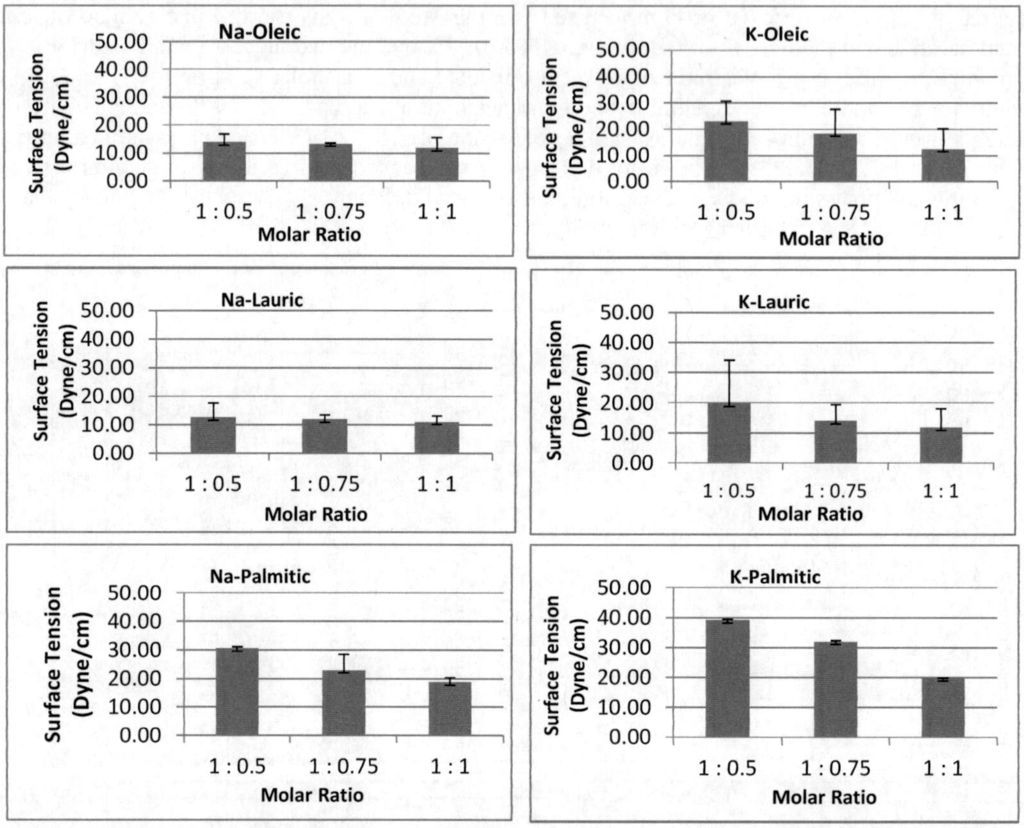

Figure 3. Histogram of the analysis of surface tension of sodium oleic, lauric, palmitic and potassium oleic, lauric, palmitic

3.2.4. Interfacial Tension

Interfacial tension is a force per unit length which occurs at interfaces of two phases which are not immiscible. Interfacial tension is always lower than facial tension as the adhesive force between two liquids which are not immiscible is bigger that the adhesive force between liquid and air. Results in this study showed that the interfacial tension of sodium oleic, lauric, palmitic was 0.93–3.93 dyne/cm and of potassium oleic, lauric, palmitic was 1.32–9.85 dyne/cm. The histogram of the analysis of interfacial tension of sodium oleic, lauric, palmitic and potassium oleic, lauric, palmitic is depicted in Figure 4. There was a tendency that wider molar ratio resulted in lower interfacial tension.

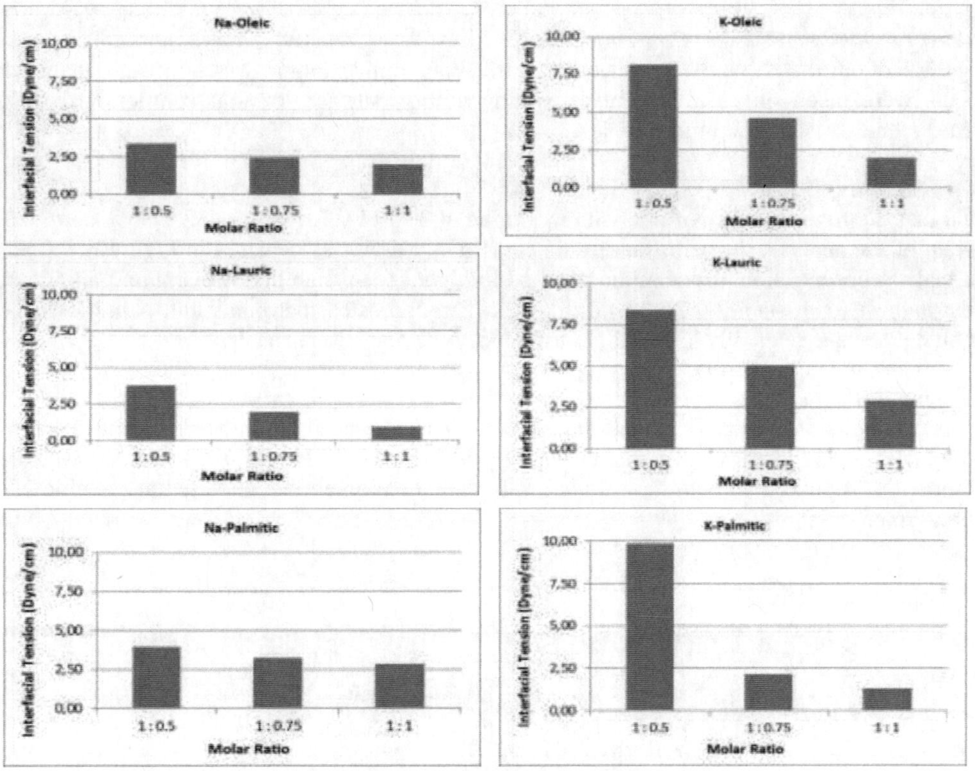

Figure 4. Histogram of the analysis of interfacial tension of sodium oleic, lauric, palmitic and potassium oleic, lauric, palmitic.

It was shown that potassium palmitic had the best results as shown in its foam stability of 82% on day 3, foaming ability of 162,2%, surface tension of 19,79 dyne/cm, and interfacial tension of 1,32 dyne/cm. Compared to the other type of surfactant (commercial), the potassium palmitic foams were more stable. On the first day, the stability of commercial surfactant foams was found to drastically drop which continued to the second day when it was found to be below 40%. The comparison between the stability of potassium palmitic foams and that of commercial surfactant foams is shown in Figure 5.

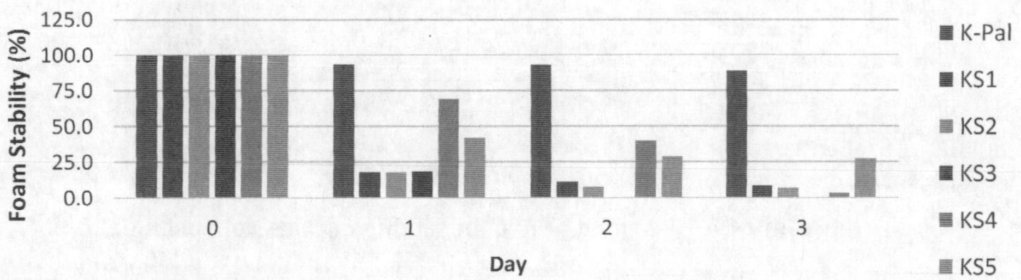

Figure 5. Results of the stability test of potassium palmitic and commercial surfactant foams

3.3. Selection of Potassium Palmitic Concentration

In this stage, selection of 3, 5, and 7% concentration of potassium palmitic was done and analyzed. Measurements were taken on foam stability, foaming ability, surface tension, solution viscosity, solution density, and wettability (contact angle).

3.3.1. Foam Stability

It was found that the foam stability of potassium palmitic at 3, 5, 7 % gave was 32.9-85% on day 3. The histogram of the analysis of foam stability of potassium palmitic is depicted in Figure 6. It was shown that higher concentration of potassium resulted in higher foam stability. Potassium palmitic of 7% gave the highest foam stability (85.2%) on day 3. This indicated that potassium palmitic at the concentration of 7% gave relatively longer foaming effect to be applied on firefighting.

3.3.2. Foaming Ability

It was found that foaming ability of potassium palmitic at the concentration of 3, 5, and 7% was 192,68 to 214,63%. The histogram of foaming ability is depicted in Figure 7. It was shown that higher consntration of potassium palmitic resulted in higher foaming ability. Potassium palmitic 7% had the highest foaming ability.

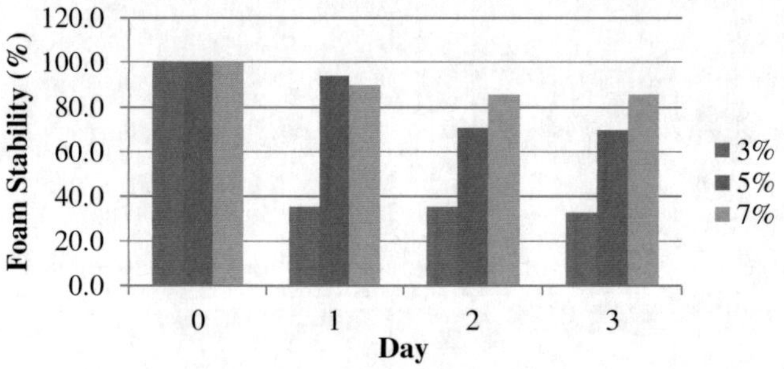

Figure 6. Histogram of foam stability of potassium palmitic

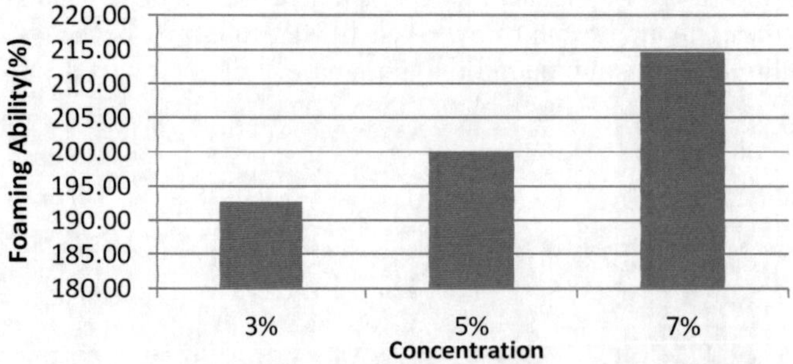

Figure 7. Histogram of foaming ability of potassium palmitic

3.3.3. Surface Tension

The surface tension of potassium palmitic at the concentration of 3, 5, and 7% was 39,25 – 55,40 dyne/cm. The histogram of surface tension of potassium palmitic is depicted in Figure 8. Higher concentration of potassium palmitic was found to result in lower surface tension. Potassium palmitic 7% had the lowest surface tension. Low surface tension of soap indicates its better ability to dissolve air. In other words, it can be said that low surface tension made the emulsion of air and soap solution easier to occur which could lead more foam formation

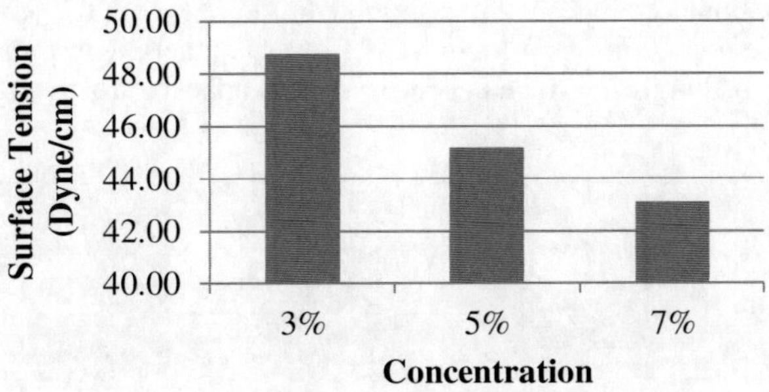

Figure 8. Histogram of surface tension of potassium palmitic

3.3.4. Viscosity

The viscosity of potassium palmitic at the concentration of 3, 5, and 7% was 1.18–1.87 cP. The histogram of viscosity of potassium palmitic is depicted in Figure 9. Higher concentration of potassium palmitic was found to result in higher viscosity. Potassium palmitic 7% had the highest viscosity (1.87 cP).

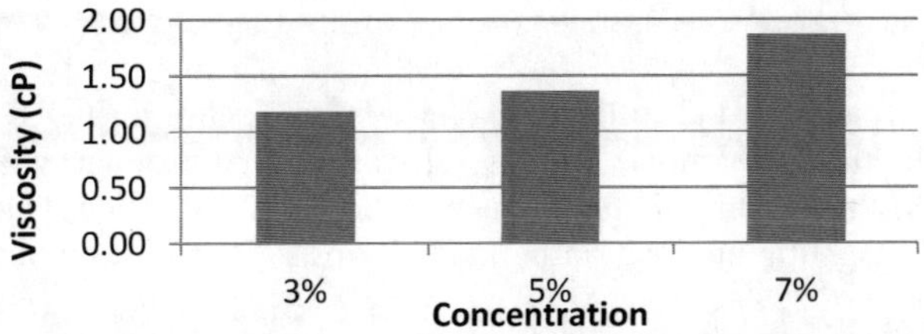

Figure 9. Histogram of viscosity level of potassium palmitic

3.3.5. Density

The density levels of potassium palmitic at the concentration of 3, 5, and 7% was 0.99734 – 0.99863 g/cm^3. The histogram of density level of potassium palmitic is depicted in Figure 10. Higher concentration of potassium palmitic was found to result in higher density. Potassium palmitic 7% had the highest density level (0.99863 g/cm^3).

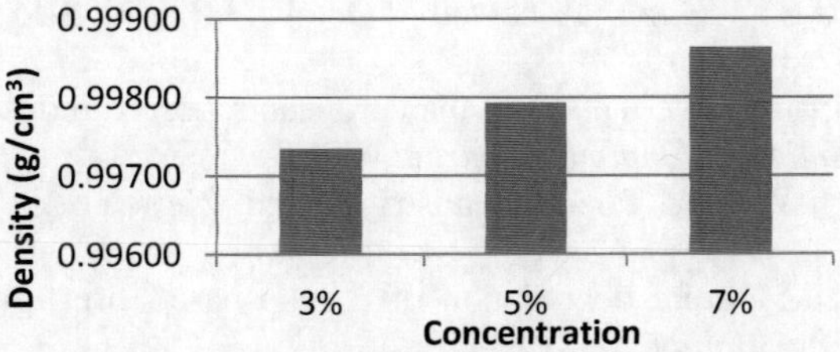

Figure 10. Histogram of density level of potassium palmitic

3.3.6. Wettability (Contact Angle)

Contact angle test was done at hydrophilic and hydrophobic solid surfaces. The contact angle of potassium palmitic at the concentration of 3, 5, and 7% at minute 0 was 60.31–65.56° which reduced to 21.15 – 33.64° at minute 10. The hostogram of contact angles of potassium palmitic is depicted in Figure 11. It was shown that at minute 10, higher concentration of potassium palmitic resulted in higher reduction of contact angles. Potassium palmitic 7% had the lowest contact angle and the highest contact angle reduction of up to 44.41°. Lower contact angle indicated that the resulted soap was hydrophilic and had higher surface energy so that its adhesive force to the surface was stronger.

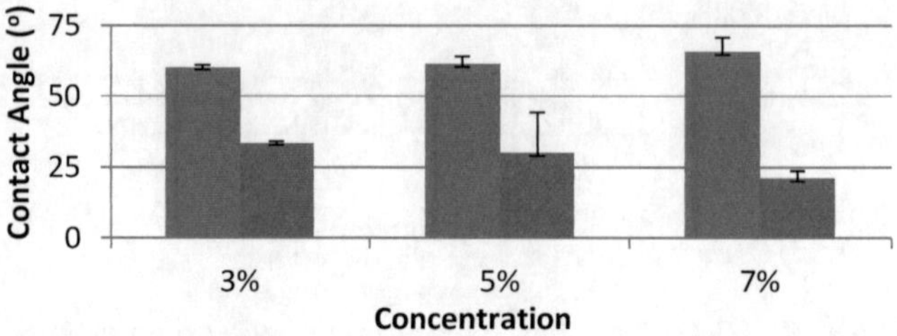

Figure 11. Histogram of contact angle

The selection process of potassium palmitic concentration showed that higher concentration of potassium palmitic increased foam stability, foaming ability, viscosity, and density and reduced surface tension and contact angle at minute 10. Potassium palmitic 7% had the highest foam stability (85.2%), foaming ability (214,63%), surface tension (43 dyne/cm), viscosity (1.87 cP), density (0.99863 g/cm^3), and contact angle reduction (44.41°).

4. Conclusions

Sodium oleic, lauric, palmitic and potassium oleic, lauric, palmitic had foam stability of about 2-3 days. Potassium palmitic with a molar ratio of 1:1 was the best product with a foam stability on day 3 of up to 82%. The best potassium plamitic concentration was 7%. It was suggested that formulation of foaming agent for firefighting application be made by using potassium palmitic with a molar ratio of 1:1.

Acknowledgements

Research funding assistance from The Indonesian Oil Palm Estate Fund (BPDP) and laboratory facility assistance from Surfactant and Bioenergy Research Center IPB were acknowledged.

References

[1] Rais F, Baati R, Damak N and Kamaun A 2008 The use of a eutetic mixture of olive pomace oil fatty amides to easilly prepared sulfated amides applied as lime soap *Dispersant* **85** 869-77
[2] Spitz L 1996 *Soap and Detergents, A Theoretical and Practical Review* (Illinois : AOCS Press)
[3] Ophardt C E 2003 *Soap* Accessed October 27 2015 (http://elmhurst.edu/-hm/vchembook/554soap.html)
[4] Simeh M A 2004 Comparative advantage of the european rapeseed industry vs other oils and fats producers *Oil Palm Industry Economic Journal* **4** 14-22
[5] Belitz H D and Grosch W 2004 Food Chemistry 3rd ed (New York : Springer Verlag Berlin Heidebers)
[6] Mizuki H *et al.* 2010 Microbial degradation of a soap-based fire-fighting agent in activated sludge *J. Environ. Eng. Manage.* **20** 109-113.
[7] Kawahara T, Hatae S, Kanyama T, Ishizaki Y, and Uezu K 2016 Development of eco-friendly soap-based firefighting foam for forest fire *Environ Control Biol.* **54** 75-78
[8] Tadros T F 2005 *Applied Surfactant : Principles and Applications* (Weinheim : Wiley-VCH Verlag GmbH & Co. KgaA) pp 259-63
[9] Exerowa D and Kruglyakov PM 1998 *Foam and Foam Films: Theory, Experiment, Application* (Netherlands : Elsevier) **Vol** 494 pp 1-3

[10] Schramm L L 2005 *Emulsions, Foams, and Suspensions : Fundamentals and Applications* (Weinheim : Wiley-VCH Verlag GmbH & Co. KgaA)

IOP Conf. Series: Earth and Environmental Science **65** (2017) 012048 doi:10.1088/1755-1315/65/1/012048

Technical problems of wastewater treatment plant in crude palm oil industry
A case study in PT Socfin Indonesia-Kebun Sungai Liput, Nanggroe Aceh Darussalam Province

Y V Paramitadevi[1] and Rahmatullah[1]

[1] Diploma Program of Bogor Agricultural University, Kumbang Street no 14, Bogor, Indonesia,

Email:vega_paramitadevy@yahoo.com

Abstract. Crude palm oil produced in Indonesia has already been known as the largest in the world. Unfortunately many of palm oil factories (CPOF) spread out in Indonesia have not good wastewater treatments (WWTP) yet. PT Socfin Indonesia, as an example, which is located in Aceh Tamiang Regency, still has BOD contained in its final effluent of the waswater treatment plant more than 150 ppm. In fact, the capability and capacity of WWTP in PT Socfin are 192 m^3 per day. Because of improper operational and maintenance of the WWTP, the technical prob lem are accumulated, such as, increasing the deposition of sludge and decreasing the retention time. The following affect is the treatment process is not going well and the quality of effluent is getting worse. The objective of this paper is to solve the technical problems by means remov ing the deposition of sludge periodically and fixing floating aeration in the aerobic pond. Method using for this paper is survey and wastewater sampling. A recommendation of the was tewater treatment system has been proposed after average BOD from WWTP outlet is defined 158 mg/L. The recommendation has seven procesess including oil separation, neutralization, closed tank anaerobic digester equipped with gas holder, extended aeration, settling tank, sand drying bed and land application.

1. Introduction

Oil palm (*Elaeis guineensis*) is a versatile monoecious oil bearing plant, its distribution is resricted to tropical regions of the world, particularly in Indonesia and Malaysia [1]. The global production of palm oil has been more than doubled every 10 years, and by 2020, it is expected to increase to 78 million tonnes [2]. The different steps of the palm oil extraction process generate various types of waste by-products including solid waste such as oil palm shell (OPS), fibres, empty fruit bunches (EFB), palm oil clinker (POC) and palm oil fuel ash (POFA) accompanied by palm oil mill effluent (POME) the most significant pollutant in the form of wastewater, if not managed properly, it could be disastrous impact on the environment [3,4].

Raw POME is a colloidal suspension containing 95-96% water, 0.6-0.7% oil and 4-5% total solids including 2-4% suspended solids [5]. The suspended solids are mainly consist of debris from palm fruit mesocarp generated from three main sources, (1). sterilizer condensate, (2). clarification wastewater, and (3). hydrocyclone waste [3,4,6,7,8]. Consequently, about 0.9, 1.5 and 0.1 m^3 of

Content from this work may be used under the terms of the Creative Commons Attribution 3.0 licence. Any further distribution of this work must maintain attribution to the author(s) and the title of the work, journal citation and DOI.
Published under licence by IOP Publishing Ltd

POME will be generated from sterilizer condensate, clarification wastewater and hydrocyclone waste for each tonne (1.13 m^3) of crude palm oil processed [5,8]. Thus, it was estimated that in year 2014, 21.35 million m^3 tonnes of POME was generated from Indonesian palm oil mills base on the total CPO production of 28.9 million m^3 tonnes. Figure 1 shows percentage of Indonesian CPO production in year 2014.

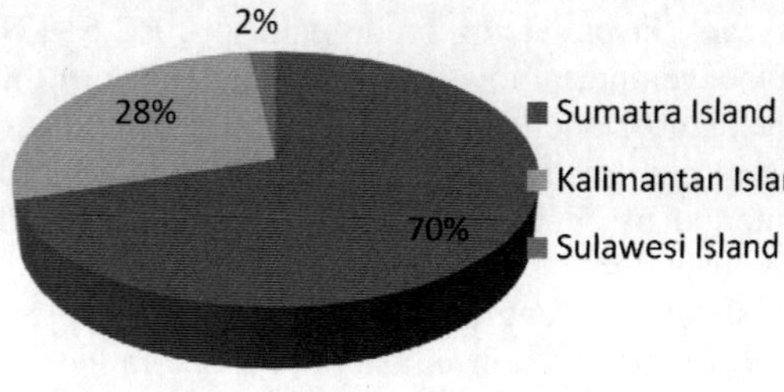

Figure 1 Indonesian CPO Production in year 2014 Source: BPS [9]

Although POME is a non-toxic liquid waste with unpleasant smell, its COD and BOD value are high enough to cause serious pollution and environmental problem to the rivers. Table 1 shows a typical characteristics of POME and discharge limits for POME discharge into water sources in Indonesia. Nevertheless, many palm oil mills are still unable to adhere to the wastewater discharge limits and thus resulting to a dramatic increase in the number of polluted rivers [5,8]. Furthermore, if the POME is discharged without prior purification, the BOD produced daily was estimated to be 9.12 million kg [10]. In other words, if each human being is assumed to produce 14.6 kg annually, this BOD value will be equivalent to the waste generated by 45.6 million people which is nearly 18% from the current population in Indonesia.

Table 1. Characteristics of wastewater to produce the POME[a] and discharge limits of treated POME in Indonesia[b].

Parameters	Typical POME	Discharge Limits (mg/ L)	Maximum Value of Pollution Loading (kg/ tonnes)
BOD_5	25 000-29 000	100	0.25
COD	51 000-64 000	350	0.88
TSS	18 000-23 000	250	0.63
Oil and Grease	6000-7000	25	0.063
Total Nitrogen	750-1200	50	0.125
pH	4.5	6-9	

[a] Lam and Lee [5], Liew *et al.* [6], Wu *et al. [8]*

[b] Decree of Minister of Environment and Forestry [14]

Ponding system is the most conventional method for treating POME [6,11], but other processes such as aerobic and anaerobic digestions, physicochemical treatments and membrane filtration may also provide the palm oil industries with a possible insight into the improvement of current POME trea tment process. More than 85% of palm oil mills in Indonesia have adopted the ponding system for PO ME treatment [12,13]. However, few have been put into full scale operation because of the unsatisfact ory performance, high capital investment and maintenance cost. The open lagoon POME treatment sys tem covers 10-12 acres of land, along with this; it has various drawbacks as long hydraulic retention ti me (HRT), foul smell, greenhouse gas emissions, and large occupied area. The exsisting technologies are unable to consistenly meet the proposed stringent biological oxygen demand (BOD) regulatory dis charge standard of 100 mg/L to be imposed by the Ministry of Environment and Forestry, Indonesia [1 4].

PT Socfin Indonesia Kebun Sungai Liput (PT SI-KSL), located in Aceh Tamiang regency, has been operated for more than 15 years. In producing crude palm oil, PT SI-KSL has undergone several processes of rehabilitation and development towards perfection. However those improvements are still not optimum evidenced by the number of oil content contained in POME is still high.The fatfit pond is unable to recover POME from clarification process. As the result, the BOD loading into POME treatment plant has reached more than 25 000 ppm. In certain circumsances, the output of wastewater, especially BOD is still above goverment quality standard. Thus the problems which occurs in POME treatment plant in PT SI-KSL should be identified and solved. The ultimate goal of this paper are : (1). identifying technical problems of POME treatment plant of PT Socfin Indonesia in Kebun Sungai Lipu t, Aceh Tamiang regency, (2). giving best practices of POME treament plant for PT Socfin and Aceh Tamiang goverment district.

2. Methods

Methods used in this paper according these following steps :

2.1. Literature Review

In purpose to describe the problems of POME treatment plant at oil palm plantation, it takes at least references about the process of producing crude palm oil (CPO). References on some wastewater treatment system that has already exsisted in Indonesia and Malaysia are also still needed for comparison.

2.2. Field Survey

The field survey was conducted on February 6th, 2015 in PT SI-KSL, Aceh Tamiang Regency. Secondary data of POME treatment plant was collected for 3 months, from February to April 2015. The survey was also carried out grab sampling of wastewater generated, sampling of the output of POME treatment plant from PT SI-KSL using Indonesia National Standard method no 6989.59:2008 and measuring the depth of ponds.

2.3. Analysis

Primary data obtained from the results of laboratory analysis using Indonesia National Standard method for pH, COD, BOD, TSS, oil and fat parameters and direct observation of wastewater treatment plant are revealed and discussed to identify exsisting problems in PT SI-KSL. When key parameter of BOD has reached > 150 mg/L, solution or appropriate alternative for PT SI-KSL's WWTP can be decided.

3. Results

3.1. Field Survey

PT SI-KSL has maximum production capacity 20 tonnes of fresh fruit bunches (FFB) an hour. This plant operates on an intermittent basis, when the demand for CPO increased and the price suited, the plant is operated for 24 hours. In reality, this plant often operates for 16 hours or 2 shifts. CPO production process begins from the sterilization process and continues with threshing palm fruit. Then the fruit has been crushed separately and pressed. After pressing, the impurified liquid oil is accomodated in the clarification unit. Clarification unit consists of three processes, namely filtration, deposition and evaporation. In the deposition unit, oil contained in the upper layer flows into evaporation process. The clarified oil reached certain standard of CPO is stored in storage tanks. Those process and maintaining machines generate wastewater in huge amount.

3.1.1. POME Quantity and Quality

The amount of wastewater produced about 60 percent of plant's capacity. When plant capacity is 20 tonnes FFB per hour, the number of POME is about 12 tonnes per hour. Production process is running in 2 shift, which is 16 hours per day. So, the number of wastewater is 192 m^3 per day.

The influent quality of wastewater (before fatpit and pre-sedimentation) and effluent quality (aerobic pond) is based on the laboratory analysis in Table 2.

Table 2. Characteristics of wastewater in PT SI-KSL (April 2014-March 2015)

Parameters	Average Influent Discharge (ppm)	Average Effluent Discharge (ppm)	Discharge Limits[a] (ppm)	Effectivity (%)
BOD_5	31 000	158	100	99.5
COD	25 000	320	350	98.7
TSS	15 000	78	250	99.5
Oil and Grease	13 500	24	25	99.8
Total Nitrogen	550	6	50	99
Ph	4.5	8.7	6-9	48.3

[a] [14]

3.1.2. POME Exsisting Treatment Plant

POME exsisting treatment plant was built in 1996 considering untreated wastewater with BOD/COD ratio of 0.5 and greater can be treated easily by biological means. Aerobic and anerobic treatment is suitable for POME treatment. In comparison of these 2 treatment methods, the anaerobic treatment can be regarded to be more suitable for POME treatment due to its lower energy consumption. Figure 2 shows flow diagram of POME treatment plant in PT SI-KSL.

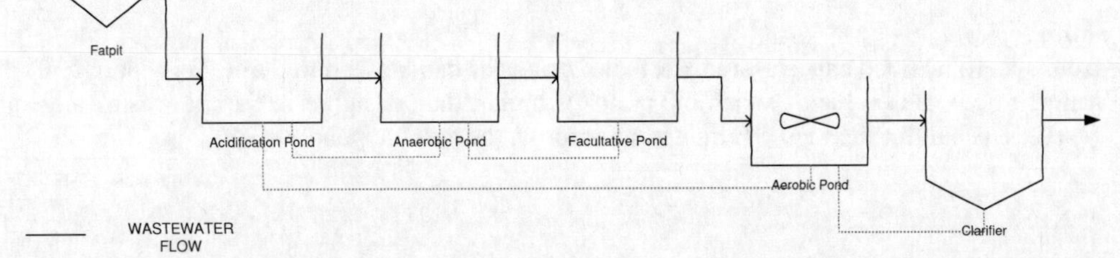

Figure 2. Flow diagram of exsisting POME treatment plant in PT SI-KSL

Wastewater from all CPO production process which flows into fatfit tank still has acidic characteristic (pH 4-4.7) and medium temperature (50-75°C). The oil contained in the wastewater will be easily separated in that condition. The fatfit dimension is 3 x 50 m^2 with 0.8 m depth. Hydraulic retention time (HRT) is 15 hours with remained BOD value is 25 000-35 000 ppm. Wastewater from fatpit tank flows into cooling tank. Capacity of cooling tank is 15 m^3 with 1.5 hours HRT. The temperature of wastewater decreases become 38°C which suitable for mesophilic bacteria.

The second wastewater process is anaerobic ponds which are consisted of 3 ponds and operated in sequence. The first is acidification anaerobic pond that has HRT 62 days and dimesion 78 x 38 x 4 m^3. BOD outlet value from this pond is 1300-2500 ppm. There are 2 circulation pipe coming from anaerobic and aerobic ponds in acidfication anaerobic pond. The second is called anaerobic pond with dimension 61 x 32 m^2 and 4 m depth. This pond HRT is 41 days, BOD outlet level is 500-1100 ppm. Circulating pump in this pond is streamed back to adification anaerobic pond. The third pond is facultative pond. This pond is an intermediate between anaerobic and aerobic pond with dimension 61 x 32 x 4 m^3. The HRT of this pond is up to 41 days. Estimated BOD outlet level is 200-300 ppm with pH is 5 to 6.

The third is aerobic process with dimension of 110 x 100 x 1.5 m^3 and HRT 86 days. Aerobic process implemented with floating aerator that served to supply oxygen into the wastewater and switched on for minimal 15 hours a day. BOD outlet level is 150 to 200 ppm.

The last pond is settling pond with dimension of 50 x 32 m^2 and 2.5 m depth. The HRT is 21 days with BOD outlet level is about 130-150 ppm. Wastewater discharge directly to secondary class of river based on Goverment Regulation no 82 year 2011, Tamiang River.

3.2. Technical Problems of POME Treatment in PT SI-KSL

3.2.1. Fatpit Pond

Fatpit pond should runs well with HRT of 15 hours. Only little amount of oil that can be refined. Operational process of fatpit definetely has been many of irregularities, based on observation revealed the following things :

- Although operating temperature is about 60°C in fatpit pond, the oil refinery process is still unoptimal. Wastewater flow rate through the pipeline is too fast, so many impurities had not enough time to settle. The real HRT become far below 15 hours. As a result, a lot of emulsified oil in the bottom layer flows into acidification pond.
- Reduced HRT occurs because of hardened sludge on the bottom of fatpit pond. The depth of fatpit is reduced to 50 cm. This condition led to reduce the volume of effective chamber and to decrease the settling process. Treatment management in fatpit pond has not done properly. Moreover, cleaning hardened sludge should be done periodically.

3.2.2. Anaerobic Pond

Pointing from anaerobic pond capacity, the HRT of 3 ponds as can be seen in **Table 3** is enough. Even organic loading rate (OLR) value of anaerobic ponds is higher than design criteria, the wastewater can be directly serves as bait for next POME treatment process, the aerobic pond.

Table 3. Comparation of design criteria and exsisting design of Anaerobic and Aerobic Pond in PT SI-KSL

Problems	Anaerobic Ponds			Design Criteria for Anaerobic Ponds[a]	Aerobic Ponds	Design Criteria for Aerobic Ponds[a]
	Acidification	Anaerobic	Facultative			
The Depths	4 m	4 m	4 m	5-7 m for anaerobic, 1-1.5 m for facultative	1.5 m	0.5-1 m
HRT	62 days	41 days	41 days	30-45 days for anaerobic, 15-20 days for facultative	86 days	24 days
COD removal Efficient	98.8 %			97.8%	50%	55%
OLR	1.67 kg COD/ m^3 day			1.4 kg COD/ m^3 day	-	-

[a][5]

3.2.3. Aerobic Pond
Aerobic pond made with 1.5 depth and considered for contacting air to wastewater. The deposition of sludge on all the corner of the pond will reduce HRT. Floating aerator at the time of observation is damaged and need to be fixed. In Table 3, HRT of aerobic pond 86 days exceed design criteria which is 24 days. Prolonged 86 days of HRT made treated wastewater become anaerobic. Aerobic process will be optimal using biofilter as media.

However, the aerobic pond was filled with scum which led to decrease HRT. Creation of scum on the surface and body be subject to buid-up at the bottom of the pond is main difficulties with this system, which lowered efficiency. So, regular desludging need to do either using submersible pumps or excavator to continue the preferred efficiency.

The 'dead space' will lead to volume reduction in each pond. The POME treatment plant of PT SI KSL is not need of energy, which means without stirring, pumping and desludging. The gases produced also smell sting of each the pond. Lack of biogas and CH_4 control is disadvantage of ponding system.

Aerobic ponds made up of earthen structures with no lining at the bottom. This condition will led to groundwater contamination. Aerobic ponds should be constructed with geotextile liner and used biofilter as a means of bacteria-growth.

3.3. Recommendation of POME Treatment in PT SI-KSL
According to advantages and disadvantages of the ponding system, it is necessary to redesign the process and plant at PT SI-KSL. The process will be dominated by biological processes, although there is physical and chemical process as well.

The first step is oil separation tank which has similar method with fatpit system. Because POME has acid characteristic, the second step is neutralization. The third step is closed tank anaerobic digester. Closed tank anaerobic digester are more efficient in removing COD (>95%) at a lower HRT of 17 days [15]. Closed system also generate biogases which are eventually either converted to elecricity in a biogas engine (gas holder) or burned into gas flare. This system is also equipped with circulation tank for optimizing and controlling the amount of sludge in digester.

Supernatant from the recirculation tank will flow to extended aeration tank and settling tank. Effluent from settling tank is discharged to land application, mainly for irrigation. Sludge treatment plant is using sand drying bed treatment.

Through digestion and extended aeration processes, the nutrient content particularly nitrogen and potassium will go down through leaching and settling of sludge solids at the plant underneath. This mechanism breaks down complex organic solids for enhancing nutrient uptakes by plants. BOD level will be decrease with these system.

4. Conclusion

Based on field survey and data analysis at PT SI-KSL, the aerobic ponding system processing capacity has decreased due to damaged floating aerator and sludge accumulation. This condition led to decrease COD and BOD removal efficiency, which is only 50%. BOD average is always higher than effluent discharge, 158 mg/L. The ponding system is not operated and maintained well. PT SI-KSL recommend to redesign their POME treatment plant.

References

[1] United Nations Environment Programme UNEP Global Environmental Alert Service GEAS 2011 Oil Palm Plantations: treats and opportunities for tropical ecosystems (USA: Center, EROS (Ed.), US Geological Survey (USGS))
[2] Mielke T 2013 Proc 2nd palm oil internet seminar or POINTERS (Kuala Lumpur: MPOC2013)
[3] Ahmed Y, Yaakob Z, Akhtar P, Sopian K 2015 J of Renewable and Sustainable Energy Reviews **42** 1260-78
[4] Tabassum S, Zhang Y, Zhang Z 2015 J of Cleaner Prod JCLP 02(2015)056
[5] Lam KM, Lee TK 2011 J of Biotechnology Advances **29** 124-141
[6] Liew LW, Kassim AM, Muda K, Loh KS, Affam CA 2015 J of Env Management **149** 222-35
[7] Poh EP, Chong MF 2009 J of Bio Resource Tech **100** 1-9
[8] Wu YT, Mohammad WA, Jahim JMd, Anvar N 2010 J of Env Management **91** 1467-90
[9] Biro Pusat Statistika 2014 Pertanian Indonesia dalam Angka tahun 2012-2014 (Jakarta: BPS) p 134
[10] Ahmad AL, Chan CY 2009 J Appl Sci **9** 3074-9
[11] Gobi K, Vadivelu VM 2013 J of Renewable and Sustainable Energy Reviews **28** 788-803
[12] Rahardjo PN 2005 Jurnal Air Indonesia 1 **1** 43-51
[13] Rahardjo PN 2006 Jurnal Air Indonesia 2 **1** 66-72
[14] Decree of Indonesian Ministry of Environment and Forestry No 5/ 2014 attached III about Wastewater Quality Standards
[15] Yacob S, Hassan MA, Shirai Y, Wakisaka M, Subash S 2006 J of Sci Total Env 366 **1** 187-196

IOP Conf. Series: Earth and Environmental Science **65** (2017) 012049 doi:10.1088/1755-1315/65/1/012049

The design of financial recording system in industrial bio-briquette of Ramie (*Boehmeria nivea*) *decortication waste with design thinking approach*

R Irianto, D Purnomo [2], Prima S [3], A Wulandari[4]

[1] Department of Agricultural Engineering and Bio-system, Faculty of Industrial Technology of Agriculture, Padjadjaran University, Bandung, Indonesia
[2] Department of Agroindustrial of Technology, Faculty of Industrial Technology of Agriculture, Padjadjaran University, Bandung, Indonesia
[3] Department of Accounting, Faculty of Economic and Business, Padjadjaran University, Bandung, Indonesia
[4] Department of Biology, Faculty of Mathematic and Natural Science, Padjadjaran University, Bandung, Indonesia

Email : ryaniriantoo@gmail.com

Abstract. The production process of ramie (*boehmeria nivea*) fibers generates waste which contents 5.95 to 7.83% ash; 1.88 to 2.87% silicate; 30.67 to 31.08% lignin; 33.81 to 35.99% alpha cellulose; 62.95 to 63.78% holoselulosa; 17.43 to 18.14% pentosan, which can be used as raw material of bio-briquette. Those potential can be used to generate a business opportunity, such as industrial bio-briquette of ramie decortication waste. The purpose of this research is to create accounting information which could present an income statement that is easily applied on industrial bio-briquette of ramie decortication waste. This research use descriptive analysis method with design with design thinking approach to gather the information through depth observation on human being as the object to achieve the purpose. The result in this research is financial recording system of industrial bio-briquette of ramie decortication waste in a desktop application. The system is integrated with production activities according to the needs of accounting information particularly at managerial production. The existing applications creates information in the form of financial operations which can be used as a factor in decision-making.

1. Introduction

Nowadays, businesses agribusiness ramie plant (*Boehmeria nivea*) utilize its fiber as the raw subtitute material for cotton. On production process generates decorication waste which data is based on Saroso *et al.* [1] has ash content 5,95-7,83%; silicate 1.88 to 2.87%; lignin from 30.67 to 31.08%; alpha cellulose from 33.81 to 35.99%; holoselulosa 62.95 to 63.78%; pentosan 17.43 to 18.14% in which based on Prawirohatmojo [2] can be used as a raw material for bio-briquette because it contains high lignin and celulose to produce carbon. The potential use of such waste can be used as ramie bio-briquette industry and to integrate the business of ramie into ramie industry center.

Content from this work may be used under the terms of the Creative Commons Attribution 3.0 licence. Any further distribution of this work must maintain attribution to the author(s) and the title of the work, journal citation and DOI.
Published under licence by IOP Publishing Ltd

To develop ramie bio-briquette industry, one of the pillars that can be built for business fundamental is by fixing the financial management and systems that is used in these industries. The basic problem in management on a small and medium enterprises (SMEs) is about the owner which is not using the proper financial recording system in their industry [3]. Financial recording when conducted with carefulness and discipline, will help the industry in terms of financial management so that the industry will run according to its business model development.

The process of designing activities according to financial records in Ervillia [4] should include: characteristics of the raw materials used; company's assets; as well as the characteristics of the users of financial reports. These things should be known in depth so that financial recording system can be designed properly using design thinking approach. Design thinking is a mindset based on the designer glasses when solving the problem focused on human oriented approach [5]. The principle of design thinking by Glinski in Purnomo [5], Collaborate in the processes of systematic human-centered as a user through a planned process that produces behavioral changes and conditions in line with expectations. Stages of the design process with design thinking is described as follows.

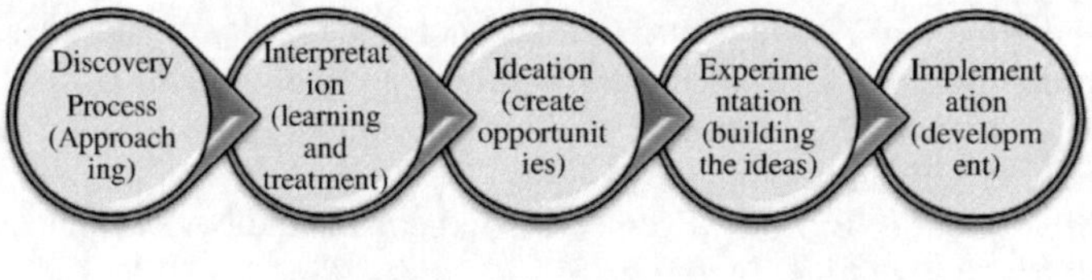

Figure 1. The Design Thinking Process
Source: Purnomo [5]

2. Method

The method used is descriptive analysis with design thinking approach to collect data by performing in-depth observation on the characteristics of SMEs in conducting financial records. Design thinking approach to collaborate intuitive data and analytics data to design the appropriate financial recording system for ramie bio-briquette industry.

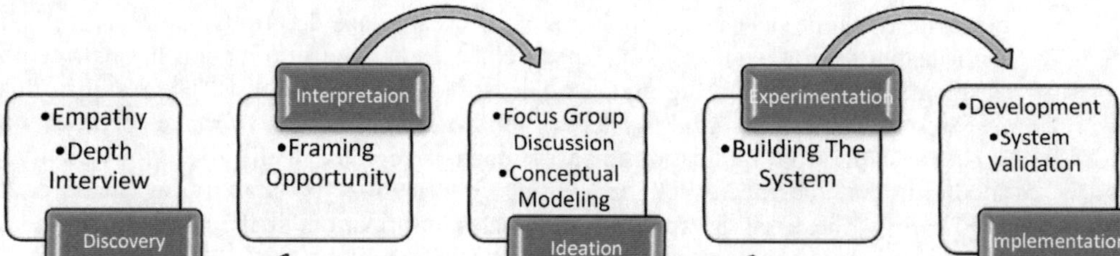

Figure 2. Procedure of Research using Design Thinking Approach

3. Results and Discussion

3.1. Discovery
Discovery stage is done by identifying the overall ramie business activities ranging from the treatment, ramie fiber decortication process to produce decortication waste and utilization to become ramie bio-briquette. Discovery stage requires an approach to build empathy so the result is in the following:

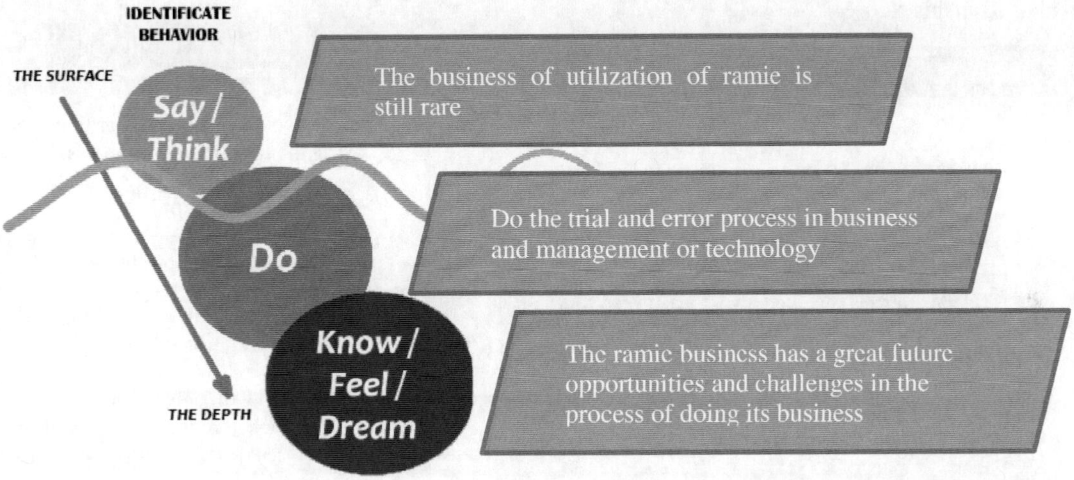

Figure 3. The Result Empathy Build of Utilization of Plant Operations Ramie to Bio-briquette

In the result of empathy building is known that what is felt and dreamt from the ramie businesses, resulting in its big efforts as well as the learning process. It covers the process technology, tools and machinery, product quality and governance to marketing products based on ramie.

On discovery process also found analytics and intuitive facts which can be used as a as a reference to find solution by collaborating these facts.

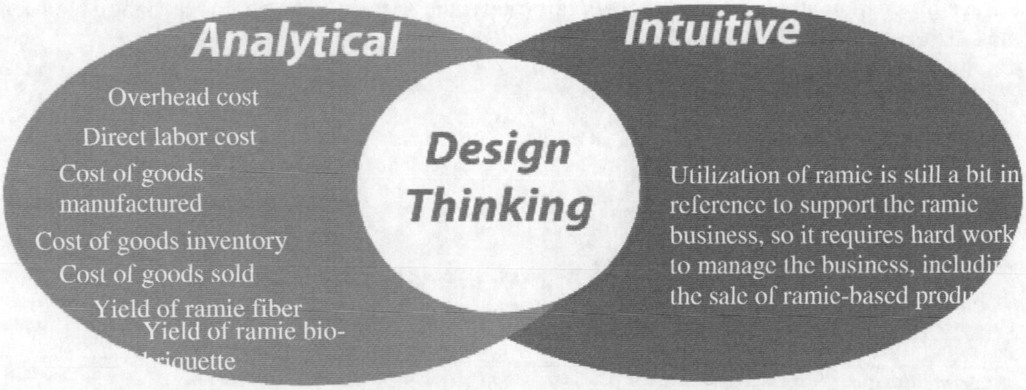

Figure 4. Facts Found of Ramie Bio-briquette industry

3.2. Interpretation
On the discovery process is obtained the characteristics of ramie bio-briquette industry which caused on the recording activities, including: 1) Business management utilization of ramie is not structured

yet which made the businesses spend a lot of time in seeking ramie bio-briquette industry so it is not motivated to do a proper financial record activities; 2) Tools are needed for recording that is able to become a trigger to awareness for the businesses in recording business activities. 3) The needed tools should be able to meet the variable of success in design thinking.

3.3. Ideation

The development of ideas is done by considering the variable of success in design thinking which explained in this figure.

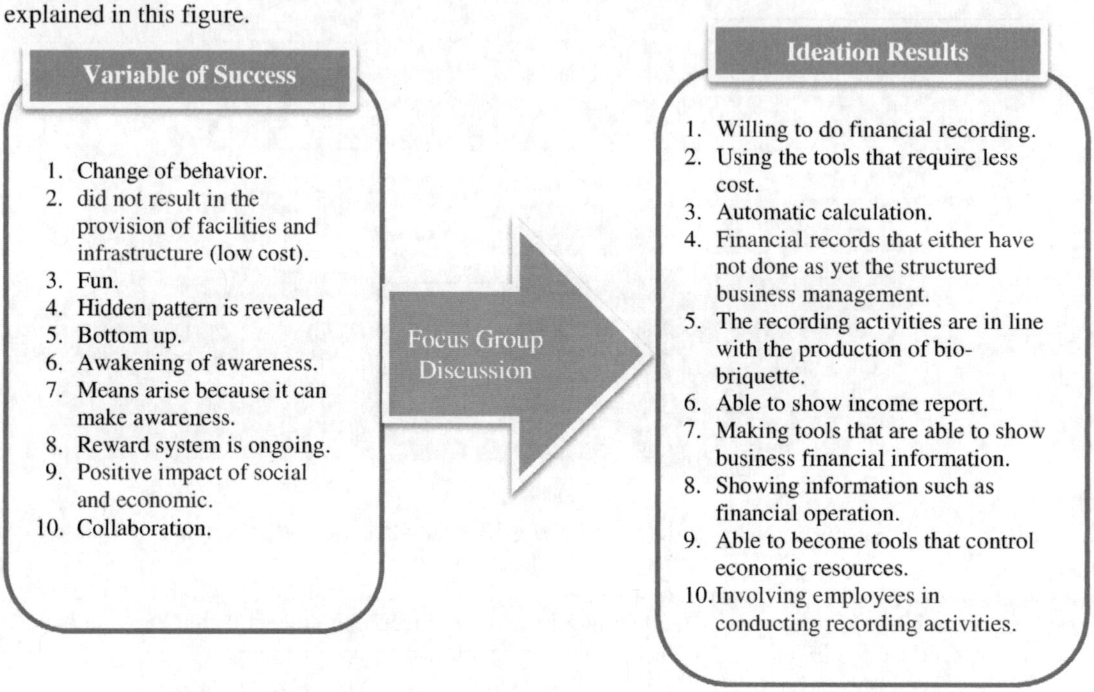

Figure 5. Ideation stage using the variable of success in design thinking

At the stage of ideation also make the process of prototyping is to visualize the ideas that have been built next to the workings of the tools that is understandable.

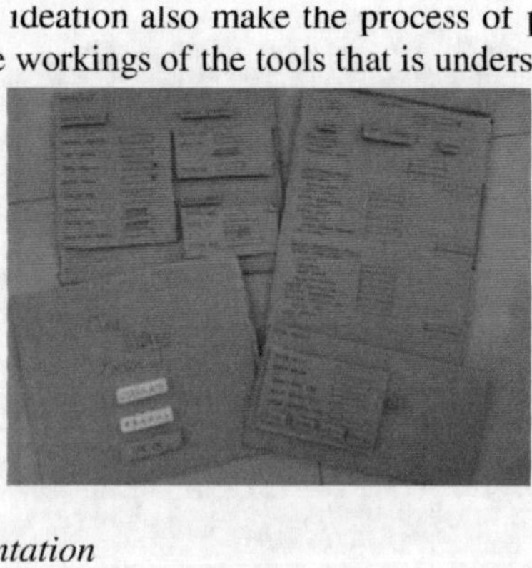

Figure 6. Prototyping for Financial Recording Tools

3.4. Experimentation

Experimentation stages is realizing the result of prototyping in the ideation stage. Prototyping is built with Microsoft visual studio as the document flow mechanism and Microsoft SQL Server as the access in relational data management system. Prototyping is built into the system in the financial records to production and sales, and culminate in the presentation of income statement.

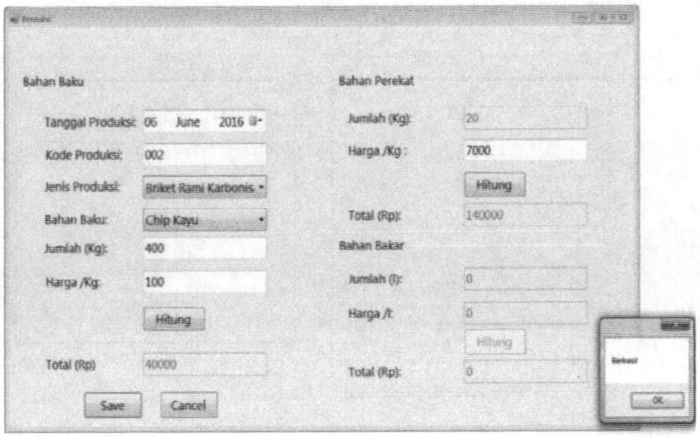

(a)

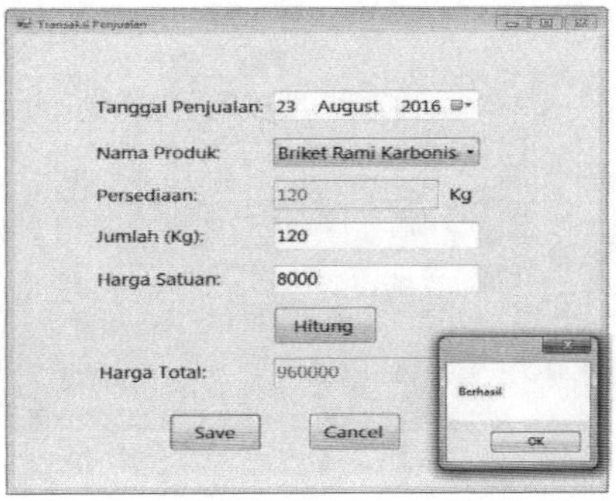

(b)

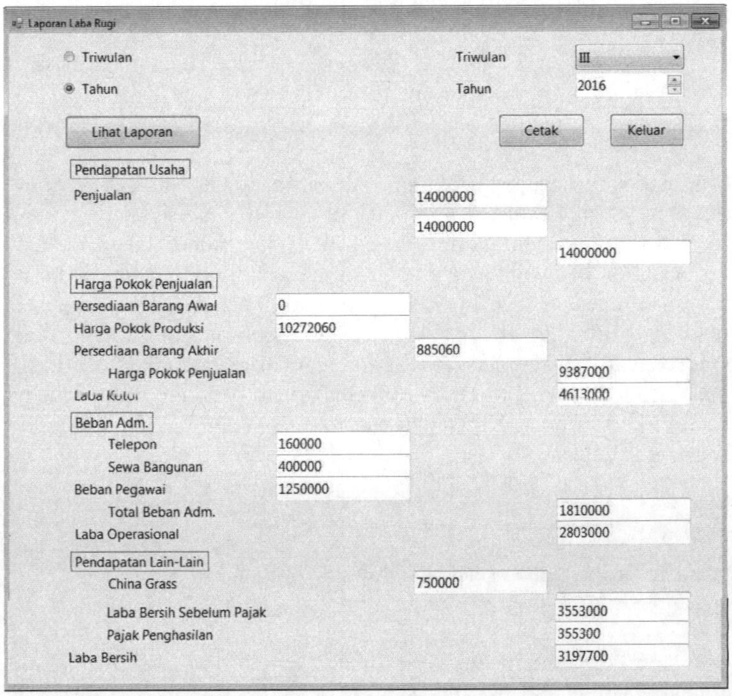

(c)

Figure 7. The Result of Experimentation from Prototyping System: (a) Recording on production, (b) The result of recording sales (c) income statement on specified period.

3.5. Implementation

The systems have been built and then are implemented on ramie bio-briquette industry to determine appropriate parameters that suitable to variable of success in design thinking. The system was implemented in record business activities in the second quarter period of June 2016 and produce variables that have been classified in the following table.

Table 1. The Parameter of success in implementation of financial recording system

Change	Deficiency
1. There has been a tools as a means to financial recording activities. 2. The emergence of collaboration in recording activities with employee. 3. The existence of a bottom-up process that groove recording in conjunction with the production flow. 4. Development of awareness in knowing the information income statement operating.	1. In overall the system is not making many positive impacts on economic because the system is unable to show the financial position of the business. 2. The absence of significant changes in behavior because it has not lead to positive economic impact. 3. The reward system is not running because there is no balance sheet features that present financial information business. 4. The revealed hidden pattern is the businesses need a financial recording system that is capable to show information about financial position (balance sheet). 5. The financial recording system has not been pleasant because of its hidden pattern requirement's is not met.

4. Conclusion

The financial recording system in industrial bio-briquette of ramie decortication waste that is built based on the characteristic of the user using design thinking approach have produced a system such as desktop application. The financial recording system is integrated with the ramie bio-briquette production activities and generate information of income statement. In the system implementation, it is not able to change the behavior in awareness to do financial recording. This happened because some of the variable of success in design thinking are not fulfilled yet, such as not showing the overall positive impact on economic; reward system is not ongoing; and the financial recording system that have not been fun because of the hidden pattern that has been revealed is not right. Such as the businesses of ramie bio-briquette industry need a financial recording system that is capable of presenting information of financial position (balance sheet).

References

[1] Saroso B A, Sastrosupadi B W, Winarto and Sugesty 1999 *Pemanfaatan Limbah Dekortikasi Rami untuk Pulp Kertas* (Indonesia : Report of Researched Balittas, Malang)
[2] Prawirohatmodjo S 2004 *Sifat-sifat Fisika Kayu* (Indonesia : Forestry faculty of Gadjah Mada University, Yogyakarta)
[3] Haryani E 2012 accounting system for small business in Indonesia *Journal of Faculty Information Technology* (Indonesia : Universitas Kristen Satya Wacana)
[4] Ervillia P 2009 *Analisis Perumusan dan Penerapan Sistem Akuntansi pada Usaha Kecil Menengah Studi Kasus UKM Waroeng Cokelat Bogor* (Indonesia : Bogor Agricultural University)
[5] Purnomo D 2013 Konsep design thinking bagi pengembangan rencana program dan pembelajaran kreatif dalam kurikulum berbasis kompetensi *Journal of Konferensi Nasional "Inovasi dan Technopreneurship* (Indonesia : RAMP Bogor Agricultural University)

IOP Conf. Series: Earth and Environmental Science **65** (2017) 012050 doi:10.1088/1755-1315/65/1/012050

The Potential of Palm Oil Waste Biomass in Indonesia in 2020 and 2030

E Hambali[1,2] and M Rivai[1]

1) Surfactant and Bioenergy Research Center, Bogor Agricultural University
Jl. Raya Pajajaran No. 1, Bogor, Jawa Barat, Indonesia
2) Department of Agroindustrial Technology, Faculty of Agricultural Technology, Bogor Agricultural University
Jl. Raya Darmaga Kampus IPB Darmaga, Bogor, Jawa Barat, Indonesia
Email: erliza.h@gmail.com

Abstract. During replanting activity in oil palm plantation, biomass including palm frond and trunk are produced. In palm oil mills, during the conversion process of fresh fruit bunches (FFB) into crude palm oil (CPO), several kinds of waste including empty fruit bunch (EFB), mesocarp fiber (MF), palm kernel shell (PKS), palm kernel meal (PKM), and palm oil mills effluent (POME) are produced. The production of these wastes is abundant as oil palm plantation area, FFB production, and palm oil mills spread all over 22 provinces in Indonesia. These wastes are still economical as they can be utilized as sources of alternative fuel, fertilizer, chemical compounds, and biomaterials. Therefore, breakthrough studies need to be done in order to improve the added value of oil palm, minimize the waste, and make oil palm industry more sustainable.

1. Background to the Study

Oil palm (*Elaeis Guineensis*) is a plant producing oil mostly produced and consumed all over the world. In Indonesia, this plant was first introduced in 1911 and since 1970 it has been developed fast as one of the most important commodities. Data of Directorate General of Estate Crops, Ministry of Agriculture of Republic Indonesia showed that in 1970, oil palm plantations were run only by state-owned and private enterprises while small-scale farmer plantations were found to operate in 1979. In 1970, oil palm plantation area was only 133,298 hectares and 11.3 million hectares in 2015. Compatible climate and soil condition have made oil palm plantation extensively distributed in 22 provinces spreading from the western to the eastern parts of Indonesia. In 2015, with an oil palm plantation area of 11.3 million hectares, Indonesia produced 37.5 million tons of palm oil consisting of 31.3 million tons of crude palm oil (CPO) and 6.2 million tons of palm kernel oil (PKO) [1]. Provinces in Sumatra and Kalimantan share the largest oil palm areas in Indonesia.

As an estate crop commodity, oil palm is very productive and able to produce oil seven times higher than rapeseeds (*Brassica napus)* and eleven times higher than soybean per hectare. Since 2004, palm oil has shared the highest in vegetable oil world market with a total production of 30 million tons and an average growth rate of 8% per year. These figures were higher than soybean production of 25 million tons and average growth rate of 3.8% per year.

Content from this work may be used under the terms of the Creative Commons Attribution 3.0 licence. Any further distribution of this work must maintain attribution to the author(s) and the title of the work, journal citation and DOI.
Published under licence by IOP Publishing Ltd

During replanting activity in oil palm plantation, biomass including palm frond and trunk are produced. During harvest, palm bunches and fronds are cut off. In palm oil mills, during the conversion process of fresh fruit bunches (FFB) into crude palm oil (CPO), several kinds of waste including empty fruit bunch (EFB), mesocarp fiber (MF), palm kernel shell (PKS), palm kernel meal (PKM), and palm oil mills effluent (POME) are produced. The production of these wastes is abundant and can be found in 65% of provinces in Indonesia as oil palm plantation area.

In addition to CPO and PKO as the main products, wastes resulted from palm oil processing industry still have economical values as they can be used as sources of alternative fuel, fertilizer, chemical compounds, and biomaterials. Based on this notion, this review aims to assess the potential of wastes resulted from national oil palm plantations and palm oil processing industries and providing information on the opportunity of further utilization of them.

1.1. Methods of Data Collection and Analysis

Secondary data were used in this study. Data were collected from various related institutions including The Indonesian Central Bureau of Statistics (BPS). Ministry of Agriculture, Regional Development Planning Board (BAPPEDA), Indonesian Palm Oil Association (GAPKI), research institutions, and other related institutions.

Descriptive and forecasting data analysis methods were used. Descriptive analysis was used to identify, explain, and describe current condition and potential of oil palm development and oil palm waste availability. This analysis was done by providing explanation and detailed projection by using graphic and cross tabulation techniques and data processing by using statistical methods. Forecasting method was used to estimate the availability of oil palm wastes (MF, PKS, EFB, POME, and PKM) in 2016, 2020, and 2030. The technique used in this method was trend analysis which created a mathematical equation line to make projection in the future. A regression model was developed.

2. Area Coverage and Production of National Oil Palm Plantations

During the period of 1980 to 2015, the area of oil palm plantations in Indonesia grew by, on average, 14.26% per year from 294,560 to 11,300,370 hectares. In the same period, the national palm oil production increased by, on average, 17.46% per year from 849,121 tons (721,172 tons of CPO and 127,949 tons of PKO) to 37,541,167 tons (31,284,306 tons of CPO and 6,256,861 tons of PKO). Detailed area coverage and production are showed in figure 1.

In an oil palm plantation, there are three types of plants, namely (a) unproductive plants (UPP) which are not in their productive age, (b) productive plants (PP) which are fruit producing plants, and (c) old/damaged plants (OP/DP) which already pass their productive age or no longer produce fruits. In 2002-2015, UPP was found to occupy 35.11% of oil palm plantation areas. In the future, there will be an increased oil palm production as more plants will be in their productive age. The development of oil palm plantation area, CPO production, and productivity of national oil palm plantations in 2002-2015 is listed in table 1.

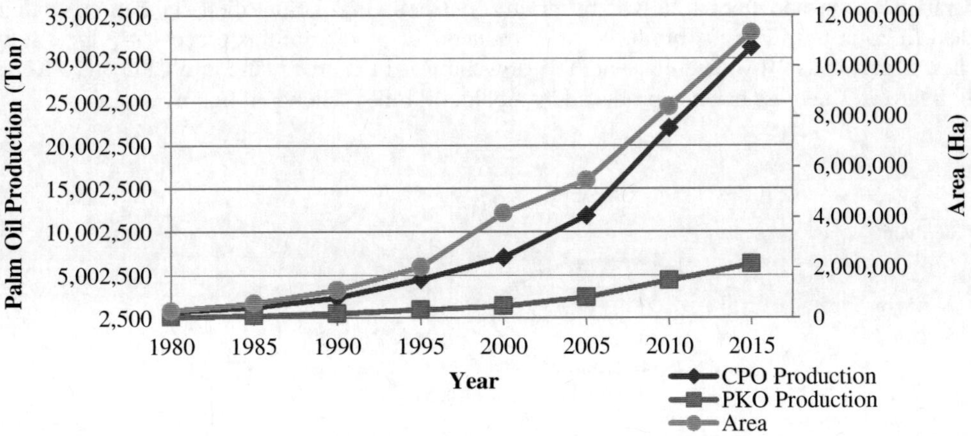

Figure 1. Graphic of development of plantation area coverage and production of palm oil (CPO and PKO) in 1980 – 2015
Source: [1]

Table 1. Area coverage, production, and productivity of oil palm plantations in Indonesia in 2002-2015

Year	Area Coverage (Ha)			Palm Oil Production/CPO (Tons)	Productivity (Kg CPO/Ha)
	PP	UPP	OP/DP		
2002	3,307,419	1,669,376	90,267	9,622,345	2,909
2003	3,428,580	1,780,088	74,889	10,440,834	3,045
2004	3,823,324	1,393,614	67,785	10,830,389	2,833
2005	4,054,683	1,330,454	68,680	11,861,615	2,925
2006	4,960,529	1,564,186	70,199	17,350,848	3,498
2007	4,881,335	1,791,095	94,407	17,664,725	3,619
2008	5,122,275	2,142,425	99,497	17,539,788	3,424
2009	5,541,422	2,229,469	102,402	19,324,293	3,487
2010	6,108,275	2,169,042	108,078	21,958,120	3,595
2011	6,550,800	2,332,886	109,138	23,096,541	3,526
2012	6,989,653	2,472,631	110,432	26,015,518	3,722
2013	7,856,254	2,454,080	154,685	27,782,004	3,536
2014	8,129,570	2,430,276	194,956	29,278,190	3,601
2015*	8,503,784	2,567,633	228,953	31,284,306	3,679

Source: [1]

By comparing national palm oil production (CPO) to the area coverage, it was known that in general there was an increased national oil palm productivity from 2,909 kg/ha in 2002 to 3,679 kg/ha in 2015.

3. Oil Palm Processing and Distribution of National Palm Oil Mill Location

Palm oil mill is a place where FFB was processed into crude palm oil (CPO) and palm kernel oil (PKO). The processing of FFB into CPO involves cooking, threshing, digestion, pressing, settling, purification, and drying. Pressing process results in a waste product, namely nut/fiber. Further process is the separation of nut and fiber followed by drying, nut cracking, kernel/shell separation, drying, pressing, and filtering until PKO is produced. By-products obtained from this process include empty fruit bunches, fiber, shell, PKM, and effluent. The flowchart of FFB processing into CPO and PKO is depicted in figure 2. The mass balance produced by a palm oil mill is shown in figure 3.

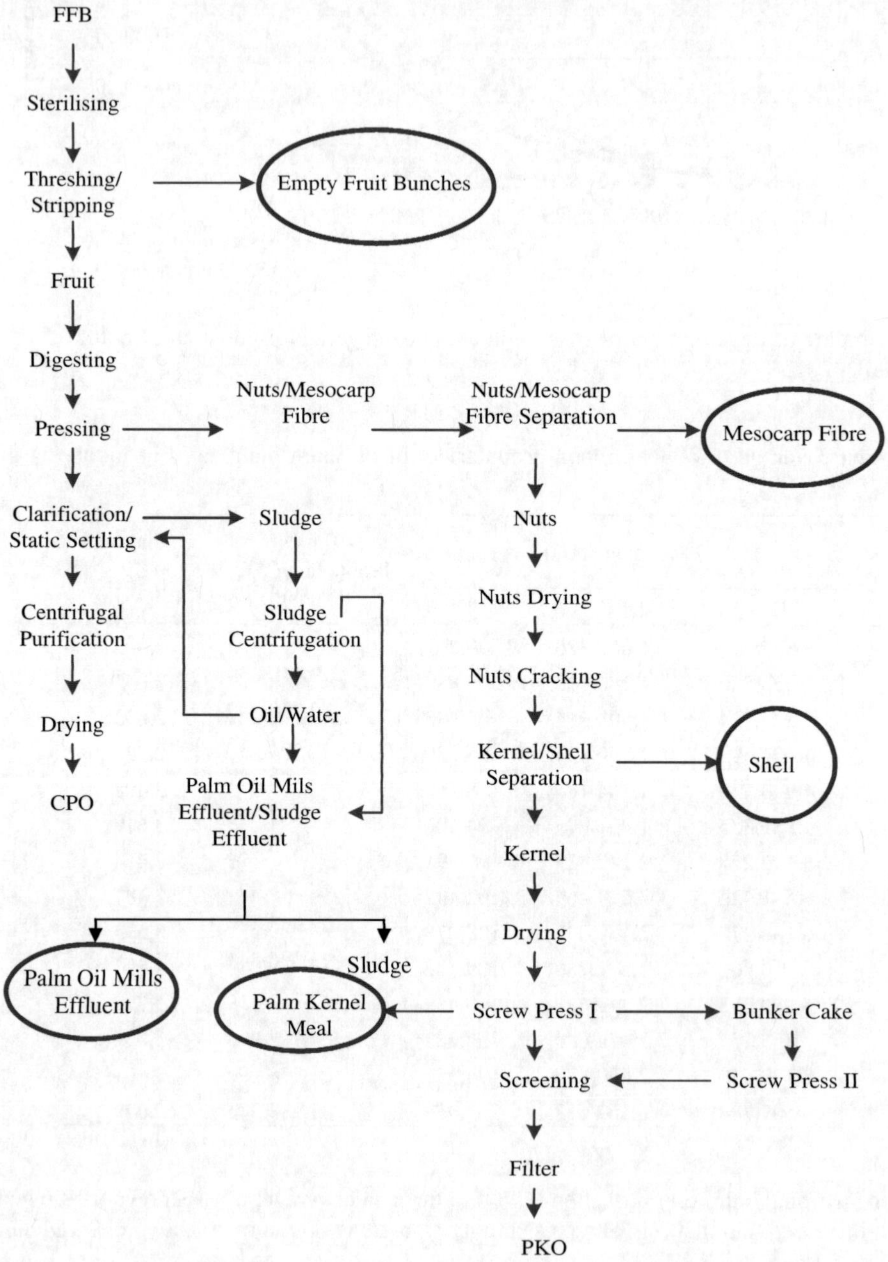

Figure 2. Flow chart of FFB processing into CPO and PKO

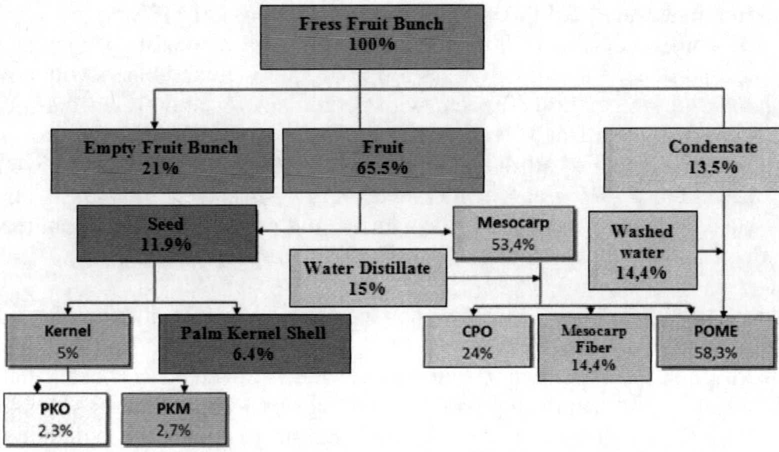

Figure 3. Mass balance of oil palm processing [2]

Similar to the distribution of oil palm plantations, national palm oil mills are distributed all over 22 provinces. Riau province has the most palm oil mills of 140 units with production capacity of 6660 tons FFB/hour followed by North Sumatra with 92 units and production capacity of 3815 tons FFB/hour and West Kalimantan with 65 units and production capacity of 5475 tons FFB/hour. In all, there are 608 units of national palm oil mills with production capacity of 34,280 tons FFB/hour. The number of national palm oil mills and their production capacity in 2014 are listed in table 2.

Table 2. National Palm Oil Mills and Their Production Capacity in 2014

No	Province	Number of Palm Oil Mills (Unit)	Production Capacity (Tons FFB/hour)
1	NAD	25	980
2	North Sumatra	92	3,815
3	West Sumatra	26	1,645
4	Riau	140	6,660
5	Riau Islands	1	40
6	Jambi	42	2,245
7	South Sumatra	58	3,555
8	Bangka Belitung	16	1,235
9	Bengkulu	19	990
10	Lampung	10	375
11	West Java	1	30
12	Banten	1	60
13	West Kalimantan	65	5,475
14	Central Kalimantan	43	3,100
15	South Kalimantan	15	770
16	East Kalimantan	29	1,545
17	Central Sulawesi	7	590
18	South Sulawesi	2	150
19	West Sulawesi	6	260
20	Southeast Sulawesi	3	260
21	Papua	3	140
22	West Papua	4	360
	Total	**608**	**34,280**

Source: [3]

4. Waste Production in National Oil Palm Plantations and Palm Oil Mills

Oil palm frond (OPF) consists of about 100 leaflets and each leaflet consists of lamina and midrib, central rachis, petiole, and petiole base. Leaflet has a size of 55-65 cm and tapers with a width of 2-4 cm. A mature frond has a size of up to 7.5 cm with petiole size of about one-fourth of the frond length. Spikes are found all over a frond. A frond has a convex shape and the base part is wider and harder than the tip part. The length of a frond can reach 5-9 m. OPF is a lignocellulosic material high in fiber consisting of cellulose (51%) and hemicellulose (15%) [4]. OPF biomass is obtained from replanting activity when oil palm plants are 25-30 years old and pruning activity. From replanting and pruning activities, the resulted OPF biomass is 14.4 and 10.4%, respectively [5].

Oil palm is a monocotyledon plant having a single cotyledon and therefore, its trunk has no cambium. In general oil palm plants grow with no branches. Branches are found only in plants growing with abnormality. Oil palm trunk grows straight (phototrophic) and is covered with leaf sheath. The base part of the trunk is bigger in size than the top part. Until the age of 3 years, the trunk is not visible as it is still wholly covered with leaf sheath. The height of oil palm trunk grows at the rate of about 45 cm per year depending on plant age, nutrient availability, soil condition, climate, and plant genetic. Cultivated oil palm plant can reach a maximum height of 15-18 m while wild oil palm plant can be up to 30 m height. Tenera is palm varieties commonly grown in Indonesia with average trunk length (in rejuvenation phase) of about 17-20 m, diameter 60-85 cm, volume 0.9-3 m^3, and a number of plants per hectare 143. The tenera has shell 0.5-4mm thick, 60-90% mesocarp and a fiber ring. Beirnaert and Vanderweyen defined the tenera as fruit with a shell thickness of 0.5-2mm [6]. Considering the life cycle of oil palm plant with about 25 years productive period, 4%/year rejuvenating rate, and 11.3 million hectares oil palm plantation area all over Indonesia (2015), it can be estimated that there will be about 34.13 million tons/year of trunk waste production. In old oil palm plants, the trunk base has a bigger size. The trunk is usually covered with leaf sheath which will fall off when the plant reaches the age of ten years.

Table3. Total national FFB and oil palm waste production in 2002-2015

Year	Production (Tons)				
	FFB	EFB	MF	PKS	POME
2002	40,093,104	9,417,663	5,369,331	2,587,503	22,718,497
2003	43,503,475	10,218,742	5,826,053	2,807,599	24,650,962
2004	45,126,621	10,600,011	6,043,428	2,912,353	25,570,707
2005	49,423,396	11,609,301	6,618,858	3,189,655	28,005,447
2006	72,295,200	16,981,770	9,681,886	4,665,741	40,965,606
2007	73,603,021	17,288,970	9,857,032	4,750,144	41,706,674
2008	73,082,450	17,166,691	9,787,316	4,716,548	41,411,696
2009	80,517,888	18,913,237	10,783,081	5,196,412	45,624,939
2010	91,492,167	21,491,038	12,252,774	5,904,663	51,843,443
2011	96,235,588	22,605,243	12,888,020	6,210,790	54,531,271
2012	108,397,992	25,462,129	14,516,828	6,995,720	61,423,019
2013	115,758,350	27,191,040	15,502,539	7,470,738	65,593,718
2014	121,992,458	28,655,400	16,337,421	7,873,071	69,126,235
2015	130,351,275	30,618,843	17,456,846	8,412,527	73,862,704

IOP Conf. Series: Earth and Environmental Science **65** (2017) 012050 doi:10.1088/1755-1315/65/1/012050

Meanwhile, other types of waste including oil palm frond and trunk are produced in oil palm plantation activities. Frond waste is produced every time the frond part is pruned during harvest time or biyearly maintenance. Oil palm frond biomass is produced by 6.3 tons per year. In 2015, with 11.3 million hectares of oil palm plantation area, 71.2 million tons frond biomass was produced. Total national oil palm frond production as a waste during the period of 2000-2015 is listed in table 4.

Table4. Total National Oil Palm Frond Production in 2000-2015

Year	Area Coverage (Ha)	Frond Production (Tons)
2002	5,067,058	55,616,029
2003	5,283,557	57,992,322
2004	5,284,723	58,005,120
2005	5,453,817	59,861,095
2006	6,594,914	72,385,776
2007	6,766,836	74,272,792
2008	7,363,847	80,825,585
2009	7,873,294	86,417,275
2010	8,385,394	92,038,085
2011	8,992,824	98,705,236
2012	9,572,715	105,070,120
2013	10,465,020	114,864,060
2014	10,754,801	118,044,696
2015	11,300,370	124,032,861

Oil palm trunk biomass is produced from the regeneration of old plants (25-30 years old) which are classified as old/damaged and unproductive plants. By using a rejuvenation rate of 4% of total plantation area, it was estimated that in 2015, the rejuvenation area was 452,015 hectares. Based on an assumption that there were 143 plant populations per hectare of oil palm plantation area, in 2015, the potential oil palm trunk waste was estimated to reach 34,13 million tons trunks.

Table 5. Oil palm trunk biomass production during plant rejuvenation

Year	Rejuvenation Area (Ha)	TrunkBiomass Production (Tons)
2002	202,682	15.302.515
2003	211,342	15.956.342
2004	211,389	15.959.863
2005	218,153	16.470.527
2006	263,797	19.916.640
2007	270,673	20.435.845
2008	294,554	22.238.818
2009	314,932	23.777.348
2010	335,416	25.323.890
2011	359,713	27.158.328
2012	382,909	28.909.599
2013	418,601	31.604.360
2014	430,192	32.479.499
2015*	452,015	34.127.117

5. Projection National Oil Palm Waste Production in 2016, 2020, and 2030

Projected production of national EFB, MF, PKS, POME, oil palm frond, and oil palm trunk was made based on data obtained from 2002 to 2015. By using a trend analysis, the growth trend of EFB, MF, PKS, POME, oil palm frond, and oil palm trunk production in several years to come was estimated. This projection was used to determine the potential national production of oil palm wastes in 2016, 2020 and 2030.

Therefore, in 2016, there would be a production of 31,380,742 tons EFB, 17,891,231 tons MF, 8,621,859 tons PKS, 75,700,657 tons POME, 128,914,621tons frond, and 59,722,455tons trunk. In 2020, there would be a production of 37,816,105 tons EFB, 21,560,251 tons MF, 10,389,975 tons PKS, 91,224,865 tons POME, 128,914,621tons frond, and 59,722,455tons trunk. In 2030 a total production of 53,904,512 tons EFB, 30,732,801 tons MF, 14,810,264 tons PKS, 130,035,387 tons POME, 128,914,621tons frond, and 59,722,455tons trunk was estimated (Table 29). The same estimated yearly production of frond and trunk from 2016 to 2020 and 2030 is caused by the moratorium of land use extension for oil palm plantations as instructed by the President of Republic Indonesia in 2016.

The potential production of solid and liquid wastes from oil palm industry in the future is huge. Therefore, efforts to utilize them as product with economic value need to be done. POME can be utilized to produce biogas/biomethane for power generator. The electricity produced can then be used fulfill the power need of the mill and people living around the mill location. The liquid waste resulted from the use of POME for biogas/biomethane processing is still rich in nutrients which can be further used as fertilizer, land application, and source of nutrient-rich-medium for microalgae growth. CO_2 resulted from this process can also be used for microalgae growth. Palm Kernel Shell (PKS) and Mesocarp Fiber (MF) are commonly used as boiler fuel to drive production machinery and material to surface road around a palm oil mill. PKS can be processed into biopellet to be used as fuel for self-use or export. Oil palm frond, trunk, and empty bunch can be used as mulching material to help maintain soil humidity and inhibit weed growth, raw material for composting, material for covering plant vegetation, and a source of fiber for various composite products.

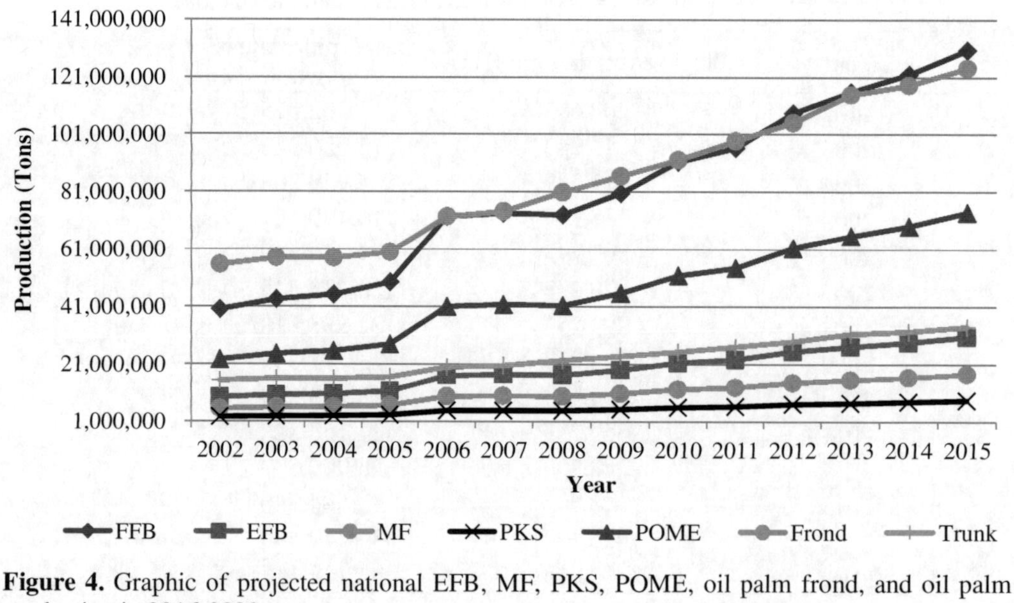

Figure 4. Graphic of projected national EFB, MF, PKS, POME, oil palm frond, and oil palm trunk production in 2016-2030

Oil palm biomass in the form of oil and its wastes are in abundance in Indonesia. Therefore, breakthrough studies need to be conducted in order to improve the added value of oil palm, minimize the waste, and make oil palm industry more sustainable

6. Conclusions

In 2030, it was estimated that there would be a production of 54 million tons EFB, 31 million tons MF, 15 million tons PKS, 130 million tons POME, 115 million tons oil palm frond, and 59.7 million tons oil palm trunk. Further research and technology development need to be done in order to improve the added value of oil palm waste biomass.

Acknowledgements
Funding, facility and equipment assistance from Surfactant and Bioenergy Research Center IPB were acknowledged.

References

[1] Directorate General of Estate Crops Ministry of Agriculture 2016 Tree crop estate statistics of indonesia 2015-2016 palm oil (Jakarta: Directorate General of Estate Crops)
[3] Directorate General of Estate Crops Ministry of Agriculture of Republic Indonesia 2015 Tree crop estate statistics of indonesia 2014-2015 palm oil (Jakarta: Directorate General of Estate Crops)
[2] Erliza H 2010 Role of agro-industry process technology in the development of oil palm downstream industry (Bogor: PT Penerbit IPB Press)
[4] Khalil H P S A, Jawaid M, Hassan A, Paridah M T and Zaidon A 2012 Oil palm biomass fibres and recent advancement in oil palm biomass fibres based hybrid biocomposit Composites and their aplication *INTECH* Chapter 8 187-220
[5] Muzzammil M A 2010 OIL Palm trunk (opt) as an alternative cellulosic material for brown paper production (Malaysia: Faculty of Chemical & Natural Resources Engineering, Universiti Malaysia Pahang.)
[6] Beirnaert A and Vanderweyen R 1941 Contribution à l'étude génétique et biométrique des variétés d'Elaeis guineensis Jacq *Pub. de l'INEAC Série Scientifique* **27**

IOP Conf. Series: Earth and Environmental Science **65** (2017) 012051 doi:10.1088/1755-1315/65/1/012051

Torrefaction study for energy upgrading on Indonesian biomass as low emission solid fuel

R Alamsyah[1], N C Siregar [1] and F Hasanah[1]

1 Center for Agro-Based Industry-CABI , Jl. Ir. H. Juanda No. 11, Bogor, Indonesia

Email : rizalams@yahoo.com

Abstract. Torrefaction is a pyrolysis process with low heating rate and temperature lower than 300°C in an inert condition which transforms biomass into a low emission solid fuel with relatively high energy. Through the torrefaction process biomass can be altered so that the end product is easy to grind and simple in the supply chain. The research was aimed at designing torrefaction reactor and upgrading energy content of some Indonesian biomass. The biomass used consist of empty fruit bunches of oil palm (EFB), cassava peel solid waste, and cocopeat (waste of coconut fiber). These biomass were formed into briquette and pellet form and were torrified with 300°C temperature during 1.5 hours without air. The results of terrified biomass and non-torrefied biomass were compared after burning on the stove in term of energy content and air emission quality. The result shows that energy content of biomass have increased by 1.1 up to 1.36 times. Meanwhile emission air resulted from its combustion was met with Indonesian emission regulation.

1. Introduction

Depletion of oil reserves and global warming are problems faced by the world today. Many of the possible solutions are already in place, but they have not been widely adopted [1,2]. The best possible solution is to reduce the world's dependence on non-renewable resources and to improve overall conservation efforts [3]. Biomass is one of good alternative energy sources over fossil fuel. It is renewable and it is generally commonly available [4]. Biomass uses also can be a solution on dependence on fossil or petroleum energy and fluctuations price of petroleum energy which became an obstacle for the development of industry [5,6]. Biofuels can help reduce carbon emissions, greenhouse gas build up and dependence on fossil fuels. Biofuels are commonly used to power vehicles and cooking stoves [7].

The problem of biomass when used for fuel or solid fuel are low energy density and low heating value, low fixed carbon content, high moisture content which can reach 50% , high volatile matter content, easiness to absorbs moisture (hygroscopic), low ash content, high alkali metal content (Na, K), low Chloride content compared to herbaceous biomass, less air for stoichiometric combustion, low combustion efficiency and smoking during combustion, wide variations in sizes, shapes and types (difficulties on handling and storage), and too bulky, not economical to transport over long distances

Content from this work may be used under the terms of the Creative Commons Attribution 3.0 licence. Any further distribution of this work must maintain attribution to the author(s) and the title of the work, journal citation and DOI.
Published under licence by IOP Publishing Ltd

[8]. Technology barriers to their utilization as energy sources is one of the main constrains of empowering biofuels [9].

Torrefaction is a pyrolysis process with temperature of 200-300°C near atmospheric pressure, absence of oxygen or air, low heating rate with residence time of 30 minutes up to 2 hours [10,11].

Currently available torrefaction technologies are basically designed and tested for woody biomass so further research is required to address on utilization of the agricultural biomass with technically and economically viable. Indonesia is the biggest producer of biomass such as oil palm empty fruit bunches (EFB), cocopeat, cassava peel waste [12]. The opportunity to develop non-fossil energy renewable is quite large this is due to: 1) the availability of a variety of renewable energy resources, 2) economic growth the better which will increase the demand in energy, 3) purchasing power of private investment in industrial development, and 4) the potential market for national, regional and international levels which are still open [13,14,15].

This preliminary research was aimed at designing bench scale of torrefaction reactor in upgrading energy content of some Indonesian biomass such as empty fruit bunches (EFB) of oil palm, cassava peel solid waste, and cocopeat or waste from coconut fiber. This study also was addressed to know emission quality of torrefied biomass and non-torrefied biomass. It is hope that combustion of biomass characteristics before and after torrefaction, its emission quality data can be recommendations for further study or scale up biomass torrefaction reactor

2. Materials and Methods

2.1. Torrefaction reactor

The bench scale torrefaction unit is shown schematically in Figure 1. The length, wide, and height of the reactor were 100 cm, 80 cm, and 100 cm respectively. Meanwhile the capacity of torrefaction process is 10 kg. The reactor was made from stainless steel, and it is equipped with the heater of 3000 Watt. The reactor was also was equipped with thermometer control (0-400°C) to observe biomass temperature during torrefaction. During the torrefaction process, absence of oxygen or air was controlled by a blade at the back of the reactor.

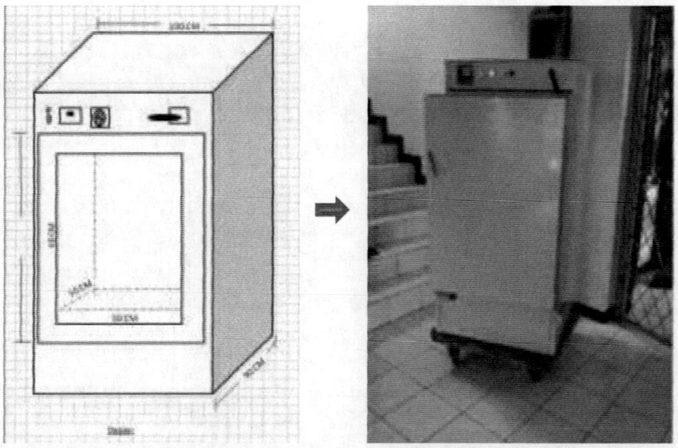

Figure 1. Bench scale biomass torrefaction reactor

2.2. *Materials*
Biomass waste such as empty fruit bunches (EFB) of oil palm trees, cassava peel waste, and cocopeat or waste from coconut fiber were used as solid fuel in this experiments. Tapioca starch with concentration of 5 % from biomass mixture was used as binding agent for making cocopeat pellet and cassava peel briquette, and EFB briquette. To make appropriate rheological properties of mixture of pellet when pelletizing it was added waste cooing oil with concentration 5 % of mixture

2.3. *Pellet and briquette making*
Cocopeat pellet, EFB briquette, and cassava peel briquette were produced by the following steps. Firstly biomass were grinded with 2 horse power (hp) milling unit and continued by sortation of biomass to purge foreign materials and to produce raw materials, and to make uniform size. The mesh used for sortation was 60. Drying biomass was done in order to get the level of dryness biomass with of 14% maximum (wet basis) and to obtain the quality of good structures pellet as well as to produce low emission pellet or briquette (Lee et al, 200; Amin et al, 2014). Agitation of mixing started by the addition of grinded biomass (90 %), tapioca starch (5 %), and waste cooking oil (5 %) respectively for 10 minutes. Briquetting or Pelletizing was intended to form a biomass fuel in the form of briquettes (3m diameter) and pellets (2 cm long and 8 mm diameter). Making briquettes and pellets were done by manually press machine and pelletizer (10 hp). Finally briquette and pellet were then cooled to evaporrate moisture of their surface.

2.4. *Torrefaction*
Torrefaction took place without air access and the temperature in the reactor was gradually increased to 300 ºC. The temperature in the reactor was read from the display on the reactor's control unit. After fitting and sealing the reactor windows, and reaching the temperature of 300ºC, 5 kilograms (kg) biomass was torrefied during 1.5 hours. Samples were taken inside torrefaction reactor unit and it was stopped when the color of the pellet or briquette was changed into dark shiny color [17].

2.5. *Analysis and emission test*
In general, testing of biomass before torrefction and torrified biomass were to determine some parameters such as volatile matter (%), moisture content (%), ash content (%), fixed carbon (%),density (kg/m^3), energy content (kJ/kg) of biomass, briquette and pellet (before and after torrrefaction), and air emissions quality of combustion pellet and briquette (CO, O_2, HC, and CO_2) during combustion on the stove

2.6. *Combustion*
To know the pellet and briquette properties and combustion performance, experiments were carried out by burning such torrefied biomass on the stove. A half (0.5) kilogram pellet an briquette put in stove or bed furnace. Observation was conducted for combustion time which start from flame appear until coals .from pellet or briquettes disappeared. The same experiments also were revealed by non-torrefied pellet and briquettes.

3. Results and Discussion

3.1. *Torrefaction reactor performance*
To test the performance of torrefaction reactor, torrefaction of biomass pellet and briquette was conducted at 300°C without air. This observation was carried out to know optimum torrefaction time of biomass. These works were based on observations each interval time in 15 minutes, 30 minutes, 45 minutes, 60 minutes, 75 minutes, 90 minutes, 105 minutes and 120 minutes. Indicators used of quality biomass was glossy color of biomass [16,17]. In 90th minute observation, biomass looks black and shiny and when it reached 120 minutes biomass started to burn and turn into coal. Thus, the length of the torrefaction process done for 90 minutes is the optimum time for biomass torrefaction.

Based on the optimum time the experimental research on some biomass and measurements will be done during 90 minutes or 1.5 hours. Figure 2 shows torrefied pellet and briquettes.

Non-torrefied Torrefied pellet Briquette

Figure 2. Non - torrefied and terrified pellet/briquette

The use of torrefied wood or agricultural biomass in commercial industry is still in development, even though torrefaction is not a new process. Several companies are developing commercial torrefaction equipment based on one methods obtained. For instance it is used a cylindrical reactor equipment with a device for removal of the liquid products and measurement of the gaseous products [18]. It would be better torrefaction process should also used others method in order to have optimum energy and minimum operational cost. The technologies under current development use a variety of combinations of temperature and residence time for processing woody biomass into torrefied wood [19]. For this research, a temperature used about 250°C with reaction times around 1 hour is recommended for torrefaction. However, at same operating conditions, mass and energy yields will vary for different biomass, as the polymeric composition and reactivity may differ. Consequently, each biomass will have its own set of operating conditions to achieve the same product quality. It would be better also for a good design and operation to use a biomass with pellet form and further understanding for optimisation of the process is required.

3.2. Chemical composition and energy

Table 1 shows the characteristics of torrefied biomass and non-torrefied biomass obtained in the performed experiments as well as the ash and the moisture contents, volatile matter, ash content, density, fixed carbon, and caloric value of the torrefied biomass and non-torrefied biomass.

Table 1. Chemical and energy of torrefied biomass

PARAMETERS	Biomass					
	Non-torrefied cocopeat pellet*)	Torrefied cocopeat pellet*)	Non-torrefied EFB briquette	Torrefied EFB briquette	Non-torrefied Cassava peel briquette	Torrified Cassava peel briquette
Moisture (%)	8.490	6.745	1.668	1.643	9.943	1.299
Volatile matter (%)	73.283	48.996	75.997	54.210	67.767	54.83
Ash content (%)	10.565	13.996	6.527	7.782	9.512	7.751
Density (kg/m3)	850	1052	494	517	411	494
Fixed carbon (%)	19.451	37.996	17.746	37.697	22.721	37.376
Calorific value (cal/g)	4083	5565	4289	4555	4494	5610
Energy ratio (increase of energy)	1.36		1.1		1.25	

*) Alamsyah *et al.* [17]

The results in Table 1 above shows that there is an increasing calorific values for the three types of torrefied biomass. The increase of calorific value of torrefied biomass were 1.36 times for torrefied cocopeat pellet, 1.1 times for torrefied EFB briquettes, and 1.25 times for torrefied cassava peel briquette. This is due to the loss of substances such as celullose, lignin, and volatile matter, and relatively smaller loss of calorie value. This increase is quite significant, especially for torrefied cocopeat pellet and cassava pell briquette so torrefaction process is expected to be a promising alternative to increase the added value of biomass as a renewable energy or solid fuel.
The type of biomass influenced the product characteristics. The torrefied cassava peel briquette and torrefied EFB briquette gave lower moisture content than torrefied cocopeat pellet. On the other hand the content of volatile matter which is owned cassava peel and briquettes briquettes EFB smaller than the same substance found in cocopeat pellet.

From the perspective of the agro-based industry logging and timber industry, literature indicates that raw material for torrefaction can vary in size and can include thin and thick chips, briquette, pellet, and even larger wood chunks. Bjorck [20] employed wood pellets of similar sizes. It was Cylindrical pellets that has 20 mm in length, and 16 mm and 6 mm diameter respectively and it was selected for torrefaction process. The energy densification resulted range from 0.85 up to 1. It can be shown that the characteristics of terrified biomass will be depend on the pretreatment before process.
Such torrefaction results it depend on the equipment design, and considering characteristics such as pre-drying, processing temperature and reaction time; it appears that performance of the torrefaction process may be vary and also types of woods, biomass, and their form [21]. However, research studies to support implementation commercial industry have been done partially. Research carried out are still complete yet and therefore need to be complementary to an integrated manner so that the result will be more efficient and maximum

3.3. Combustion and emission test

Figure 3 shows the result of terrified biomass combustion using stove. This figure shows the beginning of the combustion with full of flame (1), the middle combustion with half full of flame (2) and the end of combustion with full of coal (3). Table 2 shows biomass combustion time of terrified pellet and briquette biomass.

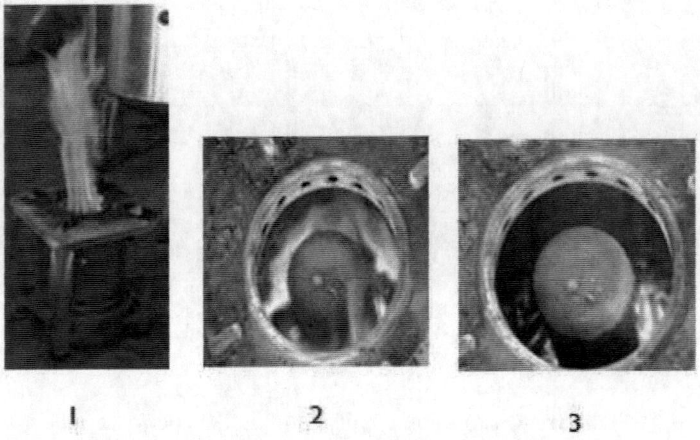

Figure 3. Combustion of torrefied biomass on the stove

The results shows that the biomass obtained from the process torrefaction have greater energy. This is advantageous because the stored energy becomes greater than the non-torrefied biomass, then the experiments shows that the best combustion time in term of flame time until the coal disappear was shown by torrefied cocopeat pellet for 120 minutes, followed by torrefied cassava peel briquettes cassava peel for 75 minutes, and torrefied EFB briquettes for 70 minutes.
The emissions test was done to find out whether the exhaust gases produced from the burning of torrefied biomass fulfill the requirements according the national regulations. Biomass pellet or briquettes combustion either torrtified or non torrefied ones were presented in Figure 4, 5, and 6. Based on the emission regulation specified in the Decree of the Minister of Environment of Republic of Indonesia No. KEP-13 / MENLH3 / 1965, all results of exhaust gas quality parameters from torrefied biomass meet the requirements. Thus, torrefaction treatment of exhaust emissions aspects is better and superior than non-torrefied biomass. From this table, are also found in combustion exhaust emissions test based on biomass feedstock on the stove in of a small quantity (0.5 kg) for all exeriments either torrefied biomass and non-torrfied biomass. It is still need to do experiments with the large quantity of torrified biomass to know wether the emission of hazardous gases in the pellet combustion such as production dangerous gas (especially CO) and the concentrations levels should be in safe limit.

Table 2. Biomass combustion time

PARAMETERS	BIOMASS					
	Non-torrefied cocopeat pellet	Torrefied cocopeat pellet	Non-torrefied EFB Briquettes	Torrefied EFB Briquettes	Non-torrefied Cassava skin briquette	Non-torrefied Cassava skin briquette
Weight (kg)	0.5	0.5	0.5	0.5	0.5	0.5
Flame and coals time after burned (min)	20	120	35	70	30	75

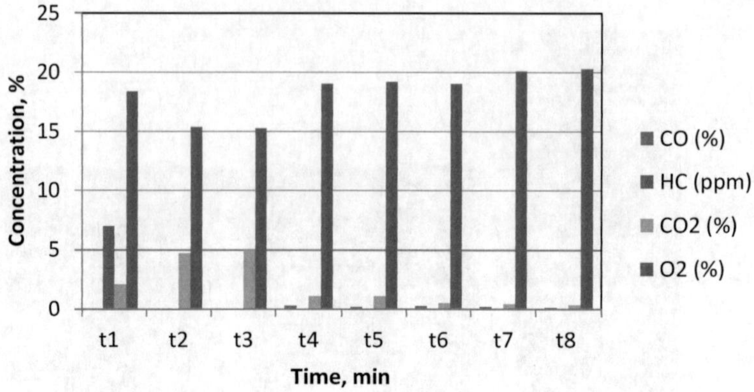

Figure 4. Emission quality of torrefied cocopeat pellet
(Standard gas limit CO: 4.5 %, HC: 1200 %, CO_2: 20 %)

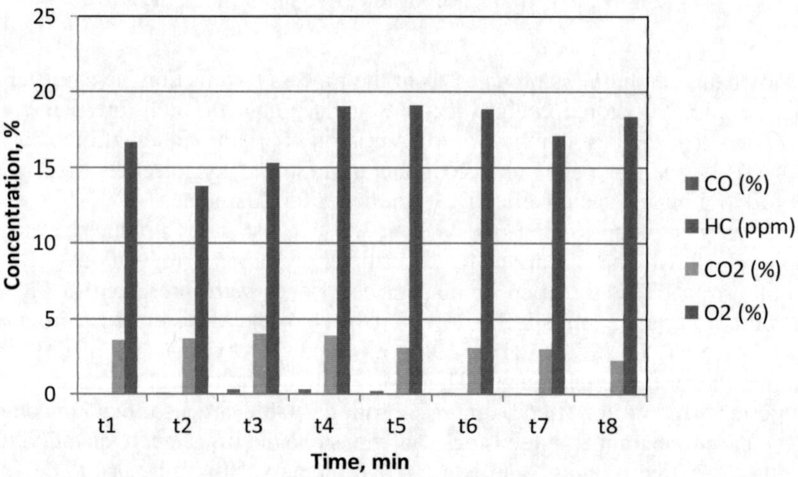

Figure 5. Emission quality of cassava peel briquette
(Standard gas limit CO: 4.5 %, HC: 1200 %, CO_2: 20 %)

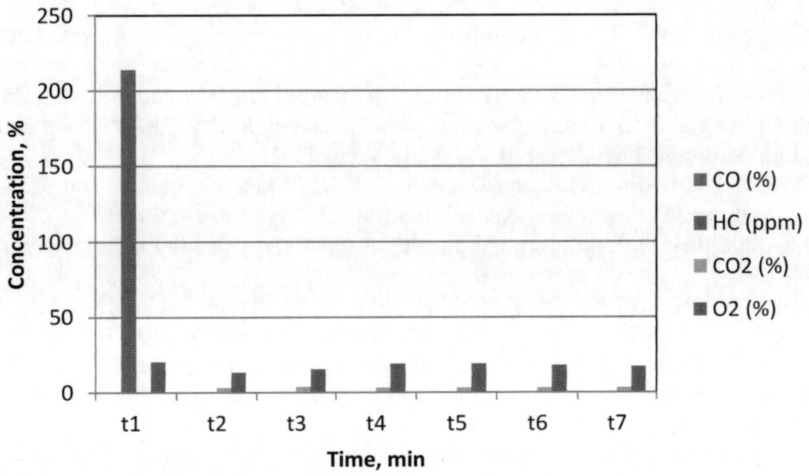

Figure 6. Emission quality of EFB briquette
(Standard gas limit CO: 4.5 %, HC: 1200 %, CO_2: 20 %,)

However, little attention has been paid to the influence of torrefaction of biomass processing for solid fuel. A series of experiments were performed to examine the characteristics of the torrefaction process, the properties of torrefaction. products, and the effects of torrefaction on gas composition, cold gas efficiency and gasification efficiency. The results showed that during the torrefaction process the moisture content of biomass were reduced, and the wood fiber structure of the material was destroyed. Torrefaction is one of attractive option of upgrading biomass energy and the combustion result of terrified biomass give low emission [11]. From the observation of the smoke is concluded that there is only very little smoke. This findings was similar to those conducted by Azhar and Rustamaji [22] which indicates that the process torrefaction of bamboo have eliminated most of the volatile on bamboo. On the other hand observations of the burning briquettes can be concluded that the ash formed showed in a very small amount [22].

4. Conclusion

Energy of biomass can be upgraded and used as a fuel by torrefaction process. It was shown by the energy ratio of caloric value of biomass pellet after torrefaction. For cassava peel briquette, caloric value increase from 4494 cal/g to 5610 (or energy ratio was 1.25), for EFB briquette caloric value increase from 4289 to 4555 cal/g (or energy ratio was 1.1), and for cocopeat pellet caloric value increase from 4083 to 5565 cal/g (or energy ratio was 1.36)

Cocopeat pellet treated with torrefaction process reveals flame and coals time after burned is longer when combustion which means the energy stored in the biomass pellet was greater and pellet shape of biomass has greater influence on combustion. Meanwhile emission air resulted from its combustion of all torrefied bioamass (pellet and briquette) were met with Indonesian emission regulation standard. For a good design and operation of a biomass torrefaction reactor plant, further experiments for optimization of the process is required. It is suggested that future work should focus on biomass in Indonesia for optimum torrefaction conditions, as well as investigations of the combustion behaviour of these fuels.

Acknowlwdgement

The authors would like to acknowledge Center for Agro-Based Industry (CABI), Ministry of Industry for supporting budget in this research, Local Department of highway traffic (DLLAJR) and Bogor for air emission quality test.

References

[1] Höök M and Tang X 2013 Depletion of fossil fuels and anthropogenic climate change e a review *Energy Policy* **52** 797- 809

[2] Ridha M 2014 New Renewable Energy and Energy Conservation Nreec: Existing Program & Breakthrough Policies Toward Achieving National Energy Policy Target Paper presented on The 3rd Indonesia EBTKEConEx JCC Jakarta 4 June 2014

[3] Naik S N, Goud V V, Rout P K and Dalai A K 2010 Production of first and second generation biofuels: a comprehensive review *Renew. Sustain. Energy Rev.* **14** 578-97

[4] Demirbas A 2009 Biofuels securing the planet's future energy needs *Energy Convers. Manag.* **50** 2239-49

[5] Liew W H, Hassim MH Ng DKS 2014 Review of Evolution Technology and Sustainability Assessments of Biofuel Production *Journal of Cleaner Production* **17** 11-29

[6] Ariyati R 2007 National Policy on Biofuel Development *Works. Mini Scale Biodiesel Plant Design BRDST BPPT Research and Technology DJLPE EMR and PT Astra Agro Lestari Tbk* November 13-15 2007 Indonesia-Serpong

[7] Egebäck K-E and Westerholm R 1997 *Environmental Potential of the Alternative Fuels* (Italy : Biogas Ethanol Methanol Natural Gas RME and DME Swedish Transport and Communications Research Board KFB Baveno)

[8] Bergman P C A, Boersma A R and Kiel J H A 2005 *Torrefaction for Biomass Co-Firing in Existing Coal-Fired Power Stations "Biocoal"* Report ECN-C- 05-013 (Belanda : ECN)

[9] Cheng J J and Timslina G R 2011 Status and barriers of advanced biofuel technologies: A review *Renewable Energy* **36** 3541-49

[10] Acharya B, Sule I and Dutta A 2012 A review on advances of torrefaction technologies for biomass processing *Journal of Biomass Conversion and Biorefinery* **2** 349-369

[11] Chen W L and Kuo P G 2011 Torrefaction and Co-Torrefaction Characterization of Hemicelluloses Cellulose and Lignin as Well as Torrefaction of Some Basic Constituents in Biomass *Energy* **36** 803-11

[12] Alamsyah R and Siregar N C 2013 Study for production of low emission pellet from Coco Peat Waste Jakarta International *Journal on Coconut R & D - CORD APCC* **29** 1-12

[13] Belonio A 2012 Cocopeat Pellet Gas Stove (http://biomass-eventscom /promo-pellet_characteristics) Accessed august 21 2016

[14] Hambali E, Thahar A and Komarudin 2010 The potential of oil palm and rice biomass as bioenergy feedstock *7th Biomass Asia Workshop* November 29 – December 01 Jakarta Indonesia

[15] Abdullah K 2009 *Sustainable Parameters in Introducing Renewable Energy Technology ISESCO Science and Technology Vision* Vol 5 pp 7-10

[16] Pelheat 2010 Biomass pellet production guide accessed on April 27 2016 from http://wwwPelHeatcom

[17] Alamsyah A, Siregar N C dan HasanahF 2016 Improvement of caloric value of cocopeat biomass pellet as renewable energy fuel by torrefaction application warta *ihp Journal of Agro-based Industry* **33** 18-25

[18] Pach M, Zanzi R A V and Bjornbom E 2002 Torrefied Biomass a Substitute for Wood and Charcoal paper on 6th Asia-Pacific *Int. Sym. on Combustion and Energy Utilization6th Asia-Pacific International Symposium on Combustion and Energy Utilization*

[19] Kongkeaw N and Patumsawad S 2011 Thermal upgrading of biomass as a fuel by torrefaction *2nd Int. Conf. on Environmental Engineering and Applications IPCBEE IACSIT* (Singapore :Press Singapore) **vol** 17

[20] Bjorck C 2012 Torrefaction of biomass assignment mini-project report 18th May 2012 (Britania Raya : The University of Sheffield)

[21] James J 2010 Creating power from biomass: creating power from biomass: torrefaction torrefaction - a key facilitating technology paper *US Department of Energy Biomass 2010 Conference Agri-Tech Producers LLC* April 31

[22] Azhar and Rustamaji H 2009 solid fuels from biomass bamboo with torefaksi and densification process *Jurnal Rekayasa Proses* **3** 26-29

[23] Ministrial Decree 1995 Minister of Environment Decree No KEP-13 / MENLH3 / 1995 concerning quality standards stationary sources of emissions

Two-steps microwave-assisted treatment on acid hydrolysis of sago pith for bioethanol production

T C Sunarti[1], S D Yanti[1], E Ruriani[2]
[1]Department of Agroinustrial Technology, Faculty of Agricultural Technology, Bogor Agricultural University
[2]Department of Technology of Agricultural Products, Faculty of Agricultural Technology, University of Jember

Email : titi-cs@ipb.ac.id

Abstract. Sago is a genus of palm that can be utilized to produce fermentable sugars as substrate for bioethanol. Sago pith is a heterogeneous substrate consists of starch and fiber. Acid hydrolysis by microwave heating radiation can break down starch and fibers together in a very short time, so it is considered to be very efficient process. The use of microwave energy (as power level) and variation of heating time can produce fermentable sugar with certain characteristics. This study included the preparation and analysis of sago pith flour; process of acid hydrolysis (0.3 M and 0.5 M H_2SO_4) using two steps microwave heating, first with power level 30% (1, 2 and 3 min) and second with power level 70% (3 min); and ethanol production. The conventional treatment (autoclaving at 121°C for 15 min) was carried for the comparison. The highest fermentable sugar (105.7 g/l) was resulted from microwave heating with power level 30% for 2 min followed by the power level 70% for 3 min. This hydrolyzate then used as substrate for bioethanol fermentation and partially neutralized (pH 3, 4, 5) by using yeast *Issatchenkia orientalis*, and the highest ethanol (2.8 g/l) was produced in pH 5.

1. Introduction

Sago palm (*Metroxylon sp.*) is a potential commodity for agroindustrial raw material, and considered as an extremely sustainable plant becuase of its ability to thrive in most soil conditions. This crop has a significant value to be utilized for some commercial products, such as modified starches, lactic acid, cyclodextrins, and ethanol [1]. It was also predicted that approximately 60 million tons of sago starch will be produced per year at the end of the twentieh century [2]. In addition, it has also been reported that Indonesia covered 1.25 million ha area of sago forest, and assumed its productivity 100 kg per stem with the population of 30 stems per ha, then 6.48 millions ton per year of sago starch could be produced [3].

Sago starch is derived from the pith of sago and accumulated in the pith of sago trunk. It has a high content of starch (81-85%) on the pith [4]. It was also stated that sago pith starch have a high potency as a raw material of ethanol production, since sugar and starch-based biomass are currently predominant and economically material in industry [5]. In addition, our previous study [6] found that

Content from this work may be used under the terms of the Creative Commons Attribution 3.0 licence. Any further distribution of this work must maintain attribution to the author(s) and the title of the work, journal citation and DOI.
Published under licence by IOP Publishing Ltd

sago pith also containing fiber (cellulose and hemicelulose), and this lignocellulose was also able to be converted into ethanol. The conversion of starch, cellulose and hemicellulose into fermentable sugars using energy efficient, economic and faster way was the greatest concern for commercial fuel ethanol production [7]. The conversion includes two processes: hydrolysis of starch, cellulose and hemicellulose to fermentable sugar, and fermentation of the sugars to ethanol by yeast or bacteria. Hydrolysis is a critical stage in fermentable sugars production, before the fermentation process. Chemical and enzyme hydrolysis have been widely employed to breakdown the starch and cellulose into fermentable sugar [8, 9, 10], however the approaches still tend to be slow, expensive and of high dilutions that give poor yields of glucose [11]. Microwave irradiation is an alternative method for hydrolysis of biomass to simple sugars. Microwave heating presents a potentially faster, efficient and selective method for thermal treatment of biomass, compared with conventional heating [12], the reaction rate of starch hydrolysis to glucose was accelerated 100 times under microwave irradiation [13]. Therefore, it was considered as energy efficient approach for biomass pretreatment under low pressure and temperature [14, 15], and the microwave heating even improved the glucose selectivity [16]. Previously, it was reported that microwave treatment for saccharification in water and acid hydrolysis can produce oligosaccharides [17, 18].

In this study, a breakthrough system developed by assisting the acid hydrolysis with two-steps microwave treatment to improve the more efficient saccharification process of starch and fiber from sago pith. This research investigated the effect of microwave treatment and optimal acid concentation to sugar characteristics resulted by modifying the acid concentration, power level (energy) of radiation and radiation period. The hydrolisates then be fermented into bioethanol by using *Issatchenkia orientalis*. The typical yeast has a high tolerance in low pH [19] and this characteristic was important to against the acid condition due to the acid hydrolysis. In addition, the yeast also has an ability to convert both hexsose and pentose sugar into bioethanol [20, 21], and this characteristic was significantly needed in bioethanol production from lignocellulose biomaterial that consisted of the two types of sugar.

2. Methods

2.1. Materials

The raw material used in this research was sago pith from *Metroxylon sp*., provided from local sago starch industry in Cimahpar, Bogor, Indonesia. The hydrolysis used sulphuric acid (p.a), then assisted by heating of radiation from microwave oven Sharp type R-348 C output 1000 W High power and 32 L capacity. The *Isaatchenkia orientalis* was a collection of Biology Department, Faculty of Mathematic and Science, Bogor Agricultural University. This isolate was obtained from rotten fruits and has been identified and characterized its performance in bioethanol fermentation.

2.2. Preparation and Characterization of Sago Pith Flour

Native sago pith was cut into smaller size, rasped and dried in sun drying. The crude flour then pulverized and passed 35 mesh screens. Some chemical characteristics of the dried flour were analyzed as moisture, ash, protein, fat, crude fiber, and carbohydrate (by differences) contents, starch content (*Luff Schoorl*), and fibre component.

2.3. Acid Hydrolysis Assisted by Two-Steps Microwave Heating

Sago pith mainly consists of starch and limited amount of fiber. Sago pith slurries (10%) hydrolysed in aqueous sulphuric acid (0.3 M and 0.5 M) were poured into glass jar to get the first heating from microwave oven on 30% of power level (output 350 Watt) for a various time (1, 2 and 3 minute). The second heating was continued on higher power level (70%, output 650 Watt) for 3 minute. After that, the hydrolizat was filtered and neutralized by NH_4OH, and the fermentable sugar resulted was then analyzed its total sugar amount, reducing sugar amount, dextrose equivalent (DE), degree polimerization (DP), filtrat volume, syrup clarity, and HMF and furfural content. Microscopic analysis of the filtrat and residue was also observed. A conventional heating by autoclaving was carried out as a comparison.

2.4 Production of Bioethanol
Fermentable sugar produced from the best treatment was fermented by using *Isaatchekia orientalis* (10% starter) in a various pH (3, 4, and 5). The carbondioxide volume as the indicator of by-product in bioethanol production was measured in every 3h for the first 12 h fermentation period, then every 6h in the next 36 h period, and every 12 h in the last 36 h. After 72h of fermentation, the ethanol content was determined by using *Gas Chromatography*. The substrate of autocalving treatment and technical glucose were used as comparison.

3. Results and Discussion

3.1. Characteristics of Sago Pith Flour
Proxymate data (Table 1) explained that the major component of sago pith was carbohydrate (81.6%) and 55.86% of it was starch. It has been verified that sago pith, inner portion of the trunk after the removal of the outer bark-like layer, contained the starch which was accumulated in the pith of sago palm stem from the base upwards [1]. The starch content is influenced by the maturity of sago plant. The trunk is fully saturated with starch almost to the crown at the optimal maturity [22].

The second component of sago pith was fiber (7.93%), consisted of cellulose (2.93%) and hemicellulose (4.72%). The starch and two component of fiber are potential substrates to be hydrolized into fermentable sugar (glucose and xylose), since glucose is major components of starch and cellulose, whereas hemicelluose contains xylose and glucose as the major components, together with arabinose and galactose, and lower levels of rhamnose, mannose, fucose and uronic acids [23]. The content of lignin (6.86%) is less than reported by Ref. [1] that was about 9% to 22%, therefore the delignification process was not needed to be carried out in this study.

However, the moisture contents was high (14.5%), this will decrease its quality because of the microorganism growth. To solve the problem the sample has been dried in order to minimize the moisture content into safe condition for storage (≤10%).

Table 1. Chemical characteristics of sago pith flour.

Components	Percentage (%)
Moisture (db)	14.52
Ash (db)	5.16
Fat (db)	3.72
Protein (db)	1.59
Crude fiber (db)	7.93
Carbohydrate (db)	81.6
Starch (db)	55.86
NDF (neutral detergent fiber)	14.51
ADF (acid detergent fiber)	9.79
Cellulose	2.93
Hemicellulose	4.72
Lignin	6.86

IOP Conf. Series: Earth and Environmental Science **65** (2017) 012052 doi:10.1088/1755-1315/65/1/012052

3.2. Acid Hydrolysis Assisted by Two-Steps Microwave Heating

The approach of hydrolysis can be chemically or enzimatically. The first technique wis wide applied in industry because relative low cost, fast, and incomplex process, although it produces by-product that could be inhibit the next process. Acid hydrolysis using sulphuric acid was usually used to convert starch or lignocellulose into intermediate products such as maltodextrin, oligosaccharide, and glucose syrup that have a wide application in food, feed, fuel, textile and pharmaceutical industries. The existence of acid could break down starch molecules at random and, theoritically it produced 60% glucose [24]. Furthermore, the glucose decomposition resulted some components that were identified as 50-hydroxymethylfurfural (HMF), 1,6-anhydroglucose, levulinic acid, and formic acid [25].

However, since the sago pith consisted starch and fiber which could not soluble in cold water, the pre-treatment before hydrolysis step was important to be developed. Microwave heating is one of the pre-treatment which has an important role in degrading some polysaccharides into their monomer and oligomers [18]. The heating energy of microwave penetrate directly and uniformly into materials, and then absorbed by polysaccharide causing the cleavage of the glycosidic polysaccharide linkages. Figure 1 shown the change of fiber microscopic structure after microwave heating and acid hydrolysis treatment.

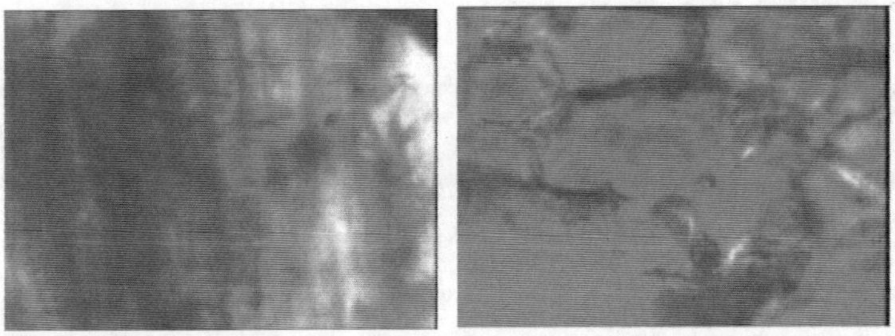

Figure 1. The microscopic fiber structure observed by light-polarized microscope: (1a) native sago pith starch and (2a) hydrolizates of sago pith starch after two-stages microwave heating (power level 30% in 3'and power level 70% in 3') and hydrolysis (0.5 M sulphuric acid)

It can be seen that the two combined treatments convert a compact fibre (Fig 1a) into amorphous structure (Fig 1b), and most of the native starch granules disappeared, and converted into gel forms products. The blue color of fiber fraction also indicated not completely degradation of crystalline parts. The heating pretreatment using autoclaving and microwave could increase the solubility and break down the crystallinity of starch and fiber, and caused the starch gelatinized and hydrolyzed directly into depolymerized products [6]. This degradation process can proved by analyze the soluble sugars contents in the hydrolyzates (Table 2).

Table 2. Parameter of hydrolizates after acid hydrolysis assisted by two-step microwave treatment of sago pith flour.

Treatment	Reducing sugar (g/l)	Total sugar (g/l)	Average DE value	Average DP value	HMF (mg/l)	Furfural (mg/l)	Filtrat clarity (%T)	Filtrat volume (ml)	Residue weight (%)
A1B0	83.17	90.50	91.90	1.08	8.10	0.06	89.85	75.75	24.01
A1B1	18.85	68.38	27.57	3.63	0.28	0.00	79.35	59.75	47.10
A1B2	24.40	80.81	30.20	3.32	0.45	0.00	87.05	56.50	42.48
A1B3	26.87	81.20	33.09	3.03	0.48	0.00	89.50	54.00	37.00
A1B4	32.24	82.93	40.08	2.49	0.63	0.00	76.30	49.00	35.25
A2B0	106.20	111.61	95.15	1.05	65.64	1.71	87.15	80.00	20.63
A2B1	23.34	80.69	28.93	3.46	0.96	0.04	65.60	67.50	30.62
A2B2	30.33	92.83	32.67	3.06	1.46	0.05	72.15	65.50	29.42
A2B3	48.86	105.56	46.24	2.16	1.56	0.05	76.15	64.00	28.14
A2B4	37.41	102.39	37.51	2.74	1.88	0.06	61.30	61.00	26.43

A1 : 0.3M sulphuric acid
A2 : 0.5M sulphuric acid
B0 : autoclaving heating (comaprison)
B1 : 70% power level (control)
B2 : 30% power level (1') + 70% power level (3')
B3 : 30% power level (2') + 70% power level (3')
B4 : 30% power level (3') + 70% power level (3')

Table 2 described detail chemical characteristics of hyrdolizates products. Total sugar is the important parameter of hydrolysis efficiency, since it determines the amount of soluble reducing sugar (monomer), dimer and its oligomers those the products of degradation process of sago pith starch and fiber. The amount of total sugar ranged from 80.81 g/L to 111.61 g/L was resulted by two-steps heating rather than one-step heating (68.38 g/L to 80,69 g/L) from microwave treatment in dilute acid. This due to the radiation energy obtained by the material was higher in two step heating (30% power level continuing to 70% power level) than the other, and made the more release of polysaccharides linkages. It drove to the increase of total sugar and reducing sugar amount, indicated by the decrease of average DP values (2.49-3.32). DP (degree of polymerization) expressed the number of monomer unit in the molecule, the more few number of DP value, the more high the content of reducing sugar. DP value up to one meaning there are much more monomers rather than disaccharides or oligosaccharides (DP value above 2), since glucose and xilose only contains 1 unit molecule.

However, autoclaving treatment produced the highest total sugar and reducing sugar (111.61 g/l and 106.20 g/l) with the lowest DP value (1.05). It described that the hydrolysis process occuring perfectly, but the undesired product HMF (hydroxy methyl furfural) also accumulated in much more high amount as 65.64 mg/L rather than microwave heating (0.28-1.28). This by-products could inhibit the fermentation of bioethanol, therefore microwave heating was the better treatment than others.
In general, the acid concentration influence the degradation process. The hydrolysis in dilute acid of 0.5 M resulted a higher total sugar amount than 0.3 M in the same condition of microwave heating. The best combination treatment of acid concentarion, hydrolysis time, and total amount of energy was obtained from 0.5 M acid concentration, first heating on 30% power level for 2 minute in the first and continuing to second heating on 70% of power level for 3 minutes. This best treatment resulted the high amount of total sugar (105.56 g/L) and reducing sugar (48.86 g/L), and also the minimal DP value (2.16). These three parameters could indicate the sufficient fermentable sugar to be converted into ethanol.

3.3. Production of Bioethanol

The fermentable sugar resulted from best treatment of hydrolysis assisted with two stages microwave was determined by the content of total sugar, reducing sugar and HMF. The high amount of total sugar and reducing sugar indicated the high availiblity of fermentable sugar as the substrate of bioethanol. In contrast, the HMF content should be minimized in order to avoid the inhibition of ethanol production, since this by product could decrease the yield of ethanol. The hydrolized sago starch from the best treatment then was converted into ethanol by *Isaatchenkia orientalis*. The typical hexose sugar, including glucose, galactose and mannose can be fermented by numbers of wild type microorganisms, yet pentose sugar such as xylose and arabinose can only be fermented by only few number of wild type microorganisms, and frequently resulting low ethanol yield. *Isaatchenkia orientalis* was one of a specific yeast strain, because it was low pH tolerance [20, 21, 26], . It is important to be noticed that the fermentable sugar from acid hydrolysis is acid condition. It was also underlined that its superior ability to ferment glucose to ethanol under high stress conditions, such as acid (H_2SO_4), salt (Na_2SO_4), or heat stress [27].

The observation toward CO_2 volume used as additional data for indicating the fermentation was running well. It was also suggested that the volume of CO_2 were equivalent with the ethanol content, shown by the chemical reaction as follows:

$$C_6H_{12}O_6 \rightarrow 2C_2H_5OH + 2CO_2 \quad (1)$$

$$3C_5H_{10}O_5 \rightarrow 5C_2H_5OH + 5CO_2 \quad (2)$$

From the reaction above, the ethanol content which are resulted from either glucose (i) or xylose (ii) are equivalent with the yielded glucose. The alteration of CO_2 volume are not directly used for determining bioethanol content from hydrolysate [28]. It was mainly determined by multiplying the volume of CO_2 with coefficient 1.045 as Gay-Lussac equation. Figure 2 explained that as the time of fermentation increased, the level of CO_2 was also escalated, yet after 12 h of incubation the increasing level of CO_2 were not significant. This condition only occured in pH 4 and 5. It can be confirmed by measuring the ethanol content using GCMS (Table 3).

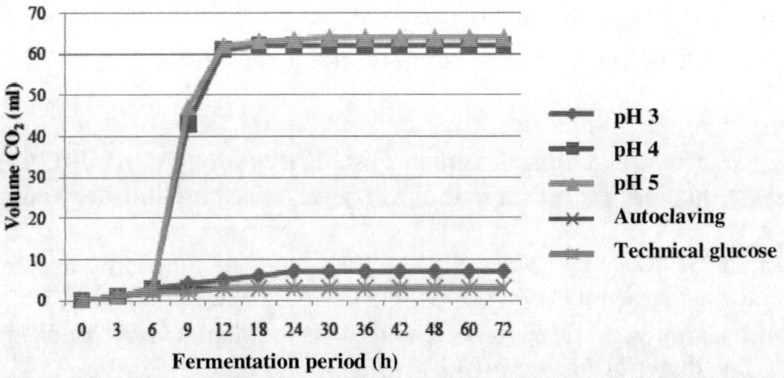

Figure 2. The accumulation of augmented CO_2 derived from 72 hours of sago pith hydrolysate fermentation in any various pH and the comparison (autoclaving and technical glucose).

Table 3. Ethanol yield and fermentation kinetics parameter of fermentable sugar resulted by acid hydrolysis assisted by two-step microwave treatment of sago pith flour.

Substrate	Parameter			
	Ethanol yield (%)	Ethanol yield (g/l)	Δs/s	Yp/s
Sample (pH 3)	0.220	1.783	0.126	0.022
Sample (pH 4)	0.314	2.481	0.180	0.032
Sample (pH 5)	0.361	2.852	0.250	0.036
Autoclaving pH 5 (comparison)	1.840	14.536	0.167	0.186
Technical glucose pH 5 (control)	1.611	12.727	0.118	0.175

Table 3 shown the higher yield of ethanol (2.852 g/l) resulted by fermentation in pH 5, rather than pH 3 and 4. This can be assumed that optimal condition for *Issatchenkia orientalis* growth was in pH 5. However, the ethanol yield produced was still lower compared to autoclaving treatment. This could be suggested that autoclaving has a sufficient time and energy to hydrolysed the sago pith. The average DP values of this treatment was 1.05 (Table 2). It means that the starch and fiber had been hydrolized perfectly into their monomers. The content of reducing sugar was about 106.20 g/l, and it was two times higher than microwave heating (48.86 g/l). The higher concentration of reducing sugar made an available substrate for fermentation, and increased ethanol yield.

4. Conclusion

Two-steps microwave-assisted on sulphuric acid hydrolysis of sago pith flour produced a better characteristics of fermentable sugar rather than one-step treatment. The higher acid concentration and longer hydrolysis time increased the total sugar and reducing sugar amount of the hydrolizates. It also improved the hydrolysis process, indicated by the raise of DE value and the decrease of DP value. In addition, the sugar resulted has been proven able to be a substrate of bioethanol (0.22-0,36%) by *Issatchenkia orientalis* in pH 5. For further research, some improvements should be carried out to increase the hydrolysis efficiency and ethanol yields, such as : size reduction of sago pith, detoxsification of fermentable sugar, and isolation of indigenous yeast.

References

[1] Singhal RS, Kennedy JF, Gopalakrishnan GM, Kaczmarek A, Knill CJ and Akmar PF 2008 Industrial production, processing, and utilization of sago palm-derived products *Carbohydr. Polym.* **72** 1-20

[2] Wee LL, Annuar MSM and Ibrahim S 2011 Energetics of glucoamylase-catalyzed hydrlysis of commercial sago starch *Asia Pas. J. Mol. Biol. Biotechnol.* **19** (4) 117-120

[3] Syakir M and Karmawati E 2013 Potency of sago plant (*Metroxylon spp.*) as a source of bioenergy raw material *Perspektif* **12** (2) 57-64

[4] Fujii S, Kitahara S and Komoto M 1985 Studies on improvement of sago starch quality in: *Proc. of the 3rd International Sago Symposium.* Tokyo-Japan, May 20-23

[5] Subashini D, Ejilane J, Radha A, Jayasri MA and Suthindhiran K 2011 Ethanol production from sago waste using Saccharomyces cerevisiae Vits-M1 *Curr. Res. J. Biol. Sci.* **3** (1) 42-51

[6] Sunarti TC, Dwiko M, Richana N and Meryandini A 2011 Effect of microwave treatment on acid hydrolysis of sago pith for the preparation of fermentable sugar in bioethanol production *Proc. of the International Conference on Chemical Innovation.* Kemaman-Malaysia, May 23-24

[7] Thangavellu SK, Ahmed AS and Ani FN 2014 Bioethanol production from sago pith waste using microwave hydrothermal hydrolysis acelerated by carbondioxie *Appl. Energy* **128** 277-283

[8] Balat M, Balat H and Cahide O 2008 Progress in bioethanol processing *Prog. Energy Combust. Sci.* **34** 551
[9] Demirbas A 2012 Competitive liquid biofuels from biomass *Appl. Energy* **88** 17-28
[10] Limayem A and Ricke S 2012 Lignocellulosic biomass for bioethanol production: current perspective, potential issues and future prospects *Prog. Energy Combust. Sci.* **38** 449
[11] Fan J, Zhu Z, Budarin V, Gronnow M, Gomez JD, Macquarrie D and Clark J 2013 Microwave-enhanced formation of glucose from cellulosic waste C*hemic. Engineer. Process: Process Intensific.* **71** 37-42
[12] Chen WH, Tu YJ and Sheen HK 2011 Disruption of sugarcane bagasse lignocellulosic structure by means of dilute sulfuric acid pretreatment with microwave-assisted heating *Appl. Energy* **88** 2726
[13] Kunlan L, Lixin X, Jun L, Jun P, Guaying C and Zuwei X 2011 Salt-assisted hydrolysis of starch to D-glucose under microwave irradiation *Carbohydr. Res.* **331** 9-12
[14] Lu X, Xi B, Zhang Y and Angelidaki I 2011 Microwave pretreatment of rape straw for bioethanol production: foccus on energy efficiency *Bioresour. Technol.* **102 7**937.
[15] Kannan TS, Ahmed AS and Ani FN 2013 Energy efficient microwave irradiation of sago bark waste (SBW) for bioethanol production *Adv. Material Res.* **701** 249.
[16] Tsubaki S, Oono K, Onda A, Yanagisawa K and Azuma JI 2012 Microwave-assisted hydrothermal hydrolysis of cellobiose and effects of additions of halide salts *Bioresour. Technol.* **123** 703
[17] Matsumoto A, Tsubaki S, Sakamoto M and Azuma J 2011 A novel saccharification method of starch using microwave irradiation with addition of activated carbon *Biores. Technol.* **102** 3985-2988
[18] Warrand J and Janssen HG 2007 Controlled production of oligosaccharides from amylose by acid-hydrolysis under microwave treatment: comparison with conventional heating *Carbohydr. Polym.* **69** 353-362
[19] Seo SH, Rhee CH and Park HD 2007 Degradation of malic acid by Issatchenkia orientalis KMBL 5774, an acidophilic yeast strain isolated from Korean grape wine pomace *J. Microbiol.* **45** 521- 27
[20] Hisamatsu M, Furubayashi T, Karita S, Mishima T and Isono N 2006 Isolation and identification of a novel yeast fermenting ethanol under acidic condition *J. Appl. Glycosci.* **53** 111-113
[21] Ruriani E, Meryandini A and Sunarti TC 2012 Yeast isolation for bioethanol production *Hayati J. Biosci.* **19** (3) 145-149
[22] Lang AT, Mohamed AMD and Karim AA 2006 Sago starch and composition of associated components in palms of different growth stages *Carbohydr. Polym.* **63** 283-286.
[23] RungCang S, Jones GL, Tomkinson J and Bolton J 1999 Fractional isolation and partial characterization of non-starch polysaccharides and lignin from sago pith *Industr. Crops. Prod.* **9** 211-220
[24] Woiciechowski AL, Nitsche S, Pandey A and Soccol CR 2012 Acid and enzymatic hydrolysis to recover reducing sugars from cassava bagasse: an economic study *Braz. Arch. Biol. Technol.* **45** (3) 393-400
[25] Xiang Q, Yong Y, Lee and Torget RB 2004 Kinetic of glucose decomposition during dilute acid hydrolysis of lignocellulosic biomass *App. Biochem. Biotechnol.* **115** 1-3 1127-38
[26] Mosier N *et al.* 2005. Features of promising technologies for pretreatment of lignocellulosic biomass. *Biores. Technol.* **96** 673–686
[27] Naoto I, Hayakawa H, Usami A, Mishima T and Hisamatsu M 2012 A comparative study of ethanol production by Issatchenkia orientalis strains under stress conditions *J. Biosci. Bioengineer.* **113** (1) 76-78
[28] Bonciu C, Cristiana T and Gabriela B 2010 Yeast isolation and selection for bioethanol production from inulin hydrolysates *Innovat.*

Utilization of porous carbon from waste palm kernel shells on carbon paper as a supercapacitors electrode material

O N Tetra[1], Syukri[2], A Alif[1], R Fristina[1] and H Aziz[1]

[1] Laboratory of Photo/Electrochemistry, Department of Chemistry, Faculty of Mathematics and Natural Sciences, Andalas University, Indonesia
[2] Laboratory of Material, Department of Chemistry,Faculty of Mathematics and Natural Sciences, Andalas University, Indonesia

Email : haziz13@yahoo.com

Abstract. A porous carbon prepared from waste palm kernel shells has been applied as a new supercapacitor electrode material. Activated carbon was synthesized through a carbonization process at a temperature of 400 °C which has a particle size of 90 µm and provide a surface area of 301.482 $m^2 g^{-1}$. Carbon from waste palm kernel shells was distributed on a sheet of carbon paper surface with a variation of carbon paper and palm kernel shells carbon mass ratio. From the results reported that the capacitance value was increased to 1331.8 µF in rolling method with addition of carbon palm shells mass have higher capacitance values than which was obtained in plate or sandwich method.

1. Introduction

Supercapacitors have been attracted much attention as an electrochemical energy storage device due to their high power density and long durability compared to batteries. Activated carbon material has been widely used as electrode materials of supercapacitors because of high surface area, chemical inertness, good thermal and mechanical stability. Generally highly developed surface area, which is characteristic for microporous carbons is necessary for formation of electrical double layer but the presence of mesopores are crucial if a high propagation of charge is demanded specially for transportation system still in acceleration where is the power peak required as well as during braking for energy recovery [1]. Various types of activated carbon materials, for example coffee grounds [2], ginkgo shells [3] waste tea-leaves [4] and palm kernel shell [5] were used in a fabrication of supercapacitor.

The development of oil palm kernel shell biomass as a source of bioenergy and also in reducing the environmental impact must be accompanied by another development of a new electrode material as an important aspect in supporting the energy storage technology development. Regarding for next utilization of waste palm kernel shells biomass to produce an activated carbon to pursuing a licensing strategy of the future supercapacitor electrode [6,7].

The common methods used in the present study is in order to increase the capacitance of a new carbon electrode material is actually our main research objective especially in the electrode surface modification or fabrication of a new carbon fiber paper-based supercapacitors [8], a fabrication of

Content from this work may be used under the terms of the Creative Commons Attribution 3.0 licence. Any further distribution of this work must maintain attribution to the author(s) and the title of the work, journal citation and DOI.
Published under licence by IOP Publishing Ltd

Polyaniline (PANI) nanofibers supported by carbon paper [9]. In this paper, we choose palm kernel shell carbon which was coated on the commercial carbon paper surface as a template in a new super-capacitor electrodes in order to increase supercapacitor capacitance.

2. Materials and Methods

2.1. Equipments and materials

Equipments which were used are LCR-Meter (Tonghui *Electronic* TH2820-LCR), Multimeter (Heles UX-838TR), petridish, and other laboratory glasses equipment's.

Waste palm kernel shells materials were obtained from Agam regency, West Sumatra, carbon rods of waste batteries, H_3PO_4 (Merck), carbon paper (Munix Kangaro) and distilled water.

2.2. Preparation of carbon palm kernel shell

Palm kernel shells are dried and reduced into smaller size. The formation of carbon through carbonization process at a temperature of 400 °C for 60 min with a certain particle size of 90 μm [5].

2.3. Preparation of carbon paper as electrode materials

Supercapacitors were made with such a roll and sandwich form (Fig 1). Carbon rod originally from waste batteries with a length of 5 cm and a diameter of 0.3 cm. Carbon paper electrode plate of palm kernel shells are rolled up circling the carbon rods from the battery. On the first roll is the front of the paper carbon that acts as a positive electrode (sheet 1) and then continue rolling the rear carbon paper acting as a negative electrode (sheet 2). The same treatment is done for variations in particle size and surface area variations of carbon paper [5].

In the sandwich plate method, carbon paper was cut with a variety of sizes. Each carbon paper arranged in layers as Figure 2 and as a current collector is a copper plate that is placed flanking the two electrodes.

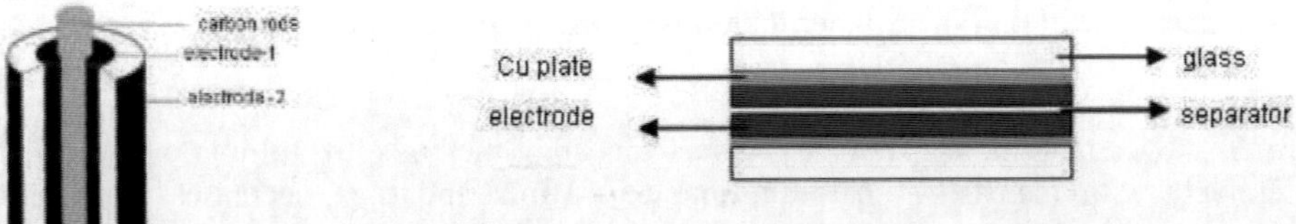

A. B.

Figure 1. Schematic illustration of the supercapacitor preparation (Rolling Method (A) and Sandwich Method (B)

2.4. Preparation of palm kernel shell carbon on the carbon paper

The sheets paper spread up with glue and sprinkle the carbon of palm kernel shells on its surface. Plate electrode of carbon which had been formed then dried at room temperature and the carbon paper weighed before and after being coated with palm kernel shell carbon.

3. Results and Discussions

Characterization of the carbon palm kernel shell obtained at the combustion temperature of 400 °C has been reported previously. From the results of EDX, the percentage of atomic weight of carbon at the combustion temperature of 400° C is 72.70%, and 26.62% are oxygen atoms [5] and is based on the measurement results obtained at the BET 90μm particle size provides the largest surface area is 301 482 m^2 / g , Therefore for further research of carbon used is from carbon palm shells with the size of 90 μm.

3.1. Effect of of carbon paper size

Figure 2 indicates that the larger the surface area of an electrode, the greater the capacitance value. This is because the larger the surface area of the electrode, the greater the ability of the electrodes to store charge. Large surface area electrodes will provide a place for the storage of electrical charge layer formed on the electrode surface.

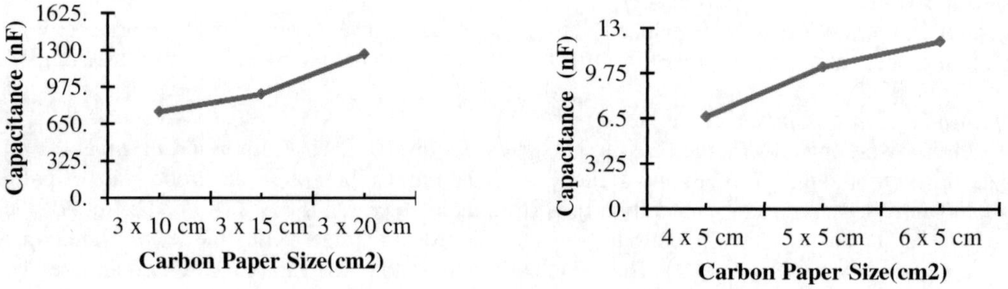

a b

Figure 2. Effect of the size of the carbon paper on the capacitance using 0.1 N H_3PO_4 electrolyte and the charging time of 60 minutes (a) the method of rolling, (b) the method of sandwich.

Comparison of capacitance values in both methods with the same surface area (30cm^2) showed the differences in the ability of supercapacitors charge storing. Supercapacitor with the rolling method has a capacitance value of 63 times higher than it is in the sandwich method.

3.2. Effect of charging time

Figure 3 shows that the capacitance value for both methods of rolling or sandwich increased progressively with increasing charging time to 60 min and then decreases with increasing time charging

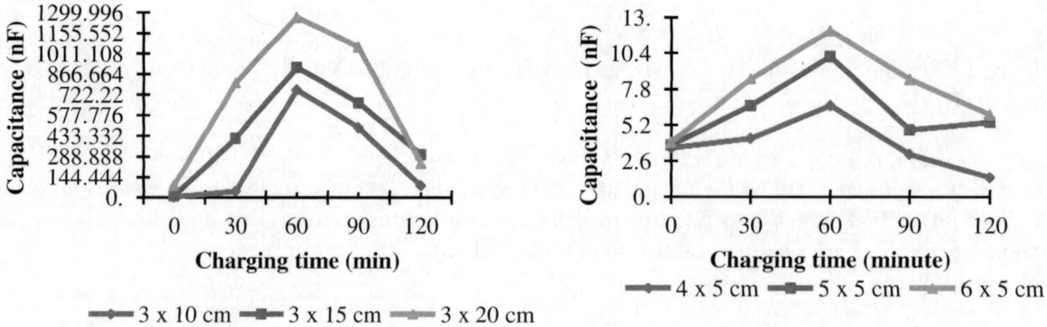

a b

Figure 3. Effect of time of charging at the electrode capacitance size variation carbon paper (a) the method of rolling, (b) the method of sandwich.

In the method of scrolling obtained capacitance was 105 time higher than it is in the sandwich method at the optimum charging time of 60 min.

3.3. Effect of time charging on the conductivity electrode

Conductivity is calculated based on the measured resistance value of supercapacitor by rolling and sandwich methods of a surface area of the carbon paper as 3 x 20 cm^2 on the rolling method and carbon paper size of 6 x 5 cm^2 on the sandwich method.

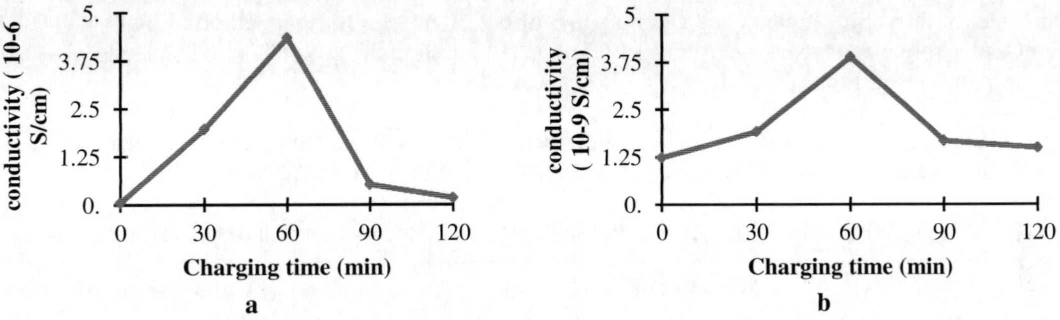

Figure 4. Effect of charging time to the conductivity of the electrode (a) the method of scrolling (b) the method of sandwich.

Figure 4 shows the influence of the charging time on the conductivity where is the conductivity increased by the charging time of 60 min, and then decreased after 60 min.

3.4. Effect of H_3PO_4 concentration

H_3PO_4 electrolyte concentration variations performed on carbon paper sizes 3 x 20 cm^2 on the rolling method and carbon paper size 6 x 5 cm^2 on the sandwich method with a charging time of 60 min.

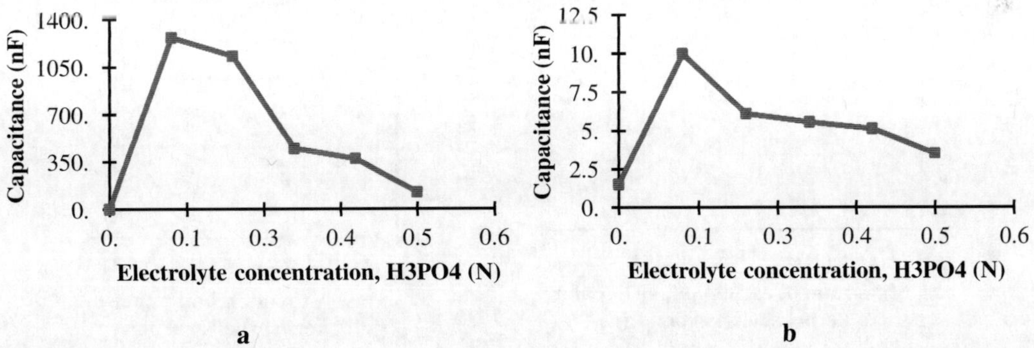

Figure 5. Effect of H_3PO_4 concentration on the capacitance supercapacitor (a) the method of rolling, (b) the method of sandwich.

Figure 5 shows higher concentration of H_3PO_4, lower capacitance of the supercapacitor. Greater concentration of electrolyte, electrolyte ions will be accumulated at electrolyte and separator interface during charging. The great number of ions will disturb the discharge process because of the bulk ions makes the ions or the charge is difficult back to its initial position.

3.5. The effect of addition of carbon palm shells on carbon paper on the value of the supercapacitor capacitance

The addition of carbon in the carbon electrode supercapacitor will improve the ability of charge storage [2]. Therefore, in this study, the addition of palm kernel shells carbon on carbon paper has been considered. Carbon palm shells are added in the variation mass ratio between carbon paper and palm kernel shells carbon.

The optimum addition of carbon of the rolling and the sandwich method is on a carbon paper of 3 x 20 cm^2 size for the rolling method and carbon paper of 6 x 5 cm^2 size for the method of plate with 0.1 N H_3PO_4 electrolyte and the charging time of 60 min. The addition of carbon from palm shells was made by variations of mass ratio as 1: 1, 1: 2, 1: 3, 1: 4 and 1: 5.

In the rolling method by addition of oil palm shell carbon on the carbon paper electrodes show the capacitance values as 32 times higher than carbon electrode made of palm kernel shells.

Table 1. Differences in capacitance of the supercapacitor electrode material on a rolling method

Treatment	Capacitance (μF)
Without addition of palm kernel shells carbon	1.2665
With addition of palm kernel shell carbon in ratio of 1: 2	1331.8

Table 1 shows that the addition of carbon palm kernel shell can increase the capacitance equal to 1052 times greater than without the addition of the carbon. Carbon has pores that will absorb and store incoming charge. By addition another carbon than carbon paper will increase the amount of carbon occuring in the carbon paper surface so that more charge will be stored by these new electrodes.

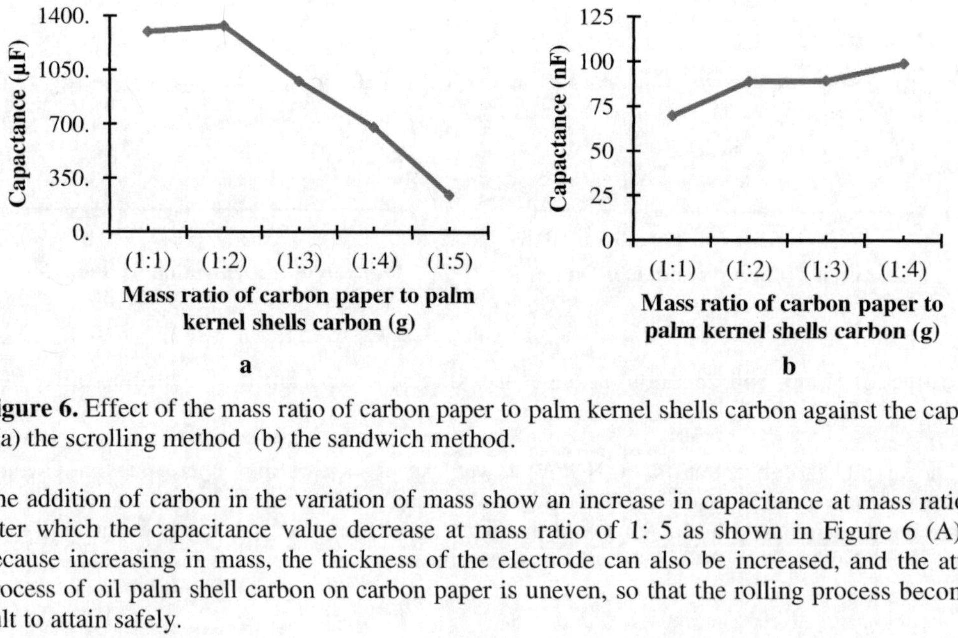

Figure 6. Effect of the mass ratio of carbon paper to palm kernel shells carbon against the capacitance : (a) the scrolling method (b) the sandwich method.

The addition of carbon in the variation of mass show an increase in capacitance at mass ratio of 1: 2, after which the capacitance value decrease at mass ratio of 1: 5 as shown in Figure 6 (A). This is because increasing in mass, the thickness of the electrode can also be increased, and the attachment process of oil palm shell carbon on carbon paper is uneven, so that the rolling process becomes difficult to attain safely.

Figure 6 (B) shows that the capacitance will increases with the addition of palm kernel shells carbon mass. On the sandwich method, palm kernel shell carbon attachment to the carbon paper is more easily prepared because that paper used smaller carbon size.

Table 2. The capacitance value of supercapacitors without and with addition of palm kernel shell carbon.

Treatment	Capacitance (nF)
Without addition of palm kernel shells carbon	12.043
With addition of palm kernel shell carbon in ratio of 1: 4	99.05

Table 2. shows that the addition of carbon palm shells can increase the capacitance value of 8.22 times higher than without the addition of this carbon.
Based on the capacitance value it can be seen that the method of rolling has higher capacitance values than the sandwich method.

4. Conclusion

In this present study show that the carbon paper made from palm kernel shells is capable to increase supercapacitor capacitance by using both methods, namely the rolling method and sandwich method. The rolling method has higher capacitance values (1331.8 μF on carbon paper size of 3 x 20 cm^2) than the sandwich method (1.2665 μF on carbon paper size of 6 x 5 cm^2), the optimum charging time found to be in 60 min and optimum concentration of electrolyte found as 0.1 N H_3PO_4. The addition of palm kernel shell carbon on the carbon paper can increase the capacitance value where is found respectively as 1052 times in the rolling method and 8.22 times in the sandwich method which were higher than values obtained without the addition of palm kernel shells carbon.

References

[1] Frackowiak E 2006 supercapacitors based on carbon materials and ionic liquids *J. Braz.Chem. Soc.* **17** 1074-82
[2] Kamikuri N, Hamasuna Y, Tashima D, Fukuma M, Kumagai S, John D W, and Madden W 2014 Low cost activated carbon materials produced from used coffee grounds for electric double-layer capacitors *International journal of engineering science and innovative technology (IJESIT)* **3** 492-501
[3] Jiang L, Yan J, Hao L, Xue R, Sun G, and Yi B 2013 High rate performance activated carbons prepared from ginkgo for electrochemical supercapacitors *Carbon* **56** 146-54
[4] Peng C, Yan X B, Wang R T, Lang J W, Qu Y J and Xue Q J 2013 Promising activated carbons derived from waste tea-leaves and their application in high performance supercapacitors electrodes *Electrochimica Acta.* **87** 401-8
[5] Hermansyah A, Tetra O N, Alif A and Syukri W R 2016 Electrical properties of supercapacitor electrode - based on activated carbon from waste palm kernel shells *Der Pharma Chemica* **8** 227-232
[6] Gultom E M and Lubis M T 2014 Aplikasi karbon aktif dari cangkang kelapa sawit dengan aktivator h3po4 untuk penyerapan logam berat Cd dan Pb *Jurnal Teknik Kimia* **3** 5-10
[7] Kurniati E 2008 Pemanfaatan cangkang kelapa sawit sebagai arang aktif *J. Penelitian Ilmu Teknik* **8** 96-103
[8] Danga W, Dong C, Zhang Z, Chen G, Wang Y and Guan H 2016 Self-grown MnO_2 nanosheets on carbon fiber paper as high-performance supercapacitors electrodes *Electrochimica Acta* **217** 16–23

[9] Wang G, Zhang Y, Zhou F, Sun Z, Huang F, Yi Y, Chen L and Mu P 2016. Simple and fast synthesis of polyaniline nanofibers/carbon paper composites as supercapacitor electrodes *Journal of Energy Storage* **7** 99-103

IOP Conf. Series: Earth and Environmental Science **65** (2017) 012054 doi:10.1088/1755-1315/65/1/012054

Varieties of sweet sorghum Super-1 and Super-2 and its equipment for bioethanol in Indonesia

M B Pabendon[1], R Efendi[1], S B Santoso[1] and B Prastowo[2]

[1]Indonesia Cereals Research Institute, Maros, Indonesia
[2]Indonesia Center for Estate Crops Research and Development, Bogor, Indonesia

Email : marcia.pabendon@gmail.com

Abstract. The demands for alternative sources of energy are currently growing because people now are more aware of the many negative impacts fossil fuel gives to the environment. Plant based renewable energy provides potential sources of energy with advantages of cleaner fuel effect and capability of integration with food crop production. Sorghum have been considered to be a highly potential source of food, feed and fuel, especially sweet stalk sorghum that posses both functions as source of food from its grain and fuel made from its stalk juice. Sorghum varieties are well known to have excellent adaptability in marginal areas, especially drought prone areas where other food crops are unable to thrive. The current paper aimed to share ongoing research on many functional uses of sweet stalk sorghum varieties released at Indonesian Cereals Research Institute (ICERI). Among many varieties that had been released were two sweet stalk sorghum varieties SUPER-1 and SUPER-2 released in 2013 that stands out in biomass yield and bioethanol production. Based on various researches conducted at different location and planting season, yield potential of biomass ranged at 30 - 40 t/ha with higher yield occurred during dry season. Stalk juice sugar content in brix were found to be higher in dry season ranged at 13.6 % to 18.4 %, and the amount of juice stalk was obtained at about 30-50 % from total biomass yield. Furthermore, bioethanol production from stalk juice after fermentation was at the range of 8 to 10 % from total stalk juice volume. Modification of processing equipment of bioethanol have also been carried out and was able to increased the concentration of ethanol being distilled from 85% -92% to 90% -95%. Another result obtained was able to decreased fermentation time from 14-21 days to 6-10 days. Furthermore, the yield of ethanol from juice was also from an average of 4.95% to 6.75%.

1. Introduction

Sorghum is a cereal type of plat that has multiple functions, it could be used as food, feed and in recent trend it also use as a source of ethanol (fuel). The potential of utilizing sorghum as a main source for ethanol production has drawn many research and studies performed in many Indonesian research institutions. In Indonesia especially, production of ethanol mainly utilize sugarcane molasses and maize kernel as its raw materials. And the demand from the consumer to use ethanol as the new source of renewable energy grows rapidly more than the supply can fulfill [1]. Consequently, with the increasing of demand, ethanol production will be utilizing more toward sustainable raw materials that do not compete with major commodities such as food crop both in term of land use and supply of materials. It is also important not to risk national food security in deciding which crop that is suitable as a source of ethanol. According to Schaffert *et al.* [2], crisis of fuel oil occurred in 1970 was the reason that caused sweet stalk sorghum began to be used for ethanol to substitute fuel oil. And in the last few years sweet sorghum has increasingly been used to produce free-fermentation sugar, which

Content from this work may be used under the terms of the Creative Commons Attribution 3.0 licence. Any further distribution of this work must maintain attribution to the author(s) and the title of the work, journal citation and DOI.
Published under licence by IOP Publishing Ltd

could potentially derived to substitute fuel oil, food, feed and various other product [3]. Because of these advantages, sorghum is the crop that has promising feature as a source of biofuel, food and feed. The function of sweet stalk sorghum is equal to what sugarcane offer, because it contains high sugar content which can produce ethanol through fermentation process [4].

Sweet sorghum (*Sorghum bicolor* [L.] Moench), which later can be divided into grain sorghum and sweet stalk sorghum are highly water efficient crop that requires low amount of water intake for its biomass growth and grain production. It can be cultivated on marginal land, mainly drought prone areas where other crops cannot survive [5]. The morphology of sweet sorghum is relatively different from grain sorghum, which is generally have taller plant height and possess high biomass [6]. Furthermore, stalk of sweet sorghum produce juice that has high sugar content, which can be use as a main source of ethanol production [7,8], whereas grain of sorghum can also be used as raw materials for food, feed and functional food [9]. Beside for bioethanol, sorghum grain and juice can also be use to produce other kind of industrial products such as bioplastic, beverages and also syrup.

Factors affecting biomass and juice yield are genetic of the varieties and the environment of the cultivation areas. Sweet stalk sorghum have higher biomass than grain sorghum, however yield of biomass also depend on the precipitation rate during anthesis and drought stage [10]. Another factor that can influence sweet sorghum is growing season, where longer growing season in temperate climate can affect and improve biomass and stalk juice yield [11]. Brix content of stalk juice is strongly controlled by the genetic each varieties posses. The amount of Nitrogen fertilizer added does not significantly affect brix content of sweet sorghum [10]. Beside genetic, factor that likely affects level of brix content is water consumption which can either be by rain precipitation of by irrigation during cultivation [12].

Among many varieties that have been released by Indonesian Department of Agriculture, almost all of those varieties were grain sorghum aimed for food. However recent release of Super-1 and Super-2 inbred varieties in 2013 were the first sweet stalk sorghum ever release in Indonesia. These varieties specifically breed, select and aimed to become potential source of bioethanol. Nevertheless, its grain can still be utilizes as a food source and the remains of the stalk biomass (bagasse) can be use as silage for cattle feed or compos manure.

Researches on the utilization of sorghum for biofuel in the form of bioethanol have been intensively carried out by Indonesian Cereals Research Institute (ICERI) as one of their mandate crop. Various data used in this review were mainly obtained from several research conducted at ICERI, ranges from research on grain yield, biomass potential and sorghum juice brix content for bioethanol.

2. Results and analysis

2.1. *Prospect of Sweet Stalk Sorghum Varieties Super-1 and Super-2 used for Bioethanol*

The quality feature of sorghum that suitable to be use for bioethanol production is determined by its high sugar content in stalk or the amount of carbohydrate in grain. Whereas the protein content in grain and lignocellulosic in the stalk is an added value [13]. However, the most secure and practical parts to produce ethanol is to use the juice from the stalk. The grain can be used for food, while bagasse containing cellulose can be used for silage (fodder) or compost. Higher rate of ethanol production derived from stalk juice is determined by the amount of stalk biomass weight, stalk juice volume and sugar content.

Among many parts of sorghum plant, the stalk gives highest contribution to produce juice as raw material for bioethanol [14]. Stalk biomass production of sorghum in average ranged between 30-50 t/ha, and leaf biomass gained 20-40 t/ha. To produce 1 liter of bioethanol requires 22-25 kg sorghum stalk [13]. Result of research conducted at Indonesian Cereals Research Institute in 2006 showed that ethanol production from stalk juice ranges between 415-2.576 l/ha. This result was similar to what had been done by Sumaryono [9] which gained 2.300 l/ha bioethanol. Another study carried out by Singgih [15] and Pabendon *et al.* [16] stated that ethanol production was determined by the production of biomass and sugar content of the stalk.

Sorghum is a food crop that has high response in its growth and yield due changes in climate, especially in Indonesia that have two main seasons, which are rainy season (MH) and dry season (MK). Response of Super-1 and Super-2 varieties is presented in Figure 1, it shows production potential of stalk biomass from both varieties was higher in dry season compare to rainy season. Even though average yield of biomass gained in dry season of Super-1 (25.20 t/ha) and Super-2 (28.75 t/ha) were relatively below their potential, nevertheless the true potential of these variety ranges from 38.7 t/ha and 39.30 t/ha. Lower biomass yield in rainy season can be resulted from high leaching of fertilizer applied, which eventually reduced N uptake available for plant [10]. Juice volume was relatively higher in rainy season for Super-1, while juice yield of Super-2 gained similar result in both seasons. Sugar content in brix was higher in dry season than rainy in both Super-1 (15.10 %), and Super-2 (13.90 %). Study also obtained data that during dry season, brix percentage tends to increase at 17 – 21 % in both varieties. Similar finding occurred in sugarcane where sugar level in stalk decreased during rainy season. Furthermore, ethanol production was higher in dry season compare to rainy, and ethanol yield obtained higher in Super-1 (1.199,33 l/ha) while Super-2 at 1.187,92 l/ha. Similar study obtained by Smith and Buxton [17] in non-irrigated sorghum sugar concentration increased versus irrigated location. Furthermore, Pabendon *et al.* [18] also reported that stalk biomass, brix level in stalk juice had high correlation to sweet sorghum ethanol production.

Sorghum plant is sensitive to photoperiodism or relative length of day and night. Sorghum requires at least 12 hours to trigger the internal mechanism that initiate growth, reproduction, flowering and seed development [19]. This is because sorghum is a type of plant that has high efficiency to convert CO_2 to become sugar compare to other crop such as sugarcane or maize.

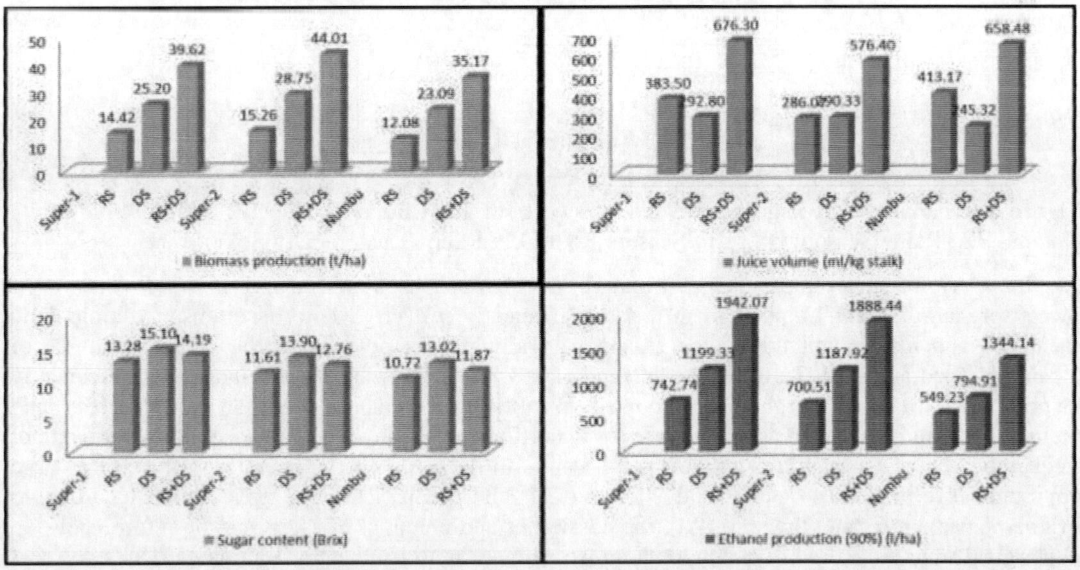

Figure 1. Diagram of average yield stalk biomass, juice volume, brix percentage and ethanol production of sweet stalk sorghum Super-1, Super-2 and control variety Numbu . RS : Rainy Season; DS : Dry Season; RS + DS : Rainy Season + Dry Season.

The grain of Super-1 and Super-2 can be use as functional food, therefore there are advantages in added value by cultivating sweet sorghum. Because not only the main purpose of bioethanol production can be achieved, addel value of obtaining food security can also be achieved. At Indonesian Cereals Research Institute (ICERI), the grain of Super-1 and Super-2 are commonly consumed as "sorghum rice" and can be made into other kind delicacies such as cookies and cakes. Average grain yield in Figure 2 showed that during dry season the grain yield was higher compare to rainy season, and Super-2 variety obtained higher yield than Super-1. One weakness of Super-2 is its susceptibility to sorghum anthracnose disease, which can infect on the stalk and affect the quality and quantity of stalk juice. However, this disease is specific to small region so in order to develop Super-2 in vast area requires vigilant selection of location.

In general the direction in developing both sweet sorghum varieties are mostly suitable in arid and dry climate region, because in addition to higher ethanol and grain yield in both quantity and quality, it can also avoid the high percentage of pest (Aphids) and anthracnose infection. Another reason is that to prevent the decrease of sugar content in stalk, it is best cultivating when the generative growth of plant occurred during dry season. Because with higher amount of rainfall, usually sugar content decrease below 10 % Brix, whereas cultivating at dry season can increase sugar content at value of 18-21 % Brix.

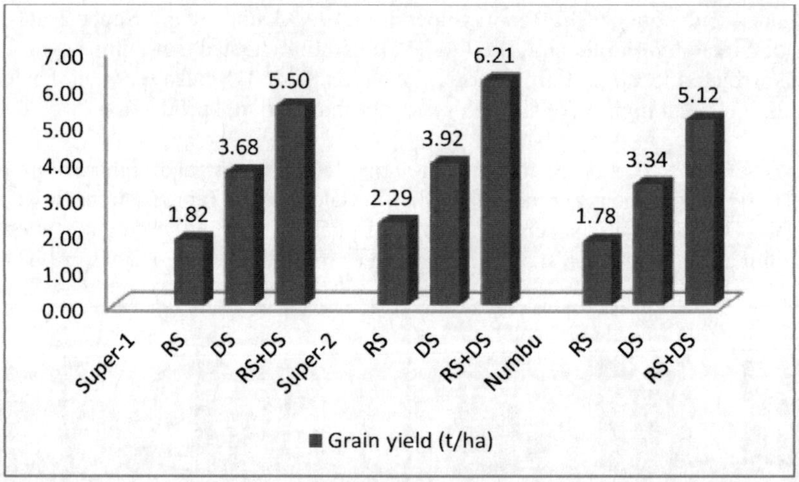

Figure 2. Diagram of average grain yield of sweet sorghum Super-1, Super-2 and control variety Numbu. RS : Rainy Season; DS : Dry Season; RS + DS : Rainy Season + Dry Season.

2.2. *The Potential in Developing Super-1 and Super-2 Varieties as a Source of Sustainable Energy*

Sweet sorghum is special type of plant that can efficiently convert CO_2 to become sugar compare to sugarcane and maize, and this is the reason why sorghum is potential to be use as a source of bioenergy, food and feed. To fulfill the demand of raw materials in bioethanol industry, consequently planting areas of sweet sorghum are suppose to cultivate throughout the year on vast areas, and preferably not to interfere areas that are use for major food crops. In the areas susceptible to drought of frequently get waterlogged, sorghum plant can still be cultivated. Therefore, there are large opportunities to increase sorghum production through land extensification of agriculture. According to Trikoesoemaningtyas and Suwarto [20], the possibility and potential of sweet sorghum to be develop in marginal areas are still vastly open, both in wet climate areas (Sumatera, Kalimantan, Sulawesi and Papua) and dry climate areas (Nusa Tenggara Timur, Sulawesi Tenggara, and parts of Sumatera and Java). Dry climate areas that have land topography waving slope (slope < 8 %) are also suitable for sorghum plantation. Previous data showed that Super-1 and Super-2 varieties are suitable and

IOP Conf. Series: Earth and Environmental Science **65** (2017) 012054 doi:10.1088/1755-1315/65/1/012054

excellent to be developed in dry areas. Bennet and Anex [21] stated that juice production from sweet sorghum that is sustainable can be performed by on-farm method.

Another advantage of cultivating sorghum is its ability to grow new shoot (ratoon) from previously harvested stalk, and the shoot will grow into new plant that can be harvested for the second time [22,23]. According to Tsuchihashi and Goto [24], sorghum plant can be harvested twice or third times which include primer plant and its ratoon. Therefore, it can supply the demand for raw materials for carbohydrate, green fodder or bioethanol in sustainable ways. By utilizing the high ability of ratoon in sorghum plant, cultivating sorghum can increase in its efficiency because it can reduce labor cost, planting time, land tillage, seed usage and energy. Cultivation sorghum by ratoon system has been carried out in India, Hawaii, Australia, Philippine, Indonesia, USA, and Africa [25,26]. In ratoon system, cultivation of sorghum started by planting seed for first harvest, followed by maintaining ratoon plant through second and third harvest, which is a cultivation system that can meet the demand of sustainable raw materials for biomass or seed.

Study of ratoon ability perfomed by Effendi *et al.* [27] in rainy season on several sorghum genotypes showed that Super-2 and Super-1 had low ratoon ability, with ratoon percentage at 33-44 % and also low brix *percentage* in stalk (9 %). However, in order to understand the effect of ratoon in dry season there should be another study performs on those two varieties.

2.3. *Sweet Sorghum Processing Machine Modification*

Simple mechanical device that was manufactured to produce ethanol currently was able to obtained ethanol at 1.850-2.500 liters with alcohol percentage of around 85-92 %. The period time of fermentation roughly 14 – 21 days by adding only bread yeast (*Saccharomyces cerevisiae*). The amount of ethanol produced was highly related and affected by the method being used to ferment stalk juice [28-32]. And the hours needed for distillation usually take about 9-10 hours. Modification applied on ethanol processing machine was aimed to increase ethanol productivity and quality, such as effort in increasing ethanol percentage at 90 – 95 % with less fermentation period (4-6 days) and shorter distillation time. Modification that performed were include changing fermentation process from mechanically into fully automated, it was done by stirring the juice from previous manual stirring every 3 days for 5-10 minutes to become automated stirrer attached on to the machine. Result of the modification showed that by stirring for 24 hours at stirring speed of 30 rpm, fermentation period was able to decrease from 22 days to only 5 days with remaining percentage of brix at 4.9 – 6.2 % (Figure 2).

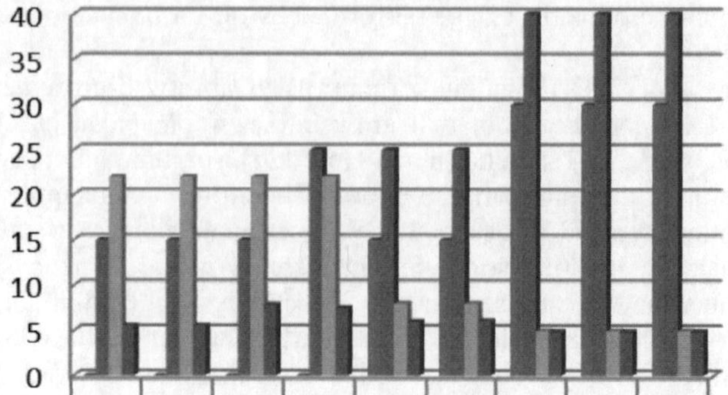

	1	2	3	4	5	6	7	8	9
Rotation speed (rpm)	0	0	0	0	15	15	30	30	30
Juice volume (L)	15	15	15	25	25	25	40	40	40
Fermentation time (days)	22	22	22	22	8	8	5	5	5
Brix final level (%)	5.6	5.6	7.9	7.5	6.0	6.1	4.9	4.9	4.9

Figure 3. Fermentation period of sweet stalk sorghum based on stirring speed level.

Result of ethanol distillation showed that there was an increase in ethanol percentage at >90%. The time period and volume of ethanol obtained cannot be known with certainty because the volumes of stalk juice were lower compare to previous research at 40 liters. Therefore, the modification will still in progress in order to gain expected and optimum results

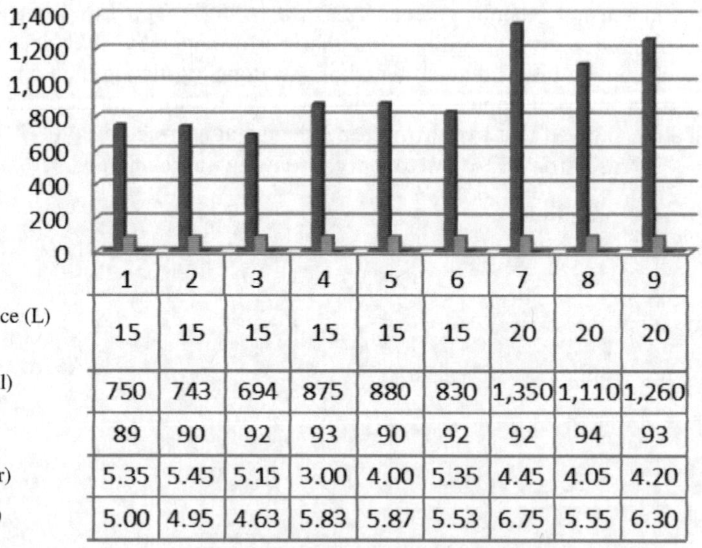

	1	2	3	4	5	6	7	8	9
Fermented stalk juice (L)	15	15	15	15	15	15	20	20	20
Ethanol volume (ml)	750	743	694	875	880	830	1,350	1,110	1,260
Ethanol (%)	89	90	92	93	90	92	92	94	93
Distillation time (hr)	5.35	5.45	5.15	3.00	4.00	5.35	4.45	4.05	4.20
Brix final level (%)	5.00	4.95	4.63	5.83	5.87	5.53	6.75	5.55	6.30

Figure 4. Ethanol production by distillation of stalk juice after modification; No. 1-3, distillation results before modification

2.4. *Challenges in developing Super-1 and Super-2 Varieties*

Challenges in developing these sweet sorghum varieties consist of cultivation technology aspect, post harvest technology and market and demand guarantee. Other challenges are dissemination and socialization of both varieties as a source of renewable energy. Even though cultivation technology of specific location for sorghum is not yet available, nevertheless number of sweet stalk sorghum varieties is still needed to be increase, especially sorghum for bioenergy. Major challenges that constraint sustainability of renewable energy is the guarantee of market demand, and raw material supply, so that farmers would be interested in cultivating sorghum. Market for sorghum both for grain and biomass is not yet conducive at regional or national level. Moreover, dissemination of information about sorghum potential and training in the social economy of sorghum farming at farmer level is not yet intensive. One problem often become constrain is the availability of seed in term of variety, amount, quality, time and location is low. Even when in time that many farmers are willing to cultivate and grow sorghum.

In developing sorghum for bioindustry, close connection are required between government policies, farmer producer, and other industry related starting from research (technology assembly), development (technological transfer), production (provision of production means), agribusiness practitioner (collecting, storing, marketing and processing), and utilization of goods. Supporting data on sweet sorghum demand for bioethanol production cannot appropriately presented because the productivity of sweet sorghum in Indonesia fluctuate between seasons, and also is yet to become national priority. This is due to several reasons such as : (a). Sorghum market is still very limited, and the reality on the field shows that sorghum has not been widely known to the public; (b). Sorghum cultivation is highly influence by current issues and trend occurred in public; (c). Level of planting areas have not yet achieve stable number because there is no guarantee of utilizing sorghum for specific purposes; and (d). Whenever issues and trend of alternative biofuel are brought up, farmers are very eager to cultivate sorghum. However, when the price of fuel oil decrease and bioenergy issues become less discussed, the farmer become harder to sell their sorghum products. This is the reason they would be reluctant to grow sorghum for the next season.

The reality found on the field showed that sorghum plant has yet publicly known by the people, especially farmers. One of many factor that cause underdeveloped sweet sorghum is that sorghum market is still very limited to small number of regions. Farmers will be interested in developing sweet sorghum if there are other benefits that can be obtained, such as utilizing sorghum grain for food or functional food. However, some areas have long known as the center for sorghum production in Indonesia, these districts of sorghum producer are presented in Table 1.

Table 1. Province and district of sorghum producer in Indonesia based on Directorate General of Food Crop, Department of Agriculture (2007)

Province	District of sorghum development
West Java	Indramayu, Cirebon, Kuningan, Ciamis, Garut, Cianjur dan Sukabumi
Central Jawa	Tegal, Kebumen, Kendal, Demak, Grobogan, Boyolali, Sukoharjo dan Wonogiri
DI. Yogyakarta	Kulon Progo, Sleman, Bantul dan Gunung Kidul
East Java	Pacitan, Bojonegoro, Tuban, Lamongan, Bangkalan, Pamekasan, Sampang, Sumenep, Pasuruan, Probolinggo, Malang dan Lumajang
West Nusa Tenggara	Lombok Tengah, Sumbawa, Dompu dan Bima
East Nusa Tenggara	Sumba Barat, Sumba Timur, Manggarai, Ngada, Ende, Sikka, Flores Timur, Lembata, Alor, Timor Tengah Utara, Kupang, Belu, Timor Tengah Selatan dan Rote Ndao

Areas in East Nusa Tenggara (NTT) with dry climate and many other drought prone and marginal areas in NTT are highly potential to develop sorghum, especially Super-1 and Super-2 which are suitable varieties for dry climate. By developing sorghum in these particular areas, source of renewable energy, food and feed can be fulfill in NTT, hence requires priority.

3. Conclusion

Sweet stalk sorghum potential for bioethanol production was proven to be highly suitable to be develop in many parts of Indonesia. Breeding varieties of Super-1 and Super-2 was an effort to socialize and disseminate sweet stalk sorghum for its multiple functions for food, feed and fuel. From various researches conducted at ICERI the results obtained were high biomass yield of both varieties in two different seasons, and both varieties was able to thrive and gained more yield during dry season. This is an indication that both varieties are excellent crop to be cultivated in drought prone areas of Indonesia especially in East Nusa Tenggara. Brix content of sweet sorghum was strongly influence by genetic, however environment factor of water intake affect sugar content of stalk juice. During the dry season result obtained brix content was higher compared to rainy season. On the contrary, the amount of stalk juice was relatively higher in rainy season for Super-1 as compare to dry season. Machine modification during fermentation through addition of stirring component effectively increased fermentation from 22 days to only 5 days. Furthermore, distilled ethanol was improved with purity percentage > 90 %, this mean higher efficiency occurred during distillation.

References

[1] Kennedy D and Turner J A 2004 Sustainable hydrogen production *Science* **305** 72- 974.

[2] Schaffert RE 1992 Sweet sorghum substrate for industrial alcohol in: Utilization of sorghum and millets *Proceedings of the International workshop on policy practice and potential relating to uses of sorghum and millets* 8-12 Feb 1988 ICRISAT Center Bulawayo Zimbabwe International Crops Research Institute for the Semi-Arid Tropics p 131-137

[3] Chiaramonti D G, Grassi A, Nardi and Grimm HP 2004 *ECHI-T: large bioethanol project from sweet sorghum in China and Italy* Energia Trasporti Agricoltura Florence Italy

[4] FAO 2002 *Sweet Sorghum in China Agriculture and consumer protection* Food agricultural organization of United Nations Department

[5] McLaren JS N Lakey and J Osborne 2003 *Sorghum as a bioresources platform for future renewable resources Proc 57th Corn and Sorghum Research Conference* CD ROM (Alexandria :American Seed Trade Association)

[6] Turhollow A F, Webb E G and Downing M E 2010 Review of Sorghum Production Practices: Applications for Bioenergy (America : U.S. Department Of Energy) Available from http://infoornlgov/sites/publications/files/Pub22854pdf
[7] Hons F M, R F Moresco, R P Wiedenfeld and JT Cothren 1986 Applied Nitrogen and phosphorus effects on yield and nutrient uptake by high-energy sorghum produced for grain and biomass *Agronomy Journal* Vol **76** 1069-1078
[8] Reddy N and Yang Y 2005 Biofibers from agricultural by products for industrial applications *Trends in Biotechnology* **23** 22-27
[9] Sumaryono W 2006 Kajian komprehensif dan teknologi pengembangan bioetanol sebagai bahan bakar nabati (BBN) *Bioenergy Seminar : Prospek Bisnis dan Peluang Investasi* Agency for the Assessment and Application of Technology December 6, Indonesia
[10] Maw M J W, Houx III J H and F B Fritschi 2016 Sweet sorghum ethanol yield component response to nitrogen fertilization *Industrial Crops and Products* **84** : 43-49
[11] Houx III J H and Fritschi F B 2013 Influence of midsummer planting dates on ethanol production potential of sweet sorghum *Agronomy Journal* **105** : 1761-1768
[12] Tamang P L, Bronson K F, Malapati A, Schwartz R and Johnson J and Moore-Kucera J 2011 Nitrogen requirements for ethanol production from sweet and photoperiod sensitive sorghums in the southern high plains *Agronomy Journal* **103** : 431-440
[13] Yudiarto M A 2006 Pemanfaatan sorgum sebagai bahan baku bioetanol Focus Group Discussion Prospek Sorgum dalam Mendukung Ketahanan Pangan dan Energi September 5, Serpong-Indonesia
[14] Almodares A and Hadi M R 2009 Production of bioethanol from sweet sorghum: A review *African J. Agri.* **4** 772-780
[15] Singgih S 2006 Kajian pendahuluan berbagai plasma nutfah sorgum sebagai bahan bioteanol Laporan akhir tahun 2006: koleksi rejuvenasi karakterisasi dan evaluasi plasma nutfah serealia Balai Penelitian Tanaman Serealia Maros
[16] Pabendon M B, Mas'ud S Sarungallo R S dan Nur A 2012 Penampilan fenotipik dan stabilitas sorgum manis untuk bahan baku bioetanol *Penelitian Pertanian Tanaman Pangan* **31** 60-69
[17] Smith G A and Buxton D R 1993 Temperate zone sweet sorghum ethanol production potential *Bioresource Technol.* **43** : 71-75
[18] Pabendon M B, Sarungallo R S and Mas'ud S 2013 Pemanfaatan Nira Batang Bagas dan Biji Sorgum Manis sebagai Bahan Baku Bioetanol *Penelitian Pertanian Tanaman Pangan* **31** 180-187
[19] Qingshan L and J Dahlberg 2001 Chinese sorghum genetic resources *Economic Botany* **55** 401-425
[20] Trikoesoemaningtyas and Suwarto 2006 Potensi pengembangan sorgum di lahan marginal *Paper in Focus Grup Discussion "Prospek Sorgum untuk Mendukung Ketahanan Pangan dan Energi"* MENRISTEK (Ministry of Research, Technology and Higher Education of the Republic of Indonesia) - BATAN September 26, Serpong-Indonesia.
[21] Bennett A S and Anex R P 2008 Farm-Gate Production Costs of Sweet Sorghum as a Bioethanol Feedstock *Transactions of the ASABE* **51** 603-613
[22] Duncan R R, Mrllen F R, and Bocrnolr A J 1980 Inheritance of tiller regrowth in ratooned sorghrim *Can. J. Plant Sci.* **60** 473-478
[23] Livingston S and Coffman D 2003 Ratooning grain sorghum on the Texas Gulf Coast (http://soilcrop.tamu.edu/publications/pubs/ l1568.pdf)
[24] Tsuchihashi N and Goto Y 2004 Cultivation of sweet sorghum (Sorghum bicolor (L.) Moench) and determination of its harvest time to make use as the raw material for fermentation, practiced during rainy season in dry land of Indonesia *Plant Prod. Sci.* **7** 442-448
[25] Gardner B Pearce and Mitchell RL 1991 *Physiology of crop plants* (America : The Iowa State University)

[26] Enserink H J 1995 *Sorghum agronomy in West Kenya: investigations from a farming systems perspective* (Amsterdam : Royal Tropical Institute)

[27] Efendi R M, Aqil dan Pabendon M 2013 Evaluasi genotipe sorgum manis (Sorghum bicolor (L) Moench) produksi biomas dan daya ratun tinggi *Penelitian Pertanian Tanaman Pangan* **32** 116-125

[28] Asli M S 2010 A study on some efficient parameters in batch fermentation of ethanol using Saccharomyces cerevisiae SC1 extracted from fermented siahe sardasht pomace *African Journal of Biotechnology* **9** 2906-2912

[29] Liu R and F Shen 2008 Refining sweet sorghum from stalk juice of sweet sorghum by immobilized yeast fermentation *Renewable Energy* **33**:1130-1135

[30] Sipos B J, Reczey Z and Somorai Z Kadar D Dienes and K Reczey 2010 Sweet sorghum as feedstock for ethanol production: enzymatic hydrolysis of steam-pretreated bagasse *Application of Biochemical Biotechnology* **8** 8423-8429

[31] Jones A M, Thomas K C and Ingledew W M 1994 Ethanolic fermentation of blackstrap molasses and sugarcane juice using very high gravity technology *Journal of Agricultural and Food Chemistry* **42** 1242–1246

[32] Kundiyana D K, Bellmer D D, Huhnke R L, Wilkins M W and Claypool P L 2010 Influence of temperature pH and yeast on in-field production of ethanol from unsterilized sweet sorghum juice *Biomass and Bioenergy* **34** 1481-1486

IOP Conf. Series: Earth and Environmental Science **65** (2017) 012055 doi:10.1088/1755-1315/65/1/012055

Production of gasohol from isobutanol

Z Aziz[1] and M Gozan[1]

[1]Chemical Engineering Department, Faculty of Engineering, University of Indonesia, Depok, Indonesia

Email : zarkoni_azis@yahoo.com

Abstract. Butanol is a four carbon chain alcohol compound of a non-polar used as a solvent and as an intermediate in several consumer products. It can be produced from petrochemical process route as well as biochemical process. A common petrochemical route to produce butanol is hydroformylation of propylene as primary feedstock followed by hydrogenation. The increasing demand for butanol and the scarcity of petrochemical basestock has put the urgent need for a renewable resource of butanol production. Biobutanol is butanol produced from renewable, biological feedstock through fermentation of plant materials types of food, non-food and biomass. One isomer of butanol which is more similar with gasoline property is isobutanol. The use of isobutanol is commonly as solvent in coatings industry, as lacquers, melamine or phenolic resin. Isobutanol can be potentially used as a gasoline blending component, and it is better than ethanol due to its higher heating value, lower volatility and less corrosive. A brief review of process technology for butanol production is outlined in this paper. The benefit of isobutanol over ethanol was also overviewed. In order to prove the compatibility on fuel application, the production of gasohol was carried out by mixing isobutanol and base gasoline, then evaluated according to commercial fuel specification. From the research work, it has been shown that the addition of isobutanol in base gasoline by certain volum fraction has met the typical fuel specifications used for spark-ignition engine. The physical and chemical properties of the mixture was quite similar to gasoline 88 and gasoline 92 specifications.

1. Introduction

Gasoline or petrol is petroleum derived liquid mixture having 5 to 12 carbon atoms per molecul, and it is primarily used as a fuel in spark-ignition engines. It is an aromatic, flammable liquid that mainly consist of hydrocarbon mixture of alkanes, cycloalkanes and alkenes, with their actual ratio depends on the oil refinery especially crude oil type and grade of gasoline product.

The increasing demand of gasoline fuel and the scarcity of crude oil source, it is an urgent need to produce biofuel from renewable resources to overcome the energy scarcity as well as improving the environment and global economy. Biofuel has been an alternative substitution for current petroleum fuels because it is

Content from this work may be used under the terms of the Creative Commons Attribution 3.0 licence. Any further distribution of this work must maintain attribution to the author(s) and the title of the work, journal citation and DOI.
Published under licence by IOP Publishing Ltd

renewable and can be produced continuously. Indonesia energy policy has set the target for new and renewable energy development by 2025 is 23% minimum of the total energy mix, wherein bioenergy shares about 10%, hydro 3%, geothermal 7% and other NRE of 3% [1].

Biogasoline is gasoline fuel produced by biochemical process using renewable feedstock. Gasohol is a blend of petroleum gasoline with etanol or butanol. Indonesia has experienced to produce gasohol from bioethanol in 2006 until 2010, but it was stopped due to inconsistent supply and price volatility, frequently resulting in a purchase price below cost of production. The portion of ethanol in gasoline strictly limited and not more than 10% because of its lower energy content, higher volatility and hygroscopics. To overcome the lack of ethanol quality, butanol will be an alternative solution to replace ethanol and it will be the next generation fuel. Butanol is a higher member of straight chain of alcohol series with each molecule of butanol ($C_4H_{10}O$) containing four carbon atoms. Butanol is commonly used industrially as a solvent which is largely produced petrochemically. The global butanol market has steadily increased in recent years and is projected to continue increasing at an annual rate of 3.2% [2].

Basically, there are two alternative process routes to produce butanol, one is petrochemical process and other is biochemical process relied on renewable resources. Oxo synthesis, reppe synthesis and crotonaldehye hydrogenation are some of petrochemical process route to produce butanol. Most of commercially butanol is produced petrochemically by low pressure oxo process by hydroformylation of propylene as primary feedstock using rhodium-based catalyst according to reaction (1) and (2) [3].

2CH3CH=CH2 + 2CO + 2H2 ⟶ CH3CH2CH2CHO + (CH3)2C(H)CHO (1)

Propylene Syngas normal butyraldehyde *iso*-butyraldehyde

n + iso – Butyraldehyde ⟶ Hydrogenation ⟶ Product refining ⟶ n-Butanol ↘ n-Isobutanol

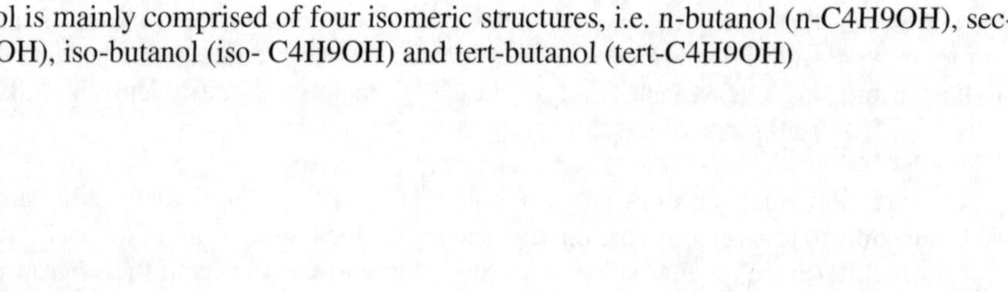

(2)

Butanol is mainly comprised of four isomeric structures, i.e. n-butanol (n-C4H9OH), sec-butanol (sec-C4H9OH), iso-butanol (iso- C4H9OH) and tert-butanol (tert-C4H9OH)

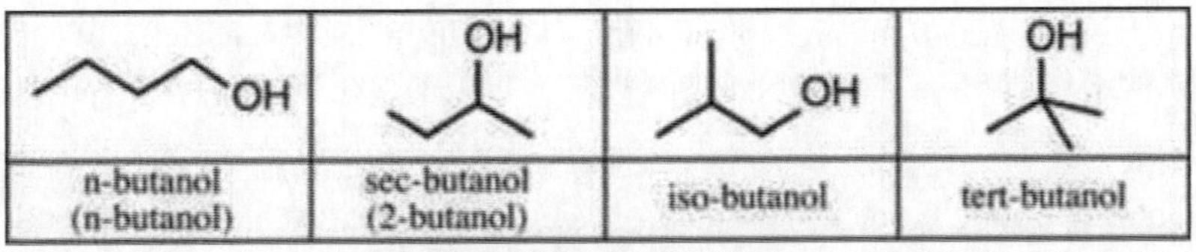

Figure 1. Butanol isomers

Table 1. Properties of ethanol, isobutanol and n-butanol [4]

No	Properties	Isobutanol	n-butanol	Ethanol
1	Density at 20 °C (g/cm^3)	0.802	0.81	0.794
2	Boiling Point at 1 atm (°C)	108	118	78
3	Water Solubility at 20 °C (g/100ml)	8	7.7	Miscible
4	Net Heat of Combustion (BTU/gal)	95000	93000	80000
5	Octane Number (R+M)/2	103.5	87	112
6	Reid Vapour Pressure (psi at 100°F)	5	4.3	18 - 22

Isobutanol or 2-methyl-1-propanol is one of butanol isomers that's more similar to gasoline and can be used as a fuel on internal combustion engine without any modification. Butanol can be transported in existing gasoline pipelines and produces more power than ethanol and providing potential cost savings, flexible and efficient access to end-user markets.

Standard specification of butanol blending with gasoline for use as automotive spark-ignition engine fuel [5] has been announced by ASTM D7862 since October 2013.

The specification covers three butanol isomers namely 1-butanol, 2-butanol, and 2-methyl-1-propanol as shown in table 2.

Table 2. Comparison of butanol isomers [6]

No	Properties	1-Butanol (n-Butanol)	2-Butanol (sec-Butanol)	2-Methyl-1-propanol (Isobutanol)
1	Density (kg/cm3)	809.8	806.3	801.8
2	Research Octane Number (RON)	96	101	113
3	Motor Octane Number (MON)	78	82	94
4	Melting Point, °C	-89.5	-114.7	-108
5	Boiling Temperature, °C	117.7	99.5	108
6	Enthalpy of Vaporation at T_{boil}, (kJ/kg)	584	550	564
7	Flash Point, °C	35	24	28
8	Auto ignition Temperature, °C	345	406	415
9	Flammability Limits:			
	- Lower Limit (Vol %)	1.4	1.7	1.2
	- Upper Limit (Vol %)	11.2	9.8	10.9

Butanol can also be produced from process of fermentation by means of bacteria of the *Clostridium species*. This process occurs under anaerobic conditions, and butanol as one of the products - called biobutanol.

Biobutanol is obtained from biochemical process using renewable or biological feedstock such as molasses, sugar cane, sugar bit, and other carbohydrate subtrats, as well as cellulosic raw materials or biomass [4]. As the cost of raw material has a significant influence on price, the main interest has recently been in low-cost substrates such as agricultural residues or industrial by-products and waste materials.

Biobutanol has not been produced commercially yet, but the process technology development has reached to pilot scale stage in several countries. Current activity of research and development are mainly focused on improving and optimizing the process condition to produce biobutanol effectively.

Upstream processing before fermentation includes pretreatment of biomass feedstock, hydrolysis, and in some cases detoxification of inhibitors formed during pretreatment. The common fermentation process is called as Acetone-Butanol-Ethanol (ABE) fermentation. The anaerobic fermentation consists of two stages: first the acidogenic phase where Clostridial bacteria produce acetic and butyric acids, carbon dioxide (CO_2) and hydrogen (H_2) from sugars, followed by the solventogenic phase where acids are converted into acetone, butanol and ethanol, typically in the ratio of 3:6:1 [7].

The final products obtained from fermentation are recovered and purified in downstream processing. Adsorption, gas stripping, liquid-liquid extraction, pervaporation, perstraction and reverse osmosis are the most used separation methods integrated with the ABE fermentation [8]. Butanol obtained from fermentation has toxic to organism so it leads to low production and the concentration in fermenter is about 1-2wt% of butanol. In situ removal of ABE by gas stripping has been reported to be one of the most important techniques of solvent removal which be able to maintain the ABE concentration in the fermentation broth below toxic levels [9]. In order to improve the butanol production from ABE fermentation and to minimize inhibition caused by butanol as well as facilitate product recovery, the use of feedstock comprising of carbohydrate source is fermented in the presence of bacteria to produce butyric acid and hydrogen. Then butanol can be directly produced by hydrogenation of butyric acid in the presence of a catalyst. The Hogdrogenation process improves upon the biomass second step of the two-step fermentation process [10].

Biomass can be converted to fermentable carbohydrates with bacteria to butyric acid, and then converts butyric acid and hydrogen to butanol by catalytic hydrogenation. The fermentation-hydrogenation process provides overall butanol yields higher than ABE fermentation yield [11].

An integrated fermentation-separation process, such as extractive fermentation, can be used to reduce product inhibition and increase process productivity and product yield. Extractive fermentation may also allow the process to produce and recover the fermentation product in one continuous step, so it reduce downstream processing load and recovery costs.

The fermentation route generally is not so competitive with petrochemical synthesis but by doing research and development continuously there will be a breakthrough to find out the newest technology overcoming all constraints that can be implemented and commercialized efficiently. Shota et al in 2007 has successed in producing 1-butanol from the engineered Escherichia Coli and opening the possibility for using non-native, easily manipulated organisms for 1-butanol production [12].

Considering all of recent technologies in bioprocess, strain development, fermentation method and efficiently fuel separation, biobutanol will be the new generation fuel that can be utilized as cost-effective fuel and enhances environmental friendly and sustainability. Whether it will be used as a transportation fuel directly or an additive component to gasoline, butanol will offer a better flexibility as an alternative fuel in the future. While awaiting for commercially production of biobutanol, the current isobutanol can be used as an attractive choice to overcome the lack of ethanol quality.

The aim of this research work is to examine and prove experimentally the effect of isobutanol addition in gasoline on the physical and chemical properties based on the standard fuel specification of gasoline 88 dan gasoline 92 for typical spark-ignition engines.

2. Material and Methods

The experimental work was carried out through the following steps:

2.1. Base gasoline preparation, by mixing of low and high octane mogas components to get bas gasoline with octane number 84.
2.2. Isobutanol sampel came from BASF product
2.3. Blending of base gasoline and isobutanol at some variate of compositions
2.4. Laboratory testing according to standard gasoline fuel parameters

Tabel 3. Typical Properties of Isobutanol

No	Propeties	Results
1	Isobutanol purity, %wt	99.7
2	Water, %wt	0.01
3	Colour	Colourless
4	Acidity as acetic aqcid, %wt	< 0.001
5	Density, g/cm3	0.802
6	Flash Point, ^{o}C	27
7	pH value	7
8	Viscosity Dinamic @ 20 ^{o}C, MPas	3.9
9	Boiling range at 1013 mbar, ^{o}C	106 -108
10	Vapour pressure @20 ^{o}C, mbar	9.5

Base gasoline was prepared by mixing low octane mogas component and high octane mogas component supplied from national oil refinery to get a certain quality with 84 octane number. The right compositions of blending components will determine the gasoline quality.

The methodology to prepare gasohol from isobutanol was by mixing process using batch stirring reactor at mixing speed 300 rpm for about 3 minutes. The composition of blending component was varied and the product was tested in laboratory according to standard test parameters for gasoline fuel. The simplified process flow diagram of gasoline-isobutanol is shown in figure 1.

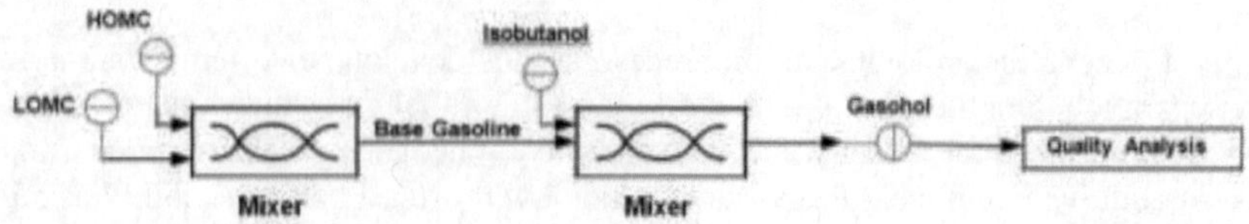

Figure 2. Simplified process flow diagram of gasohol production from isobutanol

The evaluation of gasohol from isobutanol is also done by comparing some test parameters with neat gasoline prepared by mixing of low and high octane mogas components. The test parameters evaluated at the same octane number are reid vapor pressure (RVP), density at 15°C and sulfur content.

3. Results and Discussion

The key properties for motor gasoline fuels are ignition quality, volatility, corrosivity and stability. The occurance of knocking in spark ignition engine can be reduced by increasing the gasoline's resistance to autoignition, which is expressed by its octane rating. Octane rating is measured relative to a mixture of 2,2,4-trimethylpentane (an isomer of octane) and n-heptane using CFR engine according to ASTM D2699-86 test method. The typical unit for octane ratings is research octane number (RON) and motor octane number (MON). The research octane number (RON) describes the behavior of the fuel in the engine at lower temperatures and speeds, and is an attempt to simulate acceleration behavior. Whereas motor octane number (MON) describes the behavior of the fuel in the engine at high temperatures and speeds – a full-throttle range, comparable to driving fast on a highway. The spread between the two numbers (MON & RON) is known as the fuel sensitivity. Octane rating parameter for commercial gasoline fuel specification normally use research octane number (RON).

The key gasoline characteristic for good driveability is *volatility* namely a gasoline's tendency to vaporize. Gasoline that vaporizes easily allows a cold engine to start quickly and warm up smoothly. Warm-weather gasoline is blended to vaporize less easily to prevent engine vapor lock and other hot fuel handling problems and to control evaporative emissions that contribute to air pollution.

With respect to gasoline, vapor pressure (RVP) is the single most important property for cold-start and warm-up driveability. *Cold-start* means that the engine is at ambient temperature, not that the ambient temperature is cold. Higher values of vapor pressure generally result in better cold-start performance, but lower values are better to prevent vapor lock and other hot fuel handling problems.

The normal range of vapor pressure for gasoline 88 is 69 kPa maximum and in range of 45-60 kPa for gasoline 92.

From the variation of isobutanol addition in base gasoline, it was obtained the octane rating data as shown in figure 2.

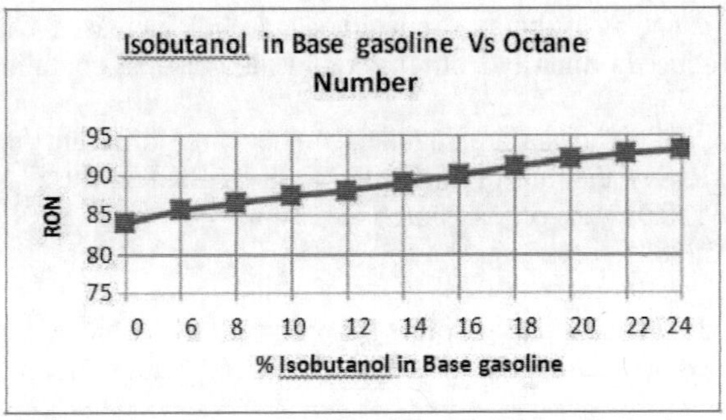

Figure 3. Effect of isobutanol addition into base gasoline on RON

From the octane testing data, gasoline 88 grade was achieved by adding isobutanol at 12%volum, gasoline 90 was achieved at 16%volum of isobutanol, whereas gasoline 92 was obtained by adding 20%volum of isobutanol into base gasoline. The sensitivity of isobutanol addition to octane rating increment in the gasoline blend is about 2.5 % per-octane unit.

The octane rating of typical commercially available gasoline varies by country. In Indonesia, RON 88 is the standard for regular unleaded gasoline named as Premium 88 or gasoline 88, whereas RON 90 is equivalent to standard gasoline of Pertalite and RON 92 is used for standard gasoline 92 called as Pertamax.

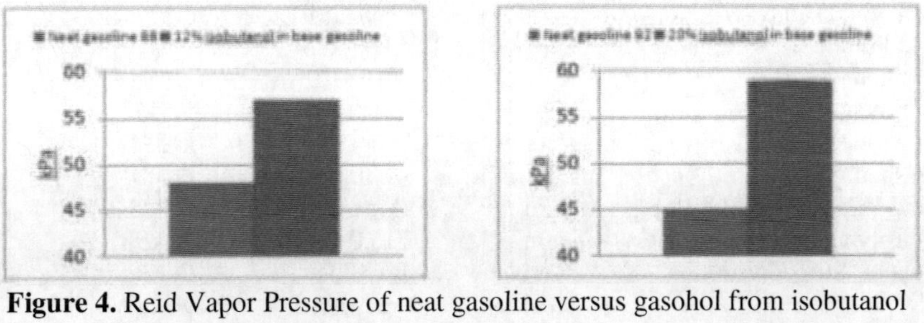

Figure 4. Reid Vapor Pressure of neat gasoline versus gasohol from isobutanol

Higher values of vapor pressure of 12% and 20% isobutanol containing gasohol result in better cold-start performance than neat gasoline 88 and 92 respectively.

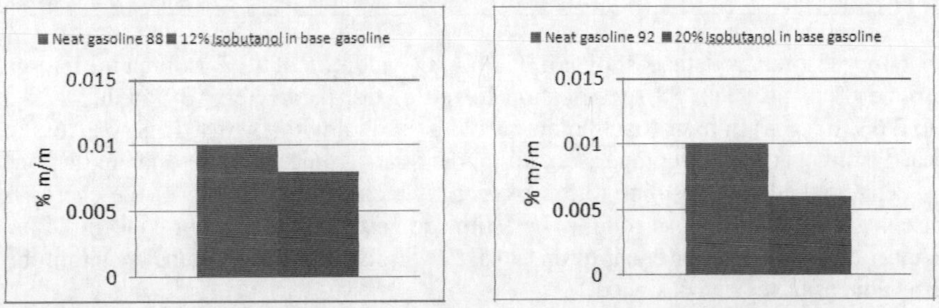

Figure 5. Sulfur content of neat gasoline versus gasohol from isobutanol

Lower values of sulfur content as isobutanol contained gasohol increased can reduce sulfur dioxide emission because SO_2 is produced primarily from the combustion of fuels containing sulfur.

Gasoline fuel specifications in Indonesia are regulated under three different regulations based on their Research Octane Number (RON), that are Premium RON 88 by SK 933/2013 as of 19 November 2013, Pertalite RON 90 by SK313/2013 as of 22 March 2013 and Pertamax RON 91 & 95 based on SK 3674/2006 as of 17 March 2006.

In the experimental work, gasoline 88 dan gasoline 92 produced by isobutanol addition are tested to the complete test parameter based on National gasoline standard specfications as shown in table 4 and 5.

Table 4. Laboratory Test Report of 12% Isobutanol in gasohol

No	Properties	Unit	Gasoline 88 Specification	Method	Results
1	Octane Number	RON	Min. 88	D 2699	88.0
2	Oxydation Stability	minutes	Min. 360	D 525	>360
3	Sulfur Content	%m/m	Max. 0.05	D 2622	0.008
4	Lead Content	g/l	Max. 0.013	D 3237	0.004
5	Metal content (Mn, Fe)	mg/l	Not detected	D 3831	-
6	Oxygen content	%m/m	Max 2.7	D 4815	2.91
7	Distillation :			D 86	
	10% vol. evap	^{o}C	Max. 74		56
	50% vol. evap	^{o}C	75 - 125		92
	90% vol. evap	^{o}C	Max. 180		138
	End Point	^{o}C	Max. 215		171
	Residue	% vol	Max. 2.0		1.0
8	Sediment	mg/l	Max. 1.0	D 5452	-
9	Unwashed gum	mg/100ml	Max. 70	D 381	-
10	Washed gum	mg/100ml	Max. 5	D 381	1
11	Reid Vapour Pressure	kPa	Max. 69	D 323	57
12	Density at 15 ^{o}C	kg/m^3	715.0-780.0	D 1298	752.7
13	Copper Strip Corrossion	Merit	Max.Class 1	D 130	Class 1
14	Mercaptane Sulfur	% m/m	Max. 0.002	D 3227	0.001
15	Appearance		Clear & Bright	Visual	Clear & bright
16	Colour		Yellow	Visual	-
17	Dye Content	gr/100 l	Max. 0.13		-
18	Odour		Marketable	Visual	Marketable

Based on laboratory test report data as shown in table 4, the addition of 12% isobutanol into gasoline can meet the gasoline 88 or premium 88 specification for all of test parameters. By adding 12% isobutanol (equivalent to 2.6% oxygen) in base gasoline may yield a gasohol with energy density is less than energy density of neat gasoline alone. For comparison, all modern cars are designed to operate on concentrations of up to 10% denatured ethanol gasoline (3.5% oxygen), which has about 96% of the energy of gasoline alone. Because the energy density can roughly be estimated based on the oxygen concentration, any blend with an oxygen content less than or equal to that of E10 will have an energy density within the range for which modern vehicles have been designed.

The extra oxygen in gasoline allows for adding more fuel to the engine and therefore making more heat energy to push the piston down in the cylinder. The extra oxygen also helps the cylinder burn all of the fuel within it more efficiently. The more fuel an engine can burn, the more power the engine can make.

The use of 12% isobutanol in gasoline blend can improve the environment impact as shown in low sulfur content 80 ppm that meets EURO III specification of 500 ppm.

The use of 20% concentration of isobutanol in base gasoline as shown in table 5 can meet mainly the gasoline 92 specification or pertamax having octane number 92 minimum.

Tabel 5. Laboratory Test Report of 20% Isobutanol in gasohol

No	Properties	Unit	Gasoline 92 Specification	Method	Results
1	Octane Number	RON	Min. 92	D 2699	92.1
2	Oxydation Stabilitty	minutes	Min. 480	D 525	>480
3	Sulfur content	% m/m	Max. 0.05	D 2622	0.006
4	Lead content	g/l	Max. 0.013	D 3237	0.005
5	Metal content (Mg, Fe)	mg/l	Not detected	D 3831	-
6	Oxygen content	% m/m	Max. 2.7	D 4815	5.3
7	Distillation :			D 86	
	10% vol. evap	^oC	Max. 70		54
	50% vol. evap	^oC	77 - 110		90
	90% vol. evap	^oC	130 - 180		138
	End Point	^oC	Max. 215		182
	Residue	% vol	Max. 2.0		1.0
8	Sediment	mg/l	Max. 1.0	D 5452	-
9	Unwashed gum	mg/100ml	Max. 70	D 381	-
10	Washed gum	mg/100ml	Max. 5	D 381	3
11	Reid Vapour Pressure	kPa	45 - 60	D 323	59
12	Density at 15 ^oC	kg/m^3	715.0-770.0	D 1298	757.5
13	Copper Strip Corrossion	Merit	Max.Class 1	D 130	Class 1
14	Doctor Test		Negative	IP 30	Negative
15	Mercaptane Sulfur	% m/m	Max. 0.002	D 3227	0.0008
16	Colour		Blue	Visual	-
17	Dye Content	gr/100 ltr	Max. 0.13		-
18	Appearance		Clear & bright	Visual	Clear & bright

The increasing of oxygen content in gsohol to more than 3.5% may reduce fuel energy content and will increase the nitrogen oxide emission during combustion, but the ignition quality as represented by octane number for this oxygenate fuel is still in the range of limit. In addition to ignition quality, the use of gasohol from 12% and 20% of isobutanol additions generally showed the normal properties in case of volatility, corrosivity and stability according to minimum requirement for typical gasoline fuel. The decreasing in sulfur content to about 60 ppm in gasohol from 20% isobutanol is approximate to EURO IV specification, so it can improve the exhaust gas emission from sulphur dioxide and minimizing the acid rain impact.

4. Conclusion

Isobutanol is an alternative oxygenate that can be used to overcome the lack of ethanol quality as gasoline blending component in spark-ignition engine. The use of 12% and 20% isobutanol in gasohol resulted in fuel quality that comply with standard fuel specification of Gasoline 88 or Premium and Gasoline 92 or Pertamax respectively. The increasing of oxygen content as isobutanol increased tend to lower energy content, but the fuel ignition quality as expressed by octane rating, as well as its volatility, corrosivity and stability has met the standard quality specifications of gasoline fuel. The decreasing of sulfur content in isobutanol-gasoline blend can reduce the formation of sulfur oxide emission during fuel combustion in engine and will be an alternative of environmental friendly fuel. The production of gasohol using isobutanol can be implemented at early stage by utilizing the existing isobutanol while awaiting for efficiently and commercially production of biobutanol.

References

[1] Peraturan Pemerintah RI No.79 2014 Kebijakan Energi Nasional (Indonesia)

[2] Kolodzie J R and Scheib J 2012 Bio-isobutanol: the next generation biofuel *Hydrocarbon Processing* (Texas: Gulf Publishing Company)

[3] Richard T and Ashley M 2007 Enhancement of industrial hydroformylation processes by the adoption of rhodium-based catalyst *Platinum metals Review* **51** 116–126

[4] Ndaba B, Chiyanzu I and Marx S 2015 n-Butanol derived from biochemical and chemical routes: A review *Biotechnology Reports* **8** 1-9

[5] Spark-Ignition Engine Fuel *D7862 − 13.* West Conshohocke US: ASTM Intl.

[6] ASTM 2013 Standard Specification for Butanol for Blending with Gasoline for Use as Automotive

[7] Niemistö J, Saavalainen P, Pongrácz E and Keiski R L2013 Biobutanol as a potential sustainable biofuel - assessment of lignocellulosic and waste-based feedstocks *Journal of Sustainable Development of Energy, Water and Environment Systems* **1** 58-77.

[8] Groot W J, Van der Lans R G J M and Luyben K C A M 1992 Technologies for butanol recovery integrated with fermentations. *Process Biochemistry* **27** 61-75

[9] Ezeji T C, Qureshi N, Blaschek H P 2003 Production of acetone, butanol and ethanol by Clostridium beijerinckii BA101 and in situ recovery by gas stripping *Microbiology &Biotechnology* **9** 595-603

[10] Panta S 2013 *Optimal Production of Biobutanol via Hydrogenation of Butyric Acid* (Undergraduate Honors Thesis) (Arkansas: University of Arkansas)

[11] Yang S T 2008 USA Patent No. US 2008/0248540 A1 : Methods of Producing Butanol.

[12] Atsumi S, Cann A F, Connor M R, Shen C R, Smith K M, Brynildsen M P, Chou K J Y, Hanai T and Liao J C 2007 Metabolic engineering of Escherichia coli for 1-butanolproduction *Metabolic Engineering* **10** 305-311

AUTHOR INDEX

AUTHOR INDEX

Institute of Physics
Dirac House, Temple Back
Bristol BS1 6BE UK

ISSN: 1755-1307
ISBN 978-1-5108-4172-7

CURRAN ASSOCIATES INC.
proceedings.com